Bioactive Components of Milk

Advances in Experimental Medicine and Biology

Editorial Board:

Nathan Back, State University of New York at Buffalo

Irun R. Cohen, The Weizmann Institute of Science

Abel lajtha, N.S. Kline Institute for Psychiatric Research

John D. Lambris, University of Pennsylvania

Rodolfo Paoletti, University of Milan

Recent Volumes in This Series:

Volume 600
SEMAPHORINS: RECEPTOR AND INTRACELLULAR SIGNALING MECHANISMS
Edited by R. Jeroen Pasterkamp

Volume 601
IMMUNE MEDIATED DISEASES: FROM THEORY TO THERAPY
Edited by Michael R. Shurin

Volume 602
OSTEOIMMUNOLOGY: INTERACTIONS OF THE IMMUNE AND SKELETAL SYSTEMS
Edited by Yongwon Choi

Volume 603
THE GENUS YERSINIA: FROM GENOMICS TO FUNCTION
Edited by Robert D. Perry and Jacqueline D. Fetherson

Volume 604
ADVANCES IN MOLECULAR ONCOLOGY
Edited by Fabrizio d'Adda di Gagagna, Susanna Chiocca, Fraser McBlane and Ugo Cavallaro

Volume 605
INTEGRATION IN RESPIRATORY CONTROL: FROM GENES TO SYSTEMS
Edited by Marc Poulin and Richard Wilson

Volume 606
BIOACTIVE COMPONENTS OF MILK
Edited by Zsuzsanna Bösze

A Continuation Order Plan is available for this series. A continuation order will bring delivery of each new volume immediately upon publication. Volumes are billed only upon actual shipment. For further information please contact the publisher.

Zsuzsanna Bösze
Editor

Bioactive Components of Milk

@ Springer

Editor
Zsuzsanna Bösze
Agricultural Biotechnology Center,
Gödöllo, Hungary

ISBN 978-0-387-74086-7   e-ISBN 978-0-387-74087-4

Library of Congress Control Number: 2007935077

© 2008 Springer Science+Business Media, LLC

All rights reserved. This work may not be translated or copied in whole or in part without the written permission of the publisher (Springer Science + Business Media, Inc., 233 Spring Street, New York, NY 10013, USA), except for brief excerpts in connection with reviews or scholarly analysis. Use in connection with any form of information storage and retrieval, electronic adaptation, computer software, or by similar or dissimilar methodology now known or hereafter developed is forbidden. The use in this publication of trade names, trademarks, service marks and similar terms, even if they are not identified as such, is not to be taken as an expression of opinion as to whether or not they are subject to proprietary rights.

Printed on acid-free paper

9 8 7 6 5 4 3 2 1

springer.com

Foreword

To the mammalian neonate, milk is more than a source of nutrients. It furnishes a broad range of molecules that protect the neonate against a more or less hostile environment. In human neonates, the incidence of digestive and respiratory diseases, which is significantly lower in breastfed infants than in those who have been formula-fed, has been attributed to the immune globulins, antimicrobial proteins, and antibacterial peptides present in maternal milk.

For humans, particularly in Western countries, milk is a source of both food and substances beneficial to growth and health in children and adults. For example, the calcium supplied by milk, acknowledged for its role in bone accretion, is also involved in controlling body weight and blood pressure.

Advances in the analysis of milk composition have led to the identification and characterization of a large number of its components. Current interest in human nutrition and health has made it possible to demonstrate that many of these components are biologically active and exert beneficial effects. Recently, however, the highly publicized negative health impact of excessive milk consumption has raised questions as to the value of dairy products. It remains necessary to clarify the different arguments advanced to support the benefits of milk and dairy products.

The purpose of this volume is to report advances in our knowledge of bioactive milk constituents, in a series of comprehensive reviews by internationally reputed scientists.

The first two parts concern the activities and properties of milk lipids and native milk proteins. An overview of the nutritional factors controlling lipogenic gene expression in ruminant mammary glands may help us to understand the nutritional importance of modifying milk fat composition to enable a positive health impact for milk. Important findings are presented on the role of the milk fat fraction as a source of lipophilic microconstituents (vitamins, phytosterols, etc.). Advances in proteomic technologies are being used to explore the protein content of the milk fat globule membrane. Another source of membrane proteins in milk (milk serum lipoprotein membrane vesicles) is also discussed. Although they only represent a small proportion of milk proteins, through their protein-protein interactions and enzymatic activities, these proteins may assume specific functions in both the mammary gland and the gastrointestinal environment of newborns.

The antimicrobial, anti-inflammatory, and anticancer activities of lactoserum proteins such as lactoferrin, CD14, α-lactalbumin, and oligosaccharide, and the immunomodulatory activities of polypeptide from colostrum, are discussed. The structure-function relationship of these molecules, the occurrence of active complexes with other molecules present in either milk or the neonatal gastrointestinal system, and their ability to promote the maturation of cells in the immune response constitute very interesting new orientations for the development of novel therapeutic approaches.

Bioactive peptides encrypted in native proteins are released following the hydrolysis of precursor proteins by specific enzymes. An overview of different *in vivo* and *in vitro* studies concerning the effects of milk peptides on the maturation of the neonatal immune system is presented in the third part of the book. Antimicrobial and antitumor peptides derived from milk during digestion may be of physiological significance in suckling neonates and also supply valuable dietary proteins and peptides that will contribute to human well-being. Genetic engineering in animals now enables the production of biologically active proteins or modifications to milk composition. The fourth part of this volume will serve as a comprehensive guide to this recent but highly active field of producing biologically active foreign proteins in the milk of livestock animals.

Colostrum and milk provide hormones and growth factors to the neonate: Their roles are exhaustively described and discussed. Part V presents information on probiotics in milk and epidemiological aspects of breastfeeding and allergic diseases.

This volume offers a very complete survey of current ideas on this active and rapidly evolving field of research. In addition to achieving an admirable summary regarding biologically active components in milk, it underlines the importance of evaluating the role of these different components as interacting molecules. It does not constitute a definitive view of the subject but reflects current thinking, providing a wealth of information and numerous suggestions for future research related to milk composition, the impact of dairy products on health, and applications for the design of dietary products or pharmaceutical preparations.

Michèle Ollivier-Bousquet
Unité UR1196 Génomique et Physiologie de la Lactation, INRA, 78352,
Jouy-en-Josas Cedex, France
E-mail: michele.ollivier@jouy.inra.fr

Contents

Foreword v

I. Milk Lipids: A Source of Bioactive Molecules

Trans **Fatty Acids and Bioactive Lipids in Ruminant Milk** 3
K. J. Shingfield, Y. Chilliard, V. Toivonen, P. Kairenius, and D. I. Givens

Expression and Nutritional Regulation of Lipogenic Genes in the Ruminant Lactating Mammary Gland 67
L. Bernard, C. Leroux, and Y. Chilliard

Lipophilic Microconstituents of Milk 109
A. Baldi and L. Pinotti

II. Biological Activity of Native Milk Proteins: Species-Specific Effects

Milk Fat Globule Membrane Components—A Proteomic Approach 129
M. Cavaletto, M. G. Giuffrida, and A. Conti

Milk Lipoprotein Membranes and Their Imperative Enzymes 143
N. Silanikove

Lactoferrin Structure and Functions 163
D. Legrand, A. Pierce, E. Elass, M. Carpentier, C. Mariller, and J. Mazurier

Milk CD14: A Soluble Pattern Recognition Receptor in Milk 195
K. Vidal and A. Donnet-Hughes

Apoptosis and Tumor Cell Death in Response to HAMLET (Human α-Lactalbum Made Lethal to Tumor Cells) 217
O. Hallgren, S. Aits, P. Brest, L. Gustafsson, A.-K. Mossberg, B. Wullt, and C. Svanborg

A Proline-Rich Protein from Ovine Colostrum: Colostrinin with
Immunomodulatory Activity 241
M. Zimecki

III. Milk Peptides

Milk Peptides and Immune Response in the Neonate 253
I. Politis and R. Chronopoulou

Protective Effect of Milk Peptides: Antibacterial
and Antitumor Properties 271
I. López-Expósito and I. Recio

Antihypertensive Peptides Derived from Bovine Casein
and Whey Proteins 295
T. Saito

IV. Induced Biologically Active Components from the Milk of Livestock Animals

Targeted Antibodies in Dairy-Based Products...................... 321
L. Hammarström and C. Krüger Weiner

Manipulation of Milk Fat Composition Through Transgenesis 345
A. L. Van Eenennaam and J. F. Medrano

Producing Recombinant Human Milk Proteins in the Milk of Livestock
Species 357
Z. Bösze, M. Baranyi, and C. B. A. Whitelaw

V. The Influence of Nutrition on the Production of Bioactive Milk Components

Insulin-Like Growth Factors (IGFs), IGF Binding Proteins, and Other
Endocrine Factors in Milk: Role in the Newborn 397
J. W. Blum and C. R. Baumrucker

Probiotics, Immunomodulation, and Health Benefits 423
H. Gill and J. Prasad

Potential Anti-inflammatory and Anti-infectious Effects of Human Milk
Oligosaccharides 455
C. Kunz and S. Rudloff

On the Role of Breastfeeding in Health Promotion and the Prevention of Allergic Diseases 467
L. Rosetta and A. Baldi

Subject Index 485

Contributors

Sonja Aits
Institute of Laboratory Medicine, Section for Microbiology, Immunology and Glycobiology, Lund, Sweden

Antonella Baldi
Department of Veterinary Sciences and Technology for Food Safety, University of Milan, Via Trentacoste, 2-20134 Milnao-Italy, Tel.: +39 0250315736, Fax: +39 0250315746
e-mail: antonella.baldi@unimi.it

Mária Baranyi
Agricultural Biotechnology Center, H-2100 Gödöllő, Szent-Györgyi A., St. 4, Hungary

Craig R. Baumrucker
Department of Dairy and Animal Science, Penn State University, University Park, PA 16802, USA, Tel.: +1-814-863-0712, Fax: +1-814-863-6042
e-mail: crb@psu.edu

L. Bernard
Adipose Tissue and Milk Lipid Laboratory, Herbivore Research Unit, INRA-Theix, 63 122 St. Genès-Champanelle, France, Tel.: +33-473-624051, Fax: +33-473-624519
e-mail: Laurence.Bernard@clermont.inra.fr

Jürg W. Blum
Veterinary Physiology, Vetsuisse Faculty, University of Bern, CH-3012 Bern, Switzerland, Tel.: +41-78-7211220, Fax: +41-31-8292042
e-mail: juerg.blum@physio.unibe.ch

Zsuzsanna Bősze
Agricultural Biotechnology Center, H-2100 Gödöllő, Szent-Györgyi A.,
St. 4, Hungary, Tel.: +36-28-526150, Fax: +36-28-526151
e-mail: bosze@abc.hu

Patrick Brest
Institute of Laboratory Medicine, Section for Microbiology, Immunology
and Glycobiology, Lund, Sweden

Mathieu Carpentier
From the Unité de Glycobiologie Structurale et Fonctionnelle, Unité Mixte de
Recherche N°8576 du Centre National de la Recherche Scientifique, Université
des Sciences et des Technologies de Lille, IFR 147, 59655 Villeneuve d'Ascq
Cedex, France

Maria Cavaletto
Biochemistry and Proteomics Section – DISAV Dipartimento Scienze
dell'Ambiente e della Vita, Department of Environmental and Life Sciences,
Universitá del Piemonte Orientale, via Bellini 25/G, 15100Alessandria, Italy,
Tel.: +390131360237, Fax: +390131360391
e-mail: maria.cavaletto@unipmn.it

Y. Chilliard
Adipose Tissue and Milk Lipid Laboratory, Herbivore Research Unit,
INRA-Theix, 63 122 St. Genès-Champanelle, France

Roubini Chronopoulou
Department of Animal Science, Agricultural University of Athens, 75 Iera
Odos, 18855 Athens, Greece

Amedeo Conti
ISPA-CNR, Institute of Science of Food Production, Colleretto Giacosa, Italy

Anne Donnet-Hughes
Nutition and Health Department, Nestlé Research Center, Nestec Ltd,
Vers-Chez-Les-Blanc, P.O. Box 44, CH-1000 Lausanne 26

A. L. Van Eenennaam
Department of Animal Science, University of California Davis, One Shields
Ave., Davis, CA 95616-8521, USA, Tel.: 530-752-7942, Fax: 530-752-0175
e-mail: alvaneenennaam@ucdavis.edu

Contributors

Elisabeth Elass
From the Unité de Glycobiologie Structurale et Fonctionnelle, Unité Mixte de Recherche N°8576 du Centre National de la Recherche Scientifique, Université des Sciences et des Technologies de Lille, IFR 147, 59655 Villeneuve d'Ascq Cedex, France

Harsharn Gill
Primary Industries Research Victoria, Department of Primary Industries, Werribee Vic 3030, Australia and School of Molecular Sciences, Victoria University PO Box 14428, Melbourne, Victoria 8001, Australia

Maria Gabriella Giuffrida
ISPA-CNR, Institute of Science of Food Production, Colleretto Giacosa, Italy

D. I. Givens
The University of Reading, Reading, UK

Lotta Gustafsson
Institute of Laboratory Medicine, Section for Microbiology, Immunology and Glycobiology, Lund, Sweden

Oskar Hallgren
Department for Experimental Medical Sciences, Section for Lungbiology, Lund, Sweden

Lennart Hammarström
Division of Clinical Immunology, Department of Laboratory Medicine, Karolinska University Hospital Huddinge, SE-141 86, Stockholm, Sweden

P. Kairenius
MTT Agriculture Food Research Finland, Jokioinen, Finland

C. Kunz
Institute of Nutritional Sciences, Justus Liebig University Giessen, Germany; Wilhlemstr 20, 35392 Giessen, Germany, Tel.: +49-641 9939041,
Fax: +49-641-9939049
e-mail: clemens.kunz@uni-giessen.de

Dominique Legrand
From the Unité de Glycobiologie Structurale et Fonctionnelle, Unité Mixte de Recherche N°8576 du Centre National de la Recherche Scientifique, Université des Sciences et des Technologies de Lille, IFR 147, 59655 Villeneuve d'Ascq Cedex, France Tel.: 33 3 20 33 72 38, Fax: 33 3 20 43 65 55
e-mail: dominique.legrand@univ-lille1.fr

C. Leroux
Adipose Tissue and Milk Lipid Laboratory, Herbivore Research Unit,
INRA-Theix, 63 122 St. Genès-Champanelle, France

Iván López-Expósito
Instituto de Fermentaciones Industriales (CSIC), Juan de la Cierva 3, 28006 Madrid, Spain

Christophe Mariller
From the Unité de Glycobiologie Structurale et Fonctionnelle, Unité Mixte de Recherche N°8576 du Centre National de la Recherche Scientifique, Université des Sciences et des Technologies de Lille, IFR 147, 59655 Villeneuve d'Ascq Cedex, France

Joël Mazurier
From the Unité de Glycobiologie Structurale et Fonctionnelle, Unité Mixte de Recherche N°8576 du Centre National de la Recherche Scientifique, Université des Sciences et des Technologies de Lille, IFR 147, 59655 Villeneuve d'Ascq Cedex, France

J. F. Medrano
Department of Animal Science, University of California Davis, One Shields Ave., Davis, CA 95616-8521, USA, Tel.: 530-752-7942, Fax: 530-752-0175
e-mail: jfmedrano@ucdavis.ed

Ann-Kristin Mossberg
Institute of Laboratory Medicine, Section for Microbiology, Immunology and Glycobiology, Lund, Sweden

Annick Pierce
From the Unité de Glycobiologie Structurale et Fonctionnelle, Unité Mixte de Recherche N°8576 du Centre National de la Recherche Scientifique, Université des Sciences et des Technologies de Lille, IFR 147, 59655 Villeneuve d'Ascq Cedex, France

Luciano Pinotti
Department of Veterinary Sciences and Technology for Food Safety, University of Milan, Via Trentacoste, 2- 20134 Milnao, Italy

Ioannis Politis
Department of Animal Science, Agricultural University of Athens, 75 Iera Odos, 18855 Athens, Greece, Tel.: 30-210-529-4408, Fax: 30-210-529-4413
e-mail: i.politis@aua.gr

Jaya Prasad
Primary Industries Research Victoria, Department of Primary Industries,
Werribee Vic 3030, Australia and School of Molecular Sciences, Victoria
University, P.O. Box 14428, Melbourne, Victoria 8001, Australia

Isidra Recio
Instituto de Fermentaciones Industriales (CSIC), Juan de la Cierva 3, 28006
Madrid, Spain, Tel.: +34 91 5622900, Fax: +34 91 5644853
e-mail: recio@ifi.csic.es

L. Rosetta
CNRS UPR 2147, 44 rue de l' Amiral Mouchez, 75014 Paris, France

S. Rudloff
Department of Pediatrics, Justus Liebig University Giessen, Germany

K.J. Shingfield
MTT Agrifood Research Finland, FIN 31600, Jokioinen, Finland,
Tel.: +358 3 41883694, Fax: +358 3 41883661
e-mail: kevin.shingfield@mtt.fi

Nissim Silanikove
Agricultural Research Organization, Institute of Animal Science, P.O. Box 6,
Bet Dagan, 50-250, Israel, Tel.: +972-8-9484436, Fax: + 972-8-9475075
e-mail: nsilanik@agri.huji.ac.il

Catharina Svanborg
Institute of Laboratory Medicine, Section for Microbiology, Immunology
and Glycobiology, Sölvegatan 23, 22362 Lund, Sweden,
Tel.: +46 46 173972, Fax: +46 46 137468
e-mail: catharina.svanborg@med.lu.se

V. Toivonen
MTT AgriFood Research Finland, Jokioinen, Finland

Karine Vidal
Nutition and Health Department, Nestlé Research Center, Nestec Ltd,
Vers-Chez-Les-Blanc, P.O. Box 44, CH-1000 Lausanne 26
e-mail: karine.vidal@rdls.nestle.com

Carina Krüger Weiner
Division of Clinical Immunology, Department of Laboratory Medicine,
Karolinska University Hospital Huddinge, SE-141 86, Stockholm, Sweden

C. Bruce A. Whitelaw
Roslin Institute, Department of Gene Function and Development, Roslin, Midlothian, EH25 9PS, Scotland, United Kingdom

Björn Wullt
Institute of Laboratory Medicine, Section for Microbiology, Immunology and Glycobiology, Lund, Sweden

Michall Zimecki
The Institute of Immunology and Experimental Therapy, Wrocllaw, Poland

Part I
Milk Lipids: A Source of Bioactive Molecules

Trans Fatty Acids and Bioactive Lipids in Ruminant Milk

K. J. Shingfield, Y. Chilliard, V. Toivonen, P. Kairenius and D. I. Givens

Introduction

There is increasing evidence that nutrition plays an important role in the development of chronic diseases in the human population, including cancer, cardiovascular disease, insulin resistance, and obesity. Developing foods that enhance human health is central to dietary approaches for preventing and reducing the economic and social impacts of chronic disease. Numerous studies in human subjects have implicated a high consumption of saturated fatty acids (SFA) and *trans* fats as risk factors for cardiovascular disease risk, with evidence that high-SFA intakes may also be related to lowered insulin sensitivity, which is a key factor in the development of the metabolic syndrome. While it is generally accepted that SFA raise plasma total and low-density lipoprotein cholesterol concentrations, atherogenic effects are confined to 12:0, 14:0, and 16:0. Consistent with the effects of individual SFA, there is some evidence to suggest that physiological responses to *trans* fatty acids (TFA) may also be isomer-dependent.

National nutritional guidelines with the target of reducing the incidence of cardiovascular disease have advocated a population-wide reduction in the intake of total fat, SFA, and TFA. Milk and dairy products are the major source of 12:0 and 14:0 in the human diet and also make a significant contribution to 16:0 and TFA intake. However, developing public health policies promoting a decrease in milk, cheese, and butter consumption ignores the value of these foods as a versatile source of nutrients. Furthermore, consumption of milk and dairy products may confer beneficial effects with respect to the prevention of osteoporosis, cancer, atherosclerosis, and other degenerative disorders (Heaney, 2000; Ness et al., 2001; Kalkwarf et al., 2003; Valeille et al., 2006). A number of minerals, proteins, peptides, and lipids in milk and fermented dairy products exhibit bioactive properties with the potential to

---

K. J. Shingfield
MTT Agrifood Research Finland, FIN 31600, Jokioinen, Finland
e-mail: kevin.shingfield@mtt.fi

improve long-term human health (Parodi, 2001; Pereira et al., 2002; Korhonen & Pihlanto, 2006; Mozaffarian et al., 2006; Tholstrup, 2006). Milk fat contains a number of components, including 4:0, branch-chain fatty acids, *trans*-11 18:1, *cis*-9, *trans*-11 conjugated linoleic acid (CLA), *trans*-9, *trans*-11 18:2 CLA, vitamins A and D, β-carotene, and sphingomyelin, that have been shown to elicit antimutagenic properties in a number of *in vitro* experiments with human cell lines and animal model studies (Parodi, 2001; Bauman et al., 2005; Collomb et al., 2006). In other respects, the putative effects of bioactive compounds in milk have to be considered in relation to the combined effects of the overall fatty acid profile, amount and duration of milk fat consumption, and other macro and micro nutrients in the diet.

Nutritional modification of milk fatty acid composition through sustainable, environmental, and welfare-acceptable means could be used as an integral component of an overall strategy for dietary disease prevention. This review provides an overview of recent evidence on the activity and properties of specific lipids in milk and the metabolic origins of fatty acids incorporated into milk fat and considers the role of nutrition in the lactating cow for modifying milk fat composition for improved long-term human health.

Altering Milk Fatty Acid Composition for Improved Human Health

During the past decade, the increased awareness of the association between diet and health has led to nutritional quality becoming an increasing important determinant of consumer food choices. A major development has been the recognition that lipids in the diet can affect health maintenance and disease prevention, resulting in the development of national health policies recommending a population-wide reduction in the intake of total fat and SFA (e.g., Committee on Medical Aspects of Food Policy, 1984, 1994). More recently, changes in legislation on the nutritional labeling of foods have been implemented in several countries with the sole purpose of reducing TFA in the human diet. The following section reviews the biological activity of lipids and fatty acids in milk with respect to human health.

Saturated Fatty Acids

The relationship between the amount and type of fat in the diet and incidence of cardiovascular disease (CVD)—coronary heart disease (CHD) in particular—has been extensively investigated, with strong and consistent associations reported across a wide body of data (Kris-Etherton et al., 2001; World Health Organization, 2003). Overall, SFA are known to raise total and low-density lipoprotein (LDL) cholesterol, but specific saturates have markedly different effects. In particular, intakes of myristic (14:0) and palmitic (16:0) acid have been associated with elevated serum LDL-cholesterol concentrations in human subjects (Katan et al., 1995; Temme et al., 1996), while the other major SFA in foods, stearic acid (18:0), has been shown to be essentially neutral (Bonanome & Grundy, 1988). Some studies suggest that lauric acid (12:0) and 14:0 exert more potent effects on plamsa cholesterol than 16:0, while others suggest that 14:0 and 16:0 are more atherogenic than 12:0.

Establishing a clear role of a specific SFA on plasma lipid profiles within the context of intakes relevant to a given population remains challenging. This is at least in part due to the effects of oils or fats rather than a specific SFA being reported; changes to a specific SFA have often been examined at levels much higher than that consumed from the habitual diet, or the substitution of other fatty acids with SFA alters the overall fatty acid profile of the diet (Wilke & Clandinin, 2005). Furthermore, there is emerging evidence that the balance of fatty acids in the diet may be a more important determinant of physiological effects than the intake of individual fatty acids *per se*. Studies in human subjects have shown that 16:0 has no effect on plasma total or LDL cholesterol in hypercholesterolemic (Clandinin et al., 1999) or normocholesterolemic (Ng et al., 1992; Sundram et al., 1995; Clandinin et al., 2000; French et al., 2002) subjects when the intake of 18:2 *n*-6 exceeds 5.0% of dietary energy and cholesterol is less than 400 mg/day. Recent comparisons of diets supplying 0.6 or 1.2% of dietary energy as 14:0 provided evidence that at moderate levels, increases in 14:0 intake were associated with significant decreases in plasma triacylglyceride and increases in HDL-cholesterol concentrations in healthy men (Dabadie et al., 2005). Higher intakes of 14:0 were also associated with significant enrichment of 20:5 *n*-3 and 22:6 *n*-6 in plasma phospholipids and an increase in 22:6 *n*-3 concentrations in plasma cholesterol esters, possibly mediated via the regulatory effects of 14:0 on Δ-6 desaturase (Dabadie et al., 2005). Further comparisons of moderate increases in 14:0 intake (1.2% vs. 1.8% dietary energy) confirmed the beneficial effects on plasma lipids reported in earlier studies but suggested that the positive changes were partially diminished when 14:0 provided 1.8% of dietary energy (Dabadie et al., 2006). Such findings highlight the challenge of developing nutritional guidelines for the prevention of chronic disease and provide support that changes in the consumption of specific classes of fatty acids have to be considered in the context of the fat composition of the diet.

The majority of 12:0 and 14:0 in the diet and a significant amount of 16:0 in the human diet are derived from whole milk, cheese, and butter (Gunstone et al., 1994). More recently, a European-wide survey on fatty acid consumption revealed that milk and dairy products were consistently the largest source of SFA in the human diet, with the highest values reported in Germany and France, where almost 60% of saturate intakes were derived from these food sources (Hulsof et al., 1999). Across the countries studied, milk and milk-derived foods were also found to contribute on average to almost 40% of the total TFA intake (Hulsof et al., 1999).

Due to the relatively high proportion of 12:0, 14:0, and 16:0 in ruminant milk fat (Table 1), consumption of whole milk, butter, and cheese would be expected to have adverse effects on serum LDL-cholesterol levels. Data from controlled intervention studies have shown that consumption of modified milk, butter, cheese, and ice cream (20% of dietary energy intake) containing lower concentrations of 10:0, 12:0, 14:0, 16:0 and increased *cis*-9 18:1 content significantly reduced total and LDL cholesterol (4.3% and 5.3%, respectively) in middle-aged men and women but had no effect on HDL cholesterol or triacylglycerides (Noakes et al., 1996). Consumption of a similarly modified butter diet (20% dietary energy) was also shown to significantly decrease plasma total and LDL cholesterol (7.9% and 9.5%, respectively) in male volunteers (Poppitt et al., 2002). In both studies, changes in milk fat and dairy product composition used in these studies were achieved using oilseeds protected from ruminal metabolism that result in decreases in SFA accompanied by higher *cis*-9 18:1 concentrations without substantial increases in TFA. Clinical trials using butter modified through the use of plant oils (lowered SFA and higher *trans*

Table 1 Typical Fatty Acid Composition of Bovine, Caprine, and Ovine Milk Fat

Fatty Acid	Composition (g/100g fatty acids)		
	Bovine [1]	Caprine [2]	Ovine [3]
4:0	3.88	2.64	2.18
6:0	2.49	2.11	2.39
8:0	1.39	2.41	2.73
10:0	3.05	9.35	9.97
10:1 *cis*-9	0.28	0.18	0.24
12:0	4.16	5.35	5.00
14:0	11.36	11.99	9.81
14:1 *cis*-9	1.11	0.24	0.18
16:0	29.36	27.47	28.23
16:1 *cis*-9	1.94	0.76	1.43
17:0	0.55	0.90	0.72
17:1 *cis*-9	0.28	0.40	0.39
18:0	11.36	6.92	8.88
18:1 *cis*-9	21.88	16.41	17.17
18:1 *trans*-11	0.28	0.72	0.78
18:2 *n*-6	1.94	1.99	3.19
18:3 *n*-3	0.55	0.96	0.42
20:0	0.00	0.22	0.15
Summary			
Σ saturates	70.08	74.68	72.38
Σ MUFA	25.76	20.41	21.99
Σ PUFA	3.32	2.93	4.31

MUFA = monounsaturated fatty acids; PUFA = polyunsaturated fatty acids.
[1] Adapted from McCance and Widdowson (1998).
[2] Derived from Bernard et al. (2005b).
[3] Adapted from Alonso et al. (1999).

18:1 content) were unable to demonstrate an improvement in plasma lipid profiles associated with CVD risk (Tholstrup et al., 1998). More recent data from a large longitudinal cohort study of 2778 black and white men and women, initially aged 18–30 years old, appear to support earlier findings (Steffen & Jacobs, 2003). Measurements of diet composition and plasma lipid profiles were assessed over a seven-year period. Plasma LDL cholesterol was found to increase by 0.078 mmol/L across all quintiles of high-fat dairy intake ($P < 0.05$), although it was postulated that the true mean increase was probably three- to sixfold higher (0.26–0.47 mmol/L) after correction for within-subject errors in diet assessments. This study also indicated that the increases in LDL cholesterol in young adults consuming high-fat dairy foods were, at least in part, balanced by an increase in plasma high-density lipoprotein- (HDL-) cholesterol concentrations compared with cohorts consuming no- or low-fat dairy foods. In other respects, an extensive meta-analysis of 60 controlled intervention trials examining the association between SFA and CHD risk reached rather different conclusions (Mensink et al., 2003). In this analysis, the relationship between the intake of fatty acids from the diet and plasma total-to-HDL cholesterol was examined based on the premise that this ratio is a more reliable predictor of coronary artery disease (CAD) than LDL-cholesterol concentrations. Overall, CAD risk was substantially reduced when SFA in the diet were replaced with *cis*-monounsaturated fatty acids (MUFA), but there were notable differences between individual SFA. Even though consumption of 12:0, 14:0, and 16:0 was associated with elevated LDL-cholesterol concentrations, higher intakes of 12:0, 14:0, and 18:0 were related to a decrease in the total-to-HDL cholesterol ratio. Higher intakes of 16:0 were found to be associated with an increase in the total-to-HDL cholesterol ratio (Mensink et al., 2003). Comparison of various fat sources in typical U.S. diets indicated that butter was predicted to give rise to the largest increase in plasma total-to-HDL cholesterol, which could be attributed in the most part to 16:0. If it is accepted that the ratio of total-to-HDL cholesterol is a more robust and reliable predictor of CAD/CHD risk than measurements of plasma LDL-cholesterol concentrations, these findings would tend to imply that nutritional strategies for reducing milk fat SFA content should be targeted toward reducing 16:0 concentrations.

Most of the biological activity of SFA in human subjects has focused on the effects on plasma lipids and associated increases in CAD/CHD risk. However, there is emerging evidence that high intakes of SFA may also be related to lowered insulin sensitivity, a key factor in the development of the metabolic syndrome (Nugent, 2004). In epidemiological studies, high intakes of SFA have been associated with a higher risk of impaired glucose tolerance and higher fasting plasma glucose and insulin concentrations (Feskens & Kromhout, 1990; Parker et al., 1993; Feskens et al., 1995). Furthermore, a three-month intervention study involving 162 healthy subjects (Vessby et al., 2001) offered diets rich in SFA (from butter and margarine) or MUFA (from high oleic sunflower oil) demonstrated that subjects fed the SFA-rich diet had significantly impaired insulin sensitivity (-10%), but no change was observed for volunteers given the

MUFA-rich diet (Table 2). It was also evident from this study that dietary supplements of long-chain n-3 fatty acids from fish oil had no effect on insulin sensitivity or insulin secretion, while the generally positive effects associated with a diet rich in MUFA were not seen in individuals consuming relatively high amounts of fat (>37% of energy intake).

Based on the evidence from human clinical and epidemiological studies, reducing the consumption of saturates—medium-chain SFA, in particular—would be expected to represent an effective nutritional strategy for reducing the impact of chronic disease. Since milk and dairy products represent a major source of SFA in the human diet, reductions in the proportions of medium-chain saturates in milk and dairy fats could be expected to contribute to improved human health.

Butyrate

Milk fat is a unique and relatively rich source of butyrate in the human diet, containing between 75 and 130 mmol/mol of butyric acid (Parodi, 1999). Butyrate is known to exhibit anticarcinogenic effects, inhibit cell growth, promote differentiation, and induce apoptosis in various human cancer cell lines (Parodi, 1999). It has also been suggested that butyrate may also prevent the invasion of tumors via inhibitory effects on urokinase (Parodi, 2001). The role of butyrate as an antimutagen has been largely focused on preventing colon cancer, since butyrate is an end product of microbial fermentation of

Table 2 Effect of Challenging with Saturated Fatty Acids or Monunsaturated Fatty Acids on Insulin Parameters, Plasma Glucose, and Serum Lipids in Healthy Men and Women[1]

Fatty Acid Challenge	Saturated Fatty Acids			Monunsaturated Fatty Acids		
	Relative Change[2]			Relative Change[2]		
Parameter	Mean	(%)	P Value	Mean	(%)	P Value
Insulin sensitivity index (SI)	−4.2	−10.3	0.032	+0.10	+12.1	0.518
Serum insulin (mU/l)	+0.25	+3.5	0.466	−0.35	−5.8	0.049
First phase insulin response (mU/l)	+3.3	+9.0	0.029	+3.8	+10.1	0.139
Plasma glucose (mmol/l)	0.00	+/− 0	0.995	−0.03	−0.60	0.413
Total cholesterol (mmol/l)	+0.14	+2.5	0.018	−0.15	−2.7	0.012
LDL cholesterol (mmol/l)	+0.15	+4.1	0.006	−0.19	−5.2	0.006

[1] Data derived from Vessby et al. (2001), where 162 health volunteers were offered an isoenergetic diet for three months containing a high proportion of monounsaturated fatty acids (derived from margarines prepared from high oleic sunflower oil) or saturated fatty acids (derived from butter and high-saturate margarine).
[2] Mean change during challenge expressed as least square mean.
LDL = low-density lipoprotein.

carbohydrates in the human gut. There is also some evidence to suggest that the physiological effects of butyrate may be enhanced in the presence of other bioactive compounds including retinoic acid (Chen & Breitman, 1994), vitamin D (Tanaka et al., 1989) and 3-hydroxy-3-methylglutaryl coenzyme A reductase inhibitors (Velazquez et al., 1996). The mode of action is not well understood, but it has been suggested that butyrate may mediate its effects by increased accessibility of DNA to transcription factors via inhibitory effects on histone deacetylase (Parodi, 2001).

Several studies using rodent animal models have established that butyrate is also effective in reducing the incidence of chemically induced mammary carcinomas. Supplementing the diet with 6% sodium butyrate was shown to decrease the occurrence of mammary papillary carcinomas and adenocarcinomas in rats (Yanagi et al., 1993). More recently, the effects of butyrate in the form of anhydrous milk fat or tributyrin on the development of nitrosomethylurea-induced mammary tumors was demonstrated in rats (Belobrajdic & McIntosh, 2000). Compared with milk fat, feeding isoenergetic diets containing fat in the form of sunflower seed oil resulted in an 88% increase in the relative risk of rats developing a tumor. Inclusion of 1% and 3% of tributyrin in the diet was found to reduce tumor incidence by 20% and 52%, respectively. Even though ingestion of milk and dairy products may not result in substantial increases in plasma butyrate concentrations, it has been argued that the physiological effects of butyrate from these sources may be enhanced several-fold via synergistic interactions with other anticarcinogenic components in milk fat and other constituents in the human diet (Parodi, 2001).

Branch-Chain Fatty Acids

Milk fat contains a diverse range of branch-chain fatty acids, with 56 specific isomers being reported, with chain lengths varying from 4 to 26 carbon atoms (Ha & Lindsay, 1990; Jensen, 2002). The major branch-chain fatty acids in milk fat can be classified into one of three classes: even-chain *iso* acids, odd-chain *iso* acids, and odd-chain *anteiso* (Vlaeminck et al., 2006). Milk fat also contains relatively minor amounts of *w*-alicyclic fatty acids with or without a substitution for double bonds or hydroxylation (Brechany & Christie, 1992, 1994). The terms *iso* and *anteiso* designate the position of the branch chain in the fatty acid moiety. For *iso*-methyl fatty acids, the position of the branch chain is located on the penultimate carbon atom, whereas the branch point is positioned on the carbon atom two from the end in *anteiso*-methyl-fatty acids. In ruminant milk fat, 15:0 *anteiso* and 17:0 *anteiso* are typically the most abundant branch-chain fatty acids in milk (Table 3).

A number of studies have demonstrated that several branch-chain fatty acids exhibit anticarcinogenic properties. Incubation of 13-methyltetradecanoic acid (15:0 *iso*) extracted from a fermented soybean product induced cell death in a wide range of human cell lines, including colon, gastric, liver, lung, and prostate

Table 3 Branch-Chain Fatty Acid Concentrations in Bovine Milk Fat

Systematic Name	Short-Hand	Concentration (mg/100 g fatty acids)
2-Methylbutanoic acid	5:0 *iso*	6.4
3-Methylpentanoic acid	6:0 *iso*	3.2
4-Methylhexanoic acid	7:0 *iso*	0.5
8-Methylnonanoic acid	10:0 *iso*	<0.1
11-Methyldodecanoic acid	13:0 *iso*	40
12-Methyltridecanoic acid	14:0 *iso*	89
13-Methyltetradecanoic acid	15:0 *iso*	224
14-Methylpentadecanoic acid	16:0 *iso*	209
15-Methylhexadecanoic acid	17:0 *iso*	272
3-Methylbutanoic acid	5:0 *anteiso*	1.1
4-Methylpentanoic acid	6:0 *anteiso*	1.7
10-Methyldodecanoic acid	13:0 *anteiso*	83
12-Methyl-tetradecanoic acid acid	15:0 *anteiso*	462
14-Methyl-hexadecanoic acid	17:0 *anteiso*	501

Source: Adapted from Ha and Lindsay (1990) and Vlaeminck et al. (2006).

carcinoma, and mammary and pancreatic adenocarcinoma and leukemia *in vitro* (Yang et al., 2000). Oral administration of 15:0 *iso* (70 mg/kg body weight) over a 40-day period was found to inhibit the growth of human prostate and liver cancer cells transplanted into the prostate and liver of mice, by 84.6% and 65.2%, respectively (Yang et al., 2000). Further studies showed that a number of branch-chain fatty acids were effective in inhibiting the fatty acid synthesis of human breast cancer cells *in vitro* (Wongtangtintharn et al., 2004). Both *anteiso* and *iso* branch-chain fatty acids inhibited tumor growth. The highest activity was observed with 16:0 *iso*, while the inhibitory effects were reduced with an increase or decrease in carbon chain length. Furthermore, the cytotoxicity of the branch chain in fatty acids was reported to be comparable to that exerted by CLA. Based on *in vitro* incubations, it was concluded that branch-chain fatty acids inhibit fatty acid synthesis of tumor cells via direct effects on fatty acid synthetase and reductions in fatty acid precursor supply (Wongtangtintharn et al., 2004).

Trans Fatty Acids

The role of TFA in chronic disease has been the stimulus for recent changes in food labeling legislation in several countries. Regulations in Canada and the United States have been implemented to reduce TFA consumption via legislation that requires a declaration on foods containing 0.5 g or more TFA per serving. In Denmark, local or imported oils and fats containing more than 2% TFA are excluded in the manufacture of processed foods, while animal-derived foods are exempt. In EU countries, Australia, and New Zealand, declarations of the TFA content of foods remain voluntary and are only required if nutritional claims are made. Canadian regulatory authorities and the U.S. Food and Drug Administration have for regulatory purposes defined TFA as "all unsaturated fatty acids that contain one or more isolated (i.e., nonconjugated) double bonds in a trans-configuration." Fatty acids with a conjugated bond, principally isomers of CLA in ruminant-derived foods and dietary CLA supplements, are exempt from TFA labeling regulations. Processing technologies have been developed for reducing the TFA content of edible oils and fats to meet legislative requirements. As a result, the contribution of industrial sources to TFA consumption is declining, with the implication that in the future, the majority of TFA in the diet will be from ruminant-derived foods (Lock et al., 2005b). It is probable that TFA in the human diet will be the subject of closer scrutiny in the future following the report on "Diet, Nutrition and the Prevention of Chronic Diseases" (WHO, 2003) that recommended that intake of TFA should not exceed 1% of total energy to reduce CVD risk, without explicitly discriminating between sources of TFA in the human diet. The following section examines the role of TFA in the diet on chronic disease and specifically emphasizes the possible implications of ruminant-derived TFA in milk on human health-related outcomes.

Over a number of years, there has been an increasing body of data indicating that high intakes of TFA are associated with a substantial increase in CHD (Willett et al., 1993; Kromhout et al., 1995; Ascherio et al., 1999a, b). Early studies in the United Kingdom identified an association between consumption of hydrogenated vegetable and marine oils and deaths from ischemic heart disease (Thomas et al., 1983; Thomas, 1992), and by the mid-1990s, it was clear that in addition to evidence of increased CHD risk from epidemiological studies, unique adverse effects of TFA on plasma lipids were also evident (Ascherio et al., 1999a, b). Even though medium-chain SFA and TFA result in comparable increases in LDL cholesterol (Mensink & Katan, 1990; Judd et al., 1994), TFA, but not SFA, lower HDL-cholesterol concentrations, resulting in an increase in the total-to-HDL cholesterol ratio (Mozaffarian et al., 2006), which has been suggested to be a robust predictor of CHD risk (Mensink et al., 2003). Based on an extensive meta-analysis, there is strong support for consumption of TFA being a greater risk factor for CHD than SFA (Mensink et al., 2003), although it should be noted that isomer-specific effects of TFA have not been considered. In addition to the effects on plasma cholesterol, there is now clear evidence that TFA also increase plasma triglycerides relative to other dietary fatty acids (Mensink et al., 2003), raise concentrations of the Lp(a) lipoprotein (Ascherio et al., 1999a), and reduce mean LDL-cholesterol particle size (Mauger et al., 2003). It is probable that these additional effects on blood lipids also contribute to the positive association between TFA and CHD risk. In support of this, the incidence of CHD identified in prospective studies is higher than would be predicted by changes in plasma cholesterol alone (Mozaffarian et al., 2006).

Recent research has also provided evidence that TFA are pro-inflammatory (Mozaffarian et al., 2006). Since inflammation is an independent risk factor for atherosclerosis, sudden cardiac death, and other aspects of chronic disease, pro-inflammatory properties may be a crucial component of the adverse effects attributed to TFA in the human diet. Data are also emerging to suggest that the physiological effects of TFA are isomer-dependent, in much the same way that carbon chain length determines metabolic responses to SFA.

Partially hydrogenated vegetable oils and ruminant-derived foods are the main sources of TFA in the human diet. Estimates of TFA intake and the contribution from industrial and ruminant fats vary considerably between countries. It has been estimated that approximately 80% of dietary TFA in the United States is derived from partially hydrogenated plant oils, with the remainder being supplied by ruminant products (U.S. Food and Drug Administration, 2003). However, the TRANSFAIR study (Hulsof et al., 1999) indicated that in all the European countries studied, *trans* isomers of 18:1 were the major TFA in the diet (Table 4), but this varied between individual populations. Due to analytical limitations, specific isomers in food ingredients were not determined, but the contribution of whole milk, cheese, and butter to total TFA was estimated (Hulsof et al., 1999). Overall, the mean contribution across all countries was 37.8% of total TFA intake, but the range varied considerably, from 16.7% (Netherlands) to 71.8% (Germany). The source of TFA in the human diet is thus an important factor when attempting to establish the role of TFA on human health-related outcomes. Ruminant-derived foods typically contain 1–8% of total fatty acids as TFA, with *trans* 18:1 being quantitatively the most important in ruminant milk (Wolff, 1995; Precht & Molkentin, 1997, 1999; Ledoux 2002; Goudjil et al., 2004) and meat (Wolff, 1995; Dannenberger et al., 2004; Nuernberg et al., 2005). In contrast to hydrogenated oils, the distribution of positional isomers in milk fat and ruminant meats is skewed, and in most cases *trans*-11 18:1

Table 4 Mean Intake of *Trans* Fatty Acids Across 14 European Countries

Trans fatty acid	Men	Women
Total intake (g/day)	3.14	2.69
% of total *trans* intake		
14:1 *trans*-9	5.7	5.7
16:1 *trans*-9	8.5	8.4
18:1[1]	67.0	68.5
18.2 *trans*-9, *trans*-12	11.2	10.9
18:3 *trans*[2] + 20:1 *trans*-11	3.5	3.6
20:2 *trans*-11, *trans*-14	2.6	1.2
22:1 *trans*-13	1.5	1.9

[1] Calculated as the sum of *trans*-6, -9, and -11 18:1.
[2] Specific mono, di, and tri *trans* isomers of 9, 12, and 15 18:3 not determined.
Data derived from Hulshof et al. (1999).

(vaccenic acid) is the major isomer (Figs. 1a and b). Partially hydrogenated vegetable and marine oils can contain up to 60% TFA, predominantly as *trans* 18:1 (Lock et al., 2005b). The isomeric profile of partially hydrogenated oils typically follows a Gaussian distribution resulting in *trans*-9, -10, -11, and -12 as the main positional isomers (Fig. 1d). Interestingly, the *trans*-18:1 profile of human milk fat is less distinct compared with ruminant milk fat (Fig. 1c), reflecting the contribution of both industrial and ruminant-derived lipids to TFA intake. Due to the marked differences in isomer profile, there is a need to distinguish between the biological effects of industrial and ruminant-derived TFA in the human diet.

Overall, the evidence from epidemiological studies examining the association between the intake of TFA from ruminant-derived foods and the risk of CHD have pointed toward rather innocuous or possibly protective effects (Jakobsen et al., 2006). In two prospective cohort studies (Willet et al., 1993; Pietinen et al., 1997), an inverse association between energy-adjusted TFA from ruminant foods and CHD risk was reported. These findings are consistent with an earlier case-control study reporting that the relative risk of myocardial infarction (MI) for the highest versus the lowest quintile of energy-adjusted intake of ruminant

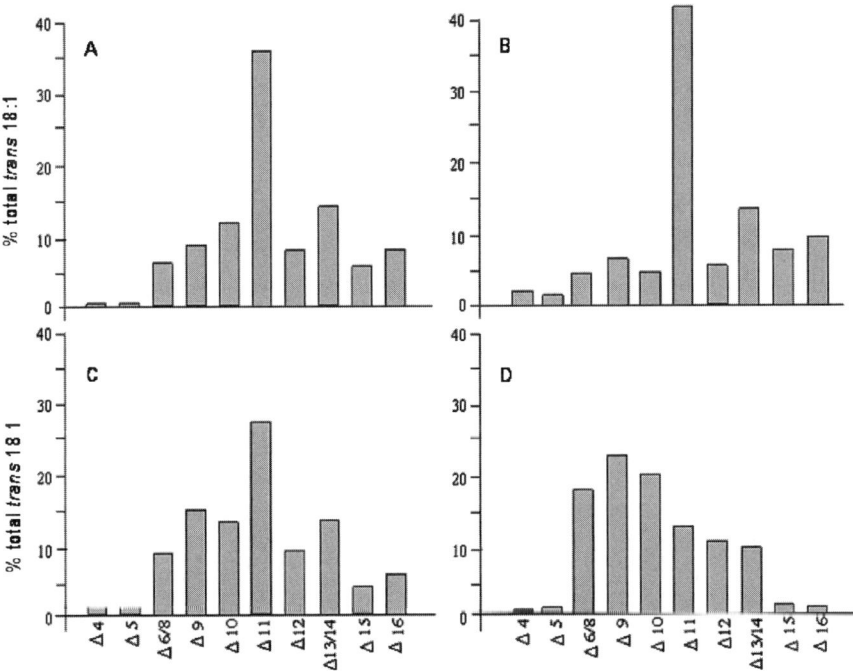

Fig. 1 Distribution of *trans* 18:1 isomers in (a) caprine milk fat, (b) bovine milk fat, (c) human milk fat, and (d) margarines. Values on the *x*-axis indicate the position of the double bond. (Data from Precht and Molkentin, 1999, and Ledoux et al., 2002.)

TFA was 1.02 (95% CI 0.43–2.41; Ascherio et al., 1994). In contrast, the findings from a prospective population study in the elderly identified a direct, but nonsignificant, association between the intake of TFA from both ruminant and industrial sources and the risk of coronary heart disease (Oomen et al., 2001). Increases of 0.5% of dietary energy from intake from ruminant TFA were associated with a relative risk of CHD of 1.17 (95% CI 0.69–1.98) compared with 1.05 (95% CI 0.94–1.17) for TFA derived from partially hydrogenated vegetable oils (Oomen et al., 2001). No study has yet found a significant positive relationship between the intake of ruminant-derived TFA and CHD risk. In a more recent prospective case-control study, the intake of milk fat based on 15:0 and 17:0 as biomarkers was found to be inversely associated with the risk of first acute MI, but adjustment for clinical risk factors removed this relationship (Warensjö et al., 2004). Most of the evidence to date suggests that increased milk consumption is associated with a reduction in CHD risk (refer to the review of Elwood et al., 2005). Mozaffarian et al. (2006) proposed that the lack of increase in CHD risk with the higher intakes of TFA from ruminant-derived foods, relative to the substantial risk associated with TFA from industrial sources, may simply reflect either the lower intake of ruminant TFA in the human diet or an isomer-specific bioactivity or be due to the activity of other compounds in dairy and meat products that negate or mitigate against any adverse effects of TFA.

Comparisons of relative effects of TFA derived from ruminant and industrial sources on CHD risk (Willett et al., 1993; Ascherio et al., 1994; Pietinen et al., 1997) have been based on quintiles of intake, which implies that associations were evaluated across different ranges in TFA between the sources. For ruminant TFA, the quintiles varied between ca. 0.5–2.5 g/day, while that for industrial TFA ranged from 0.1 to 5.1 g/day. When relationships have been examined and associations are based on absolute intakes of up to 2.5 g/day, no differences in CHD risk between dietary sources of TFA have been found (Weggemans et al., 2004). At higher daily intakes (>3 g), total and industrial TFA were associated with an increased risk of CHD, but insufficient data were available on ruminant TFA at this level of intake, leading Weggemans et al. (2004) to conclude that in light of the limited data available, there was no justification for discriminating between the sources of TFA in the human diet. Lock et al. (2005b) challenged these findings and argued that TFA from milk fat did not contribute to CHD risk based on evidence that (1) the degree of coronary artery disease in patients who underwent coronary angiograph was positively related to platelet content of *trans*-9 and *trans*-10 18:1, but not *trans*-11 18:1 (Hodgson et al., 1996), (2) *trans*-11 18:1, the major TFA in ruminant fats, is converted via the action of Δ-9 desaturase to *cis*-9, *trans*-11 CLA in humans (Adlof et al., 2000; Turpeinen et al., 2002), and (3) most of the evidence suggests that *cis*-9, *trans*-11 CLA has no adverse effects on blood lipids (Terpstra, 2004; Wahle et al., 2004; Yaqoob et al., 2006).

More recent experiments in animal models and human intervention studies have provided indirect evidence on the possible role of ruminant TFA on CVD risk. Feeding hamsters modified butter containing higher concentrations of *cis*-9, *trans*-11 CLA and *trans*-11 18:1 (69% of total *trans*-18:1) and lower concentrations of SFA was shown to significantly reduce the ratio of [very low-density lipoprotein (VLDL) + intermediate-density lipoprotein + LDL cholesterol]/HDL (0.60) compared with standard butter (1.70) or as a mixture (3:1, w/w) with standard butter (1.04; Lock et al., 2005a). The effects on plasma lipoproteins was confirmed in hamsters fed butter naturally enriched with *trans*-11 18:1 and *cis*-9, *trans*-11 CLA containing comparable SFA concentrations as the control butter (Valeille et al., 2006). Furthermore, aortic cholesterol ester concentrations were lower on diets containing CLA-enriched butter compared with control butter that contained similar concentrations of SFA and *cis*-18:1 fatty acids, leading to the conclusion that *cis*-9, *trans*-11 CLA exerts antiatherogenic effects in the hamster and possibly in humans. A similar strategy has been used to indirectly assess the effects of ruminant TFA in the hypercholesterolemic rabbit model. Rabbits were fed standard butter, butter containing enhanced concentrations of *cis*-9, *trans*-11 CLA, and *trans*-11 18:1, or butter containing *trans*-10 18:1 as the major isomer (72% of total *trans*-18:1), with both modified butters supplying the same amount of SFA. Measurements of blood lipids indicated higher plasma triacylglycerides and Apolipoprotein B concentrations in rabbits fed standard and *trans*-10 18:1-rich butter compared with butter containing enhanced *cis*-9, *trans*-11 CLA and *trans*-11 18:1 concentrations (Bauchart et al., 2007). Over a 12-week period, diets containing *trans*-10 18:1-rich butter induced a shift toward more dense LDL and increased the [VLDL + LDL]:HDL ratio 1.7–2.3-fold compared with the other butter sources (Bauchart et al., 2007). The *trans*-10 18:1-rich butter also increased VLDL, LDL, and total cholesterol, non-HDL:HDL ratio, and aortic lipid deposition compared with the *cis*-9, *trans*-11 CLA and *trans*-11 18:1-enriched butter, whereas the latter decreased HDL cholesterol and increased liver triacylglycerides compared with other treatments (Roy et al., 2007). Overall, the results suggested that higher intakes of *cis*-9, *trans*-11 CLA and *trans*-11 18:1 are neutral or reduce aortic lipid deposition, whereas increases in *trans*-18:1 similar to the isomer profile of industrial TFA sources are associated with detrimental effects on plasma lipids and lipoprotein metabolism in hypercholesterolemic rabbits (Roy et al., 2007).

The findings from experiments with animal models are in broad agreement with an intervention study examining the impact of *cis*-9, *trans*-11 CLA-enriched milk, butter, and cheese on the blood lipid profile, the atherogenicity of LDL, and markers of inflammation and insulin resistance in healthy middle-aged men compared with standard milk and dairy products (Tricon et al., 2006). Alterations in the diet of lactating cows to produce CLA-enriched milk and dairy products also reduced SFA content and induced several-fold increases in TFA concentrations (mainly *trans*-11), leading to marked differences in mean intakes of *trans*-18:1 between control and CLA treatment groups (0.78 and 6.34 g/day, respectively). Increases in *cis*-9, *trans*-11 CLA and *trans*-18:1 consumption over a six-week period had no significant effects on body weight, inflammatory markers, insulin, glucose, triacylglycerols, total, or LDL or HDL cholesterol but resulted in a small significant increase in the ratio of LDL to HDL cholesterol (mean change −0.10 and 0.11, for control and CLA-enriched products, respectively). Modified dairy products were also found to alter the fatty acid composition of LDL cholesterol but had no significant effect on LDL particle size or the susceptibility of LDL to oxidation. Overall, variables related to CVD risk were not significantly altered in volunteers consuming full-fat dairy products containing elevated concentrations of TFA (Tricon et al., 2006). The effects of CLA-enriched butter containing more than 10-fold enrichment in CLA and added coconut oil and palm stearin resulted in a significantly lower reduction in plasma total cholesterol and in HDL-to-total cholesterol ratio relative to the habitual diet than a standard control butter containing a small proportion of olive oil in overweight men (Desroches et al., 2005). No effects on abdominal accumulation of visceral or subcutaneous adipose were found. In contrast, butter enriched in *trans*-11 18:1 was reported to significantly decrease total and plasma HDL-cholesterol concentrations (6% and 9%, respectively) in healthy young men compared with a standard butter, but it had no effect on the total-to-HDL cholesterol ratio (Tholstrup et al., 2006). It was suggested that these differences may reflect the lower SFA content of the modified butter rather than a direct effect of *trans*-11 18:1 on plasma lipids. Consistent with this suggestion, comparisons of standard and modified dairy fats in human volunteers indicated positive effects on plasma LDL-to-HDL ratio and Lipoprotein (a) concentrations associated with a reduction in milk fat SFA and increase in *cis* and *trans* 18:1 (Seidel et al., 2005). In addition to differences in the amount of test fats in the diet, genotype, gender, and body mass index, variation in response to plasma lipid profiles to modified milk and dairy products may therefore be explained by changes in the overall milk fat fatty acid profile characterized as reductions in SFA and elevated TFA content that accompany the increases in CLA concentrations. Further research is required to evaluate the effects of specific *trans* 18:1 isomers to substantiate the relative risks on CHD associated with TFA from industrial or ruminant sources.

Measurements of adipose in human subjects has also indicated that *trans* isomers of 18:2 *n*-6 (collectively *cis*-9, *trans*-12 18:2, *trans*-9, *cis*-12 18:2, and *trans*-9, *trans*-12 18:2) account for up to 25% of total TFA, suggesting that mono or di *trans* octadecadienoic fatty acids are also a major source of TFA in the human diet (Lemaitre et al., 1998). Several studies have recently examined the association between the intake of *trans*-containing isomers of 18:2 on the incidence of MI. Evidence from a population-based case-control study (Lemaitre et al., 2002) revealed that higher intakes of TFA were moderately associated with an increase in MI risk. More detailed assessment of the source of TFA in the diet revealed that high consumption of *trans* 18:1 was not associated with an increase in the incidence of MI, whereas high intakes of *trans* 18:2 were associated with a threefold increase in MI risk. In a follow-up, further measurements from the same populations indicated that higher concentrations of *trans* 18:2 in plasma phospholipid were associated with an increased risk of fatal ischemic heart disease, while *trans* 18:1 concentrations above the 20th percentile were associated with a lower risk (Lemaitre et al., 2006). Analysis of individual cases of sudden cardiac death also indicated the same associations (Lemaitre et al., 2006). These findings are comparable to a case-control population study from subjects experiencing primary nonfatal MI (Baylin et al., 2003). Total adipose tissue TFA was positively associated with MI risk. After adjustment for established risk factors, the increase in MI risk was largely attributable to adipose *trans* 18:2, and to a lesser extent *trans* 16:1, but no association was attributable to *trans* 18:1. In a more recent and extensive analysis, total adipose tissue TFA was also shown to be associated with increased risk of primary nonfatal MI that was mainly related to *trans* 18:2, while no relationship with adipose *trans* 18:1 concentrations was found (Colón-Ramos et al., 2006). While these studies provide no indication of cause and effect and offer no possible mechanisms to explain these findings, the evidence emerging from human clinical studies points toward *trans* fatty acids with more than one double bond as being particularly harmful, but further research is required before definitive conclusions can be drawn.

Hydrogenated plant oils, margarines, and edible oils (Ratnayake & Pelletier, 1992; Precht & Molkentin, 1997, 2000) and milk and dairy products (Ulberth & Henninger, 1994; Precht & Molkentin, 1997) contain several methylene-interrupted *trans* 18:2, but the isomer profiles of *trans* 18:2 differ between industrial and ruminant fats (Table 5). Early data on milk fat *trans* 18:2 have to be considered tentative since the identification of a number of

Table 5 Concentration of *Trans* Octadecadienoic Acids in Bovine Milk Fat, Partially Hydrogenated Vegetable Oil-Based Spreads and Margarines Reported in the Literature (g/100 g fatty acids)

Country of Origin Isomer	Canada[1]		Germany[2]	
	Hard Margarines	Soft Margarines	Margarine	Milk Fat
cis-9, *trans*-12	0.89	0.74	0.29	0.10
cis-9, *trans*-13 + *trans*-8, cis-12	1.24	1.02	0.04	0.11
trans-9, *cis*-12	0.78	0.62	0.23	0.07
trans-10, *cis*-15 + *trans*-9, *cis* 15	0.09	0.08	–	–
trans-11, *cis*-15	–	–	–	0.33
trans-9, *trans*-12	0.34	0.26	0.03	0.09
trans-8, *cis*-13 [3]	–	–	–	0.11
trans, *trans* [4]	0.25	0.15	0.04	0.19
cis,*trans*/*trans*, *cis*[4]	–	–	–	0.16
Total *trans* 18:2	3.57	2.87	0.61	1.29

[1] Data adapted from Ratnayake and Pelletier (1992).
[2] Data adapted from Precht and Molkentin (1997).
[3] Reported to co-elute with unidentified *cis*, *cis* 18:2 isomers.
[4] Double-bond position not determined.

isomers have not been demonstrated unequivocally, a criticism that also holds true for more recent studies (Shingfield et al., 2003; Jones et al., 2005; Loor et al., 2005a, b). More detailed analysis using gas-chromatography mass spectrometry analysis of 4,4-dimethyloxazoline (DMOX) fatty acid derivatives has revealed that the isomeric profile of *trans* 18:2 in bovine milk fat is more diverse than previously thought (Shingfield et al., 2006a). Further work in our laboratory based on a combination of silver-ion thin-layer chromatography and gas-chromatography mass-spectrometry analysis of DMOX derivatives (Kairenius et al., unpublished) has allowed the major mono- and di-*trans* 18:2 isomers in milk fat to be elucidated (Fig.2) and quantified (Table 6). Further studies are required to confirm these findings, but the evidence thus far suggests that the concentration and isomer distribution of *trans* 18:2 from ruminant and industrial lipids differ.

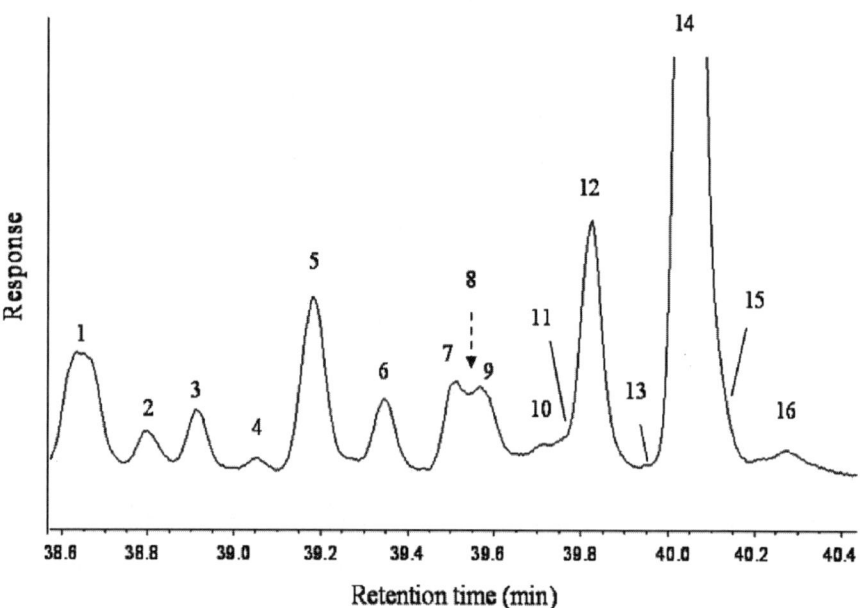

Fig. 2 Partial gas chromatogram indicating the separation of 18:2 methyl esters in milk fat from cows fed a grass silage-based diet obtained using a 100-m capillary column (CP Sil 88), temperature gradient, and hydrogen as a carrier (refer to Shingfield et al., 2003). Isomers were identified using a combination of silver-ion thin-layer chromatography and gas chromatography-mass spectrometry analysis of 4,4-dimethyloxaline derivatives (Kairenius, Toivonen, & Shingfield, unpublished). Peak identification: 1 = unresolved 19:0 and *cis*-15 18:1; 2 = *trans*-10, *trans*-14 18:2; 3 = *trans*-11, *trans*-15 18:2; 4 = *trans*-9, *trans*-12 18:2; 5 = *cis*-9, *trans*-13 18:2; 6 = 11-cyclohexyl-11:0; 7 = *cis*-9, *trans*- 14 18:2; 8 = indicates retention time of *cis*-9, *trans*-12 18:2 standard; 9 = *cis*-16 18:1; 10 = *cis*-12, *trans*-16 18:2; 11 = *trans*-9, *cis*-12 18:2; 12 = *trans*-11, *cis*-15 18:2; 13 = *cis*-7 19:1; 14 = *cis*-9, *cis*-12 18:2; 15 = unresolved *cis*-9, *cis*-15 18:2 and *cis*-9 19:1; 16 = *trans*-12, *cis*-15 18:2

Table 6 Concentration of *Trans* Octadecadienoic Acids in Milk Fat from Cows Fed Grass Silage-Based Diets (g/100 g fatty acids)[1]

Isomer	Mean	Range
cis-9, *trans*-13	0.23	0.18–0.34
cis-9, *trans*-14	0.10	0.08–0.17
trans-11, *cis*-15	0.22	0.15–0.32
trans-12, *cis*-15	0.02	0.01–0.03
trans-9, *trans*-12	0.02	0.02–0.03
trans-10, *trans*-14	0.05	0.04–0.06
trans-11, *trans*-15	0.07	0.06–0.08
Total *trans* 18:2	0.63	0.50–0.84

[1] Isomers identified based on a combination of thin-layer chromatography and GC-MS analysis of 4,4-dimethyloxazoline fatty acid derivatives (Kairenius et al., unpublished; refer to Fig. 2).

Conjugated Linoleic Acid

"Conjugated linoleic acid" is a collective term to describe a mixture of geometric and positional isomers of octadecadienoic acid containing a conjugated double bond. Dairy products are the main source of CLA in the human diet, with the *cis*-9, *trans*-11 isomer accounting for between 70–80% of total CLA intake (Lawson et al., 2001), since *cis*-9, *trans*-11 is the major isomer of CLA in ruminant milk (Palmquist et al., 2005; Luna et al., 2005; Table 7). Estimates of average CLA intake in the human diet range between 95 and 440 mg/day and differ between countries, with considerable variation between individual sectors of the population (Collomb et al., 2006).

An extensive body of data on the biological activity of CLA isomers based on studies with various cell lines, animal models, and studies with human subjects has accumulated during the last decade. A number of studies have provided strong evidence that various isomers of CLA inhibit the growth of a number of human cancer cell lines, reduce the rate of chemically induced tumor development, alter lipoprotein metabolism, modify immune function, and enhance lean body mass in animal models (Pariza, 1999; Whigham et al., 2000; Roche et al., 2001; Kritchevsky, 2003).

Early studies indicated that isomers of CLA in the diet reduced the occurrence and number of chemically induced mammary cancers in the rat in a dose-dependent manner (Ip et al., 1991, 1994). Follow-up studies also indicated that *cis*-9, *trans*-11 CLA-enriched butter was also active against incidence of tumors in a rat model of mammary carcinogenesis (Ip et al., 1999). Both *cis*-9, *trans*-11 CLA and *trans*-10, *cis*-12 CLA have been shown to be effective in reducing the formation of premalignant lesions in the rat mammary gland six weeks after carcinogen administration (Ip et al., 2002). Further studies in the rat have shown that *trans*-11 18:1, the major TFA in ruminant milk fat (Fig. 1),

Table 7 Distribution of Conjugated Linoleic Isomers in Bovine, Caprine, and Ovine Milk Fat

Isomer	Composition (g/100 g CLA)		
	Bovine[1]	Caprine[2]	Ovine[3]
cis-8, trans-10	<0.01–1.70	<0.01	NR
cis-9, trans-11	65.6–88.9	62.1–75.1	80.0–80.9
cis-11, trans-13	<0.01–0.23	0.16–0.69	NR
cis-12, trans-14	<0.01–1.06	0.00–0.13	1.69–1.83[4]
trans-7, cis-9	2.63–9.49	4.57–11.7	5.96–6.08
trans-8, cis-10	<0.01–2.33	1.85–3.48	NR
trans-9, cis-11	<0.01–3.93	<0.01–4.21	NR
trans-10, cis-12	<0.01–1.61	<0.01–0.90	0.55–0.57
trans-11, cis-13	0.06–9.33	0.22–0.48	2.14–2.38[4]
trans-6, trans-8	<0.01–1.40	0.12–1.91	<0.01
trans-7, trans-9	0.02–2.80	0.42–1.08	0.40–0.42
trans-8, trans-10	0.19–0.67	0.36–1.47	0.34–0.42
trans-9, trans-11	1.31–3.23	2.99–5.77	1.40–1.60
trans-10, trans-12	0.31–1.40	0.76–4.16	0.53–0.85
trans-11, trans-13	0.89–6.00	0.58–1.14	3.04–3.18
trans-12, trans-14	0.35–3.55	0.72–1.90	1.90–2.20
trans-13, trans-15	<0.01–0.16	<0.01	NR

[1] Derived from Piperova et al. (2000, 2002) and Shingfield et al. (2003, 2006, 2007).
[2] Shingfield, Rouel, & Chilliard, unpublished.
[3] Adapted from Luna et al. (2005).
[4] Double-bond geometry not determined; assumed configuration based on comparisons with measurements of bovine and caprine milk.
NR = not reported.

also exerts antimutagenic effects that are thought to be mediated by conversion to cis-9, trans-11 CLA via the action of Δ-9 desaturase (Corl et al., 2003; Lock et al., 2004).

In vitro studies with human tumor cells (mammary, lung, colon, prostate, and melanoma) have shown that a wide range of CLA isomers and conjugated 18:3 and 20:3 metabolites inhibit cell proliferation and induce apoptosis in a dose- and time-dependent manner (De La Torre et al., 2005; Beppu et al., 2006). Inhibitory effects were shown to be isomer-specific, with trans-9, trans-11 and 18:3 and 20:3 conjugates exhibiting the highest activity (De La Torre et al., 2005; Beppu et al., 2006). Further studies have confirmed that trans-9, trans-11 CLA exerts more potent antiprofilerative effects during incubations with human colon cancer cells than cis-9, trans-11 CLA (Coakley et al., 2006). These studies indicate that in addition to the major isomer cis-9, trans-11, minor isomers of CLA in ruminant milk fat (Table 7) may also confer antimutagenic effects in humans.

It has been proposed that the bioactivity of isomers of CLA on the development of mammary carcinomas may reflect the high concentration of adipocytes in mammary tissue (Bauman et al., 2005). Since CLA is preferentially incorporated into triacylglycerides (Banni et al., 2001), Bauman et al. (2005) argued that it would therefore accumulate in the adipocytes of the mammary gland during its development, thus providing a source of CLA that could protect against carcinogens later in life.

Isomers of CLA have also been shown to have other beneficial health effects in studies with animal models, including a decrease in plasma lipids, a reduction in the onset and severity of diabetes and obesity, immune modulation, and a change in the rate of bone formation (Whigham et al., 2000; Roche et al., 2001; Kritchevsky, 2003; Faulconnier et al., 2004; Palmquist et al., 2005). Further experimental evidence on the possible role of CLA on human health has been examined in several reviews (Wahle et al., 2004; Bauman et al., 2005; Yaqoob et al., 2006). Recent reviews have indicated that the majority of data reported in the literature have been derived from *in vitro* and animal model studies often using preparations containing a mixture of CLA isomers, containing *cis*-9, *trans*-11 and *trans*-10, *cis*-12 as major isomers. It should be noted that ruminant-derived foods contain very low amounts of *trans*-10, *cis*-12 CLA (Palmquist et al., 2005; Table 8), and therefore synthetic CLA supplements would be the main source of this isomer in the human diet. Furthermore, high doses of CLA have generally been used in experiments with animal models over short-interval studies, complicating the extrapolation of these findings to human health-related outcomes.

Epidemiological data have provided some evidence for a role of milk and dairy products in reducing breast cancer risk (Aro et al., 2000), but these findings were not supported by two subsequent studies (Chajes et al., 2002;

Table 8 Comparison of Fatty Acid Flow at the Omasum and Milk Fatty Acid Composition and Secretion in Cows Fed Grass Silage-Based Diets[1]

Fatty Acid	Omasal Flow (g/day)	Milk (g/100 g fatty acids)	(g/day)
10:0	–	2.59	15.50
10:1 *cis*-9	–	0.25	1.42
12:0	0.810	3.00	18.10
12:1 *cis*-9	–	0.07	0.42
14:0	1.880	11.70	71.60
14:1 *cis*-9	–	0.87	5.06
15:0	3.860	1.29	7.60
15:1 *cis*-9	–	0.01	0.07
16:0	37.70	30.3	187.00
16:1 *cis*-9	0.540	1.11	6.42
17:0	2.160	0.76	4.41
17:1 *cis*-9	–	0.36	2.08
18:1 *trans*-11	14.530	1.25	6.96
cis-9, *trans*-12 18:2	–	0.07	0.47
cis-9, *trans*-13 18:2	–	0.13	0.76
trans-7, *cis*-9 CLA	0.003	0.03	0.15
cis-9, *trans*-11 CLA	1.190	0.57	3.73

[1] Data derived from Shingfield et al. (2007).

Voorrips et al., 2002). More recently, an inverse relationship between milk fat intake and risk of colorectal cancer was reported (Larsson et al., 2005). In examining the literature, Yaqoob et al. (2006) concluded that studies in humans do not support the hypothesis that CLA is a protective factor in breast cancer, noting that associations between cancer risk and food consumption are difficult to establish and subject to severe bias. Extending the results from *in vitro* and animal model studies is extremely challenging due to the long latency in the development of tumors and the lack of consensus of suitable biomarkers of cancer (Bauman et al., 2005). Overall, the evidence from animal biomedical studies points toward a number of potential benefits of *cis*-9, *trans*-11 CLA in humans, including reductions in atherosclerosis and improved blood lipid profiles, in addition to potential protection against cancer.

Sphingomyelin

Milk and dairy products contain several classes of phospholipids including phosphatidylethanolamine, phosphatidylinositol, phosphatidylserine, and phosphatidylcholine. Important sphingolipids include sphingomyelin, glucosylceramide, and lactosylceramide, while lysophosphatidylcholine and phosphatidic acid are rarely detected (Christie et al., 1987). Sphingolipids are a class of phospholipids containing a phosphorylated polar group and a long-chain *N*-acetylated fatty acid component, often referred to as a sphingoid base. Long-chain bases are typically comprised of dihydroxy analogs, saturated (sphinganine) or unsaturated (sphingosine) fatty acids (Jensen, 2002). Milk fat is secreted as small lipid droplets varying in size from 0.1 to 15 µm. During the extrusion from mammary secretory cells, fat droplets are surrounded by a membrane comprised of lipid and proteins, which results in the incorporation of sphingomyelin and minor amounts of other sphingolipids into the milk fat globule membrane (Spitsberg, 2005). Typically, phospholipids account for 0.2–1.0% of total lipids in bovine milk fat (Christie et al., 1987; Bitman & Wood, 1990; Rombaut et al., 2005). The milk fat globule membrane is comprised of three major phospholipid species—sphingomyelin, phosphatidyl choline, and phosphatidyl ethanolamine—with sphingomyelin accounting for between 18–20% of total phospholipids in milk (Avalli & Contarini, 2005; Rombaut et al., 2005). The sphingomyelin content of milk fat varies between 0.65–1.27 mg/g fat, with the concentrations in whole milk typically ranging between 26.4–119 µg/g milk (Bitman & Wood, 1990; Avalli & Contarini, 2005; Rombaut et al., 2005; Graves et al., 2007).

In recent years, several studies have examined the bioactivity of phospholipids, including sphingolipids. One phospholipid in particular, sphingomyelin, has been shown to reduce the number of colon tumors and aberrant crypt foci in mice and inhibit the proliferation of colon carcinoma cell lines (Parodi, 2001; Berra et al., 2002; Schmelz, 2003; Spitsberg, 2005). Sphingomyelin has also been shown to reduce cholesterol absorption in rats, possibly due to the highly saturated long-chain fatty acyl groups inhibiting the rate of luminal lipolysis, micellar solubilization, and transfer of micellar lipids to the enterocyte (Noh & Koo, 2004). Sphingomyelin via the action of the biologically active metabolites ceramide and sphingosine is known to be important in transmembrane signal transduction and cell regulation, causing the arrest of cell growth and the induction of cell differentiation and apoptosis (Parodi, 2001).

Metabolic Origins of Milk Lipids

In order to develop effective nutritional strategies for enhancing the concentration of specific fatty acids and bioactive lipids in milk, the metabolic origins of these compounds have to be considered. In the following section, a brief overview of mammary fatty acid synthesis and ruminal lipid metabolism is provided. A more detailed consideration of the biochemistry of milk fat synthesis and lipid metabolism in ruminant animals is the subject of several reviews (Harfoot & Hazlewood, 1997; Griinari & Bauman, 1999; Lock & Shingfield, 2004; Palmquist et al., 2005).

Mammary De Novo Fatty Acid Synthesis

Bovine milk typically contains between 3–5% fat depending on diet and genotype (Givens & Shingfield, 2006). Lipids in milk are secreted as globules emulsified in the aqueous phase and contain nonpolar core lipids mainly as triacylglycerides (96–98% of total milk lipids) with small amounts of cholesteryl esters (0.02%), free fatty acids (0.22%), and retinol esters (Jensen, 2002). Milk fat globules are surrounded by a membrane comprised of phopholipids, cholesterol, and cholesterol esters (Spitsberg, 2005). Milk fat triacylglycerides are thought to contain ca. 400 fatty acids (Jensen, 2002), but it is probable that this number will increase with advances in analytical methods. Whilst there is enormous diversity, the major fatty acids in milk fat triacylglycerides include SFA 4:0–18:0, MUFA, *cis*-9 16:1, *cis*-9 18:1, *trans* 18:1, 18:2 *n*-6, and 18:3 *n*-3 (Lock & Bauman, 2004). Even though a wide range of triacylglyceride structures have been indentified, assembly of milk fat triacylglycerides is not random. Almost all 4:0 and the majority of 6:0 (ca. 90%) are preferentially esterified at *sn*-3; 10:0, 12:0, and 14:0 are esterified at all positions, with *sn*-2 being the most common; and 16:0, 18:0, and 18:1 are distributed equally between *sn*-1 and *sn*-2 (Jensen, 2002).

Fatty acids incorporated into milk fat triacylglycerides are derived from two sources: mammary *de novo* synthesis and the uptake of preformed fatty acids from peripheral circulation. Direct uptake typically contributes to about 60% of total fatty acid secretion in milk fat (Chilliard et al., 2000). Both acetate (2:0) and β-hydroxybutyrate derived from organic matter digestion in the rumen are used by mammary epithelial cells to synthesize short- and medium-chain fatty acids. Mammary *de novo* synthesis accounts for all 4:0 to 12:0, most of the 14:0 (ca. 95%), and about 50% of 16:0 secreted in milk, while all 18 carbon and longer-chain fatty acids are derived entirely from circulating plasma lipids (Lock & Shingfield, 2004).

De novo fatty acid synthesis has an absolute requirement for acetyl-CoA, two key enzymes (acetyl-CoA carboxylase and fatty acid synthetase), and a supply of NADPH-reducing equivalents (refer to Lock & Shingfield, 2004). Acetate and, to a lesser extent, β-hydroxybutyrate contribute to the initial four carbon units required for fatty acid synthesis. Acetate is converted to acetyl Co A in the cytosol and incorporated into FA via the malonyl-Co A pathway, whereas β-hydroxybutyrate is incorporated directly following activation to butyl-Co A. Conversion of acetate to acetyl-CoA via acetyl-CoA carboxylase is considered to be the rate-limiting step (Bauman & Davis, 1974). Fatty acid synthetase consists of a large enzyme complex and is responsible for chain elongation. Acetyl, butyl, and malonyl-Co A condense within the fatty acid synthetase complex, and chain elongation occurs through continual loading of additional malonyl-Co A groups. A distinctive feature of the bovine mammary gland is its ability to release fatty acids from the synthetase complex at various stages, resulting in the secretion of a wide range of short- and medium-chain fatty acids. It is now well established that increases in the supply of long-chain fatty acids to the mammary gland inhibit the synthesis of short- and medium-chained saturates (Chilliard et al., 2000).

Mammary uptake of 16:0 and all longer fatty acids are derived from the absorption of fatty acids in the small intestine and during mobilization of tissue adipose. Absorbed fatty acids derived from the diet, microbial fatty acid synthesis in the rumen, and endogenous lipids are used for the assembly of triacylglycerides in the intestinal epithelium and transported as plasma chylomicrons and VLDL (Noble, 1981). Long-chain fatty acids taken up by the mammary gland are obtained from the triacylglycerol fractions of circulating VLDL and chylomicrons via the action of mammary lipoprotein lipase, an enzyme located on the surface of mammary endothelial cells (Christie, 1981). However, fatty acids incorporated into cholesterol esters and phospholipids (PL) and transported in plasma mainly as HDL are relatively poor substrates for lipoprotein lipase. Long-chain fatty acids liberated during lipolysis of adipose triacylglycerides are transported in blood as nonesterified fatty acids (NEFA). Even though plasma triacylglycerides and NEFA represent less than 3% of total plasma lipids, these sources contribute to about 60% of the fatty acids secreted in milk (Chilliard et al., 2000).

Role of Δ-9 Desaturase

Mammary gland epithelial cells contain the Δ-9 desaturase complex (often referred to as stearoyl-CoA desaturase) that is responsible for catalyzing the oxidation of fatty acyl CoA esters that results in the introduction of a *cis* double bond between carbon atoms 9 and 10. Palmitoyl- and stearoyl-CoA are the preferred substrates for Δ-9 desaturase, leading to the formation of palmitoleoyl- and oleoyl-CoA, respectively (Ntambi, 1999). Studies of Δ-9 desaturase in ruminants are limited, and most of the data on the regulation of this enzyme has been derived from experiments with rodents and rodent cell lines. Results indicate that gene expression and the amount of enzyme are regulated by dietary factors such as glucose and PUFA and hormones such as insulin, glucagon, and thyroid hormone and that Δ-9 desaturase is not inhibited by substrate supply or product formation (Palmquist et al., 2005). Few studies have investigated the nutritional regulation of mammary Δ-9 desaturase mRNA abundance and/or protein activity in ruminants. In cows, data from four nutritional studies examining the impact of plant or marine oil supplements revealed that only treatments containing rumen-protected fish oil were shown to significantly reduce Δ-9 mRNA abundance (for a review, refer to Bernard et al., 2006). In goats fed hay-based diets, the inclusion of *cis*-9 18:1 or 18:2 *n*-6-rich sunflower oil and linseed oil were shown to have negative effects on *in vitro* Δ-9 desaturase activity (Bernard et al., 2006). Similarly, supplementing hay-based diets with formaldehyde-treated linseed decreased Δ-9 desaturase mRNA, whereas lipid supplements had no effect on Δ-9 desaturase mRNA or activity *in vitro* in goats fed maize silage-based diets (Bernard et al., 2006). Studies in late-lactation goats also demonstrated that the addition of soybeans to Lucerne hay-based diets had no effect on Δ-9 desaturase mRNA (Bernard et al, 2005a). Results from goats suggest the existence of interactions between the composition of the basal diet and lipid supplements with PUFA derived from the diet, or fatty acid metabolites formed during ruminal biohydrogenation may inhibit transcriptional or post-transcriptional regulation of Δ-9 desaturase.

Activity of Δ-9 desaturase in the mammary gland of ruminants is thought to occur as a mechanism to maintain and regulate the fluidity of milk for efficient secretion from the mammary gland. Conversion of 18:0 to *cis*-9 18:1 is the predominant precursor: product of the Δ-9 desaturase complex and results in about 40% of the 18:0 taken up by the mammary gland being converted to *cis*-9 18:1, while conversion of 16:0 is much lower, at ca. 8% (Chilliard et al., 2000). The action of Δ-9 desaturase in the bovine mammary gland is not confined to 16:0 and 18:0 fatty acids, and other SFA, including 10.0, 12:0, 14:0, and 17:0, also serve as substrates (Bauman & Davis, 1974; Fievez et al., 2003; Shingfield et al., 2006b). Studies in rat liver microsomal systems have also established that Δ-9 desaturase is also capable of converting positional isomers of *trans* 18:1 to *cis*-9, *trans* 18:2 (Mahfouz et al., 1980; Pollard et al., 1980). *Trans* 18:1 with double bonds from Δ-4 to Δ-13, other than Δ-8, 9, and 10, served as substrates, with the rate of desaturation being higher as the distance of the *trans* double bond from the Δ-9 position increased.

The important role of Δ-9 desaturase in the appearance of *cis*-9, *trans*-11 CLA in ruminant milk fat was first recognized by Griinari & Bauman (1999) following observations that supplements of 18:3 *n*-3 in the diet increased the concentration of this isomer of CLA in milk fat even though *cis*-9, *trans*-11 CLA is not an intermediate of 18:3 *n*-3 metabolism in the rumen. Strong support for the role of Δ-9 desaturase in endogenous *cis*-9, *trans*-11 CLA synthesis was gained following observations that postruminal infusions of 12.5 g *trans*-11 18:1/day resulted in a 31% increase in milk fat *cis*-9, *trans*-11 CLA content (Griinari et al., 2000). Further studies have estimated the contribution of endogenous *cis*-9, *trans*-11 CLA synthesis in the lactating cow based on (1) postruminal infusions of sterculic oil to inhibit Δ-9 desaturase and measuring the changes in milk fatty acid composition (Griinari et al., 2000; Corl et al., 2001; Kay et al., 2004) or (2) comparison of the flow of *cis*-9, *trans*-11 CLA at the duodenum (Lock & Garnsworthy, 2002; Piperova et al., 2002; Loor et al., 2004, 2005b) or omasum (Shingfield et al., 2003) and secretion of this isomer of CLA in milk. Overall, these experiments provided evidence that between 64% and 97% of *cis*-9, *trans*-11 CLA in milk fat is synthesized endogenously via the action of Δ-9 desaturase on *trans*-11 18:1. Studies in lactating cows using ^{13}C-labeled *trans*-11 18:1 have confirmed that the mammary gland is the major site of endogenous *cis*-9, *trans*-11 CLA synthesis (Mosley et al., 2006). Measurements of ^{13}C enrichment in milk and plasma lipids indicated that 83% of *cis*-9, *trans*-11 CLA was synthesized in the mammary gland and that ca. 26% of *trans*-11 18:1 taken up by the mammary gland was desaturated (Mosley et al., 2006). Further research has shown that the extent of conversion to *cis*-9, *trans*-11 CLA is independent of the supply of *trans*-11 18:1 available to the mammary gland (Shingfield et al., 2007). A constancy of *trans*-11 18:1 desaturation and transfer of *cis*-9, *trans*-11 CLA from the abomasum into milk (Shingfield et al., 2007) also explains the close linear relationship between product and substrate for Δ-9 desaturase reported for both bovine (Griinari & Bauman, 1999; Chilliard et al., 2001) and caprine (Chilliard & Ferlay, 2004) milk fat across a wide range of *cis*-9, *trans*-11 CLA and *trans*-11 18:1 concentrations.

Studies involving postruminal infusions of sterculic acid or *trans*-10, *cis*-12 CLA to inhibit Δ-9 desaturase (Corl et al., 2002) and measurements of fatty acid flow at the duodenum (Piperova et al., 2002) or omasum (Shingfield et al., 2003) have provided strong evidence that *trans*-7, *cis*-9 CLA in milk fat (Table 7) is almost exclusively synthesized endogenously from *trans*-7 18:1. Furthermore, postruminal infusions of *trans*-12 18:1 have been shown to increase milk fat *cis*-9, *trans*-12 18:2 content (Griinari et al., 2000), but the extent of desaturation is lower than *trans*-11 18:1 (5.9% vs. 28.9%; Shingfield et al., 2007). Recent comparisons of the flow of fatty acids at the omasum or duodenum and milk fat composition in cows fed high-forage diets (forage:concentrate ratio 75:25 on a dry matter basis) based on grass silage (Shingfield et al., 2007) or grass hay (Loor et al., 2004, 2005a) have provided evidence that *cis*-9 15:1 and *cis*-9, *trans*-13 18:2 are also synthesized endogenously in the mammary gland (Table 8).

Lipid Metabolism in the Rumen

Even though ruminant diets contain predominantly unsaturated fatty acids, ruminant meat and milk contain much higher levels of SFA due in part to extensive biohydrogenation of dietary unsaturated fatty acids in the rumen. It is generally considered that rumen bacteria rather than protozoa are responsible for biohydrogenation (Harfoot & Hazlewood, 1997), which serves to reduce the toxic effects of unsaturated fatty acids on bacterial growth (Palmquist et al., 2005; Wasowska et al., 2006).

Following ingestion, plant lipids become released from structural components through mastication and microbial digestive processes and are hydrolyzed by microbial lipases. The rate of hydrolysis is inversely related to the melting point (Palmquist et al., 2005). Following lipolysis, NEFA are released into the rumen and adsorbed onto feed particles and hydrogenated or directly incorporated into bacterial lipids (Demeyer & Doreau, 1999). Numerous *in vitro* and *in vivo* studies have elucidated the major pathways of ruminal biohydrogenation (refer to Harfoot & Hazlewood, 1997). Metabolism of 18:2 *n*-6 and 18:3 *n*-3 is considered to involve at least two distinct populations of ruminal bacteria that under normal conditions proceed via isomerization of the *cis*-12 double bond resulting in the formation of conjugated 18:2 or 18:3, respectively. Conjugated intermediates are transient and are subsequently reduced to 18:0 as the final end product, with *trans*-11 18:1 as a common intermediate metabolite (Fig. 3). The final reduction step is considered to be rate-limiting; therefore, *trans* 18:1 intermediates can accumulate (Harfoot

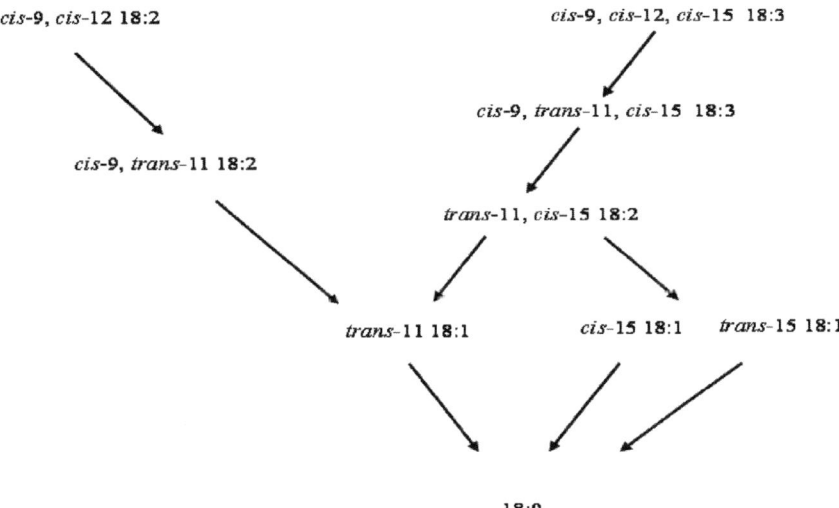

Fig. 3 Major pathways of 18:2 *n*-6 and 18:3 *n*-3 metabolism in the rumen (adapted from Harfoot & Hazlewood, 1997)

& Hazlewood, 1997; Griinari & Bauman, 1999). More recent studies have shown that biohydrogenation of dietary PUFA is more diverse than previously thought, and a wide range of fatty acid intermediates are formed in the rumen and incorporated into milk fat (Piperova et al., 2002; Shingfield et al., 2003; Loor et al., 2004, 2005a, b, c). Estimates reported in the literature suggest that on most diets ruminal metabolism of dietary 18:2 *n*-6 and 18:3 *n*-3 biohydrogenation varies between 70–95% and 85–100%, respectively (Doreau & Ferlay, 1994); therefore, with the exception of diets containing fish oil or marine lipids, 18:0 is the major fatty acid leaving the rumen.

The composition of the diet, the amount and type of lipid supplements, and interactions between these factors are known to alter the predominant ruminal biohydrogenation pathways resulting in changes in the profile of fatty acids available for absorption and incorporation into milk fat. A detailed appraisal of the impact of nutrition on ruminal lipid metabolism (Palmquist et al., 2005) concluded that (1) ruminal biohydrogenation of dietary PUFA is most extensive on low-concentrate diets based on ensiled forages, (2) incomplete metabolism of dietary PUFA to 18:0 leading to the accumulation of *trans* 18:1 intermediates occurs on diets containing high proportions of rapidly fermented carbohydrates, low amounts of fiber, and/or plant oils or oilseeds, (3) fish oil or marine lipids rich in 20:5 *n*-3 and 22:6 *n*-3 are more potent inhibitors of the reduction of *trans* 18:1 intermediates to 18:0 in the rumen than plant oils and oilseeds, (4) isolated changes in the composition of the basal ration typically have minor effects on ruminal lipid metabolism, and (5) simultaneous alterations in the carbohydrate composition and lipid content of the diet have marked effects on the supply of ruminal biohydrogenation intermediates available for absorption. The interdependency of these factors on ruminal lipid metabolism explains the challenges in accurately predicting the effects of changes in the ruminant diet on milk fatty acid composition.

Microbial Lipid Synthesis

In addition to dietary lipids, fatty acids available for absorption are also derived from rumen microbes, primarily in the form of structural lipids. Bacterial and protozoal lipids make a considerable contribution to the total flow of lipid into the duodenum (Garton, 1977). Based on an extensive evaluation of the relationship between fatty acid intake and the flow of fatty acids at the duodenum, microbial lipid synthesis has been estimated to be ca. 9 g/kg dry matter intake (Doreau & Ferlay, 1994). Bacterial lipids originate from dietary fatty acids and fatty acids synthesised *de novo*. The relative contribution of exogenous and endogenous sources to microbial lipids is dependent on dietary lipid content and bacterial species residing in the rumen (Harfoot & Hazlewood, 1997; Vlaeminck et al., 2006).

Fatty acid synthesis by bacteria in the rumen is considered to be the main source of odd and branch-chain fatty acids in milk fat (refer to Vlaeminck et al., 2006) with an anticarcinogenic potential (refer to Section 2.3). Bacterial odd and branch-chain fatty acids are, in the main, located in membranes. It is thought that the relatively high abundance of branch-chain fatty acids is, in part, related to the low melting point of these fatty acids compared with SFA of comparable chain length allowing the maintenance of membrane fluidity. *De novo* fatty acid synthesis is mediated by one of two key enzyme systems, straight-chain or branch-chain fatty acid synthetase, which differ in acyl-CoA:ACP transacylase substrate specificity (Kaneda, 1991). Branch-chain or straight-chain fatty acids are formed by the repeated condensation of malonyl-Co-A using different fatty acyl Co-A primers. Straight odd-chain fatty acids are synthesized using propyl-Co-A as a primer, whereas isovaleryl-Co A is used for 15:0 *iso* and 17:0 *iso* synthesis, and 2-methylbutyryl Co A serves as a substrate for the synthesis of 15:0 *anteiso* and 17:0 *anteiso*. Even-numbered branch-chain fatty acids (14:0 *iso* and 16:0 *iso*) are formed using isobutyryl Co A as a primer. Branch-chain fatty acyl Co-A precursors, used for fatty acid synthesis, are derived from the deamination of branch-chain amino acids in the rumen (refer to Vlaeminck et al., 2006, for a detailed account of fatty acid synthesis by rumen bacteria). It has also been suggested that methylmalonyl-Co-A derived from propionate may also serve as a substrate for the synthesis of *anteiso* fatty acids by rumen bacteria (Vlaeminck et al., 2006).

In addition to microbial lipids, there is also evidence that 15:0 and 17:0 secreted in milk are also synthesised *de novo* from propionate in ruminant tissues, including the mammary gland. In a recent evaluation of data reported in the literature, the secretion of 15:0 and 17:0 in milk was shown to be significantly higher than the flow at the duodenum or omasum, confirming the contribution of mammary synthesis to the secretion of odd straight-chain fatty acids in milk (Vlaeminck et al., 2006).

Effect of Nutrition on *Trans* Fatty Acids and Bioactive Lipids in Bovine Milk

Whole milk and dairy products are a significant source of fat in the human diet. Consequently, there has been considerable interest in developing sustainable nutritional strategies to enhance the concentrations of specific fatty acids and bioactive lipids in milk with the potential to improve long-term health. Nutrition is the major environmental factor regulating milk fat composition, stimulating an extensive number of reviews on the effect of ruminant diet on milk fatty acid composition (Chilliard et al., 2000, 2001; Chilliard & Ferlay, 2004; Lock & Bauman, 2004; Lock & Shingfield, 2004; Dewhurst et al., 2006; Givens & Shingfield, 2006). In the following sections, the role of nutrition on milk fat composition is considered in the context of developing strategies for reducing medium-chain SFA, enhancing CLA and PUFA concentrations, and dealing with the associated changes in milk fat TFA based on the most recent research. Emphasis is placed solely on the effects of nutrition on milk lipids in the lactating cow. A comprehensive evaluation of the effects of nutrition on the fatty acid composition and sensory attributes of caprine and ovine milk has been reported elsewhere (Chilliard et al., 2003; Chilliard & Ferlay, 2004; Sanz Sampelayo et al., 2007).

Effect of Forage Species

It was observed as long ago as the 1960s that when cows were turned out to pasture, the conjugated diene content of milk was increased two- to threefold (Kuzdzal-Savoie & Kuzdzal, 1961; Riel, 1963). A number of studies have confirmed that fresh pasture enhances milk fat CLA and TFA content compared with diets based on dried or ensiled forages (Chilliard et al., 2002; Elgersma et al., 2004; Ferlay et al., 2006; Table 9). Increases in milk fat CLA content at pasture have been attributed to the amount of grazed forage offered. For example, milk solely from pasture was shown to contain 1090 mg CLA/100 g fat compared with 460 mg/100 g fat in milk from mixed diets based on maize and lucerne silage (Kelly et al., 1998). Consistent with these observations, CLA concentrations in milk from grazed grass of 2210 mg CLA/100 g fat were found to be decreased to 1430 and 890 mg CLA/100 g fat when proportionately 0.33 and 0.66 of dry matter intake from grass were replaced with Lucerne hay and concentrates (Dhiman et al., 1999). Subsequent studies (White et al., 2001) also demonstrated similar effects with a C-4 grass (*Digitaria sanguinalis*; Crabgrass). Fresh grass has been shown to result in higher CLA concentrations compared with ensiled or dried maize, grass, or Lucerne (Table 9). As a consequence, under typical conditions in Switzerland, the United Kingdom, and France, milk fat CLA content is higher during the spring and summer than winter due to higher contribution of fresh grass to nutrient intake (Collomb & Buhler, 2000; Lock & Garnsworthy, 2003; Ledoux et al., 2005). In addition, alpine pastures have been shown to be particularly effective in enhancing milk fat CLA compared with diets based on conserved forages (Kraft et al., 2003) and lowland pastures (Collomb et al., 2004a). Detailed analysis also revealed that alpine pastures increased the concentration of several isomers of CLA in addition to the *cis*-9, *trans*-11 (Table 10). Increases in milk fat CLA content at pasture are accompanied by higher TFA concentrations, and milk from grazed grass often contains lower proportions of 12:0, 14:0, and 16:0 compared with ensiled forages (Table 9).

Milk fatty acid composition in grazing animals is dependent on grass maturity, with concentrations of CLA reported to be much lower six weeks after turnout compared with three weeks (0.80 vs. 1.72 g/100 g fatty acids), an effect that at least in part was associated with a reduction in the 18:3 *n*-3 content of

Table 9 Effect of Forage Species and Conservation Method on Bovine Milk Fatty Acid Composition

Forage Species	Fatty Acid Composition (g/100 g fatty acids)															Reference
	4:0	6:0	8:0	10:0	12:0	14:0	16:0	18:0	cis-9 18:1	trans 18:1	18:2 n-6	18:3 n-3	CLA[1]			
Fresh pasture (ryegrass and white clover)	NR	2.13	1.17	2.34	2.59	9.37	30.7	15.0	26.6		2.62	0.95	1.09			Kelly et al. (1998)
Maize and legume silages	NR	1.79	0.91	1.66	1.70	6.70	24.2	13.2	34.7		2.25	0.25	0.46			Dhiman et al. (1999)
Grass and legume swards	NR	NR	NR	1.80	2.33	9.10	25.1	12.1	32.6		1.40	2.02	2.21			
Lucerne hay	NR	NR	NR	2.11	2.60	9.40	24.7	15.2	31.4		4.27	0.81	0.89			White et al. (2001)
Fresh pasture (crabgrass and white clover)	1.07	1.65	1.12	2.56	3.07	10.9	31.4	13.4	21.3		1.84	0.66	0.73			
Lucerne and maize silages	1.07	1.09	1.05	2.34	2.74	9.94	31.5	15.4	22.1		2.49	0.38	0.37			Elgersma et al. (2004)
Mixed grass swards	NR	NR	NR	NR	NR	8.90	22.6	11.0	25.0	7.40	1.04	1.0	2.30			
Grass and maize silages	NR	NR	NR	NR	NR	11.7	34.8	8.80	17.9	2.70	1.08	1.1	0.37			
Lowland pasture	3.93	2.34	1.31	2.73	2.97	10.1	27.3	10.8	19.3	4.29	1.29	0.89	0.91			Collomb et al. (2002)
Mountain pasture	3.72	2.03	1.08	2.18	2.39	8.87	24.1	11.8	22.2	6.09	1.52	0.93	1.69			
Alpine pasture	3.54	1.96	1.07	2.21	2.40	8.94	23.5	10.2	20.2	7.82	1.50	1.30	2.46			
Ensiled forages (not specified)	4.28	3.04	1.57	3.44	3.79	10.7	30.9	8.50	21.7	1.68	1.84	0.37	0.31			Kraft et al. (2003)
Fresh pasture	4.21	2.89	1.52	3.14	3.19	9.90	25.5	12.0	21.8	1.55	1.71	0.97	0.98			
Alpine pasture	3.98	2.44	1.18	2.36	2.50	8.96	23.4	11.1	24.1	5.45	1.35	1.32	2.59			
Alpine pasture	40.2	2.55	1.27	2.60	2.69	9.35	24.8	9.75	20.7	6.52	1.47	1.47	3.01			

Table 9 (continued)

Forage Species	Fatty Acid Composition (g/100 g fatty acids)													Reference
	4:0	6:0	8:0	10:0	12:0	14:0	16:0	18:0	cis-9 18:1	trans 18:1	18:2 n-6	18:3 n-3	CLA[1]	
Ryegrass silage[2]	3.99	2.10	1.19	2.46	4.23	11.8	30.0	7.63	18.4	1.14	0.80	0.61	0.26	Lourenço et al. (2005)
Ryegrass silage + SPP (80:20)[2]	4.13	2.12	1.18	2.42	4.14	11.5	29.1	8.47	18.6	1.14	0.81	0.60	0.24	
Ryegrass silage + SPP (40:60)[2]	4.27	2.08	1.11	2.14	3.83	10.6	28.7	8.20	20.6	1.12	0.86	0.55	0.26	
Ryegrass silage + SPR (40:60)[2]	4.29	2.00	1.03	1.94	4.09	10.8	29.7	7.83	19.9	1.63	0.95	0.59	0.44	
Perennial ryegrass silage	4.91	2.69	1.36	2.95	3.52	11.7	32.5	11.0	20.7	1.13[3]	0.36	0.40	0.36	Dewhurst et al. (2003)
Red clover silage	5.78	2.98	1.43	2.83	3.31	11.3	30.6	11.6	20.2	1.25[3]	1.58	1.28	0.41	
White clover silage	5.16	3.04	1.57	3.47	4.16	12.7	32.9	9.70	17.9	1.06[3]	1.54	0.96	0.34	
Grass silage early cut	5.60	2.79	1.51	3.20	3.60	12.0	29.4	10.4	16.9	3.63	1.24	0.41	0.38	Vanhatalo et al. (2007)
Grass silage late cut	5.58	2.73	1.47	3.09	3.48	11.8	28.2	10.7	18.1	3.66	1.32	0.37	0.41	
Red clover silage early cut	6.17	2.82	1.46	2.79	3.01	10.4	25.5	11.2	19.9	3.98	1.80	1.34	0.36	
Red clover silage late cut	5.91	2.75	1.42	2.79	3.05	10.7	27.0	10.5	19.3	4.10	1.65	0.88	0.42	
Grass hay	2.89	2.16	1.47	3.41	3.97	13.3	34.5	9.17	15.2	3.78	1.21	0.50	0.45	Shingfield et al. (2005b)
Grass silage untreated	2.89	2.23	1.49	3.31	3.79	12.9	34.7	9.75	15.1	3.62	0.96	0.35	0.41	
Grass silage/inoculant treated	2.94	2.34	1.53	3.43	3.90	13.1	33.8	10.0	15.3	3.71	0.96	0.43	0.41	
Grass silage/formic acid treated	25.8	2.21	1.50	3.43	3.99	13.2	34.2	10.0	14.5	4.25	0.93	0.29	0.49	

Table 9 (continued)

Forage Species	Fatty Acid Composition (g/100 g fatty acids)									cis-9 18:1	trans 18:1	18:2 n-6	18:3 n-3	CLA[1]	Reference
	4:0	6:0	8:0	10:0	12:0	14:0	16:0	18:0							
Fresh natural grassland, week 3[4]	3.23	2.81		14.8			22.9	11.4		24.1	3.69	1.09	0.99	1.72	Ferlay et al. (2006)
Fresh natural grassland, week 6[4]	3.37	3.10		16.5			29.6	9.23		21.8	1.75	1.23	0.68	0.80	
Natural grassland hay	4.54	5.75		19.9			28.6	8.13		16.0	1.36	1.08	1.25	0.64	
Ryegrass hay	4.22	4.06		19.5			30.2	8.16		15.4	1.83	1.00	1.02	0.87	
Ryegrass silage	4.88	4.38		18.7			32.1	7.93		16.0	0.87	1.09	0.94	0.46	
Maize silage	4.46	4.35		19.3			31.0	7.89		16.7	1.04	1.46	0.24	0.66	
Concentrate/ Cocksfoot hay	4.28	4.75		21.9			33.5	6.65		14.1	0.62	1.77	0.46	0.39	

[1] cis-9, trans-11 conjugated linoleic acid.
[2] Fatty acid concentrations reported as g/100 g fatty acid methyl esters. SPP = silage prepared from seminatural species poor grass swards; SPR = silage prepared from botanically diverse seminatural pasture.
[3] trans-11 18:1.
[4] Refers to milk produced 3 or 5 weeks after turn-out to pasture; trans 18:1 is the sum of trans-10 18:1 and trans-11 18:1 concentrations.
NR = not reported.

Table 10 Effect of Alpine Pastures on the Concentration of Conjugated Linoleic Isomers in Bovine Milk Fat (mg/100 g fatty acids)

Pasture	Lowland[1]	Mountain[1]	Alpine[1]	Lowland[2]	Alpine[2]	L'Etivaz[2]
cis-9, trans-11	877	1587	2407	1146	2589	3014
ciss-11, trans-13	2	2	5	–	–	–
cis-12, trans-14	8	6	8	–	–	–
trans-7, cis-9	35	58	55	23	33	42
trans-8, cis-10	15	24	35	–	–	–
trans-10, cis-12	3	3	2	7	6	8
trans-11, cis-13	49	90	198	96	168	281
trans-6, trans-8	2	5	3	–	–	–
trans-7, trans-9	8	10	10	14	8	9
trans-8, trans-10	2	5	3	6	6	7
trans-9, trans-11	12	12	15	45	14	19
trans-10, trans-12	8	8	7	12	8	9
trans-11, trans-13	43	36	52	59	42	64
trans-12, trans-14	17	17	26	34	25	36

[1] Refers to milk produced from pasture at 600–650 m (lowland), 900–1210 m (mountain), or 1275–2120 (highlands) above sea level. Data adapted from Collomb et al. (2004a).
[2] Refers to milk produced from lowland pastures in Germany (ca. 500 m) or alpine pastures at various locations in Swizerland (>1200 m) or in L'Etivaz (1275–2200 m). Data adapted from Kraft et al. (2003).

grazed herbage (15 vs. 23 g/kg dry matter; Ferlay et al., 2006). Increases in TFA and CLA in milk from pasture have often been attributed to the higher intakes of PUFA compared with diets based on conserved forages that serve as substrates for cis-9, trans-11 CLA and trans-11 18:1 synthesis in the rumen. Some support for this is offered by observations that the increases in CLA and TFA and reductions in 12:0, 14:0, and 16:0 in milk from fresh compared with ensiled or dried forages are associated with increases in milk 18:3 n-3 content (Table 9). However, as Lock and Garnsworthy (2003) noted, the effects of fresh grass on milk fatty acid composition are not solely explained by differences in forage lipid concentrations, and therefore other changes in lipid metabolism must also be involved.

There is clear evidence that forage legumes enhance milk fat PUFA concentrations compared with grasses, but these changes are typically independent of changes in milk fat CLA or TFA content (Dewhurst et al., 2003; Vanhatalo et al., 2007; Table 9). Even though these forage species typically have no effect on total milk fat content, recent studies have indicated that grass silage and red clover silage result in differences in the relative abundance of several minor CLA isomers (Vanhatalo et al., 2007; Table 11). Based on data compiled in an extensive review on the role of forages in altering milk fatty acid composition (Dewhurst et al., 2006), it can be calculated that in six direct comparisons with grass silage (red studies with red clover and two with white clover), ensiled forage legumes resulted in mean increases in milk fat 18:2 n-6 and 18:3 n-3

Table 11 Effect of Forage Species on the Concentration of Conjugated Linoleic Isomers in Bovine Milk Fat (mg/100 g fatty acids)

Isomer	Grass Silage[1]		Red Clover[1]		Grass Silage/Maize Silage (DM Basis)[2]			
	Early	Late	Early	Late	100:0	84:16	66:34	50:50
cis-8, trans-10	7.5	5.4	5.2	8.0	–	–	–	–
cis,y >	375	410	361	418	476	453	438	536
cis-11, trans-13	1	1	2	2	3	8	8	8
cis-12, trans-14	4	3	7	6	–	–	–	–
trans-7, cis-9	32	37	39	47	26	28	33	44
trans-8, cis-10	6	7	6	7	10	11	10	10
trans-10, cis-12	2	3	3	4	0	0	0	3
trans-11, cis-13	16	14	8	9	20	11	6	5
trans-6, trans-8	4	6	1	2	–	–	–	–
trans-7, trans-9	8	8	6	7	–	–	–	–
trans-8, trans-10	2	2	2	2	0	1	0	0
trans-9, trans-11	11	10	13	13	31	30	23	19
trans-10, trans-12	3	4	5	5	0	4	6	10
trans-11, trans-13	15	13	30	24	15	11	10	3
trans-12, trans-14	10	8	13	10	11	10	9	1
trans-13, trans-15	1	1	2	<1	–	–	–	–

[1] Refers to milk produced from early or late cut grass silage or red clover (Vanhatalo et al., in press).
[2] Kliem et al. (unpublished).

concentrations of 0.4 and 0.6 g/100 g fatty acids, respectively. Further studies have shown that the potential of red clover for enhancing milk fat PUFA content can be further exploited by ensiling at an early stage of maturity (Vanhatalo et al., 2007; Table 9). The higher transfer of 18:3 n-3 on diets containing red clover is thought to be mediated by reductions in lipolysis in the rumen via the inhibitory effects of polyphenol oxidase on inherent plant lipases and the formation of polar lipid-phenol complexes (refer to Dewhurst et al., 2006).

Replacing silage prepared from intensively managed ryegrass with ensiled grass from seminatural grasslands has been shown to decrease milk fat 12:0, 14:0, and 16:0 content and increase trans-11 18:1 and cis-9, trans-11 CLA content (Lourenço et al., 2005; Table 9). It was also shown that silage prepared from botanically diverse grass swards resulted in a higher enrichment of trans-11 18:1, cis-9, trans-11 CLA, 18:2 n-6, and 18:3 n-3 compared with species poor pastures, effects that were attributed to the action of secondary plant compounds on ruminal lipid metabolism (Lourenço et al., 2005).

Comparison of diets based on grass silage and maize silage have shown no major differences between these forages with respect to enhancing milk fat cis-9, trans-11 CLA content, but maize silage enhances 18:2 n-6 and decreases 18:3 n-3 concentrations compared with grass silage (Ferlay et al., 2006; Table 9). Indirect comparisons between forages in diets containing 30 g/kg dry matter of a mixture of fish oil and sunflower oil (2:3 w/w) also reported no difference in milk fat *cis*-9, *trans*-11 CLA content between maize and grass silage (Shingfield et al., 2005a). However, the source of forage in the diet was found to alter the concentration and distribution of other isomers of CLA in milk, while, relative to maize silage, milk from grass silage contained higher *trans*-18:2 and lower *trans*-18:1 concentrations (Shingfield et al., 2005a). Further studies examining the impact of replacing grass silage in the diet with maize silage (from 100:0 to 50:50, on a dry matter basis) had no effect on milk fat *trans* 18:1 or total CLA content but resulted in significant linear increases in *trans*-6/8, *trans*-9, *trans*-10, and *trans*-12 18:1 from 0.09, 0.14, 0.16, and 0.20 to 0.14, 0.19, 0.31, and 0.28 g/100 g fatty acids, respectively (Kliem et al., unpublished). Increases in the proportion of maize silage in the diet were also associated with significant linear increases in *trans*-7, *cis*-9 CLA and *trans*-10, *trans*-12 CLA concentrations and decreases in milk fat *trans*-11, *cis*-13 CLA, *trans*-11, *trans*-13 CLA, and *trans*-12, *trans*-14 CLA content (Table 11). Compared with a hay-based diet, maize silage-based diets decreased milk fat branch-chain fatty acids, *trans*-11, *cis*-15 18:2, *trans*-11, *cis*-13 CLA, *trans*-11, *trans*-13 CLA, and *trans*-12, *trans*-14 CLA and increased *cis*-11 18:1, *cis*-12 18:1, *trans*-7, *cis*-9 CLA, *trans*-8, *cis*-10 CLA, *trans*-9, *cis*-11 CLA, and *trans*-10, *cis*-12 CLA (Roy et al., 2006). At least part of these changes can be attributed to differences in the profile of intermediates formed during 18:2 *n*-6 and 18:3 *n*-3 metabolism in the rumen.

Effect of Forage Conservation Method

Wilting for the production of hay, and to some extent before ensiling, is associated with decreases in forage total fatty acid and PUFA concentrations due to oxidative losses and leaf shatter, since the lipid content of leaves is higher than that in stems (Dewhurst et al., 2006). Losses of 18:3 *n*-3 in fresh grass of up to 75% can occur during drying (Doreau & Poncet, 2000; Doreau et al., 2005; Shingfield et al., 2005b). Furthermore, grass swards used for hay production are generally harvested at a relatively late stage of growth to optimize the yield of forage dry matter per hectare. Advances in maturity are associated with decreases in the lipid content of grasses (Dewhurst et al., 2006), which could also contribute to the lower fatty acid content of dried compared with fresh or ensiled grass.

Even though the concentrations of PUFA in dried grass are lower than in ensiled grass, concentrations of 18:2 *n*-6 and 18:3 *n*-3 in milk are often higher in diets containing hay than silage due to a higher efficiency of transfer (refer to Dewhurst et al., 2006). In a comparison of forage conservation methods, the transfer of 18:2 *n*-6 and 18:3 *n*-3 from the diet into milk was found to be significantly higher in diets based on hay (29% and 17%, respectively) compared with wilted silages (mean 15% and 3%, respectively) prepared from the same grass swards (Shingfield et al., 2005b). Such observations appear to be related to changes in ruminal lipid metabolism and in forage lipids during conservation. Studies in sheep have shown that biohydrogenation of PUFA in dried grass is lower than fresh or ensiled grass (Doreau et al., 2005), while a significant proportion (ca. 50%) of plant glycolipids are hydrolyzed during ensiling (Steele & Noble, 1984), resulting in the release of NEFA that, following ingestion, are immediately susceptible to biohydrogenation in the rumen. Studies in the 1970s also demonstrated that milk from hay was relatively rich in 18:3 *n*-3 (refer to Chilliard et al., 2001), while more recent research has shown that drying rather than wilting of hay to minimize losses of forage lipids can also result in milk containing similar or higher concentrations of *trans*-11 18:1 and *cis*-9, *trans*-11 CLA than grass silage (Ferlay et al., 2006). Direct comparisons of haylage (510 g DM/kg) with silage (390 g DM/kg) prepared from the same grass swards indicated only minor differences in milk fatty acid composition (Ferlay et al., 2002). Overall, the impact of the forage conservation method on milk fatty acid composition is rather limited compared with the responses to inclusion of plant oils, oilseeds, or marine lipids in the diet.

Effect of Concentrate Level and Composition

The impact of concentrate supplements in the diet on milk fatty acid composition is largely dependent on the level of inclusion. In grazing cows, increases in the proportion of concentrate in the diet from 30 to 350 g/kg dry matter resulted in an increase in milk fat 4:0 to 14:0 and total *trans* 18:1 and 18:2 *n*-6 concentrations and a reduction in *cis*-9 18:1, *trans*-11 18:1, *cis*-9, *trans*-11 CLA, and 18:3 *n*-3 content (Bargo et al., 2002, 2006; Table 12). In contrast, increases in the amount of concentrate from 35% to 65% in diets based on grass hay were found to enhance milk fat *trans*-4 to -16 18:1, *cis*-9, *trans*-11 CLA, and 18:2 *n*-6 and to decrease 16:0 and 18:0 concentrations (Loor et al., 2005a; Table 12). Similar responses have been observed with diets based on grass or legume silages (Dewhurst et al., 2003) or maize and Lucerne silage (Kalscheur et al., 1997; Piperova et al., 2002). Increases in concentrate level typically above 600 g/kg dry matter have been shown to markedly alter the profile of *trans* 18:1 in milk fat, resulting in a shift toward *trans*-10 at the expense of *trans*-11, particularly for diets containing relatively high concentrations of PUFA (Griinari et al., 1998; Piperova et al., 2002; Loor et al., 2005a). Changes in milk fat *trans* 18:1 profile toward *trans*-10 are also known to be associated with reductions in milk fat content (Bauman & Griinari, 2003). There is some evidence that alterations in milk fat *trans* 18:1 induced by high levels of concentrates may be reversed by the inclusion of mineral buffers (Kalscheur et al., 1997; Piperova et al., 2002) or vitamin E in the diet (Pottier et al., 2006).

Table 12 Effect of Increasing the Proportion of Concentrate in the Diet on the Fatty Acid Composition of Bovine Milk (g/100 g fatty acids) from Fresh Grass or Dried Hay

Basal Forage	Pasture[1]		Grass Hay[2]	
Concentrate (g/kg diet dry matter)	30	350	350	650
4:0	2.8	3.10	3.1	3.3
6:0	1.6	1.90	2.4	2.6
8:0	0.9	1.10	1.6	1.7
10:0	1.8	2.30	3.6	3.9
12:0	1.9	2.60	4.1	4.4
14:0	8.0	9.40	12.1	11.6
16:0	24.3	24.30	29.4	25.7
18:0	12.0	12.70	7.0	6.2
18:1 cis-9	29.8	26.90	15.3	14.9
18:1 trans-6/8	0.31	0.40	0.19	0.40
18:1 trans-9	0.31	0.38	0.14	0.23
18:1 trans-10	0.90	1.18	0.28	1.66
18:1 trans-11	3.58	2.85	1.12	1.32
18:1 trans-12	0.47	0.55	0.20	0.34
18:2 n-6	2.22	3.16	1.61	2.48
18:3 n-3	1.17	0.77	0.78	0.76
cis-9, trans-11 CLA	1.36	1.24	0.62	0.81

[1] Data derived from Bargo et al. (2002, 2006).
[2] Adapted from Loor et al. (2005a).

The effects of concentrate supplementation on milk fatty acid composition are also dependent on the source of starch in the diet. Replacing rapidly fermented wheat starch (300 g/kg dry matter) with more slowly degrading potato starch was shown to increase milk fat 4:0 to 16:0 and decrease cis-9 18:1 and trans 18:1 (primarily trans-10) concentrations (Jurjanz et al., 2004). It is also clear that changes in milk fat composition in response to concentrate supplementation are dependent not only on the amount but also on the composition of lipids in the diet. Decreases in the dietary forage-to-concentrate ratio from 50:50 to 20:80 were found to induce different changes in milk fat composition in diets containing SFA compared with MUFA and PUFA (Griinari et al., 1998), indicating the complex interaction between the fiber and starch composition of the basal diet and lipid supplements.

Decreases in the forage-to-concentrate ratio of the diet from 80:20 to 30:70 on a dry matter basis or increases in the proportion of maize silage in the diet have also been shown to reduce milk odd-chain and branch-chain fatty acid concentrations, resulting in linear reductions in the relative abundance of *iso* relative to *anteiso* odd-chain fatty acids in bovine milk fat (Vlaeminck et al., 2006).

Effect of Plant Lipids

The inclusion of oils or oilseeds in the diet is a well-established nutritional strategy for altering the energy metabolism of lactating cows, milk composition (Chilliard, 1993; Lock & Shingfield, 2004), and milk fatty acid composition (Chilliard et al., 2000, 2001; Chilliard & Ferlay, 2004; Lock & Bauman, 2004; Palmquist et al., 2005; Givens & Shingfield, 2006). Supplementing the diet with fish oil or marine lipids represents an effective strategy for increasing milk fat *trans*-11 18:1, *cis*-9, *trans*-11 CLA concentrations but generally has only relatively minor effects on 12:0, 14:0, and 16:0 concentrations compared with plant lipids. The following sections consider the role of plant oils and oilseeds on enhancing milk branch-chain fatty acid, 12:0, 14:0, 16:0, TFA, PUFA, and CLA concentrations. Data on the manipulation of milk sphingolipids are limited, but the most recent studies suggest that the role of nutrition to enhance milk concentrations of sphingomyelin is limited, even though concentrations are higher in summer than winter, and that genotype and stage of lactation have a greater influence (Graves et al., 2007). Several recent reviews have considered the role of fish oil or marine lipid supplements on ruminant milk fat composition (Chilliard et al., 2001; Chilliard & Ferlay, 2004; Lock & Bauman, 2004).

Changes in milk fatty acid composition to plant lipid supplements are dependent on (1) the amount of oil included in the diet, (2) the fatty acid profile of the lipid supplement, (3) the form of lipid supplement, and (4) the composition of the basal diet. Attempts to enhance the concentration of a specific fatty acid in milk invariably result in changes in other fatty acids. For example, the use of plant oils or oilseeds to decrease milk fat SFA and enhance milk *cis*-9 18:1, 18:2 *n*-6, CLA, or 18:3 *n*-3 concentrations results in an inevitable increase in TFA content (Table 13).

Milk Saturated Fatty Acids

Supplementing the diet with plant oils or oilseeds is an effective means to decrease the concentration of medium-chain SFA in bovine milk (Table 13). For example, supplementing the diet with 50 g/kg dry matter of linseed oil was shown to reduce the sum of 10:0 to 16:0 from 56 to 29 g/100 g fatty acids (Roy et al., 2006). Reductions in medium-chain fatty acids to plant lipids are also accompanied by increases in milk fat 18:0 and *cis*-9 18:1 content due to (1) increases in the amount of 18:0 available for absorption arising from extensive metabolism of unsaturated fatty acids in the rumen (Doreau & Ferlay, 1994; Loor et al., 2004), (2) increases in the flow of *cis*-9 18:1 derived from oil supplements leaving the rumen, and (3) conversion of 18:0 to *cis*-9 18:1 via Δ-9 desaturase activity in the mammary gland.

In contrast, the inclusion of lipid supplements rich in 16:0 in the diet results in an increase in milk 16:0 concentrations. Based on a comparison of six experiments, supplements of calcium salts of palm oil fatty acids (mean 762 g/day)

Table 13 Effect of Plant Oils and Oilseeds on the Fatty Acid Composition of Bovine Milk

Lipid source	Intake[1] (g/d)	Forage[2]	F:C[3]	4:0	6:0	8:0	10:0	12:0	14:0	16:0	18:0	cis-9 18:1	trans 18:1	Total 18:1	18:2 n-6	18:3 n-3	CLA	Reference
Control	0	LH/LS	44:56	2.9	2.0	1.2	2.7	3.1	9.8	30.7	9.1	21.2	2.3	24.7	3.6	0.50	0.46	Mosley et al. (2007)
Palm oil by-product	476			3.1	1.9	1.0	2.1	2.4	8.6	39.1	6.8	19.3	1.8	22.1	3.2	0.41	0.40	
Control	0	GS	50:50	2.9	2.5	1.6	3.7	4.2	12.5	30.1	11.2	19.4	1.6	21.7	1.3	0.40	0.46	Ryhänen et al. (2005)
Rapeseed oil	500			2.6	1.9	1.2	2.5	2.7	10.1	22.6	14.3	25.8	4.3	31.4	1.4	0.50	1.02	
Control	0	MS/GS	57:43	5.0	2.3	1.3	3.1	4.0	11.6	30.7	8.3	18.1	2.0	20.1	2.1	0.45	0.60	Givens et al. (2003)
Cracked rapeseed	2530			3.2	1.1	0.6	1.3	1.9	7.9	19.8	14.1	34.7	2.6	37.3	2.4	0.48	1.02	
Cracked rapeseed[1]	4100			2.7	1.0	0.4	1.0	1.4	6.0	18.0	15.8	39.3	2.0	41.3	2.8	0.60	0.74	
Control	0	Hay	59:41	3.5	2.5	1.6	4.1	5.1	13.5	35.1	6.1	13.4	2.2	15.6	2.0	0.79	0.58	Collomb et al. (2004a,b)
Ground rapeseed	920			3.4	2.6	1.6	3.7	4.3	12.8	27.7	10.3	19.5	4.0	23.5	1.8	0.79	0.70	
Ground sunflower-seed	950			3.5	2.5	1.7	3.5	4.1	12.3	28.3	9.9	18.9	4.8	23.6	2.5	0.79	0.93	
Ground linseed	1240			3.7	2.5	1.5	3.2	3.4	11.1	24.9	12.1	20.1	5.14	25.2	1.8	1.81	0.93	
Control		Pasture	5[4]	NR	1.5	0.9	2.0	2.4	9.6	24.3	11.9	22.9	4.1	27.5	1.6	0.71	1.26[5]	Rego et al. (2005)
Soyabean oil	500			NR	1.1	0.5	1.2	1.6	7.1	20.5	12.5	27.0	6.9	35.0	1.9	0.67	1.93[5]	
Control	0	LH/MS	50:50	3.9	2.5	1.5	3.5	4.0	12.1	29.4	10.4	16.1	1.8	18.3	2.6	0.54	0.40	AbuGhazaleh et al. (2002)

Table 13 (continued)

Lipid source	Intake[1] (g/d)	Forage[2]	F:C[3]	Milk Fatty Acid Composition (g/100 g fatty acids)																Reference
				4:0	6:0	8:0	10:0	12:0	14:0	16:0	18:0	cis-9 18:1	trans 18:1	Total 18:1	18:2 n-6	18:3 n-3	CLA	
Extruded soyabean[1]	2415			3.9	2.3	1.3	2.8	3.0	10.1	24.0	12.1	18.9	3.8	23.1	4.5	0.87	0.87	
Control	0	MS/GH	48:52	3.3	2.7	1.5	3.5	3.9	12.1	32.3	8.6	16.6	2.8	20.3	2.2	0.21	0.55	Roy et al. (2006)
Sunflower oil	957			2.3	1.2	0.5	1.2	1.6	7.1	18.9	13.6	28.3	11.5	41.0	2.3	0.20	0.93	
Control	0	MS	27:73	3.3	2.7	1.6	4.3	5.1	12.8	28.7	5.8	14.9	5.2	20.8	3.0	0.09	0.60	Roy et al. (2006)
Sunflower oil	755			1.8	1.0	0.5	1.2	1.8	7.4	19.1	6.3	19.4	23.7	52.0	4.6	0.15	1.17	
Control[6]	0	BS/LS/LH	60:40	4.1	2.4	1.2	2.5	2.9	11.6	30.6	9.8	17.7	4.6	23.5	1.7	0.41	0.68	Bell et al. (2006)
Safflower oil[6]	1125			2.8	1.4	0.6	1.3	1.5	8.1	18.7	11.4	17.7	17.6	38.5	2.9	0.32	4.12	
Linseed oil[6]	1066			3.2	1.6	0.7	1.4	1.6	8.5	17.9	11.1	19.2	14.3	36.6	2.0	0.73	2.80	
Control	0	GS	60:40	1.8	1.0	0.6	1.8	2.7	9.9	40.2	12.3	NR	1.1	21.0	2.0	0.72	0.16	Offer et al. (1999)
Linseed oil	250			1.8	1.0	0.6	1.5	2.1	8.8	34.0	15.6	NR	2.1	27.2	1.8	0.84	0.28	
Control	0	GH	64:36	3.0	2.3	1.4	3.5	4.2	13.1	34.9	7.0	14.2	2.1	16.8	1.6	0.74	0.54	Roy et al. (2006)
Linseed oil	1050			2.8	1.8	0.8	1.8	2.0	8.3	17.1	12.8	20.6	12.2	33.7	1.2	0.74	2.89	
Control	0	GH	65:35	3.1	2.4	1.6	3.6	4.1	12.1	29.4	7.1	15.3	2.7	20.4	1.6	0.78	0.62	Loor et al. (2005a)
Linseed oil	588			3.3	1.9	1.2	2.1	2.1	8.3	17.2	14.8	23.8	9.0	40.8	1.4	1.00	1.34	
Control	0	GH	35:65	3.3	2.6	1.7	3.9	4.4	11.6	25.7	6.2	14.9	5.0	23.3	2.5	0.76	0.81	Loor et al. (2005a)
Linseed oil	612			3.0	1.6	1.0	2.4	2.8	8.8	18.7	8.1	14.4	12.1	28.8	2.3	1.59	2.54	
Control	0	MS/GS	52:48	3.9	2.3	1.6	3.0	3.7	11.3	28.9	10.2	21.4	2.9	25.8	2.2	0.32	0.51	Offer et al. (2001)
Crushed linseed[1]	1500			3.9	2.0	1.3	2.4	2.9	9.9	23.9	12.6	26.1	3.4	31.2	2.8	0.87	0.62	

Table 13 (continued)

Lipid source	Intake[1] (g/d)	Forage[2]	F:C[3]	Milk Fatty Acid Composition (g/100 g fatty acids)															Reference
				4:0	6:0	8:0	10:0	12:0	14:0	16:0	18:0	cis-9 18:1	trans 18:1	Total 18:1	18:2 n-6	18:3 n-3	CLA	
Control	0	GS/MS	64:36	2.8	2.4	1.3	3.0	4.2	11.8	31.7	11.4	NR	2.3	23.5	2.0	0.4	0.9	Gonthier et al. (2005)
Ground raw linseed[1]	1975			2.5	1.8	0.9	1.7	2.2	7.8	20.4	19.9	NR	4.3	34.4	2.7	1.3	1.4	
Micronised linseed[1]	1930			2.5	1.9	1.0	2.0	2.6	8.3	21.7	18.3	NR	4.2	33.1	2.9	1.3	1.4	
Extruded linseed[1]	1968			2.2	1.6	0.8	1.5	2.1	8.0	21.0	16.5	NR	5.9	37.4	3.1	0.7	1.9	

[1] Oil content (g/kg) of rapeseed, soyabean, crushed linseed, and linseed, 480, 190, 280, and ca. 300, respectively.
[2] BS = barley silage; GH = grass hay; GS = grass silage; LH = Lucerne hay; LS = Lucerne silage; MS = maize silage.
[3] Forage:concentrate ratio of the diet (on a dry matter basis).
[4] Concentrate intake (kg/day).
[5] Total CLA content. In all other cases CLA refers to the concentration of cis-9, trans-11 CLA.
[6] Fatty acid concentrations reported as g/100 g fatty acid methyl esters.
NR = not reported.

were reported to increase milk fat 16:0 concentrations from 31.6 to 33.3 g/100 g fatty acids (Chilliard et al., 1993). Recent studies have confirmed the efficacy of palm oil fatty acids to increase milk fat 16:0 content, with supplements providing 1028 g/day of palmitic acid being shown to be increase milk fat 16:0 content from 30.7 to 45.6 g/100 g fatty acids (Mosley et al., 2007). Concentrations of odd- and branch-chain fatty acids in milk are decreased when oils rich in 18:2 *n*-6 or 18:3 *n*-3 are included in the diet (Vlaeminck et al., 2006), whereas milk fat 4:0, 6:0, and, to a lesser extent, 8:0 are unaffected or are marginally decreased in response to plant oils or oilseeds (Chilliard & Ferlay, 2004; Table 13).

Milk Polyunsaturated Fatty Acids

Since PUFA are not synthesized by ruminant tissues, the concentration of 18:2 *n*-6 and 18:3 *n*-3 in milk is dependent on the amounts of these fatty acids absorbed and partitioned toward the mammary gland. The supply of PUFA available for absorption is determined by both the amounts of PUFA in the diet and the extent of their metabolism in the rumen. In typical diets, the 18:2 *n*-6 concentration in milk varies between 2 to 3 g/100 g fatty acids (Table 13). Supplementing the diet with plant oils rich in 18:2 *n*-6 including soybean, sunflower, or safflower oil results in only small increases in milk fat 18:2 *n*-6 content (Chilliard & Ferlay, 2004; Dewhurst et al., 2006; Table 13). There is some evidence that extruded (AbuGhazaleh et al., 2002), micronized (Petit, 2002), or roasted (Dhiman et al., 1995) soybeans can result in higher enrichments of 18:2 *n*-6 compared with plant oils.

Other than grass or legume forages, linseed is the most common 18:3 *n*-3-rich feed ingredient. Even though rapeseed oil or rapeseeds contain 18:3 *n*-3 (ca. 7 % w/w), the use of these lipids does not result in a significant enrichment of 18:3 *n*-3 in milk fat (Chilliard & Ferlay, 2004; Givens & Shingfield, 2006; Table 13). Lipids in soybean also contain 18:3 *n*-3 (ca. 8% w/w), and increases of between 0.7 and 0.8 g/100 g fatty acids in milk fat 18:3 *n*-3 concentrations have been reported in response to roasted (Dhiman et al., 1995) or micronized (Petit, 2002) soybeans. Extruded linseeds in the diet have been shown to enhance milk fat 18:3 *n*-3 content in the range between 0.3 and 0.6 g/100 g fatty acids (Weill et al., 2002; Gonthier et al., 2005; Ponter et al., 2006), consistent with the response to linseed oil or unprocessed linseeds (Chilliard & Ferlay, 2004), but these are lower than the increases that can be achieved using fresh grass and ensiled legumes (Dewhurst et al., 2006; Table 9).

***Trans* Fatty Acids and Conjugated Linoleic Acid**

Since the majority of *cis*-9, *trans*-11 CLA in bovine milk is synthesized endogenously (Loor et al., 2005b; Palmquist et al., 2005; Mosley et al., 2006; Shingfield et al., 2007), nutritional strategies for enhancing milk fat CLA content are directed toward enhancing ruminal outflow of *trans*-11 18:1. Studies to date indicate that supplementing the diet with 18:2 *n*-6-rich lipids does not result in substantial increases in the amount of *cis*-9, *trans*-11 CLA leaving the rumen (Palmquist et al., 2005). Formulation of diets for increasing milk fat CLA content can be broadly categorized as those that contain (1) 18:2 *n*-6 or 18:3 *n*-3, both of which serve as precursors for *trans*-11 18:1 formation in the rumen, and (2) ingredients that modify ruminal biohydrogenation, resulting in an inhibition of *trans* 18:1 reduction (Palmquist et al., 2005). These attributes of the diet are not mutually exclusive, and considerable interactions occur with other constituents in the diet, including the relative proportions of starch and fiber.

Plant oils rich in 18:2 *n*-6 are effective in enhancing milk fat *cis*-9, *trans*-11 CLA content, with responses being linear to inclusion of 40 g oil/kg diet dry matter (Chilliard et al., 2000; Chilliard & Ferlay, 2004). In a recent experiment, the inclusion of 60 g safflower oil/ kg diet dry matter was shown to result in a more than fivefold increase in milk fat *cis*-9, *trans*-11 CLA content over an extended period (Bell et al., 2006). Milk fat responses to linseed oil are, in most cases, comparable to those obtained with 18:2 *n*-6-rich oils (Chilliard & Ferlay, 2004; Bell et al., 2006), confirming the importance of endogenous synthesis in the mammary gland to *cis*-9, *trans*-11 CLA incorporated into milk fat. Overall, plant oils are more effective for increasing milk fat *cis*-9, *trans*-11 CLA concentrations than oilseeds, while extruded oilseeds generally induce higher responses than unprocessed oilseeds (Chouinard et al., 1997, 2001; Bayourthe et al., 2000).

In addition to increases in *trans*-11 18:1 and *cis*-9, *trans*-11 CLA, the inclusion of plant oils in the diet also alters the profile of other *trans* 18:1, *trans* 18:2, and CLA isomers in bovine milk (Collomb et al., 2004a, b; Loor et al., 2005a, b; Rego et al., 2005; Bell et al., 2006; Roy et al., 2006; Shingfield et al., unpublished). Supplementing the diet with ground rapeseeds or rapeseed oil rich in *cis*-9 18:1 is associated with an increase in milk *trans*-4 to -10 18:1 (Table 14) and *trans*-7, *cis*-9 CLA (Table 15). Oils and oilseeds predominating in 18:2 *n*-6 typically enhance *trans*-10 18:1 and *trans*-12 18:1 (Table 14), *trans*-8, *cis*-10 CLA, *trans*-10, *cis*-12 CLA, *trans*-9, *trans*-11 CLA, *trans*-10, *trans*-12 CLA concentrations and, on high-concentrate diets, *trans*-9, *cis*-11 CLA content (Roy et al., 2006; Table 16). Linseed oil or linseeds in the diet as a source of 18:3 *n*-3 result in an enrichment of *trans*-13 to -16 18:1 (Table 14), *cis*-9, *trans*-12 18:2, *cis*-9, *trans*-13 18:2, *trans*-11, *cis*-15 18:2, *cis*-11, *trans*-13 CLA, *cis*-12, *trans*-14 CLA, *trans*-11, *cis*-13 CLA, *trans*-9, *trans*-11 CLA, *trans*-11, *trans*-13 CLA, and *trans*-12, *trans*-14 CLA (Loor et al., 2005a, b; Roy et al., 2006; Tables 15 and 16).

Changes in milk fatty acid composition in response to plant oils or oilseeds are also dependent on the composition of the basal ration, including forage species and the forage-to-concentrate ratio of the diet (Bauman & Griinari, 2003; Chilliard & Ferlay, 2004; Loor et al., 2005a; Dewhurst et al., 2006; Tables 13, 14, and 16). There is clear evidence in the literature that supplements of plant oils on low-forage diets induce changes in ruminal biohydrogenation that are typically characterized by an increase in the formation of *trans*-10 18:1 and a concomitant reduction in the amount of *trans*-11 18:1 leaving the rumen that results in corresponding alterations in milk fat 18:1 composition. Since

Table 14 Effect of Plant Oils and Oilseeds on the Distribution and Concentration of trans-18:1 Isomers in Bovine Milk Fat (g/100 g fatty acids)

Lipid Source	Intake (g/d) Oilseed	Intake (g/d) Oil	Forage[1]	F:C[2]	Δ4	Δ5	Δ6-8	Δ9	Δ10	Δ11	Δ12	Δ13/14[3]	Δ15	Δ16[4]	Reference
Control	0	0	Pasture	5[5]	NR	NR	0.37	0.27	0.59	2.40	0.43	NR	NR	NR	Rego et al. (2005)
Soybean oil		500			NR	NR	0.64	0.57	1.47	3.18	1.08	NR	NR	NR	
Control	0	0	LH/MS	50:50	NR	NR	0.19[6]	0.21	0.38[6]	1.02	NR	NR	NR	NR	AbuGhazaleh et al. (2002)
Extruded soybean	2415	454			NR	NR	0.37[6]	0.38	0.60[6]	2.41	NR	NR	NR	NR	
Total Mixed Ration	NR	NR	NR	NR	NR	NR	0.31	0.29	0.54	1.24	0.66	1.28	NR	0.51	Loor et al. (2002)
SE soybean	1675		Pasture	6.7[5]	NR	NR	0.28	0.30	0.34	4.32	0.48	0.92	NR	0.39	Collomb et al. (2004a,b)
ME soybean	2084		Pasture	6.7[5]	NR	NR	0.30	0.28	0.28	4.80	0.42	0.73	NR	0.34	
Control	0	0	GH	59:41	NR	NR	0.05	0.14	—	0.87[7]	0.10	0.27	NR	0.13	
GRrapeseed	920	472			NR	NR	0.30	0.34	—	1.60[7]	0.29	0.66	NR	0.31	
GR sunflowerseed	950	523			NR	NR	0.23	0.30	—	2.08[7]	0.40	0.79	NR	0.38	
GR linseed	1240	481			NR	NR	0.21	0.29	—	2.11[7]	0.38	1.32	NR	0.53	
Control	0	0	GS	60:40	0.02	0.02	0.20	0.21	0.17	1.12	0.27	0.47	0.38	0.34	Shingfield et al. (unpublished)
Rapeseed oil		500	GS	60:40	0.09	0.09	0.44	0.41	0.35	1.58	0.52	0.78	0.67	0.54	
Soybean oil		500	GS	60:40	0.06	0.05	0.34	0.35	0.39	1.97	0.60	1.02	0.73	0.64	
Linseed oil		500	GS	60:40	0.05	0.08	0.33	0.33	0.28	2.03	0.57	1.15	0.86	0.76	
Control	0	0	BS/LS/LH	60:40	NR	NR	0.36	0.33	0.49	1.41	0.39	0.77	0.37	0.47	Bell et al. (2006)
Safflower oil		1125			NR	NR	0.73	0.69	1.40	10.72	1.03	1.56	0.63	0.85	
Linseed oil		1066			NR	NR	0.61	0.56	0.63	6.67	1.04	2.52	1.11	1.14	
Control	0	0	GH	64:36	0.00	0.00	0.13	0.14	0.22	1.21	0.12	0.30	NR	NR	Roy et al. (2006)
Linseed oil		1350			0.04	0.03	0.70	0.55	0.70	7.49	0.87	1.87	NR	NR	
Control	0	0	GH	65:35	0.01	0.01	0.19	0.14	0.28	1.12	0.20	0.34	0.11	0.16	Loor et al. (2005a)
Linseed oil		538			0.04	0.03	0.53	0.29	0.52	3.23	0.63	2.42	0.76	0.69	
Control	0	0	GH	35:65	0.03	0.03	0.40	0.23	1.66	1.32	0.34	0.61	0.18	0.22	Loor et al. (2005a)

Table 14 (continued)

Lipid Source	Intake (g/d) Oilseed	Intake (g/d) Oil	Forage[1]	F:C[2]	Double-Bond Position Δ4	Δ5	Δ6-8	Δ9	Δ10	Δ11	Δ12	Δ13/14[3]	Δ15	Δ16[4]	Reference
Linseed oil		612			0.05	0.05	0.73	0.58	2.84	4.53	0.62	2.56	0.42	0.49	
Control	0		MS/GH	48:52	0.01	0.01	0.23	0.22	0.43	1.27	0.34	0.30	NR	NR	Roy et al. (2006)
Sunflower oil		957			0.06	0.05	0.95	0.46	7.22	1.44	0.86	0.43	NR	NR	
Control	0		MS	27:73	0.02	0.01	0.27	0.18	2.96	1.04	0.33	0.37	NR	NR	Roy et al. (2006)
Sunflower oil		755			0.05	0.09	1.43	0.59	18.62	1.36	0.74	0.80	NR	NR	
Control	0		MS/LH	60:40	NR	NR	0.05	0.10	0.26	0.54	0.23	0.43	0.13	0.14	Piperova et al. (2000)
Soybean oil		985	MS	25:75	NR	NR	1.08	0.95	9.24	1.70	0.67	1.34	0.34	0.22	

[1] BS = barley silage; GH = grass hay; GS = grass silage; LH = Lucerne hay; LS = Lucerne silage; MS = maize silage.
[2] Forage/concentrate ratio of the diet (on a dry matter basis).
[3] In some analysis may co-elute with cis 6-8 18:1.
[4] Contains cis-14 18:1 as a minor isomer.
[5] Concentrate intake (kg/day).
[6] In the original publication trans-6-8 18:1 reported as trans-6 18:1 and trans-10 18:1 misidentified as cis-6 18:1.
[7] trans-10 18:1 and trans-11 18:1 not resolved.
SE = solvent extracted; ME = mechanically extracted; GR = ground; NR = not reported.

Table 15 Effect of Plant Oils or Oilseeds on the Concentration of Conjugated Linoleic Isomers and *trans* Octadecadienoic Acids in Bovine Milk Fat (mg/100 g fatty acids)

Isomer	Ground Oilseeds[1]				Plant Oils[2]			
	Control	Rapeseed	Sunflower	Linseed	Control	Rapeseed	Soybean	Linseed
cis-9, *trans*-12/*cis*-9, *trans*-14[3]	226[4]	305[4]	350[4]	327[4]	56	81	105	122
cis-9, *trans*-13	124[5]	237[5]	282[5]	271	105	168	219	257
trans-9, *cis*-12	–	–	–	–	28	34	41	51
trans-11, *cis*-15	124[6]	135[6]	147[6]	282[6]	260	198	378	487
trans-12, *cis*-15	–	–	–	–	77	85	75	80
trans-9, *trans*-12	–	–	–	–	14	19	24	26
trans-10, *trans*-14	–	–	–	–	52	67	73	73
trans-11, *trans*-15	–	–	–	–	73	71	113	165
trans, *trans*[7]	68	90	90	102	–	–	–	–
cis-9, *trans*-11	459	617	842	547	552	727	938	848
cis-11, *trans*-13	1	1	1	2	–	–	–	–
cis-12, *trans*-14	2	4	3	14	3	6	5	23
trans-7, *cis*-9	18	74	63	37	39	99	74	58
trans-8, *cis*-10	9	12	25	14	6	9	13	8
trans-10, *cis*-12	2	4	7	3	2	2	5	2
trans-11, *cis*-13	10	12	14	26	31	17	17	41
trans-6, *trans*-8	2	2	1	1	–	–	–	–
trans-7, *trans*-9	7	7	7	7	11	8	8	8
trans-8, *trans*-10	3	3	5	2	2	2	4	2

Table 15 (continued)

Isomer	Ground Oilseeds[1]				Plant Oils[2]			
	Control	Rapeseed	Sunflower	Linseed	Control	Rapeseed	Soybean	Linseed
trans-9, trans-11	6	8	10	9	9	9	12	15
trans-10, trans-12	6	6	10	6	2	3	7	3
transs-11, trans-13	6	10	8	26	19	25	21	51
trans-12, trans-14	3	5	3	15	12	14	15	49

[1] Milk from hay-based diets (forage/concentrate ratio 59/41, on a dry matter basis) supplemented with none (control) or 1 kg/day of ground rapeseed, sunflower seed, or linseed. Data adapted from Collomb et al. (2004a, b).
[2] Milk from grass silage-based diets (forage: concentrate ratio 60:40, on a dry matter basis) supplemented with none (control) or 500 g/day of rapeseed oil, soybean oil, or linseed oil (Shingfield, Ahvenjärvi, Toivonen, Huhtanen, & Grünari, unpublished).
[3] Methyl ester of cis-9, trans-14 elutes at the same retention time as cis-9, trans-12 methyl ester standard (refer to Fig. 2).
[4] Reported to co-elute with trans-8, cis-13.
[5] Reported to co-elute with trans-8, cis-12.
[6] Rreported to co-elute with trans-9, cis-12.
[7] Double-bond position not determined.

Table 16 Effect of Inclusion of Plant Oils in Diets of Different Composition on the Concentration of Conjugated Linoleic Isomers and Major *trans* Octadecadienoic Acids in Bovine Milk Fat (mg/100 g fatty acids)

Oil supplement[2]	Basal Diet[1]					
	Grass Hay		Maize Silage		High Concentrate	
	Control	Linseed	Control	Sunflower	Control	Sunflower
cis-9, *trans*-13	80	700	150	370	130	270
trans-11, *cis*-15	70	2370	–	–	–	–
cis-9, *cis*-11	1	1	1	3	–	4
cis-9, *trans*-11	516	2771	509	769	551	1002
cis-11, *trans*-13	–	6	–	–	–	–
cis-12, *trans*-14	1	5	–	4	–	3
trans-6, *cis*-8	2	5	1	3	1	–
trans-7, *cis*-9	22	113	33	147	40	136
trans-8, *cis*-10	–	–	7	11	9	23
trans-9, *cis*-11	–	–	–	100	37	160
trans-10, *cis*-12	1	2	3	24	10	49
trans-11, *cis*-13	12	126	2	6	1	5
trans-6, *trans*-8	3	2	1	5	–	–
trans-7, *trans*-9	2	3	1	4	1	2
trans-8, *trans*-10	7	2	1	2	1	2
trans-9, *trans*-11	7	47	6	8	7	14
trans-10, *trans*-12	4	9	4	15	5	11
trans-11, *trans*-13	9	87	3	3	1	2
trans-12, *trans*-14	3	55	2	1	1	2
trans-13, *trans*-15	1	4	–	–	–	–

[1] Data adapted from Roy et al. (2006). Basal diets were comprised of grass hay (F:C ratio 64:36), maize silage (F:C ratio 48:52), or a maize silage and a high proportion of concentrate (F:C ratio 27:73).
[2] Oils included at a rate of 50g/kg diet dry matter. Concentrations reported for milk collected before and on 18 days after the start of lipid supplementation.
F:C, forage: concentrate ratio, on a dry matter basis.

trans-11 18:1 is converted to *cis*-9, *trans*-11 CLA via Δ-9 destaurase in the mammary gland and endogenous synthesis is quantitatively the most important source of this CLA isomer secreted in milk fat (refer to Section 3.2), shifts in ruminal biohydrogenation toward *trans*-10 18:1 at the expense of *trans*-11 18:1 also result in a decrease in milk fat *cis*-9, *trans*-11 CLA concentrations.

Interactions between the relative proportion of forage and concentrate in the diet and supplements of plant oils are known to alter the concentrations of TFA, CLA isomers, and other bioactive fatty acids in bovine milk fat. For example, supplements of linseed oil were found to result in higher enrichment of *trans*-10 18:1, *trans*-11, *cis*-15 18:2 and 18:3 *n*-3, smaller increases in 18:0 and *cis*-9 18:1, and lower reductions in 16:0 concentrations when included in high-concentrate diets compared with low-concentrate diets (Loor et al., 2005a). There is increasing evidence in the literature that changes in milk fatty acid composition responses to lipid supplements on high-concentrate diets are also time-dependent. Studies by Bauman et al. (2000), Dhiman et al. (2000), and Ferlay et al. (2003) were the first to demonstrate that increases in milk fat *cis*-9, *trans*-11 CLA content to sunflower, soybean, or linseed oil in the diet could be transient, typically reaching a maximum within 14 days on diet and decreasing thereafter. Supplements of a mixture (1:2 w/w) of fish oil and sunflower oil in a maize silage-based ration (45 g/kg diet dry matter) have also been shown to cause a rapid increase in milk fat *cis*-9, *trans*-11 CLA concentrations, which reached a maximum concentration of 5.37 g/100 g fatty acids within 5 days on diet, but declined thereafter to 2.35 g/100 g fatty acids by day 15 (Shingfield et al., 2006a). Further studies have shown that temporal changes in milk fat *cis*-9, *trans*-11 CLA content are dependent on the composition of the basal ration and the amount and source of lipid in the diet (Figure 4). Transient changes in milk fat *cis*-9, *trans*-11 CLA content do not occur in isolation and are also accompanied by variations in the concentrations of other isomers of CLA over time (Roy et al., 2006; Shingfield et al., 2006a; Fig. 5).

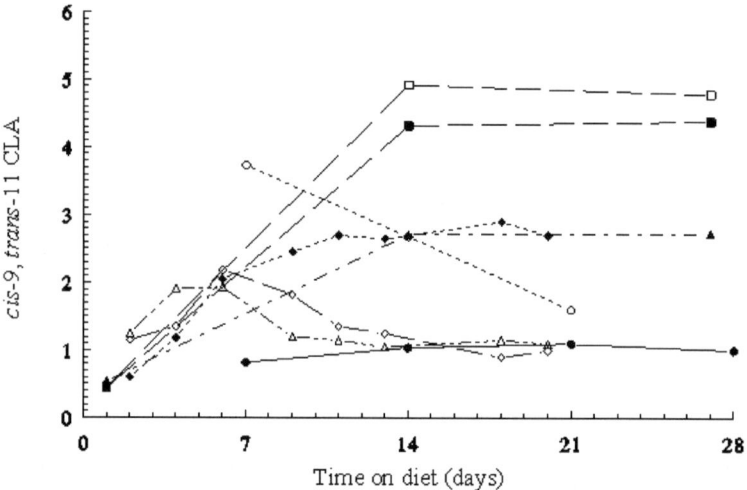

Fig. 4 Temporal changes in *cis*-9, *trans*-11 conjugated linoleic acid (CLA) concentrations (g/100 g fatty acids) in milk from cows fed diets of different compositions containing plant oils. Milk from cows fed diets based on maize silage and hay crop silage (F:C 46:54) containing 52 g sunflower oil/kg dry matter (O; Bauman et al., 2000), grass silage (F:C 50:50) supplemented with concentrates containing 50 g/kg of rapeseed oil (●; Ryhänen et al., 2005), Lucerne silage, Lucerne hay, and barley silage (F:C 60:40) containing 60 g safflower oil/kg diet dry matter (■; Bell et al., 2006), 60 g safflower oil + 150 mg dl-α-tocopheryl acetate + 24 ppm monesin/kg dry matter (□; Bell et al., 2006) or 60 g linseed oil/kg dry matter (▲; Bell et al., 2006), high-concentrate diets (F:C 27:73) containing 50 g sunflower oil/kg/dry matter (Δ; Roy et al., 2006), maize silage-based diets (F:C 48:52) containing 50 g sunflower oil/kg dry matter (♦; Roy et al., 2006) or grass hay-based diets (F:C 64:36) containing 50 g linseed oil/kg dry matter (◇; Roy et al., 2006). F:C indicates the forage: concentrate ratio of the diet, on a dry matter basis

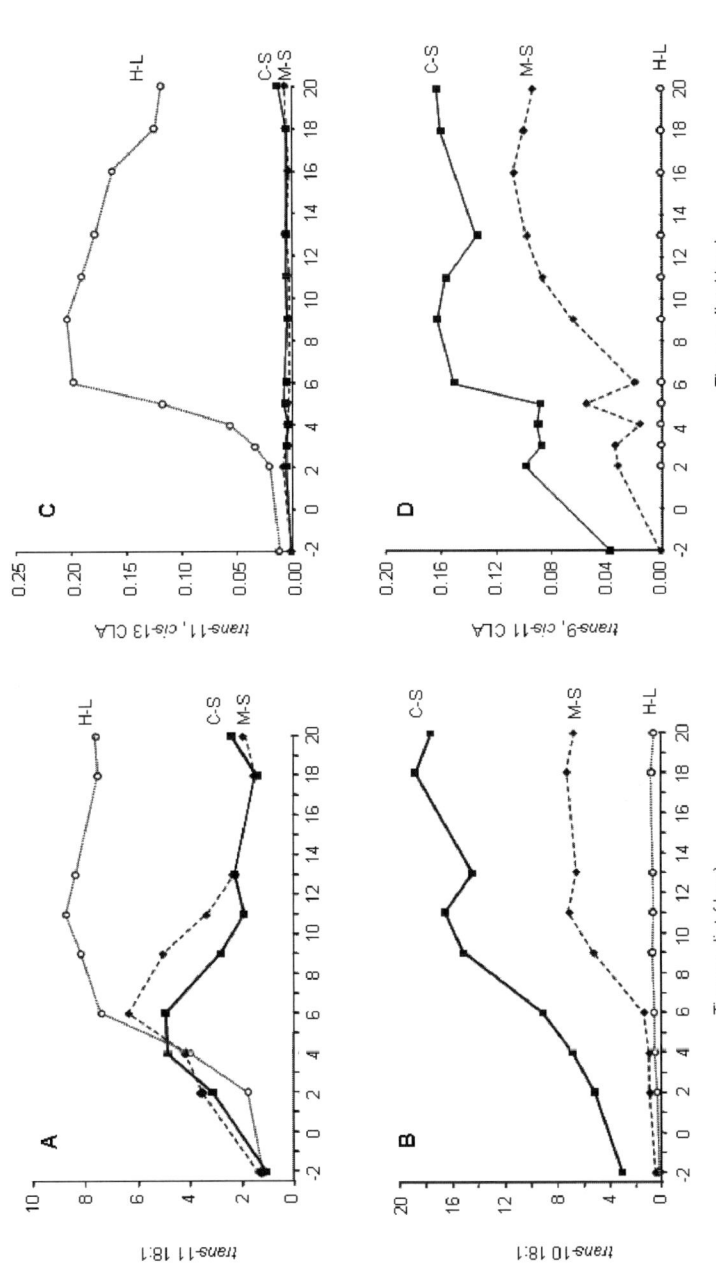

Fig. 5 Time-dependent changes in milk fat (a) *trans*-11 18:1, (b) *trans*-10 18:1, (c) *trans*-11, *cis*-13 CLA, and (d) *trans*-9, *cis*-11 CLA concentrations (g/100 g total fatty acids) in cows offered high-concentrate diets containing (g/kg diet dry matter) sunflower oil (50) (C-S); maize silage-based diets containing sunflower oil (50; M-S) or grass hay-based diets containing linseed oil (50). Values represent the mean of six cows. (Data adapted from Roy et al., 2006.)

Furthermore, after the addition of plant oil to diets containing high proportions of maize silage and/or concentrates, transient decreases in *cis*-9, *trans*-11 CLA concentrations were shown to be associated with concomitant increases in the milk fat *trans*-10 18:1 content of between 4.5 and 18.6 g/100 g fatty acids (Ferlay et al., 2003; Roy et al., 2006; Shingfield et al., 2006a). In recent studies, supplementing high-forage diets comprised of Lucerne silage, Lucerne hay, and barley silage with relatively high levels (60 g/kg diet dry matter) of safflower oil or grass hay diets with linseed oil (50 g/kg diet dry matter) have been shown to reduce milk fat 12:0, 14:0, and 16:0 content and enhance *trans*-11 18:1 and *cis*-9, *trans*-11 CLA concentrations, responses that were shown to persist over a three- (Roy et al., 2006) or eight-week period (Bell et al., 2006). The available evidence indicates that supplementing high-forage/low-starch diets with plant oils represents an effective nutritional strategy for sustainable increases in milk fat *trans*-11 18:1 and *cis*-9, *trans*-11 CLA content. Interestingly, the inclusion of safflower in combination with 150 mg dl-α-tocopheryl acetate and 24 ppm monesin/kg diet DM was shown to be particularly effective at increasing milk fat *cis*-9, *trans*-11 CLA concentrations from 0.45 to 4.75 g/100 g fatty acid methyl esters (Bell et al., 2006), but the underlying reasons to explain the persistent enrichment of *cis*-9, *trans*-11 CLA in milk fat merit further investigation.

Conclusions

Concerns about the role of the human diet on the development of chronic diseases can be expected to continue in the future. In order to reduce the social and financial burden of chronic disease and extend life expectancy, there is increased interest in changing the composition of the diet for the maintenance of human health and disease prevention. Milk and dairy products are a significant source of fat and saturated fatty acids, particularly in most Western diets. Advocating a population-wide and significant reduction in the consumption of these foods ignores the value of milk and milk products as a versatile and valuable source of nutrients, in addition to several bioactive lipid components, including butyrate, branch-chain fatty acids, *cis*-9, *trans*-11 conjugated linoleic acid, and sphingomyelin. It is possible to significantly reduce the saturate content and enhance the concentration of several bioactive lipids several-fold in milk through changes in the ruminant diet. Most strategies involve supplementing the diet with plant oils or oilseeds, which, due to the metabolism of dietary lipids in the rumen, also results in an unavoidable increase in milk *trans* fatty acid content. A high consumption of *trans* fatty acids is associated with increased cardiovascular disease risk. However, the profile of *trans* fatty acids in ruminant-derived foods and processed edible fats differs markedly, with evidence from clinical and biomedical models to suggest that the physiological effects are isomer-dependent. Overall, the evidence from human studies and animal models suggests that reductions in milk 12:0, 14:0, and 16:0 content and increases in milk fat *cis*-9 18:1, *cis*-9, *trans*-11 conjugated linoleic acid and 18:3 *n*-3, along with a decrease in dietary energy intake, have the potential to improve long-term human health.

Acknowledgment Preparation of this chapter was supported by LIPGENE, an integrated project within the EU-funded Sixth Framework Research programme (www.lipgene.tcd.ie).

References

AbuGhazaleh, A. A., Schingoethe, D. J., Hippen, A. R., Kalscheur, K. F., & Whitlock, L. A. (2002). Fatty acid profiles of milk and rumen digesta from cows fed fish oil, extruded soyabeans or their blend. *Journal of Dairy Science, 85*, 2266–2276.

Adlof, R. O., Duval, S., & Emken, E. A. (2000). Biosynthesis of conjugated linoleic acid in humans. *Lipids, 35*, 131–135.

Alonso, L., Fontecha, J., Lozada, L., Fraga, M. J., & Juárez, M. (1999). Fatty acid composition of caprine milk: Major, branched-chain, and *trans* fatty acids. *Journal of Dairy Science, 82*, 878–884.

Aro, A., Mannisto, S., Salminen, I., Ovaskainen, M. L., Kataja, V., & Uusitupa, M. (2000). Inverse association between dietary and serum conjugated linoleic acid and risk of breast cancer in postmenopausal women. *Nutrition and Cancer, 38*, 151–157.

Ascherio, A., Hennekens, C. H., Buring, J. E., Master, C., Stampfer, M. J., & Willett, W. C. (1994). *Trans* fatty acids intake and risk of myocardial infarction. *Circulation, 89*, 94–101.

Ascherio, A., Katan, M. B., Stampfer, M. J., & Willett, W. C. (1999a). *Trans* fatty acids and coronary heart disease. *New England Journal of Medicine, 340*, 1994–1998.

Ascheiro, A., Katan, M. B., Zock, P. L., Stampfer, M. J., & Willett, W. C. (1999b). *Trans* fatty acids and coronary heart disease. *Journal of the American Diet Association, 53*, 143–157.

Avalli, A., & Contarini, G. (2005). Determination of phospholipids in dairy products by SPE/HPLC/ELSD. *Journal of Chromatography A, 1071*, 185–190.

Banni, S., Angioni, E., Murru, E., Carta, G., Melis, M. P., Bauman, D. E., Dong, Y., & Ip, C. (2001). Vaccenic acid feeding increases tissue levels of conjugated linoleic acid and suppresses development of premalignant lesions in rat mammary gland. *Nutrition and Cancer, 41*, 91–97.

Bargo, F., Delahoy, J. E., Schroeder, G. F., & Muller, L. D. (2006). Milk fatty acid composition of dairy cows grazing at two pasture allowances and supplemented with different levels and sources of concentrate. *Animal Feed Science and Technology, 125*, 17–31.

Bargo, F., Muller, L. D., Delahoy, J. E., & Cassidy, T. W. (2002). Milk response to concentrate supplementation of high producing dairy cows grazing at two pasture allowances. *Journal of Dairy Science, 85*, 1777–1792.

Bauchart, D., Roy, A., Lorenz, S., Chardigny, J. M., Ferlay, A., Gruffat, D., Sébédio, J.-L., Chilliard, Y., & Durand, D. (2007). Butters varying in *trans* 18:1 and *cis*-9, *trans*-11conjugated linoleic acid modify plasma lipoproteins in the hypercholesterolemic rabbit. *Lipids, 42*, 123–133.

Bauman, D. E., & Davis, C. L. (1974). Biosynthesis of milk fat. In B. L. Larson & V. R. Smith (Eds.), *Lactation: A Comprehensive Treatise*, Vol. 2 (pp. 31–75). London: Academic Press.

Bauman, D. E., & Griinari, J. M. (2003). Nutritional regulation of milk fat synthesis. *Annual Review of Nutrition, 23*, 203–227.

Bauman, D. E., Barbano, D. M., Dwyer, D. A., & Griinari, J. M. (2000). Technical note: Production of butter with enhanced conjugated linoleic acid for use in biomedical studies with animal models. *Journal of Dairy Science, 83*, 2422–2425.

Bauman, D. E., Lock, A. L., Corl, B. A., Ip, C., Salter, A. M., & Parodi, P. M. (2005). Milk fatty acids and human health: Potential role of conjugated linoleic acid and *trans* fatty acids. In K. Serjrsen, T. Hvelplund, & M. O. Nielsen (Eds.), *Ruminant Physiology: Digestion, Metabolism and Impact of Nutrition on Gene Expression, Immunology and Stress* (pp. 529–561). Wageningen, The Netherlands: Wageningen Academic Publishers.

Baylin, A., Kabagambe, E. K., Ascherio, A., Spiegelman, D., & Campos, H (2003). High 18:2 *trans*-fatty acids in adipose tissue are associated with increased risk of nonfatal acute myocardial infarction in Costa Rican adults. *Journal of Nutrition, 133,* 1186–1191.

Bayourthe, C., Enjalbert, F., & Moncoulon, R. (2000). Effects of different forms of canola oil fatty acids plus canola meal on milk composition and physical properties of butter. *Journal of Dairy Science, 83,* 690–696.

Bell, J. A., Griinari, J. M., & Kennelly, J. J. (2006). Effect of safflower oil, flaxseed oil, monensin, and vitamin E on concentration of conjugated linoleic acid in bovine milk fat. *Journal of Dairy Science, 89,* 733–748.

Belobrajdic, D. P., & McIntosh, G. H. (2000). Dietary butyrate inhibits NMU-induced mammary cancer in rats. *Nutrition and Cancer, 36,* 217–223.

Beppu, F., Hosokawa, M., Tanaka, L., Kohno, H., Tanaka, T., & Miyashita, K. (2006). Potent inhibitory effect of *trans* 9, *trans* 11 isomer of conjugated linoleic acid on the growth of human colon cancer cells. *Journal of Nutritional Biochemestry, 17,* 830–836.

Bernard, L., Leroux, C., Bonnet, M., Rouel, J., Martin, P., & Chilliard, Y. (2005a). Expression and nutritional regulation of lipogenic genes in mammary gland and adipose tissues of lactating goats. *Journal of Dairy Research, 72,* 250–255.

Bernard, L., Rouel, J., Leroux, C., Ferlay, A., Faulconnier, Y., Legrand, P., & Chilliard, Y. (2005b). Mammary lipid metabolism and milk fatty acid secretion in alpine goats fed vegetable lipids. *Journal of Dairy Science, 88,* 1478–1489.

Bernard, L., Leroux, C., & Chilliard, Y. (2006). Characterisation and nutritional regulation of the main lipogenic genes in the ruminant lactating mammary gland. In K. Sejrsen, T. Hvelplund, & M. O. Nielsen (Eds.), *Ruminant Physiology: Digestion, Metabolism and Impact of Nutrition on Gene Expression, Immunology and Stress* (pp. 295–326). Wageningen, The Netherlands: Wageningen Academic Publishers.

Berra, B., Colombo, I., Sottocornola, E., & Giacosa, A. (2002). Dietary sphingolipids in colorectal cancer prevention. *European Journal of Cancer, 1,* 193–197.

Bitman, J., & Wood, D. L. (1990). Changes in milk fat phospholipids during lactation. *Journal of Dairy Science, 73,* 1208–1216.

Bonanome, A., & Grundy, S. M. (1988). Effect of dietary stearic acid on plasma cholesterol and lipoprotein. *New England Journal of Medicine, 318,* 1244–1248.

Brechany, E. Y., & Christie, W. W. (1992). Identification of the saturated oxo fatty acids in cheese. *Journal of Dairy Research, 59,* 57–64.

Brechany, E. Y., & Christie, W. W. (1994). Identification of the unsaturated oxo fatty acids in cheese. *Journal of Dairy Research, 62,* 111–115.

Chajes, V., Lavillonniere, F., Ferrari, P., Jourdan, M. L., Pinault, M., Maillard, V., Sebedio, J.-L., & Bougnoux, P. (2002). Conjugated linoleic acid content in breast adipose tissue is not associated with the relative risk of breast cancer in a population of French patients. *Cancer Epidemiology, Biomarkers and Prevention, 11,* 672–673.

Chen, Z.-X., & Breitman, T. R. (1994). Tributyrin: A prodrug of butyric acid for potential clinical application in differentiation therapy. *Cancer Research, 54,* 3494–3499.

Chilliard, Y. (1993). Dietary fat and adipose tissue metabolism in ruminants, pigs, and rodents: A review. *Journal of Dairy Science, 76,* 3897–3931.

Chilliard, Y., & Ferlay, A. (2004). Dietary lipids and forages interactions on cow and goat milk fatty acid composition and sensory properties. *Reproduction Nutrition Development 44,* 467–492.

Chilliard, Y., Doreau, M., Gagliostro, G., & Elmeddah, Y. (1993). Protected (encapsulated or calcium soaps) lipids in dairy cow diets. Effects on production and milk composition. *Productions Animales, 6*, 139–150.

Chilliard, Y., Ferlay, A., Mansbridge, R. M., & Doreau, M. (2000). Ruminant milk fat plasticity: Nutritional control of saturated, polyunsaturated, *trans* and conjugated fatty acids. *Annales de Zootechnie, 49*, 181–205.

Chilliard, Y., Ferlay, A., & Doreau, M. (2001). Effect of different types of forages, animal fat or marine oils in cow's diet on milk fat secretion and composition, especially conjugated linoleic acid (CLA) and polyunsaturated fatty acids. *Livestock Production Science, 70*, 31–48.

Chilliard, Y., Ferlay, A., Loor, J., Rouel, J., & Martin, B. (2002). *Trans* and conjugated fatty acids in milk from cows and goats consuming pasture or receiving vegetable oils or seeds. *Italian Journal of Animal Science, 1*, 243–254.

Chilliard, Y., Ferlay, A., Rouel, J., & Lamberett, G. (2003). A review of nutritional and physiological factors affecting goat milk lipid synthesis and lipolysis. *Journal of Dairy Science, 86*, 1751–1770.

Chouinard, P. Y., Levesque, J., Girard, V., & Brisson, G. J. (1997). Dietary soybeans extruded at different temperatures: Milk composition and in situ fatty acid reactions. *Journal of Dairy Science, 80*, 2913–2924.

Chouinard, P. Y., Corneau, L., Butler, W. R., Chilliard, Y., Drackley, J. K., & Bauman, D. E. (2001). Effect of dietary lipid source on conjugated linoleic acid concentrations in milk fat. *Journal of Dairy Science, 84*, 680–690.

Christie, W. W. (1981). The effect of diet and other factors on the lipid composition of ruminant tissues and milk. In W. W. Christie (Ed.), *Lipid Metabolism in Ruminant Animals* (pp. 193–226). Oxford: Pergamon Press.

Christie, W. W., Noble R. C., & Davies, G. (1987). Phospholipids in milk and dairy-products. *Journal of the Society for Dairy Technology, 40*, 10–12.

Clandinin, M. T., Cook, S. L., Konrad, S. D., Goh, Y. K., & French, M. A. (1999). The effect of palmitic acid on lipoprotein cholesterol levels and endogenous cholesterol synthesis in hyperlipidemic subjects. *Lipids, 34* (Suppl), S121–S124.

Clandinin, M. T., Cook, S. L., Konard, S. D., & French, M. A. (2000). The effect of palmitic acid on lipoprotein cholesterol levels. *International Journal of Food Science and Nutrition, 51* (Suppl), S61–S71.

Coakley, M., Johnson, M. C., McGrath, E., Rahman, S., Ross, R. P., Fitzgerald, G. F., Devery, R., & Stanton, C. (2006). Intestinal bifidobacteria that produce *trans*-9, *trans*-11 conjugated linoleic acid: A fatty acid with antiproliferative activity against human colon SW480 and HT-29 cancer cells. *Nutrition and Cancer, 56*, 95–102.

Collomb, M., & Buhler, T. (2000). Analysis of the fatty acid composition of milk fat. I. Optimization and validation of a high resolution general method. *Mitteilungen aus Lebensmitteluntersuchung und Hygiene, 91*, 306–332.

Collomb, M., Sieber, R., & Bütikofer, U. (2004a). CLA isomers in milk fat from cows fed diets with high levels of unsaturated fatty acids. *Lipids, 39*, 355–364.

Collomb, M., Sollberger, H., Bütikofer, U., Sieber, R., Stoll, W., & Schaeren, W. (2004b). Impact of a basal diet of hay and fodder beet supplemented with rapeseed, linseed and sunflowerseed on the fatty acid composition of milk fat. *International Dairy Journal, 14*, 549–559.

Collomb, M., Schmid, A., Sieber, R., Wechsler, D., & Ryhänen, E.-L. (2006). Conjugated linoleic acids in milk fat: Variation and physiological effects. *International Dairy Journal, 16*, 1347–1361.

Colón-Ramos, U., Baylin, A., & Campos, H. (2006). The relation between *trans* fatty acid levels and increased risk of myocardial infarction does not hold at lower levels of *trans* fatty acids in the Costa Rican food supply. *Journal of Nutrition, 136*, 2887–2892.

Committee on Medical Aspects of Food Policy (1984). Diet and cardiovascular disease. Department of Health and Social Security report on health and social subjects. No. 28. HMSO, London.

Committee on Medical Aspects of Food Policy (1994). Nutritional aspects of cardiovascular disease. Department of Health and Social Security Report on Health and Social Subjects. No. 46. HMSO, London.

Corl, B. A., Baumgard, L. H., Dwyer, D. A., Griinari, J. M., Phillips, B. S., & Bauman, D. E. (2001). The role of Δ^{-9} desaturase in the production of *cis*-9, *trans*-11 CLA. *Journal of Nutritional Biochemestry, 12*, 622–630.

Corl, B. A., Baumgard, L. H., Griinari, J. M., Delmonte, P., Morehouse, K. M., Yurawecz, M. P., & Bauman, D. E. (2002). *Trans*-7, *cis*-9 CLA is synthesized endogenously by Δ9- desaturase in dairy cows. *Lipids, 37*, 681–688.

Corl, B. A., Barbano, D. M., Bauman, D. E., & Ip, C. (2003). *Cis*-9, *trans*-11 CLA derived endogenously from *trans*-11 18:1 reduces cancer risk in rats. *Journal of Nutrition, 133*, 2893–2900.

Dannenberger, D., Nuernberg, G., Scollan, N., Schabbel, W., Steinhart, H., Ender, K., & Nuernberg, K. (2004). Effect of diet on the deposition of n-3 fatty acids, conjugated linoleic and C18:1 *trans* fatty acid isomers in muscle lipids of German Holstein bulls. *Journal of Agriculture and Food Chemistry, 52*, 6607–6615.

Dabadie, H., Peuchant, E., Bernard, M., LeRuyet, P., & Mendy, F. (2005). Moderate intake of myristic acid in sn-2 position has beneficial lipidic effects and enhances DHA of cholesteryl esters in an interventional study. *Journal of Nutritional Biochemistry, 16*, 375–382.

Dabadie, H., Motta, C., Peuchant, E., LeRuyet, P., & Mendy, F. (2006). Variations in daily intakes of myristic and alpha-linolenic acids in sn-2 position modify lipid profile and red blood cell membrane fluidity. *British Journal of Nutrition, 96*, 283–289.

De La Torre, A., Debiton, E., Durand, D., Chardigny, J. M., Berdeaux, O., Loreau, O., Barthomeuf, C., Bauchart, D., & Gruffat, D. (2005). Conjugated linoleic acid isomers and their conjugated derivatives inhibit growth of human cancer cell lines. *Anticancer Research, 25*, 3943–3949.

Demeyer, D., & Doreau, M. (1999). Targets and procedures for altering ruminant meat and milk lipids. *Proceedings of the Nutrition Society, 58*, 593–607.

Desroches, S., Chouinard, P. Y., Galibois, I., Corneau, L., Delisle, J., Lamarche, B., Couture, P., & Bergeron, N. (2005). Lack of effect of dietary conjugated linoleic acids naturally incorporated into butter on the lipid profile and body composition of overweight and obese men. *American Journal of Clinical Nutrition, 82*, 309–319.

Dewhurst, R. J., Fisher, W. J., Tweed, J. K. S., & Wilkins, R. J. (2003). Comparison of grass and legume silages for milk production. 1. Production responses with different levels of concentrate. *Journal of Dairy Science, 86*, 2598–2611.

Dewhurst, R. J., Shingfield, K. J., Lee, M. R. F., & Scollan, N. D. (2006). Increasing the concentrations of beneficial polyunsaturated fatty acids in milk produced by dairy cows in high-forage systems. *Animal Feed Science and Technology, 131*, 168–206.

Dhiman, T. R., Zanten, K. V., & Satter, L. D. (1995). Effect of dietary fat source on fatty acid composition of cow's milk. *Journal of the Science of Food and Agriculture, 69*, 101–107.

Dhiman, T. R., Anand, G. R., Satter, L. D., & Pariza, W. (1999). Conjugated linoleic acid content of milk from cows fed different diets. *Journal of Dairy Science, 82*, 2146–2156.

Dhiman, T. R., Satter, L. D., Pariza, M. W., Galli, M. P., Albright, K., & Tolosa, M. X. (2000). Conjugated linoleic acid (CLA) content of milk from cows offered diets rich in linoleic and linolenic acid. *Journal of Dairy Science, 83*, 1016–1027.

Doreau, M., & Ferlay, A. (1994). Digestion and utilisation of fatty acids by ruminants. *Animal Feed Science and Technology, 45*, 379–396.

Doreau, M., & Poncet, C. (2000). Ruminal biohydrogenation of fatty acids originating from fresh or preserved grass. *Reproduction, Nutrition, Development, 40*, 201.

Doreau, M., Lee, M. R. F., Ueda, K., & Scollan, N. D. (2005). Métabolisme ruminal et digestibilité des acides gras des fourrages. In *12iemes Rencontres Recherches Ruminants, Paris (France), 12*, 101–104.

Elgersma, A., Ellen, G., van der Horst, H., Boer, H., Dekker, P. R., & Tamminga, S. (2004). Quick changes in milk fat composition from cows after transition from fresh grass to a silage diet. *Animal Feed Science and Technology, 117,* 13–27.

Elwood, P., Hughes, J., & Fehly, A. (2005). Milk, heart disease and obesity: An examination of the evidence. *British Journal of Cardiology, 12,* 283–290.

Faulconnier, Y., Arnal, M. A., Patureau Mirand, P., Chardigny, J. M., & Chilliard, Y. (2004). Isomers of conjugated linoleic acid decrease plasma lipids and stimulate adipose tissue lipogenesis without changing adipose weight in post-prandial adult sedentary or trained Wistar rat. *Journal of Nutritional Biochemistry, 15,* 741–748.

Ferlay, A., Andrieu, J. P., Pomies, D., Martin-Rosset, W., & Chilliard, Y. (2002). Effet de l'ensilage enrubanné d'herbe de demi montagne sur la composition en acides gras d'intéret nutritionnel du lait de vache. In *9iemes Rencontres Recherches Ruminants, Paris (France), 9,* 365.

Ferlay, A., Capitan, P., Ollier, A., & Chilliard, Y. (2003). Interactions between nature of forage and oil supplementation on cow milk composition. 3. Effects on kinetics and percentages of milk CLA and *trans* fatty acids. In Y. van der Honing (Ed.), *Abstracts of the 54th Annual Meeting of European Association for Animal Production,* Rome, 31 August–3 September 2003 (p. 120). Wageningen, The Netherlands: Wageningen Academic Publishers.

Ferlay, A., Martin, B., Pradel, P., Coulon, J. B., & Chilliard, Y. (2006). Influence of grass-based diets on milk fatty acid composition and milk lipolytic system in Tarentaise and Montbeliarde cow breeds. *Journal of Dairy Science, 89,* 4026–4041.

Feskens, E. J. M., & Kromhout, D. A. (1990). Habitual dietary intake and glucose tolerance euglycaemic men: The Zutphen Study. *International Journal of Endocrinology, 19,* 953–959.

Feskens, E. J. M., Virtanen, S. M., Räsänen, L., Tuomilehto, J., Stengard, J., Pekkanen, J., Nissinen, A., & Kromhout, D. A. (1995). 20-Year follow-up of the Finnish and Dutch cohorts of the Seven Countries Study. *Diabetes Care, 18,* 1104–1112.

Fievez, V., Vlaeminck, B., Dhanoa, M. S., & Dewhurst, R. J. (2003). Use of principal component analysis to investigate the origin of heptadecenoic and conjugated linoleic acids in milk. *Journal of Dairy Science, 86,* 4047–4053.

French, M. A., Sundram, K., & Clandinin, M. T. (2002). Cholesterolaemic effect of palmitic acid in relation to other dietary fatty acids. *Asian Pacific Journal of Clinical Nutrition, 11 (Suppl 7),* S401–S407.

Garton, G. A. (1977). Fatty acid metabolism in ruminants. In T. W. Goodwin (Ed.), *Biochemistry of Lipids,* Vol. 14 (pp. 337–370). Baltimore: University Park Press.

Givens, D. I., & Shingfield, K. J. (2006). Optimising dairy milk fatty acid composition. In C. Williams & J. Buttriss (Eds.), *Improving the Fat Content of Foods* (pp. 252–280). Cambridge: Woodhead Publishing Ltd.

Gonthier, C., Mustafa, A. F., Ouellet, D. R., Chouinard, P. Y., Berthiaume, R., & Petit, H. R. (2005). Feeding micronized and extruded flaxseed to dairy cows: Effects on blood parameters and milk fatty acid composition. *Journal of Dairy Science, 88,* 748–756.

Goudjil, H., Fontecha, J., Luna, P., de la Fuente, M. A., Alonso, L., & Juárez, M. (2004). Quantitative characterization of unsaturated and *trans* fatty acids in ewe's milk fat. *Lait, 84,* 473–482.

Graves, E. L. F., Beaulieu, A. D., & Drackley, J. K. (2007). Factors affecting the concentration of sphingomyelin in bovine milk. *Journal of Dairy Science, 90,* 706–715.

Griinari, J. M., Dwyer, D. A., McGuire, M. A., Bauman, D. E., Palmquist, D. L., & Nurmela, K. V. (1998). *Trans*-octadecenoic acids and milk fat depression in lactating dairy cows. *Journal of Dairy Science, 81,* 1251–1261.

Griinari, J. M., Corl, B. A., Lacy, S. H., Chouinard, P. Y., Nurmela, K. V. V., & Bauman, D. E. (2000). Conjugated linoleic acid is synthesized endogenously in lactating dairy cows by Δ^9-desaturase. *Journal of Nutrition, 130,* 2285–2291.

Griinari, J. M., & Bauman, D. E. (1999). Biosynthesis of conjugated linoleic acid and its incorporation into meat and milk in ruminants. In M. P. Yurawecz, M. M. Mossoba, J. K.

G. Kramer, M. W. Pariza, & G. Nelson (Eds.), *Advances in Conjugated Linoleic Acid Research* (pp. 180–200). Champaign, IL: AOCS Press.

Gunstone, F. D., Harwood, J. L., & Padley, F. P. (1994). Occurrence and characteristics of oils and fats. In F. D. Padley, F. D. Gunstone, & J. L. Harwood (Eds.), *The Lipid Handbook* (pp. 47–224). Cambridge: Cambridge University Press.

Ha, J. K., & Lindsay, R. C. (1990). Method for the quantitative analysis of volatile and free branched-chain fatty acids in cheese and milk fat. *Journal of Dairy Science, 73*, 1988–1999.

Harfoot, C. G., & Hazlewood, G. P. (1997). Lipid metabolism in the rumen. In P. N. Hobson & C. S. Stewart (Eds.), *The Rumen Microbial Ecosystem*, 2nd ed. (pp. 382–426). London: Blackie Academic & Professional.

Heaney, R. P. (2000). Calcium, dairy products and osteoporosis. *Journal of the American College of Nutrition, 19*, 83S–99S.

Hodgson, J. M., Wahlqvist, M. L., Boxall, J. A., & Balazs, N. D. (1996). Platelet *trans* fatty acids in relation to angiographically assessed coronary artery disease. *Atherosclerosis, 120*, 147–154.

Hulsof, K. F. A. M., van Erp-Baart, M. A., Anttolainen, M., Becker, W., Church, S. M., Couet, C., Hermann-Kunz, E., Kesteloot, H., Leth, T., Martins, I., Moreiras, O., Moschandreas, J., Pizzoferrato, L., Rimestad, A. H., Thorgeirsdottir, H., van Amelsvoort, J. M. M., Aro, A., Kafatos, A. G., Lanzmann-Petithory, D., & van Poppel, G. (1999). Intake of fatty acids in Western Europe with emphasis on *trans* fatty acids: The TRANSFAIR study. *European Journal of Clinical Nutrition, 53*, 143–157.

Ip, C., Chin, S. F., Scimeca, J. A., & Pariza, M. W. (1991). Mammary cancer prevention by conjugated dienoic derivative of linoleic acid. *Cancer Research, 51*, 6118–6124.

Ip, C., Singh, M., Thompson, H. J., & Scimeca, J. A. (1994). Conjugated linoleic acid suppresses mammary carcinogenesis and proliferative activity of the mammary gland in the rat. *Cancer Research, 54*, 1212–1215.

Ip, C., Banni, S., Angioni, E., Carta, G., McGinley, J., Thompson, H. J., Barbano, D., & Bauman, D. (1999). Conjugated linoleic acid-enriched butterfat alters mammary gland morphogenesis and reduces cancer risk in rats. *Journal of Nutrition, 129*, 2135–2142.

Ip, C., Dong, Y., Ip, M. M., Banni, S., Carta, G., Angioni, E., Murru, E., Spada, S., Melis, M. P., & Sæbo, A. (2002). Conjugated linoleic acid isomers and mammary cancer prevention. *Nutrition and Cancer, 43*, 52–58.

Jakobsen, M. U., Bysted, A., Andersen, N. L., Heitmann, B. L., Hartkopp, H. B., Leth, T., Overvad, K., & Dyerberg, L. (2006). Intake of ruminant *trans* fatty acids and risk of coronary heart disease. *Atherosclerosis Supplements, 7*, 9–11.

Jensen, R. G. (2002). The composition of bovine milk lipids: January 1995 to December 2000. *Journal of Dairy Science, 85*, 295–350.

Jones, E. L., Shingfield, K. J., Kohen, C., Jones, A. K., Lupoli, B., Grandison, A. S., Beever, D. E., Williams, C. M., Calder, P. C., & Yaqoob, P. (2005). Chemical, physical and sensory properties of dairy products enriched with conjugated linoleic acid. *Journal of Dairy Science, 88*, 2923–2937.

Jurjanz, S., Monteils, V., Juaneda, P., & Laurent, F. (2004). Variations of *trans* octadecenoic acid in milk fat induced by feeding different starch-based diets to cows. *Lipids, 39*, 19–24.

Judd, J. T., Clevidence, B. A., Muesing, R. A., Wittes, J., Sunkin, M. E., & Podczasy, J. J. (1994). Dietary *trans* fatty acids: Effects on plasma lipids and lipoproteins of healthy men and women. *American Journal of Clinical Nutrition, 59*, 861–868.

Kalkwarf, H. F., Khoury, J. C., & Lanphear, B. P. (2003). Milk intake during childhood and adolescence, adult bone density, and osteoporotic fractures in U.S. women. *American Journal of Clinical Nutrition, 77*, 257–265.

Kalscheur, K. F., Teter, B. B., Piperova, L. S., & Erdman, R. A. (1997). Effect of dietary forage concentration and buffer addition on duodenal flow of *trans*-C18:1 fatty acids and milk fat production in dairy cows. *Journal of Dairy Science, 80*, 2104–2114.

Kaneda, T. (1991). *Iso-* and *anteiso-*fatty acids in bacteria: Biosynthesis, function, and taxonomic significance. *Microbial Reviews, 55*, 288–302.

Katan, M. B., Zock, P. L., & Mensink, R. P. (1995). Dietary oils, serum lipoproteins, and coronary heart disease. *American Journal of Clinical Nutrition, 61*, 1368S–1373S.

Kay, J. K., Mackle, T. R., Auldist, M. J., Thomson, N. A., & Bauman, D. E. (2004). Endogenous synthesis of *cis*-9, *trans*-11 conjugated linoleic acid in dairy cows fed fresh pasture. *Journal of Dairy Science, 87*, 369–378.

Kelly, M. L., Kolver, E. S., Bauman, D. E., van Amburgh, M. E., & Muller, L. D. (1998). Effect of intake of pasture on concentrations of conjugated linoleic acid in milk of lactating cows. *Journal of Dairy Science, 81*, 1630–1636.

Korhonen, H. J. T., & Pihlanto, A. (2006). Bioactive peptides: Production and functionality. *International Dairy Journal, 16*, 945–960.

Kraft, J., Collomb, M., Mockel, P., Sieber, R., & Jahreis, G. (2003). Differences in CLA isomer distribution of cow's milk lipids. *Lipids, 38*, 657–664.

Kris-Etherton, P. M., Daniels, S. R., Eckel, R. H., Engler, M., Howard, B. V., Krauss, R. M., Lichtenstein, A. H., Sacks, F., St. Jeor, S., & Stampfer, M. (2001) Summary of the scientific conference on dietary fatty acids and cardiovascular health: Conference summary from the nutrition committee of the American Heart Association. *Circulation, 103*, 1034–1039.

Kritchevsky, D. (2003). Conjugated linoleic acids in experimental atherosclerosis. In J.-L. Sebedio, W. W. Christie, & R. O. Adlof (Eds), *Advances in Conjugated Linoleic Acid Research*, Vol. 2. (pp. 293–301). Champaign, IL: AOCS Press.

Kromhout, D., Menotti, A., Bloemberg, B., Aravanis, C., Blackburn, H., Buzina, R., Dontas, A. S., Fidanza, F., Giampaoli, S., Jansen, A., Karvonen, M., Katan, M., Nissinen, A., Nedeljkovi, S., Pekkanen, J., Pekkarinen, M., Punsar, S., Räsänen, L., Simic, B., & Toshima, H. (1995). Dietary saturated and *trans* fatty acids and cholesterol and 25-year mortality from coronary heart disease: The Seven Countries Study. *Preventive Medicine, 24*, 308–315.

Kuzdzal-Savoie, S., & Kuzdzal, W. (1961). Influence de la mise à l'herbe des vaches laitières sur les indices de la matière grasse du beurre et sur les teneurs en différents acides gras polyinsaturés. *Annals of Biology, Animal Biochemistry and Biophysiology, 1*, 47–69.

Larsson, S. C., Bergkvist, L., & Wolk, A. (2005). High-fat dairy food and conjugated linoleic acid intakes in relation to colorectal cancer incidence in the Swedish Mammography Cohort. *American Journal of Clinical Nutrition, 82*, 894–900.

Lawson, R. E., Moss, A. R., & Givens, D. I. (2001). The role of dairy products in supplying conjugated linoleic acid to man's diet: A review. *Nutrition Research Reviews, 14*, 153–172.

Ledoux, M., Rouzeau A., Bas, P., & Sauvant, D. (2002). Occurrence of *trans*-C18:1 fatty acid isomers in goat milk: Effect of two dietary regimens. *Journal of Dairy Science, 85*, 190–197.

Ledoux, M., Chardigny, J. M., Darbois, M., Soustre, Y., Sebedio, J. L., & Laloux, L. (2005). Fatty acid composition of French butters, with special emphasis on conjugated linoleic acid (CLA) isomers. *Journal of Food Composition and Analysis, 18*, 409–425.

Lemaitre, R. N., King, I. B., Patterson, R. E., Psaty, B. M., Kestin, M., & Heckbert, S. R. (1998). Assessment of *trans*-fatty acid intake with a food frequency questionnaire and validation with adipose tissue levels of trans-fatty acids. *American Journal of Epidemiology, 148*, 1085–1093.

Lemaitre, R. N., King, I. B., Raghunathan, T. E., Pearce, R. M., Weinmann, S., Knopp, R. H., Copass, M. K., Cobb, L. A., & Siscovick, D. S. (2002). Cell membrane *trans*-fatty acids and the risk of primary cardiac arrest. *Circulation, 105*, 697–701.

Lemaitre, R. N., King, I. B., Mozaffarian, D., Sotoodehnia, N., Rea, T. D., Kuller, L. H., Tracy, R. P., & Siscovick, D. S. (2006). Plasma phospholipid *trans* fatty acids, fatal ischemic heart disease, and sudden cardiac death in older adults. *Circulation, 114*, 209–215.

Lock, A. L., & Bauman, D. E. (2004). Modifying milk fat composition of diary cows to enhance fatty acids beneficial to human health. *Lipids, 39*, 1197–1206.

Lock, A. L., & Garnsworthy, P. C. (2002). Independent effects of dietary linoleic and linolenic fatty acids on the conjugated linoleic acid content of cows' milk. *Animal Science, 74*, 163–176.

Lock, A. L., & Garnsworthy, P. C. (2003). Seasonal variation in milk conjugated linoleic acid and Δ-9 desaturase activity in dairy cows. *Livestock Production Science, 79*, 47–59.

Lock, A. L., & Shingfield, K. J. (2004). Optimising milk composition. In E. Kebreab, J. Mills, & D. E. Beever (Eds.), *UK Dairying: Using Science to Meet Consumers' Needs* (pp. 107–188). Nottingham, UK: Nottingham University Press.

Lock, A. L., Corl, B. A., Barbano, D. M., Bauman, D. E., & Ip, C. (2004). The anticarcinogenic effect of *trans*-11 18:1 is dependent on its conversion to *cis*-9, *trans*-11 CLA by delta 9 desaturase in rats. *Journal of Nutrition, 134*, 2698–2704.

Lock, A. L., Horne, C. A. M., Bauman, D. E., & Salter, A. M. (2005a). Butter naturally enriched in conjugated linoleic acid and vaccenic acid alters tissue fatty acids and improves the plasma lipoprotein profile in cholesterol-fed hamsters. *Journal of Nutrition, 135*, 1934–1939.

Lock, A. L., Parodi, P. W., & Bauman, D. E. (2005b). The biology of *trans* fatty acids: Implications for human health and the dairy industry. *Australian Journal of Dairy Technology, 60*, 134–142.

Loor, J. J., Herbein, J. H., & Polan, C. E. (2002). *Trans* 18:1 and 18:2 isomers in blood plasma and milk fat of grazing cows fed a grain supplement containing solvent-extracted or mechanically extracted soybean meal. *Journal of Dairy Science, 85*, 1197–1207.

Loor, J. J., Ueda, K., Ferlay, A., Chilliard, Y., & Doreau, M. (2004). Biohydrogenation duodenal flow, and intestinal digestibility of *trans* fatty acids and conjugated linoleic acids in response to dietary forage: Concentrate ratio and linseed oil in dairy cows. *Journal of Dairy Science, 87*, 2472–2485.

Loor, J. J., Ferlay, A., Ollier, A., Doreau, M., & Chilliard, Y. (2005a). Relationship among *trans* and conjugated fatty acids and bovine milk fat yield due to dietary concentrate and linseed oil. *Journal of Dairy Science, 88*, 726–740.

Loor, J. J., Ferlay, A., Ollier, A., Ueda, K., Doreau, M., & Chilliard, Y. (2005b). High-concentrate diets and polyunsaturated oils alter *trans* and conjugated isomers in bovine rumen, blood, and milk. *Journal of Dairy Science, 88*, 3986–3999.

Loor, J. J., Ueda, K., Ferlay, A., Chilliard, Y., & Doreau, M. (2005c). Intestinal flow and digestibility of *trans* fatty acids and conjugated linoleic acids (CLA) in dairy cows fed a high-concentrate diet supplemented with fish oil, linseed oil, sunflower oil. *Animal Feed Science and Technology, 119*, 203–225.

Lourenço, M., Vlaeminck, B., Bruinenberg, M., Demeyer, D., & Fievez, V. (2005). Milk fatty acid composition and associated rumen lipolysis and fatty acid hydrogenation when feeding forages from intensively managed or semi-natural grasslands. *Animal Research, 54*, 471–484.

Luna, P., Fontecha, J., Juarez, M., & de la Fuente, M. A. (2005). Changes in the milk and cheese fat composition of ewes fed commercial supplements containing linseed with special reference to the CLA content and isomer composition. *Lipids, 40*, 445–454.

Mahfouz, M. M., Valicenti, A. J., & Holman, R. T. (1980). Desaturation of isomeric *trans*-octadecenoic acids by rat liver microsomes. *Biochim Biophys Acta, 618*, 1–12.

Mauger, F. M., Lichtenstein, A. H., Ausman, L. M., Mensink, R. P., & Katan, M. B. (1990). Effect of dietary *trans* fatty acids on high-density and low-density lipoprotein cholesterol levels in healthy subjects. *New England Journal of Medicine, 323*, 439–445.

Mauger, J. F., Lichtenstein, A. H., Ausman, L. M., Jalbert, S. M., Jauhiainen, M., Ehnholm, C., & Lamarche, B. (2003). Effect of different forms of dietary hydrogenated fats on LDL particle size. *American Journal of Clinical Nutrition, 78*, 370–375.

Mensink, R. P., & Katan, M. B. (1990). Effect of dietary *trans* fatty acids on high-density and low-density lipoprotein cholesterol levels in healthy subjects. *New England Journal of Medicine, 323*, 439–445.

Mensink, R. P., Zock, P. L., Kester, A. D., & Katan, M. B. (2003). Effects of dietary fatty acids and carbohydrates on the ratio of serum total to HDL cholesterol and on serum lipids and apolipoproteins: A meta-analysis of 60 controlled trials. *American Journal of Clinical Nutrition, 77,* 1146–1155.

Mosley, E. E., Shafii, B., Moate, P. J., & McGuire, M. A. (2006). *Cis*-9, *trans*-11 conjugated linoleic acid is synthesized directly from vaccenic acid in lactating dairy cattle. *Journal of Nutrition, 136,* 570–575.

Mosley, S. A., Mosley, E., Hatch, B., Szasz, J. I., Corato, A., Zacharias, N., Howes, D., & McGuire, M. A. (2007). Effect of varying levels of fatty acids from palm oil on feed intake and milk production in Holstein cows. *Journal of Dairy Science, 90,* 987–993.

Mozaffarian, D., Katan, M. B., Ascherio, A., Stampfer, M. J., & Willett, W. C. (2006). *Trans* fatty acids and cardiovascular disease. *New England Journal of Medicine, 354,* 1601–1613.

Ness, A. R., Smith, G. D., & Hart, C. (2001). Milk, coronary heart disease and mortality. *Journal of Epidemiology and Community Health, 55,* 379–382.

Ng, T. K., Hayes, K. C., DeWitt, G. F., Jegahesan, M., Satgunasingam, N., Ong, A. S., & Tan, D. (1992). Dietary palmitic acids and oleic acids exert similar effects on serum cholesterol and lipoprotein profiles in normocholesterolemic men and women. *Journal of the American College of Nutrition, 11,* 383–390.

Noakes, M., Nestel, P. J., & Clifton, P. M. (1996). Modifying the fatty acid profile of dairy products through feedlot technology lowers plasma cholesterol of humans consuming the products. *American Journal of Clinical Nutrition, 63,* 42–46.

Noble, R. C. (1981). Digestion, absorption and transport of lipids in ruminant animals. In W. W. Christie (Ed.), *Lipid Metabolism in Ruminant Animals* (pp. 57–93). Oxford: Pergamon Press.

Noh, S. K., & Koo, S. L. (2004). Milk sphingomyelin is more effective than egg sphingomyelin in inhibiting intestinal absorption of cholesterol and fat in rats. *Journal of Nutrition, 134,* 2611–2616.

Ntambi, J. M. (1999). Regulation of stearoyl-CoA desaturase by polyunsaturated fatty acids and cholesterol. *Journal of Lipid Research, 40,* 1549–1558.

Nugent, A. P. (2004). The metabolic syndrome. *Nutrition Bulletin, 29,* 36–43.

Nuernberg, K., Nuernberg, G., Endera, K., Dannenberger, D., Schabbel, W., Grumbach, S., Zupp, W., & Steinhart, H. (2005). Effect of grass vs. concentrate feeding on the fatty acid profile of different fat depots in lambs. *European Journal of Lipid Science and Technology, 107,* 737–745.

Offer, N. W., Marsden, M., Dixon, J., Speake, B. K., & Thacker, F. E. (1999). Effect of dietary fat supplements on levels of n-3 polyunsaturated fatty acids, *trans* acids and conjugated linoleic acid in bovine milk. *Animal Science, 69,* 613–625.

Offer, N. W., Marsden, M., & Phipps, R. H. (2001). Effect of oil supplementation of a diet containing a high concentration of starch on levels of *trans* fatty acids and conjugated linoleic acids in bovine milk. *Animal Science, 73,* 533–540.

Oomen, C. M., Ocke, M. C., Feskens, E. J., van Erp-Baart, M. A., Kok, F. J., & Kromhout, D. (2001). Association between *trans* fatty acid intake and 10-year risk of coronary heart disease in the Zutphen Elderly Study: A prospective population-based study. *Lancet, 357,* 746–751.

Palmquist, D. L., Lock, A. L., Shingfield, K. J., & Bauman, D. E. (2005). Biosynthesis of conjugated linoleic acid in ruminants and humans. In S. L. Taylor (Ed.), *Advances in Food and Nutrition Research,* Vol. 50 (pp. 179–217). San Diego: Elsevier/Academic Press.

Pariza, M. W. (1999). The biological activities of conjugated linoleic acid. In M. P. Yurawecz, M. M. Mossoba, J. K. G. Kramer, M. W. Pariza, & G. J. Nelson (Eds.), *Advances in Conjugated Linoleic Acid Research,* Vol. 1 (pp. 12–20). Champaign, IL: AOCS Press.

Parker, D. R., Weiss, S. T., Troisi, R., Cassano, P. A., Vokonas, P. S., & Landsberg, L. (1993). Relationship of dietary saturated fatty acids and body habitus to serum insulin concentrations: The Normative Aging Study. *American Journal of Clinical Nutrition, 58,* 129–136.

Parodi, P. W. (1999). Conjugated linoleic acid and other anticarcinogenic agents of bovine milk fat. *Journal of Dairy Science, 82,* 1339–1349.

Parodi, P. W. (2001). Cows' milk components with anti-cancer potential. *Australian Journal of Dairy Technology, 56*, 65–73.

Pereira, M. A., Jacobs, D. R., Jr., Van Horn, L., Slattery, M. L., Kartashov, A. I., & Ludwig, D. S. (2002). Dairy consumption, obesity, and the insulin resistance syndrome in young adults: The CARDIA Study. *Journal of the American Medical Association, 287*, 2081–2089.

Petit, H. V. (2002). Digestion, milk production, milk composition, and blood composition of dairy cows fed whole flaxseed. *Journal of Dairy Science, 85*, 1482–1490.

Pietinen, P., Ascherio, A., Korhonen, P., Hartman, A. M., Willett, W. C., Albanes, D., & Virtamo, J. (1997). Intake of fatty acids and risk of coronary heart disease in a cohort of Finnish men: The Alpha-Tocopherol, Beta-Carotene Cancer Prevention Study. *American Journal of Epidemiology, 145*, 876–887.

Piperova, L. S., Teter, B. B., Bruckental, I. Sampugna, J., Mills, S. E., Yurawecz, M. P., Fritsche, J., Ku, K., & Erdman, R. A. (2000). Mammary lipogenic enzyme activity, *trans* fatty acids and conjugated linoleic acids are altered in lactating dairy cows fed a milk fat-depressing diet. *Journal of Nutrition, 130*, 2568–2574.

Piperova, L. S., Sampugna, J., Teter, B. B., Kalscheur, K. F., Yurawecz, M. P., Ku, Y., Morehouse, K. M., & Erdman, R. A. (2002). Duodenal and milk *trans* octadecenoic acid and conjugated linoleic acid (CLA) isomers indicate that postabsorptive synthesis is the predominant source of *cis*-9-containing CLA in lactating dairy cows. *Journal of Nutrition, 132*, 1235–1241.

Pollard, M. R., Gunstone, F. D., James, A. T., & Morris, L. J. (1980). Desaturation of positional and geometric isomers of monoenoic fatty acids by microsomal preparations from rat liver. *Lipids, 15*, 306–314.

Ponter, A. A., Parsy, A. E., Saade, M., Mialot, J. P., Ficheux, C., Duvaux-Ponter, C., & Grimard, B. (2006). Effect of a supplement rich in linolenic acid added to the diet of post partum dairy cows on ovarian follicle growth, and milk and plasma fatty acid compositions. *Reproduction, Nutrition, Development, 46*, 19–29.

Poppitt, S. D., Keogh, G. F., Mulvey, T. B., McArdle, B. H., MacGibbon, A. K. H., & Cooper, G. J. S. (2002). Lipid-lowering effects of a modified butter fat: A controlled intervention trial in healthy men. *European Journal of Clinical Nutrition, 56*, 64–71.

Pottier, J., Focant, M., Debier, C., De Buysser, G., Goffe, C., Mignolet, E., Froidmont, E., & Larondelle, Y. (2006). Effect of dietary vitamin E on rumen biohydrogenation pathways and milk fat depression in dairy cows fed high-fat diets. *Journal of Dairy Science, 89*, 685–692.

Precht, D., & Molkentin, J. (1997). *Trans*-geometrical and positional isomers of linoleic acid including conjugated linoleic acid (CLA) in German milk and vegetable fats. *Fett/Lipid, 99*, 319–326.

Precht, D., & Molkentin, J. (1999). C18:1, C18:2 and C18:3 *trans* and *cis* fatty acid isomers including conjugated *cis* Δ9, *trans* Δ11 linoleic acid (CLA) as well as total fat composition of German human milk lipids. *Nahrung, 43*, 233–244.

Precht, D., & Molkentin, J. (2000). Recent trends in the fatty acid composition of German sunflower margarines, shortenings and cooking fats with emphasis on individual C16:1, C18:1, C18:2, C18:3 and C20:1 *trans* isomers. *Nahrung, 44*, S. 222–228.

Ratnayake, W. M. N., & Pelletier, G. (1992). Positional and geometric isomers of linoleic acid in partially hydrogenated oils. *Journal of the American Oil Chemists Society, 69*, 95–105.

Rego, O. A., Rosa, H. J. D., Portugal, P. V., Franco, T., Vouzela, C. M., Borba, A. E. S., & Bessa, R. J. B. (2005). The effects of supplementation with sunflower and soybean oils on the fatty acid profile of milk fat from grazing dairy cows. *Animal Research, 54*, 17–24.

Riel, R. R. (1963). Physico-chemical characteristics of Canadian milk fat. Unsaturated fatty acids. *Journal of Dairy Science, 46*, 102–106.

Ritzenthaler, K. L., McGuire, M. K., Falen, R., Shultz, T. D., Dasgupta, N., & McGuire, M. A. (2001). Estimation of conjugated linoleic acid intake by written dietary assessment methodologies underestimates actual intake evaluated by food duplicate methodology. *Journal of Nutrition, 131,* 1548–1554.

Roche, H. M., Noone, E., Nugent, A., & Gibney, M. J. (2001). Conjugated linoleic acid: A novel therapeutic nutrient? *Nutrition Research Reviews, 14,* 173–187.

Rombaut, R., Camp, J. V., & Dewettinck, K. (2005). Analysis of phospho- and sphingolipids in dairy products by a new HPLC method. *Journal of Dairy Science, 88,* 482–488.

Roy, A., Chardigny, J.-M., Bauchart, D., Ferlay, A., Lorenz, S., Durand, D., Gruffat, D., Faulconnier, Y., Sébédio, J.-L., & Chilliard, Y. (2007). Butters rich either in *trans*-10-C18:1 or in *trans*-11-C18:1 plus *cis*-9, *trans*-11 CLA differentially affect plasma lipids and aortic fatty streak in experimental atherosclerosis in rabbits. *Animal, 1,* 467–476.

Roy, A., Ferlay, A., Shingfield, K. J., & Chilliard, Y. (2006). Examination of the persistency of milk fatty acid composition responses to plant oils in cows given different basal diets, with particular emphasis on *trans*-C18:1 fatty acids and isomers of conjugated linoleic acid. *Animal Science, 82,* 479–492.

Ryhänen, E. L., Tallavaara, K., Griinari, J. M., Jaakkola, S., Mantere-Alhonen, S., & Shingfield, K. J. (2005). Production of conjugated linoleic acid enriched milk and dairy products from cows receiving grass silage supplemented with a cereal-based concentrate containing rapeseed oil. *International Dairy Journal, 15,* 207–217.

Sanz Sampelayo, M. R., Chilliard, Y., Schmidely, P., & Boza, J. (2007). Influence of type of diet on the fat constituents of goat and sheep milk. *Small Ruminant Research, 68,* 42–63.

Schmelz, E. M. (2003). Dietary sphingolipids in the prevention and treatment of colon cancer. In B. F. Szuhaj & W. van Nieuwenhuyzen (Eds.), *Nutrition and Biochemistry of Phospholipids* (pp. 80–87). Champaign, IL: AOCS Press.

Seidel, C., Deufel, T., & Jahreis, G. (2005). Effects of fat-modified dairy products on blood lipids in humans in comparison with other fats. *Annals of Nutrition and Metabolism, 49,* 42–48.

Shingfield, K. J., Ahvenjärvi, S., Toivonen, V., Ärölä, A., Nurmela, K. V. V., Huhtanen, P., & Griinari, J. M. (2003). Effect of fish oil on biohydrogenation of fatty acids and milk fatty acid content in cows. *Animal Science, 77,* 165–179.

Shingfield, K. J., Reynolds, C. K., Lupoli, B., Toivonen, V., Yurawecz, M. P., Delmonte, P., Griinari, J. M., Grandison, A. S., & Beever, D. E. (2005a). Effect of forage type and proportion of concentrate in the diet on milk fatty acid composition in cows fed sunflower oil and fish oil. *Animal Science, 80,* 225–238.

Shingfield, K. J., Salo-Väänänen, P., Pahkala, E., Toivonen, V., Jaakkola, S., Piironen, V., & Huhtanen, P. (2005b). Effect of forage conservation method, concentrate level and propylene glycol on the fatty acid composition and vitamin content of cows' milk. *Journal of Dairy Research, 72,* 349–361.

Shingfield, K. J., Reynolds, C. K., Hervás, G., Griinari, J. M., Grandison, A. S., & Beever, D. E. (2006a). Examination of the persistency of milk fatty acid composition responses to fish oil and sunflower oil in the diet of dairy cows. *Journal of Dairy Science, 89,* 714–732.

Shingfield, K. J., Toivonen, V., Vanhatalo, A., Huhtanen, P., & Griinari, J. M. (2006b). Indigestible markers reduce the mammary Δ9-desaturase activity index and alter the milk fatty acid composition in cows. *Journal of Dairy Science, 89,* 3006–3010.

Shingfield, K. J., Ahvenjärvi, S., Toivonen, V., Vanhatalo, A., & Huhtanen, P. (2007). Transfer of absorbed *cis*-9, *trans*-11 conjugated linoleic acid into milk is biologically more efficient than endogenous synthesis from absorbed vaccenic acid in the lactating cow. *Journal of Nutrition, 137,* 1154–1160.

Spitsberg, V. L. (2005). Bovine milk fat globule membrane as a potential nutraceutical. *Journal of Dairy Science, 88,* 2289–2294.

Steele, W., & Noble, R. C. (1984). Changes in lipid composition of grass during ensiling with or without added fat or oil. *Proceedings of the Nutrition Society, 43,* 51A.

Steffen, L. M., & Jacobs, D. R. (2003). Relation between dairy food intake and plasma lipid levels: The CARDIA Study. *Australian Journal of Dairy Technology, 58,* 92–97.

Sundram, K., Hayes, K. C., & Siru, O. H. (1995). Both dietary 18:2 and 16:0 may be required to improve serum LDL/HDL cholesterol ratio in normocholesterolemic men. *Journal of Nutritional Biochemistry, 6*, 179–187.

Tanaka, Y., Bush, K. K., Klauck, T. M., & Higgins, P. J. (1989). Enhancement of butyrate-induced differentiation of HT-29 human colon carcinoma cells by 1,25-dihydroxyvitamin D3. *Biochemical Pharmacology, 38*, 3859–3865.

Temme, E. H. M., Mensink, R. P., & Hornstra, G. (1996). Comparison of the effects of diets enriched in lauric, palmitic, or oleic acids on serum lipids and lipoproteins in healthy women and men. *American Journal of Clinical Nutrition, 63*, 897–903.

Terpstra, A. H. M. (2004). Effect of conjugated linoleic acid on body composition and plasma lipids in humans: An overview of the literature. *American Journal of Clinical Nutrition, 79*, 352–361.

Tholstrup, T. (2006). Dairy products and cardiovascular disease. *Current Opinion in Lipidology, 17*, 1–10.

Tholstrup, T., Sandstrom, B., Hermansen, J. E., & Hølmer, G. (1998). Effect of modified dairy fat on postprandial and fasting plasma lipids and lipoproteins in healthy young men. *Lipids 33*, 11–21.

Tholstrup, T., Raff, M., Basu, S., Nonboe, P., Sejrsen, K., & Straarup, E. M. (2006). Effects of butter high in ruminant *trans* and monounsaturated fatty acids on lipoproteins, incorporation of fatty acids into lipid classes, plasma C-reactive protein, oxidative stress, hemostatic variables, and insulin in healthy young men. *American Journal of Clinical Nutrition, 83*, 237–243.

Thomas, L. H. (1992). Ischaemic heart disease and consumption of hydrogenated marine oils in England and Wales. *Journal of Epidemiology and Community Health, 46*, 78–82.

Thomas, L. H., Winter, J. A., & Scott, R. G. (1983). Concentration of 18:1 and 16:1 *trans* unsaturated fatty acids in the adipose body tissue of decedents dying of ischaemic heart disease compared with controls: Analysis by gas liquid chromatography. *Journal of Epidemiology and Community Health, 37*, 16–21.

Tricon, S., Burdge, G. C., Jones, E. L., Russell, J. J., El-Khazen, S., Moretti, E., Hall, W. L., Gerry, A. B., Leake, D. S., Grimble, R. F., Williams, C. M., Calder, P. C., & Yaqoob, P. (2006). Effects of dairy products naturally enriched with *cis*-9, *trans*-11 conjugated linoleic acid on the blood lipid profile in healthy middle-aged men. *American Journal of Clinical Nutrition, 83*, 744–753.

Turpeinen, A. M., Mutanen, M., Aro, A., Salminen, I., Basu, S., Palmquist, D. L., & Griinari, J. M. (2002). Bioconversion of vaccenic acid to conjugated linoleic acid in humans. *American Journal of Clinical Nutrition, 76*, 504–510.

Ulberth, F., & Henninger, M. (1994). Quantitation of *trans* fatty acids in milk fat using spectroscopic and chromatographic methods. *Journal of Dairy Research, 61*, 517–527.

U.S. Food and Drug Administration (2003). Questions and answers about *trans* fat nutrition labeling; www.cfsan.fda.gov/~dms/qatrans2.html.2003.

Valeille, K., Ferezou, J., Parquet, M., Amsler, G., Gripois, D., Quignard-Boulange, A., & Martin, J. C. (2006). The natural concentration of the conjugated linoleic acid, *cis*-9, *trans*-11, in milk fat has antiatherogenic effects in hyperlipidemic hamsters. *Journal of Nutrition, 136*, 1305–1310.

Vanhatalo, A., Kuoppala, K., Toivonen, V., & Shingfield, K. J. (2007). Effects of forage species and stage of maturity on milk fatty acid composition. *European Journal of Lipid Science and Technology, 109*, 856–867.

Velazquez, O. C., Jabbar, A., De Matteo, R. P., & Rombeau, J. L. (1996). Butyrate inhibits seeding and growth of colorectal metastases to the liver in mice. *Surgery, 120*, 440–448.

Vessby, B., Uusitupa, M., Hermansen, K., Riccardi, G., Rivellese, A. A., Tapsell, L. C., Nälsén, C., Berglund, L., Louheranta, A., Rasmussen, B. M., Calvert, G. D., Maffetone, A.,

Pedersen, E., Gustafsson, L.-B., & Storlien, L. H. (2001). Substituting dietary saturated for monounsaturated fat impairs insulin sensitivity in healthy men and women. *Diabetologia, 44,* 312–319.

Vlaeminck, B., Fievez, V., Cabrita, A. R. J., Fonseca, A. J. M., & Dewhurst, R. J. (2006). Factors affecting odd- and branched-chain fatty acids in milk: A review. *Animal Feed Science and Technology, 131,* 389–417.

Voorrips, L. E., Brants, H. A. M., Kardinaal, A. F. M., Hiddink, G. J., van den Brandt, P. A., & Goldbohm, R. A. (2002). Intake of conjugated linoleic acid, fat, and other fatty acids in relation to postmenopausal breast cancer: The Netherlands Cohort Study on Diet and Cancer. *American Journal of Clinical Nutrition, 76,* 873–882.

Wahle, K. W., Heys, S. D., & Rotondo, D. (2004). Conjugated linoleic acids: Are they beneficial or detrimental to health? *Progress in Lipid Research, 43,* 553–587.

Warensjö, E., Jansson, J. H., Berglund, L., Boman, K., Ahrén, B., Weinehall, L., Lindahl, B., Hallmans, G., & Vessby, B. (2004). Estimated intake of milk fat is negatively associated with cardiovascular risk factors and does not increase the risk of a first acute myocardial infarction. A prospective case-control study. *British Journal of Nutrition, 91,* 635–642.

Wasowska, I., Maia, M., Niedźwiedzka, K. M., Czauderna, M., Ramalho Ribeiro, J. M. C., Devillard, E., Shingfield, K. J., & Wallace, R. J. (2006). Influence of fish oil on ruminal biohydrogenation of C18 unsaturated fatty acids. *British Journal of Nutrition, 95,* 1199–1211.

Weggemans, R. M., Rudrum, M., & Trautwein, E. A. (2004). Intake of ruminant versus industrial *trans* fatty acids and risk of coronary heart disease—What is the evidence? *European Journal of Lipid Science and Technology, 106,* 390–397.

Weill, P., Schmitt, B., Chesneau, G., Daniel, N., Safraou, F., & Legrand, P. (2002). Effects of introducing linseed in livestock diet on blood fatty acid composition of consumers of animal products. *Annals of Nutrition and Metabolism, 46,* 182–191.

Whigham, L. D., Cook, M. E., & Atkinson, R. L. (2000). Conjugated linoleic acid: Implications for human health. *Pharmacological Research, 42,* 503–510.

White, S. L., Bertrand, J. A., Wade, M. R., Washburn, S. P., Green, J. T., & Jenkins, T. C. (2001). Comparison of fatty acid content of milk from Jersey and Holstein cows consuming pasture or a total mixed ration. *Journal of Dairy Science, 84,* 2295–2301.

Wilke, M. S., & Clandinin, M. T. (2005). Influence of dietary saturated fatty acids on the regulation of plasma cholesterol concentration. *Lipids, 40,* 1207–1213.

Willett, W. C., Stampfer, M. J., Manson, J. E., Colditz, G. A., Speizer, F. E., Rosner, B. A., Sampson, L. A., & Hennekens, C. H. (1993). Intake of *trans* fatty acids and risk of coronary heart disease among women. *Lancet, 341,* 581–585.

Wolff, R. L. (1995). Content and distribution of *trans*-18:1 acids in ruminant milk and meat fats. Their importance in European diets and their effect on human milk. *Journal of the American Oil Chemists Society 72,* 259–272.

Wongtangtintharn, S., Oku, H., Iwasaki, H., & Toda, T. (2004). Effect of branched-chain fatty acids on fatty acid biosynthesis of human breast cancer cells. *Journal of Nutritional Science and Vitaminology, 50,* 137–143.

World Health Organization (2003). Diet, nutrition and the prevention of chronic diseases. Report of a Joint WHO/FAO Expert Consultation. WHO Technical Report Series, No. 916; www.who.int/dietphysicalactivity/publications/trs916.

Yanagi, S., Yamashita, M., & Imai, S. (1993). Sodium butyrate inhibits the enhancing effect of high fat diet on mammary tumorigenesis. *Oncology, 50,* 201–204.

Yang, Y., Shangpei, L., Chen, X., Chen, H., Huang, M., & Zheng, J. (2000). Induction of apoptotic cell death and *in vivo* growth inhibition of human cancer cells by a saturated branched-chain fatty acid, 13-methyltetradecanoic acid. *Cancer Research, 60,* 505–509.

Yaqoob, P., Tricon, S., Burdge, G. C., & Calder, P. C. (2006). Conjugated linoleic acids (CLAs) and health. In C. Williams & J. Buttriss (Eds.), *Improving the Fat Content of Foods* (pp. 182–209). Cambridge: Woodhead Publishing Ltd.

Expression and Nutritional Regulation of Lipogenic Genes in the Ruminant Lactating Mammary Gland

L. Bernard, C. Leroux and Y. Chilliard

Abstract The effect of nutrition on milk fat yield and composition has largely been investigated in cows and goats, with some differences for fatty acid (FA) composition responses and marked species differences in milk fat yield response. Recently, the characterization of lipogenic genes in ruminant species allowed *in vivo* studies focused on the effect of nutrition on mammary expression of these genes, in cows (mainly fed milk fat-depressing diets) and goats (fed lipid-supplemented diets). These few studies demonstrated some similarities in the regulation of gene expression between the two species, although the responses were not always in agreement with milk FA secretion responses. A central role for *trans*-10 C18:1 and *trans*-10, *cis*-12 CLA as regulators of milk fat synthesis has been proposed. However, *trans*-10 C18:1 does not directly control milk fat synthesis in cows, despite the fact that it largely responds to dietary factors, with its concentration being negatively correlated with milk fat yield response in cows and, to a lesser extent, in goats. Milk *trans*-10, *cis*-12 CLA is often correlated with milk fat depression in cows but not in goats and, when postruminally infused, acts as an inhibitor of the expression of key lipogenic genes in cows. Recent evidence has also proven the inhibitory effect of the *trans*-9, *cis*-11 CLA isomer. The molecular mechanisms by which nutrients regulate lipogenic gene expression have yet to be well identified, but a central role for SREBP-1 has been outlined as mediator of FA effects, whereas the roles of PPARs and STAT5 need to be determined. It is expected that the development of *in vitro* functional systems for lipid synthesis and secretion will allow future progress toward (1) the identification of the inhibitors and activators of fat synthesis, (2) the knowledge of cellular mechanisms, and (3) the understanding of differences between ruminant species.

Keywords: nutrition · gene expression · lipogenesis · mammary gland · lactating ruminant

---

L. Bernard
Adipose Tissue and Milk Lipid Laboratory, Herbivore Research Unit, INRA-Theix,
63 122 St Genès-Champanelle, France
e-mail: laurence.bernard@clermont.inra.fr

Introduction

In ruminants, the major constituents of milk (lipids, proteins, carbohydrates, and salts) and their concentrations are linked to intrinsic or extrinsic (nutritional and environmental) factors (Coleman et al., 2000). Among these, the genetic factor, through the animal species, and the nutritional factor are the two major factors determining milk composition. Indeed, nutrition has a considerable effect on the composition of the lipids (Jensen, 2002), conversely to the protein fraction, which generally is only marginally affected by this factor (Coulon et al., 2001). Moreover, milk fat is an important component of the nutritional quality of dairy products, with the saturated fatty acids (FA) (mainly C12, C14, and C16) commonly considered to have a negative effect on human health when consumed in excess (Williams, 2000), whereas other FA, such as oleic and linolenic acids, have positive effects by a direct vascular antiatherogenic action (Massaro et al., 1999). Besides this, *cis*-9, *trans*-11 C18:2, the major conjugated linoleic acid (CLA) isomer found in ruminant products including milk, was shown both in animals studies and *in vitro* experiments to exert a number of advantageous physiological effects (Pariza et al., 2001). In addition, milk fat content and composition is one of the most important components of the technological and sensorial qualities of dairy products. Thus, modification of milk fat content and FA composition by dietary manipulation has been investigated in cows and goats, with particular attention on the effects of fat supplementation of the diet.

In bovines, the consequences of lipid supplementation on the milk yield and fat and protein contents have been well described, with an increase in milk production (for most lipid supplements) and a slight but systematic reduction in the protein and casein contents. In dairy cows, due to important interactions between dietary forages and concentrates and their components (fibers, starch, lipids), the supplements given do not all have the same efficiency regarding fat content modulation. Thus, concentrate-rich diets, concentrate-rich diets supplemented with vegetable oils, and diets supplemented with fish oil lead to a decrease in fat content, while encapsulated lipids lead to a large increase in fat content (Chilliard & Ferlay, 2004; Palmquist et al., 1993). Conversely, in goats almost all types of lipid supplements induce a marked increase in the milk fat content without systematic modification of milk production or protein content (Chilliard et al., 2003a). These modifications in the yield of fat are observed together with an important modification of milk FA composition, which is well documented both in cows (Chilliard et al., 2000, 2001; Palmquist et al., 1993) and in goats (Chilliard et al., 2003a, 2006a).

The mechanisms underlying these intra- and inter-species-specific responses are not yet well understood. Nevertheless, recently, thanks to the characterization of the lipogenic genes involved in milk synthesis and secretion and the development of molecular biology tools, few studies have been undertaken to relate the effects of diet on the milk FA profile to mammary gland lipid metabolism. These studies mainly considered lipogenic genes, in particular genes for the enzymes involved in the uptake, *de novo* synthesis, desaturation, and esterification of FA

Expression and Nutritional Regulation of Lipogenic Genes

in order to relate the effects of the diet on the abundance of their transcripts and/or enzymatic activities. This chapter reviews the present knowledge on the main lipogenic genes, in particular the known effects of nutritional factors, especially those of fat supplementation, on ruminant mammary lipogenic gene expression, together with milk fat content and FA composition. In addition, we present the putative molecular mechanisms underlying these regulations.

Milk Fatty Acid Origin

Milk fat is composed of ca. 98% triglycerides, of which ca. 95% is FA and less than 1% is phospholipids, with small amounts of cholesterol, 1,2-diacylglycerol, monoacylglycerol, and free FA. Milk FA have a dual origin: (1) They are either *de novo* synthesized in the mammary gland (Fig.1) from acetate and

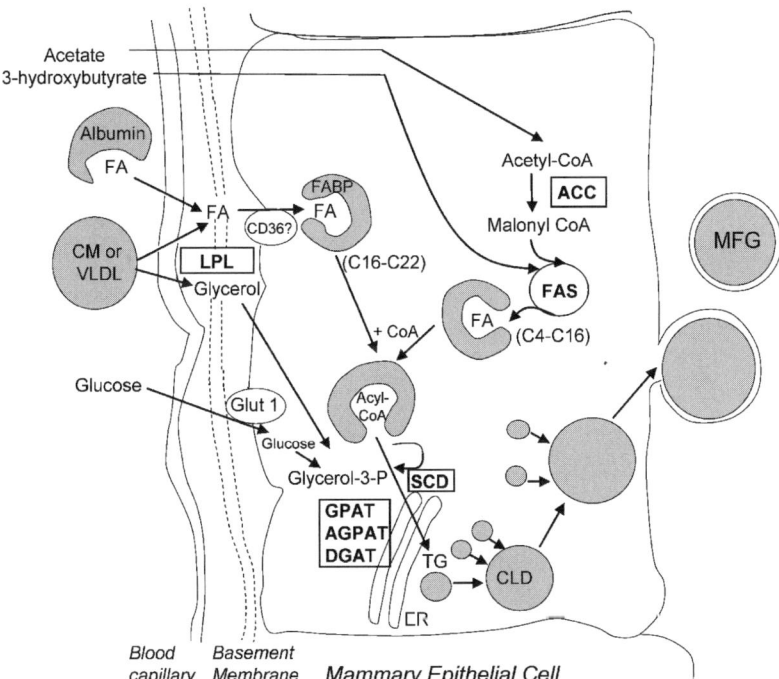

Fig. 1 Milk fat synthesis in the ruminant mammary epithelial cell. Abbreviations used: ACC = acetyl-CoA carboxylase; AGPAT = acyl glycerol phosphate acyl transferase; CD36 = cluster of differentiation 36; CLD = cytoplasmic lipid droplet; CoA = coenzyme A; CM = chylomicron; DGAT = diacyl glycerol acyl transferase; ER = endoplasmic reticulum; FA = fatty acid; FABP = fatty acid binding protein; FAS = fatty acid synthase; Glut 1 = glucose transporter1; GPAT = glycerol-3 phosphate acyl transferase; LPL = lipoprotein lipase; MFG = milk fat globule; SCD = stearoyl-CoA desaturase; TG = triglyceride; VLDL = very low density-lipoprotein

3-hydroxybutyrate, produced by ruminal fermentation of carbohydrates and by rumen epithelium from absorbed butyrate, respectively, thus resulting in short- and medium-chain FA (C4:0 to C16:0) that represent 40–50% of the FA secreted in milk, or (2) they are imported from the plasma, where they are either released by the enzyme lipoprotein lipase (LPL) (Barber et al., 1997) from triglycerides circulating in chylomicra or very low-density lipoprotein (VLDL), or derived from the plasma nonesterified fatty acids (NEFA) that circulate bound to albumin, for long-chain FA ($\geq$C18) as well as ca. one-half of the C16:0, depending on the diet composition. These long-chain FA originate mainly from dietary lipid absorption from the digestive tract (with the dietary FA undergoing total or partial hydrogenation in the rumen) and from body reserves mobilization (especially at the beginning of lactation). Commonly, mobilization of body fat accounts for less than 10% of milk FA, with this proportion increasing in ruminants in negative energy balance in direct proportion to the extent of the energy deficit (Bauman & Griinari, 2001). Furthermore, FA may be desaturated, but not elongated, in the secretory mammary epithelial cells (MEC) (Chilliard et al., 2000).

Gene Characterization and Mechanisms of Mammary Lipogenesis

The acetyl-CoA carboxylase (ACC) and fatty acid synthase (FAS) enzymes (encoded by the *ACACA* and *FASN* genes, respectively) are involved in the metabolic pathway for *de novo* FA synthesis in the mammary tissue, whereas the LPL enzyme is involved in the uptake of plasma FA. These FA could be desaturated by the stearoyl-CoA desaturase (SCD), resulting in synthesis of *cis*-9 unsaturated FA, and then esterified to glycerol sequentially via glycerol-3 phosphate acyl transferase (GPAT), acyl glycerol phosphate acyl transferase (AGPAT), and diacyl glycerol acyl transferase (DGAT). Then the triglycerides are secreted as milk fat globules (Fig. 1). The genes specifying these enzymes, implicated in the key processes of lipogenesis within the mammary gland, are candidate genes whose regulation has been studied first.

In ruminants, thanks to the recent knowledge of the cDNA sequences of several lipogenic genes (see Fig. 2)—*LPL* (Bonnet et al., 2000a; Senda et al., 1987), *ACACA* (Barber & Travers, 1998; Mao et al., 2001), *FASN* (Leroux et al., submitted; Roy et al., 2005), SCD (Bernard et al., 2001; Keating et al., 2005; Ward et al., 1998), *DGAT1* (Winter et al., 2002), *AGPAT* (Mistry & Medrano, 2002), and recently *GPAT* (Roy et al., 2006b)—molecular tools for studying their expression have been developed to allow the quantification of their mRNA by northern blot or real-time RT-PCR and/or the activity of the corresponding enzymes.

In addition, the recent development of high-throughput techniques such as microarrays allows us to complete this candidate gene approach. These new methods simultaneously provide data for the expression of thousands of genes

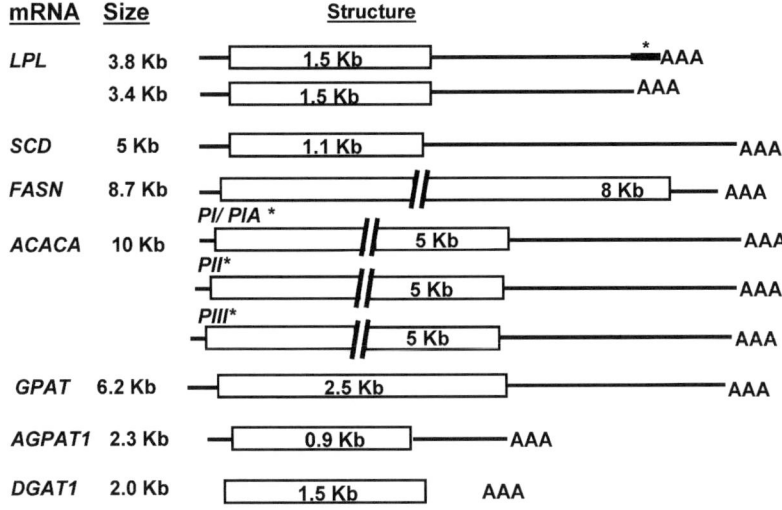

Fig. 2 Characterization (size and structure) of the transcripts of the main lipogenic genes in ruminant species, lipoprotein lipase (*LPL*; Senda et al., 1987; Bonnet et al., 2000), stearoyl-CoA desaturase (*SCD*; Bernard et al., 2001; Keating et al., 2005), fatty acid synthase (*FASN*; Roy et al., 2005; Leroux et al., in preparation), acetyl-CoA carboxylase (*ACACA*; Barber et al., 1998, 2005), glycerol-3 phosphate acyl transferase (*GPAT*; Roy et al., 2006b), acyl glycerol phosphate acyl transferase (*AGPAT*; Mistry & Medrano, 2002), and diacyl glycerol acyl transferase (*DGAT*; Grisart et al., 2002; Winter et al., 2002)

and allow a larger understanding of different mammary functions including milk lipid synthesis and secretion.

Uptake of Fatty Acids

The hydrolysis of lipoprotein triacylglycerol is catalyzed by LPL, which selectively releases FA esterified at the *sn*-1 (-3) position. In bovines, mammary tissue expresses three *LPL* transcripts, which are 1.7, 3.4, and 3.6 Kb in size (Bonnet et al., 2000a; Senda et al., 1987) due to the alternative use of the polyadenylation site. In this species, a predominance of the 3.4-Kb transcript in the mammary gland (Senda et al., 1987) as in ovine adipose tissue and of the 3.6-Kb transcript in muscles (Bonnet et al., 2000b) has been reported. In ovine species, the total cDNA sequence has been reported (Bonnet et al., 2000a). A complex regulation, by dietary and hormonal factors, modulates LPL activity via transcriptional, posttranscriptional, and posttranslational mechanisms. Immediately prior to parturition, mammary LPL activity increases markedly, remains high throughout lactation, and is simultaneously downregulated in adipose tissue in cows (Shirley et al. 1973) and goats (Chilliard et al., 2003a).

From immunochemical and biochemical studies, it was shown that LPL is located both on (or near) the surface and within the cell of the major cell types of the different tissues as well as on the luminal surface of vascular endothelial cells. Mammary tissue contains various cell types in addition to parenchymal secretory MEC, including adipocytes, in varying proportions to MEC depending on the developmental and physiological status of the mammary gland. Discrepancies concerning LPL localization were reported from histological studies made on the rodent mammary gland. Thus, some studies (Camps et al., 1990) demonstrated the presence of LPL mRNA and protein in MEC using *in situ* hybridization and immunofluorescence techniques, respectively, and concluded that the origin of mammary LPL is the secretory MEC. Conversely, others (Jensen et al., 1991) demonstrated that LPL protein and mRNA are located in mammary depleted adipocytes or adipocyte precursors located in interstitial cells, suggesting that mammary LPL could originate partly from mammary adipocytes to be subsequently secreted and transported by cellular uptake and transcytosis, both to its final site of action on the capillary endothelial cell and through the secretory MEC into milk (Jensen et al., 1991; Neville & Picciano, 1997).

Arteriovenous difference measurements in lactating goats have shown that the utilization of triglycerides and NEFA for milk lipid synthesis is related to their plasma concentrations (Annison et al., 1968). Otherwise, experiments demonstrated that the availability of the substrate determines its utilization by the mammary gland with either a large NEFA utilization when plasma triglycerides are low and NEFA are high, as in fasting animals (West et al., 1972), or no net utilization of NEFA at plasma concentrations below 0.2 mM (Nielsen & Jakobsen, 1994). Similarly, triglyceride utilization increases when its plasma concentration increases. Thus, duodenal lipid infusion (Gagliostro et al., 1991) increased plasma triglyceride concentration and apparent mammary uptake, and there was simultaneously a net production of NEFA by the mammary gland, due to their release in the vascular bed during LPL action (Fig. 1).

The mechanism by which FA crosses the capillary endothelium and interstitial space to reach the MEC has not yet been identified. After arriving at the MEC, FA could cross the plasma membrane by diffusion or via a saturable transport system. In mammals, the acyl-CoA binding proteins (ACBP) (Knudsen et al., 2000) that bind long-chain acyl-CoA have an important role in regulating the FA transport and concentration in the cytosol. Nevertheless, in ruminants, Mikkelsen and Knudsen (1987) found lower concentrations of these ACBP in the mammary gland and muscles compared to the liver cytosol. Elsewhere it has been suggested in rodents and ruminants that another FA binding protein, CD36 (cluster of differentiation 36), expressed in the lactating MEC, and found in the milk fat globule membrane, heart, platelets, and adipocytes, may function as a transporter of long-chain FA (Abumrad et al., 2000). The detection of CD36 in mammary tissue could be linked to the presence of adipocytes. Nevertheless, the presence of *CD36* mRNA in rodent MEC lines has been shown, with a slight enhancement of its gene expression after the addition of lactogenic hormones (Aoki et al., 1997).

Furthermore, fatty acid binding proteins (FABP; a family of intracellular lipid binding proteins found across numerous species) are involved in the uptake and intracellular trafficking of FA in many tissues (Lehner & Kuksis, 1996). In the bovine mammary gland, co-expression of FABP and CD36 has been shown, which increases during lactation and decreases during involution (Spitsberg et al., 1995), demonstrating that their expression is related to physiological variations of lipid transport and metabolism within the cell. In the same way, simultaneous elevations of *CD36* mRNA expression, of cytosolic TAG, and of lipid droplets were observed in primary bovine MEC (Yonezawa et al., 2004b). Barber et al. (1997) proposed a role for CD36 in the transport of FA across the secretory MEC membrane, working in conjunction with intracellular FABP. In the bovine lactating mammary gland, the presence of two forms of FABP has been demonstrated (Specht et al., 1996), identified as A-FABP and H-FABP, thus named according to the tissue of their first detection, adipose tissue and heart, respectively. Studying the proteins' cellular location, Specht et al. (1996) showed that A-FABP and H-FABP were present in myoepithelial cells and MEC, respectively. The expression of *FABP* types is generally interpreted in terms of specialized functions in FA metabolism, with *H-FABP* predominantly found in cells where FA are used as an energy source and probably involved in their β-oxidation. The significance of its abundance in mammary tissue in which active triglyceride synthesis and low FA oxidation occur during lactation remains to be understood.

In addition, ATP-binding cassette (ABC) transporters (a family of membrane proteins) are involved in the transport, against concentration gradients, of a wide variety of compounds, including ions, peptides, sugars, and lipids, at the cost of ATP energy (Klein et al., 1999). The *ABCG5* and *ABCG8* members of this family play an important role in cholesterol homeostasis and have been described specifically in intestine and liver cells in rodents (Mutch et al., 2004; Yu et al., 2002). The recent identification and expression of *ABCG5* and *ABCG8* transporters in the bovine mammary gland open a wide range of future investigations on their potential role in lipid trafficking and excretion during lactation and control of sterol concentrations in milk (Viturro et al., 2006).

De Novo Fatty Acid Synthesis

Acetyl-CoA Carboxylase Gene (*ACAC*)

The *ACACA* and *ACACB* genes are distinct genes that respectively encode the isoenzymic ACC proteins ACCα and ACCβ. The *ACACA* gene is expressed in all cell types but is found at its highest levels in the lipogenic tissues

(Lopez-Casillas et al., 1991), where its protein product ACCα provides cytoplasmic malonyl-CoA for FA synthesis. The *ACACB* gene is the major form expressed in heart and skeletal muscles (Abu-Elheiga et al., 1997), where its protein product ACCβ is implicated in the regulation of the β-oxidation of FA in the mitochondria (Abu-Elheiga et al., 2000). Expression of the ACCα isoenzyme is regulated in a complex fashion in the short term, through allosteric mechanisms with cellular metabolites possessing a positive (citrate) or negative effect (malonyl-CoA and long-chain acyl-CoA), and reversible phosphorylation on a number of specific serine residues, as well as chronically, through the regulation of transcription of the gene (Kim, 1997).

In ruminants, the *ACACA* cDNA sequence was first reported by Barber and Travers (1995) and corresponds to the synthesis of a protein with 2,346 amino acids. More recently, the *ACACA* gene was characterized in sheep (Barber & Travers, 1998) and cattle (Mao et al., 2001). Initially, the existence of three promoters, PI, PII, and PIII, was demonstrated, and their use together with alternative splicing of the primary transcripts from promoters I and II results in the generation of a heterogeneous population of transcripts differing in the sequence of their 5'UTR. These promoters are used in tissue-differential fashion. PIII use is limited to lung, liver, kidney, brain, and predominantly the lactating mammary gland in bovine (Mao et al., 2002) and ovine (Barber et al., 2003) species. PII expression is ubiquitous, with an elevated expression in the lactating mammary gland. PI is preferentially used in adipose tissue and liver under lipogenic conditions and in lactating bovine mammary gland. PI generate ~30% of *ACACA* mRNA (Mao et al., 2001), while in the ovine mammary gland, its contribution is low (2%; Molenaar et al., 2003). The reason for this difference of promoter usage between the bovine and the ovine is unknown. In addition, Barber et al. (2005) demonstrated the existence of a fourth promoter in human, rodent, and ruminant species and mainly expressed in brain, which again underlines the complexity of the structure and regulation of the *ACACA* gene.

Fatty Acid Synthase Gene (*FASN*)

The *FASN* gene encodes the protein FAS, which, under a complex homodimeric form, is responsible for the synthesis of short- and medium-chain FA (C4-C16) in the mammary gland during lactation (Wakil, 1989). In ruminants, the FAS enzyme contains six catalytic activity domains on a single protein of 2,513 amino acids. Contrary to what is observed in rodent mammary gland and duck uropygial gland, ruminant FAS synthesizes medium-chain FA without the implication of a thioesterase II (Barber et al., 1997). In addition to being able to load acetyl-CoA, malonyl-CoA, and butyryl-CoA, ruminant FAS contains a loading acyltransferase whose substrate specificity extends to up to C12, with the result that it is able to load and also release these medium-chain FA (Knudsen & Grunnet, 1982). This way of medium-chain FA synthesis is specific to the lactating ruminant mammary gland, whereas the product of FAS in other ruminant tissues is predominantly C16:0, as in non-ruminant tissues (Christie, 1979). *FASN* mRNA ranges in size from 8.4 to 9.3 Kb, depending on the species: In several human tissues (Jayakumar et al., 1995), bovine mammary gland (Beswick & Kennelly, 1998), and ovine (Bonnet et al., 1998) and porcine (Ding et al., 2000) adipose tissues, only one transcript has been detected by northern blot. Conversely, two mRNA, generated by the use of two alternative polyadenylation signals, have been detected in rat adipose tissue (Guichard et al., 1992) and mammary gland (Schweizer et al., 1989). The gene, termed *FASN*, has recently been cloned in bovine (Roy et al., 2005), and an alternate transcript was discovered without part of exon 9 (minus 358 bp). In caprine, the cDNA was recently characterized with only one transcript described (Leroux et al., submitted). In addition, Roy et al. (2006c) identified several single-nucleotide polymorphisms (SNPs) in the bovine *FASN* gene, and the analysis of two of them, located respectively in exon1 and 34, suggested an association of these polymorphisms with variations in milk fat content.

Stearoyl-CoA Desaturase

The *SCD* gene encodes a protein of 359 amino acid residues located in the endoplasmic reticulum that catalyzes the Δ-9 desaturation, introducing a *cis* double bond, of a spectrum of fatty acyl-CoA substrates, mainly from C14 to C19. In rodents, *SCD* relies on different genes whose expression and regulation by polyunsaturated FA (PUFA) are tissue-specific (Ntambi, 1999). Conversely, in ruminants there is only one *SCD* gene (Bernard et al., 2001), generating a 5-Kb transcript that was characterized in sheep (Ward et al., 1998), cows (Chung et al., 2000), and goats (Bernard et al., 2001). In goats, the 3'-UTR sequence derives from a single exon and is unusually long (3.8 Kb), as observed for humans (Zhang et al., 1999), rats (Mihara, 1990), and mice (Ntambi et al., 1988). In addition, the caprine 3'-UTR is characterized by the presence of several AU-rich elements, which could be mRNA destabilization sequences, and presents a genetic polymorphism with the presence or absence of a triplet nucleotide (TGT) in position 3178-3180 (Bernard et al., 2001, and GenBank accession number AF325499). Immediately after parturition, *SCD* mRNA in ovine (Ward et al., 1998) and activity in bovine (Kinsella, 1970) increased in the mammary gland. In lactating goats, the *SCD* gene is highly expressed in the mammary gland and subcutaneous adipose tissue, compared to perirenal adipose tissue (Bernard et al., 2005b). In the lactating mammary gland, palmitoleoyl-CoA and oleyl-CoA are synthesized from palmitoyl-CoA and stearoyl-CoA by the action of the SCD enzyme (Enoch et al., 1976). In addition, in bovine mammary gland, SCD is responsible for the synthesis of the major part of *cis*-9, *trans*-11- (Corl et al., 2001; Griinari et al., 2000; Loor et al., 2005d; Shingfield et al., 2003) and of *trans*-7, *cis*-9- (Corl et al., 2002) CLA isomers.

The promoter region of the bovine *SCD* gene has recently been characterized (Keating et al., 2006), and a region of critical importance, designated stearoyl-CoA desaturase transcriptional enhancer element (STE) and containing three binding complexes, was identified in the MAC-T cell. In addition, this STE region was shown to play a key role in the inhibitory effect on *SCD* gene transcription of *trans*-10, *cis*-12 CLA and, to a lesser extent, of *cis*-9, *trans*-11 CLA (Keating et al., 2006).

Esterification of FA to Glycerol

In mammals, FA are not distributed randomly on the *sn*-1, *sn*-2, and *sn*-3 positions of the glycerol backbone of the milk triglycerides; this nonrandom distribution determines functional and nutritional attributes (German et al., 1997). In bovine, a description of the stereospecific position of the major FA in TAG has been reviewed by Jensen (2002). A high proportion (56–62%) of FA esterified at positions *sn*-1 and *sn*-2 of the glycerol backbone are medium- and long-chain saturated FA (C10:0 to C18:0), with C16:0 equally distributed among *sn*-1 and *sn*-2, C8:0, C10:0, C12:0, and C14:0 more located at *sn*-2, and C18:0 more located at *sn*-1. In addition, about 24% of FA esterified at position *sn*-1 is C18:1. Finally, a high proportion of FA esterified at position *sn*-3 is short-chain FA (C4:0, C6:0, C8:0; 44% on a molar basis) and oleic acid (27%). When consumed by humans, milk TAG are hydrolyzed by pancreatic lipase specifically in the *sn*-1 and *sn*-3 positions, allowing the FA present at position *sn*-2 to be preferentially absorbed because they remain in the monoacyl glycerol form (Small, 1991). The first step in triglyceride biosynthesis is the esterification of glycerol-3-phosphate in the *sn*-1 position, which is catalyzed by the glycerol-3 phosphate acyl transferase (GPAT). Two isoforms of GPAT have been identified in mammals, which can be distinguished by subcellular localization (mitochondrial vs. endoplasmic reticulum) and sensitivity to sulfhydryl group modifying agent N-ethylmalmeimide (NEM). The mitochondrial isoform is resistant to NEM, and the endoplasmic reticulum isoform is sensitive to NEM. In rodents, both isoforms have a role in the TAG synthesis in the liver and adipose tissue (Coleman & Lee, 2004). Regarding the mitochondrial *GPAT* gene, the genomic structure and cDNA sequence were recently determined in ruminants, with the presence of two transcripts differing in their 5'-UTR (Roy et al., 2006b).

The second step of triglyceride synthesis is committed by AGPAT (or lysophosphatidic acid acyltransferase, LPAAT). AGPAT has a greater affinity for saturated fatty acyl-CoA (Mistry & Medrano, 2002) in the order C16 > C14 > C12 > C10 > C8 (Marshall & Knudsen, 1977), which is in accordance with the observed high proportion of medium- and long-chain saturated FA at the *sn*-2 position in milk, with palmitate as the major FA (representing 43% of the total palmitate found in triacylglycerol). Consequently, a possible regulation of substrate specificity of the enzymes of FA esterification should be of major importance for both the mammary cell and human nutrition. In addition, substrate availability in the bovine and ovine mammary gland is also a factor for the *sn*-2 position FA composition, allowing its manipulation to some extent by nutritional factors, in interaction with the aforementioned substrate affinity. Bovine and ovine *AGPAT* genes were characterized, cloned, and located on bovine chromosome 23 (Mistry & Medrano, 2002). Bovine and ovine AGPAT are proteins made up of 287 amino acids that differ by only one amino acid residue.

The third enzyme, DGAT, is located on the endoplasmic reticulum membrane. DGAT is the only protein that is specific to triacylglycerol synthesis and therefore may play an important regulatory role (Mayorek et al., 1989). However, little is known about the regulation of *DGAT* expression, whereas its gene has been particularly well studied in ruminants due to its genetic variability. The complete bovine *DGAT1* gene (Grisart et al., 2002; Winter et al., 2002) and the near-complete coding region of the caprine *DGAT1* gene (Angiolillo et al., 2006) have been sequenced. A quantitive triat loci (QTL) QTL for milk fat content has been detected in the centromeric region of cattle chromosome 14, and *DGAT1* was proposed as a positional and functional candidate for this trait (Winter et al., 2002). These studies found a nonconservative substitution of lysine by alanine (Grisart et al., 2004) in *DGAT1* caused by AA to GC dinucleotide substitution at position 10434 of the gene sequence, in exon 8 (GenBank accession number AY065621). This polymorphism was related to milk composition and yield variations. The K allele was recently shown *in vitro* to be associated with increased activity of the enzyme in agreement with its positive link with bovine milk fat percentage (Grisart et al., 2004). In addition, in the German Holstein population, Kühn et al. (2004) described five alleles at a variable number tandem repeat (VNTR) polymorphism in the *DGAT1* promoter, which showed an effect on fat content additional to the *DGAT1 K232A* mutation. Due to the presence of a potential transcription factor binding site in the 18nt element of the VNTR, the variation in the number of tandem repeats of the 18nt element might be causal for the variability in the transcription level of the *DGAT1* gene. In sheep, *DGAT1* is an obvious candidate gene for milk fat content, as a QTL was detected in chromosome 9, which is homologous to the bovine chromosome 14 region (Barillet et al., 2005).

Regulation of Mammary Lipogenic Gene Expression by Dietary Factors

The response of gene expression to nutrient changes involves the control of events that could occur at transcriptional (e.g., through transcription factors), posttranscriptional (e.g., such as mRNA stability), translational (e.g., its initiation, etc.), and posttranslational (e.g., via turnover or activation of enzymatic protein) levels. However, it is often unclear whether the regulatory factors are the dietary components themselves or their metabolites or are hormonal changes produced in response to the nutritional changes. Moreover, in most of the studies, as in those reported in this chapter, it is difficult to determine the level of regulation involved since, for a given gene, measurements of the relevant mRNA, the enzyme protein content, and the activity are not often studied simultaneously. Only few data are indeed available in ruminants on the nutritional regulation of mammary lipogenic gene expression either *in vivo* or *in vitro*.

The *in vivo* trials have been carried out in midlactation cows and goats. Data in cows come from five studies that were undertaken mainly (four of the five studies) with milk fat-depressing (MFD) diets:

- **Study 1**: a high level of concentrate (containing 76% cracked corn, 19% heat-treated soybean meal, 1% sunflower oil) with a forage- (alfalfa hay) to-concentrate ratio of 16/84, compared to a control diet with a high-forage (composed of 83% corn silage and 17% alfalfa hay) content and a forage-to-concentrate ratio of 53/47 (Peterson et al., 2003; 3 cows),
- **Study 2**: a high level of concentrate (containing 52% ground corn and 15% soybean meal) with a forage- (corn silage) to-concentrate ratio of 25/70, supplemented with 5% of soybean oil, compared to a high level of forage (containing 76% corn silage and 24% alfalfa hay) with a forage-to-concentrate (containing mainly of 57% ground corn and 34% soybean meal) ratio of 60/40 (Piperova et al., 2000; 10 cows),
- **Study 3**: a diet based on grass silage (19%), corn silage (19%), and rolled barley (44%) supplemented, or not, with either 1.7% of glutaraldehyde-protected fish oil or 2.7% unprotected fish oil (Ahnadi et al., 2002; 16 cows),
- **Study 4**: a diet based on grass (27%), corn silage (30%), and rolled barley (22%) supplemented with 3.3% unprotected canola seeds plus 1.5% canola meal, or 4.8% formaldehyde-protected canola seeds, compared to the same diet with 4.8% of canola meal, which, for the two treatment diets, represents a supply of 0.7% and 0.6% of extra lipids compared to the control diet, respectively (Delbecchi et al., 2001; 6 cows),
- **Study 5**: a forage-based diet (containing 25% corn silage, 17% alfalfa silage, and 3% alfalfa hay) with a forage-to-concentrate (22% of ground corn and 28% of grain mix) ratio of 45/55 and supplemented with 3% of soybean oil and 1.5% of fish oil, compared to a forage-based diet (containing 35% corn silage, 23% alfalfa silage, and 5% alfalfa hay) with a forage-to-concentrate (15% of ground corn and 20% of grain mix) ratio of 65/35 (Harvatine & Bauman, 2006; 16 cows).

In goats, data are from two nutritional studies undertaken with dietary lipid supplementations varying in their nature, form of presentation, and dose:

- **Study 6**: a 54% hay diet supplemented, or not, with 3.6% of lipids from oleic sunflower oil or formaldehyde-treated linseeds, with the concentrate fraction containing 40% rolled barley, 17% dehydrated sugar beet pulp, 10% pelleted dehydrated Lucerne, and 17% soybean meal (Bernard et al., 2005c; 14 goats),

- **Study 7**: two diets with 47% of either hay (H;13 goats) or corn silage (CS;14 goats) supplemented, or not, with 5.8% of either linseed or linoleic sunflower oil, with the concentrate fraction containing 24(H)–24(CS)% rolled barley, 39(H)–34(CS) dehydrated sugar beet pulp, 12(H)–0(CS) pelleted dehydrated Lucerne, and 25(H)–42(CS)% soybean meal, respectively (Bernard et al., unpublished 27 goats).

In these two goat studies, conversely to what is generally observed in dairy cows (Bauman & Griinari, 2003), dietary lipid supplementation always increased the milk fat content and did not increase (or only slightly increased) milk yield, in accordance with previously published goat studies (Chilliard et al., 2003a, 2006a). The seven cow and goat studies presented above have been used as a database to allow us to evaluate how dietary factors may change metabolic pathways and lipogenic gene expression, in interaction with animal species peculiarities.

Regulation of Lipoprotein Lipase

Studies in cows on MFD diets reported either no change in the abundance of mammary *LPL* mRNA (study 3 with 1.7% "protected" fish oil ; Ahnadi et al., 2002) or a tendency to decrease (study 1; Peterson et al., 2003) or a strong decrease (study 5; Harvatine & Bauman, 2006, and study 3 with 2.7% unprotected fish oil; Ahnadi et al., 2002). In addition, still with an MFD diet, no change in the abundance of *FABP* has been observed (Peterson et al., 2003). However, in the two former studies (1 and 3), milk long-chain FA (> C16) yield (g/day) was significantly decreased, by 43% after "protected" fish oil supplementation (study 3) and by 28% with a high-concentrate diet (study 1). Elsewhere, in suckling beef cows fed hay supplemented with 14% high-linoleate safflower seeds, compared to 12% corn and 6% safflower seed meal, a trend toward a greater *LPL* mRNA was observed ($P = 0.09$; Murrieta et al., 2006) together with an increase in long-chain FA percentage in milk fat. Furthermore, in goats fed a hay-based diet supplemented with oleic sunflower oil (study 6; Bernard et al., 2005c), no significant effect was observed on mammary LPL activity, whereas *LPL* mRNA content was increased, together with a large increase (83%) in long-chain FA (C18) secretion. In goats fed with hay or corn silage supplemented with either linseed or sunflower oil (study 7), a large increase in the secretion of milk long-chain FA (> 100%) was observed without any effect on mammary LPL activity and mRNA except when corn silage was supplemented with sunflower oil, in which case LPL activity increased. In addition, in late-lactating goats fed an alfalfa hay-based diet, supplementation with 3.8% lipids from soybeans had no effect on mammary *LPL* mRNA abundance, while the secretion of long-chain FA (C18) in milk was increased by 58% (Bernard et al., 2005b). Thus, most of the results available in ruminants are not in agreement with those reported in rodents, where a very high dietary lipid intake (20%) enhanced both mammary gland LPL activity and lipid uptake (Del Prado et al., 1999).

The aforementioned results led to the hypothesis that either *LPL* mRNA or LPL activity measured *in vitro* in optimal conditions generally does not limit the uptake of long-chain FA by the mammary gland in ruminants. Other factors such as the availability of plasma triglyceride FA (Gagliostro et al., 1991) and the location of LPL (capillary endothelial cells, MEC, or depleted adipocytes) could play an important role. These results are also in accordance with those from previous studies in lactating goats, demonstrating that mammary utilization of plasma triglycerides and NEFA is related to their arterial concentration (see earlier section). Moreover, the existence of positive correlations between milk stearic acid percentage and milk fat content was observed (Chilliard et al., 2003b), related both to the response to dietary lipids and to individual variations within each dietary treatment (Bernard et al., 2006). This suggests that a significant availability and uptake of this FA is an important factor for milk fat secretion in goats, as in cows (Loor et al., 2005a, b).

Regulation of Genes Involved in de Novo Lipid Synthesis (ACCA and FASN)

In cows, an MFD diet induced joint reductions in *ACACA* mRNA abundance and ACC activity and in FAS activity in mammary tissue, together with a dramatic 60% decrease in C4–C16 FA secretion (g/day) (study 2; Piperova et al., 2000). Furthermore, Ahnadi et al. (2002; study 3) observed a decrease in mammary *ACACA* and *FASN* mRNA levels along with a 38% decrease in C4–C16 FA secretion with fish oil-supplemented diets. Moreover, in cows with a high-concentrate diet (study 1; Peterson et al., 2003), the observed reductions of milk fat secretion (27% decrease) and of C4–C16 (30% decrease) match a reduction of mammary mRNA abundance of several genes involved in milk lipid synthesis pathways, in particular *ACACA* and *FASN*, whereas no effect on milk κ-*casein* mRNA was observed. In addition, in cows fed an MFD diet, a downregulation of *FASN* and *ACACA* as well as other lipogenic genes such as *SCD, LPL,* and *FABP3*, was observed using a bovine oligonucleotide microarray (Loor et al., 2005c). Similarly, in cows fed a low-forage diet supplemented with soybean and fish oils, Harvatine and Bauman (2006; study 5) observed a decrease in the yield of milk fat (38%) and short- and medium chain-FA (50%), together with a significant reduction in the expression of *FASN* and other genes involved in the regulation of lipid metabolism (*SREBP1, INSIG-1, S14*). However, in suckling beef cows, Murrieta et al. (2006) observed no effect from a diet supplemented with high-linoleate safflower seeds on mammary *ACACA* and *FASN* mRNA levels, despite a 33% decrease in the weight percentage of C10–C14 FA.

Elsewhere, food deprivation for 48 hours in dairy goats was shown to change mammary transcriptome profil, using a bovine 8,379-gene microarray (Ollier et al., 2007): in particular, six genes involved in lipid metabolism and transport such as *FASN*, *ACSBG1* (acyl-CoA synthetase "bubblegum"), and *AZGP1* (zinc-alpha-2-glycoprotein precursor) were downregulated simultaneously with a decrease in milk lactose, protein, and fat secretion. The regulation of these last two genes needs further studies to evaluate their impact on mammary lipogenesis.

In goats (studies 6 and 7), the observed slight decrease of 15–22% in milk C4–C16 FA secretion (g/day) after supplementation of hay or corn silage diets with vegetable lipids was not accompanied by any significant variation of *ACACA* and *FASN* mRNA levels or activities. However, a positive relationship ($r = +0.62$) was observed (Fig. 3a) between *ACACA* mRNA variation of and C4–C16 secretion response to lipid supplementation, calculated from 81 individual values obtained by biopsies from goats fed hay or corn silage diets supplemented with 5.8% of either linseed or sunflower oil (study 7). Altogether, results from cow and goat studies demonstrated a positive relationship between both *ACACA* ($r = +0.66$; Fig. 4a) and *FASN* ($r = +0.73$; Fig. 4b) mRNA abundance and C4–C16 secretion responses to dietary treatment. These results suggest that variations of *ACACA* and *FASN* mRNA abundance play a role in the response of milk short- and medium-chain FA to dietary manipulation, especially the addition of lipids to the diet. Furthermore, the responses to lipid supplementation of *ACACA* and *FASN* mRNA were correlated ($r = +0.56$; Fig. 3b) in study 7, suggesting the existence of a similar regulatory mechanism for these genes.

The differences observed between cows (Ahnadi et al., 2002; Piperova et al., 2000) and goats (Bernard et al., 2005c) in the magnitude of responses of *ACACA* (Fig. 4a) and *FASN* (Fig. 4b) mRNA and/or activity to dietary PUFA supplementation may be partly explained by factors linked to the diet, such as the level of starchy concentrate in the diet, the nature (fish oil vs. vegetable oil) and presentation (seeds vs. free oil) of the lipid supplements, as well as species differences. The fact that lipid supplementation in goats always induces an increase in milk fat content and secretion whereas in cows it generally decreases them (see earlier section) supports species-differences implications.

Altogether, from *in vivo* studies where *ACACA* and *FASN* mRNA levels and activities in the mammary gland were measured (studies 6 and 7 and study 2 for ACC in cows), positive relationships between mRNA and activities variations in response to dietary treatment were observed for *ACACA* and *FASN* (Bernard et al., 2006). This suggests a regulation at a transcriptional level, at least, for these two genes in the ruminant mammary gland, which is in accordance with data obtained in rat adipose tissue (Girard et al., 1997) and liver (Girard et al., 1994) in response to nutritional factors. For *ACACA*, this level of regulation occurs in addition to the well-documented posttranslational

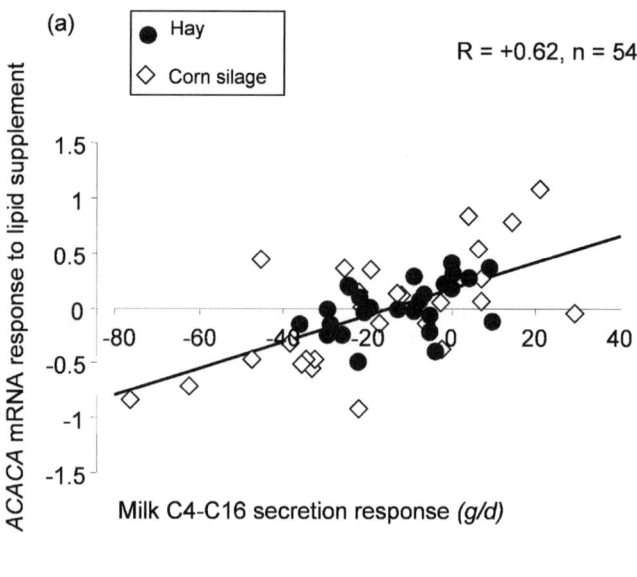

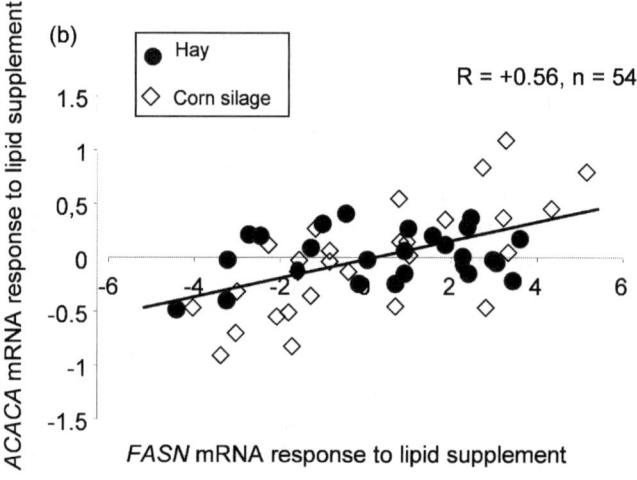

Fig. 3 Relationships between (a) milk short- and medium-chain fatty acid secretion and mammary acetyl-CoA carboxylase (*ACACA*) mRNA level responses to different dietary treatments (adapted from Bernard et al., 2006), and (b) mammary fatty acid synthase (*FASN*) and acetyl-CoA carboxylase (*ACACA*) mRNA-level responses to different dietary treatments. Results from 54 individual responses (lipid-supplemented = control) from 81 mammary biopsies in two 3 × 3 Latin squares, in 27 goats (study 7; see text). Lipid supplements were linseed oil or sunflower oil

regulation, in particular through covalent modification via phosphorylation/dephosphorylation under hormonal control as well as by allosteric activation or inhibition by cellular metabolites (see earlier section).

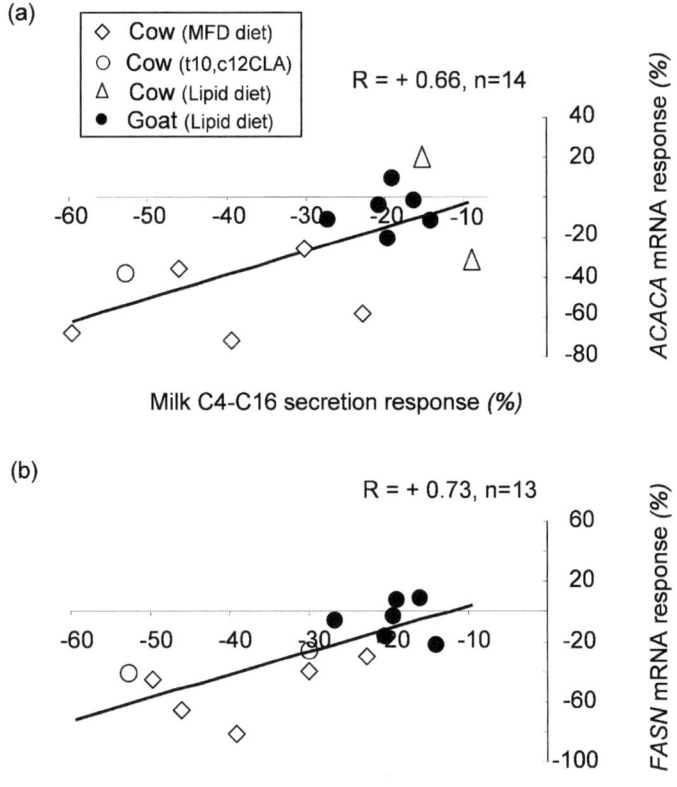

Fig. 4 Relationships between milk short- and medium-chain fatty acid secretion and mammary acetyl-CoA carboxylase (*ACACA*) (a) and fatty acid synthase (*FASN*) and (b) mRNA-level responses to different dietary treatments in goats and cows. Results are expressed as treatment mean response (vs. control group). Dietary treatments are described in the text and consisted of either lipid supplementation in goats (●; studies 6 and 7; n = 41; Bernard et al., 2005c, unpublished) and cows (△; study 4; Delbecchi et al., 2001) or milk fat-depressed (MFD) diets (◇ studies 1–3 and study 5; Peterson et al., 2003; Piperova et al., 2000; Ahnadi et al., 2002; Harvatine & Bauman, 2006), and postruminal *trans*-10, *cis*-12 CLA infusion (○ 13.6 g/day and 10 g/day, respectively, in Baumgard et al., 2002, and Harvatine & Bauman, 2006) in cows

Regulation of Stearoyl-CoA Desaturase

In rodents, nutritional regulation of SCD activity mainly occurs in the liver and has been studied extensively (Ntambi, 1999). Conversely, in the ruminant lactating mammary gland, only a few studies have investigated the nutritional regulation of *SCD* mRNA abundance and/or protein activity.

From the five nutritional studies performed in cows, four reported mammary *SCD* expression (studies 1, 3–5), and only feeding 1.7% of protected fish oil supplement (rich in long-chain n-3 FA) (Ahnadi et al., 2002) significantly decreased ($P < 0.05$) *SCD* mRNA abundance.

In goats fed hay-based diets supplemented with lipids, mammary *SCD* mRNA abundance decreased with formaldehyde-treated linseed and enzyme activity decreased with oleic sunflower oil, linseed oil, and sunflower oil, whereas lipid supplementation on *SCD* mRNA or activity had no effect on a corn silage diet (studies 6 and 7). Elsewhere, in late-lactating goats fed an alfalfa hay-based diet, mammary *SCD* mRNA was not affected by the dietary addition of 3.8% lipids from soybeans (Bernard et al., 2005b). Altogether, results from goats suggest an interaction between the basal diet and the dietary lipids used and predict a negative transcriptional or posttranscriptional regulation by dietary PUFA and/or by their ruminal biohydrogenation products may occur.

In vivo △-9 desaturase activity has often been estimated by the milk ratios for the pairs of FA that represent a product/substrate relationship for SCD (*cis*-9 C14:1/C14:0, *cis*-9 C16:1/C16:0, *cis*-9 C18:1/C18:0, *cis*-9, *trans*-11 CLA/ *trans*-11 C18:1; Bauman et al., 2001) due to the fact that the *in vitro* activity assay needs fresh materials, is laborious (Legrand et al., 1997), and is done in optimal conditions (pH, substrate, cofactors) that differ from *in vivo* conditions.

From the goat studies, we saw that the four FA pair ratios that represent a proxy for SCD activity were more or less related to the SCD activity itself, across 10 dietary groups (Bernard et al., 2005c; Chilliard et al., 2006b; Fig. 5). However, in terms of response to dietary lipids (six comparisons), the milk ratio of myristoleic acid to myristic acid (*cis*-9 C14:1/C14:0) gave the best estimation for the response of mammary SCD activity. This is due to the fact that almost all the myristoleic acid present in the milk is likely to be synthesized in the mammary gland by SCD. Indeed, myristic acid originates almost exclusively from *de novo* synthesis within the mammary gland (C14:0 is poorly represented in feedstuffs used for ruminants, including lipid supplements). Conversely, variable proportions of palmitic, palmitoleic, stearic, oleic, vaccenic, and rumenic acids come from absorption from the digestive tract and/or mobilization of body fat reserves. Then the ratios that involve these latter FA are less indicative of SCD activity. In addition, differential uptake, turnover, and use of the different FA of the four pair ratios by the mammary tissue itself may occur. Moreover, the four FA pair ratios could be influenced by other factors than SCD activity, such as a differential accuracy in the quantification of the *cis*-9 isomers in milk as well as, as stated above, by differences between *in vivo* (effective) and *in vitro* (potential) SCD activity.

Regulation of Acyltransferases

Only one study in cows reported data on genes encoding acyltransferases with an observed reduction of mammary *GPAT* and *AGPAT* mRNA abundance

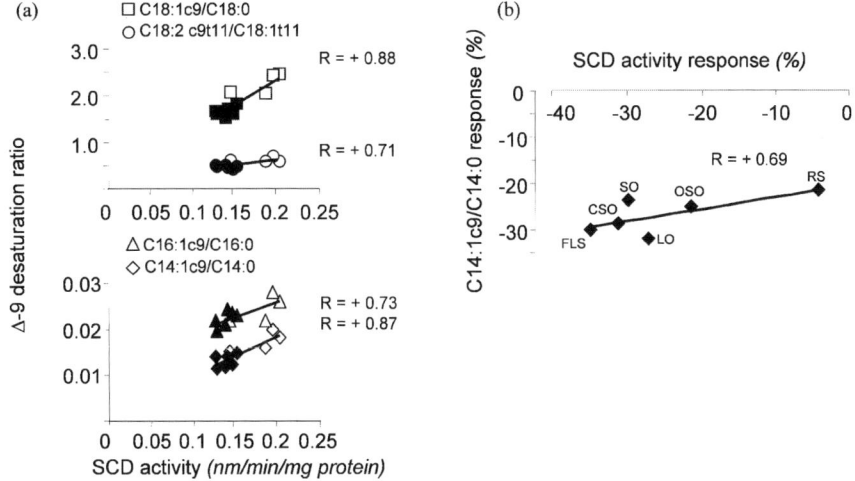

Fig. 5 Relationships between milk FA desaturation ratios and stearoyl-CoA desaturase (SCD) activity in 43 goats fed hay-based diet supplemented, or not, with lipids. The lipid supplements were either 3.6% of lipids from oleic sunflower oil (OSO) or formaldehyde-treated linseed (FLS) (study 6; Bernard et al., 2005c), or 5.8% lipids from linseed oil (LO) or sunflower oil (SO) (study 7; Bernard et al., unpublished), or 6.5% whole rapeseed (RS) or 4.5% of sunflower oil (CSO) (Chilliard et al., 2006b). (a) Results are means of the six lipid-supplemented groups (black symbols) and the four control groups (white symbols). (b) Relationship between the responses (% of the control group) of SCD activity and milk *cis*-9 C14:1/C14:0 to the six lipid supplements

together with a reduction of milk fat yield (by 27%) with a high-concentrate diet (Peterson et al., 2003). In goats, food deprivation (FD) 48 hours before slaughter was shown to increase mammary *AGPAT* and *DGAT1* mRNA content, whereas milk fat secretion decreased (Ollier et al., 2006). This apparent discrepancy could be due to a known posttranscriptional regulation of these two genes. Indeed, phosphorylation/dephosphorylation mechanisms have been suggested for AGPAT activity by Mistry and Medrano (2002), whereas a translational control has been reported for DGAT1 expression (Yu et al., 2002).

Molecular Mechanisms Involved in the Regulation of Mammary Lipogenesis by Nutrition

A Key Regulatory Role for Specific Fatty Acids

In Vivo Results in Nutritional Studies

From earlier *in vivo* studies with postruminal infusions of plant oils (e.g., Chilliard et al., 1991), a role for long-chain saturated FA and/or PUFA in decreasing mammary *de novo* FA synthesis has been suggested. Nevertheless, only recently was a central role proposed for specific *trans*-FA as potent inhibitors of mammary lipid synthesis, from studies with specific dietary conditions inducing a dramatic MFD (Bauman & Griinari, 2003). In the same way, an impairment of mammary lipid synthesis has been observed in rats fed a diet containing a mixture of *trans*-isomers (Assumpcao et al., 2002).

Indeed, the so-called low-milk fat syndrome in cows seems to be mainly due to specific PUFA biohydrogenation products formed in the rumen. Diets that induce MFD belong to three groups:

1. Diets rich in readily digestible carbohydrates and poor in fibrous components, without the addition of lipid supplements (e.g., high-grain/low-forage diets; Peterson et al., 2003), but containing a minimal amount of PUFA in dietary feedstuffs (Griinari et al., 1998),
2. Low-fiber diets associated with supplemental PUFA of plant origin (Piperova et al., 2000),
3. Diets associated with dietary supplements of marine oils (fish oils, fish meals, oils from marine mammals and/or algae) that induce MFD regardless of the level of starch or fiber in the diet (Ahnadi et al., 2002; Chilliard et al., 2001).

In the past, a number of theories have been proposed to explain diet-induced MFD, with the starting point for all these theories being an alteration in ruminal fermentations (reviews by Bauman & Griinari, 2001, 2003). One of them is that rumen production of acetate and butyrate was too low to support milk fat synthesis. Another one is that ruminal production of propionate increased, enhancing the hepatic rate of gluconeogenesis and the levels of circulating glucose and insulin and adipose tissue lipogenesis, thus inducing a shortage of nutrients available for the mammary gland. Another theory that now prevails was first proposed by Davis and Brown (1970) and, as mentioned above, suggests that mammary fat synthesis is inhibited by specific *trans*-FA, which results from alterations in rumen PUFA biohydrogenation.

In cows, *in vivo* trials support the *trans*-FA theory because a wide range of MFD diets are accompanied by an increase in the *trans*-C18:1 percentage of milk fat (review by Bauman & Griinari, 2003). An important development of this theory was the discovery that MFD was associated with a specific increase in *trans*-10 C18:1 rather than total *trans*-C18:1 isomer (Griinari et al., 1998). This finding was confirmed in several studies in cows from which a curvilinear response curve between *trans*-10 C18:1 and fat yield responses (Fig. 6) was evidenced (Loor et al., 2005a). Elsewhere, a curvilinear relationship between the decrease in milk fat percentage and small increases in milk *trans*-10, *cis*-12 CLA (Bauman & Griinari, 2003) was also observed in some, but not all, studies. For example, the reduction of milk fat (27% decrease) and of C4–C16 (30% decrease) secretion observed in cows with a high-concentrate diet (study 1; Peterson et al., 2003) was accompanied by a small but significant increase in milk fat *trans*-10, *cis*-12 CLA secretion (+0.5 g/day).

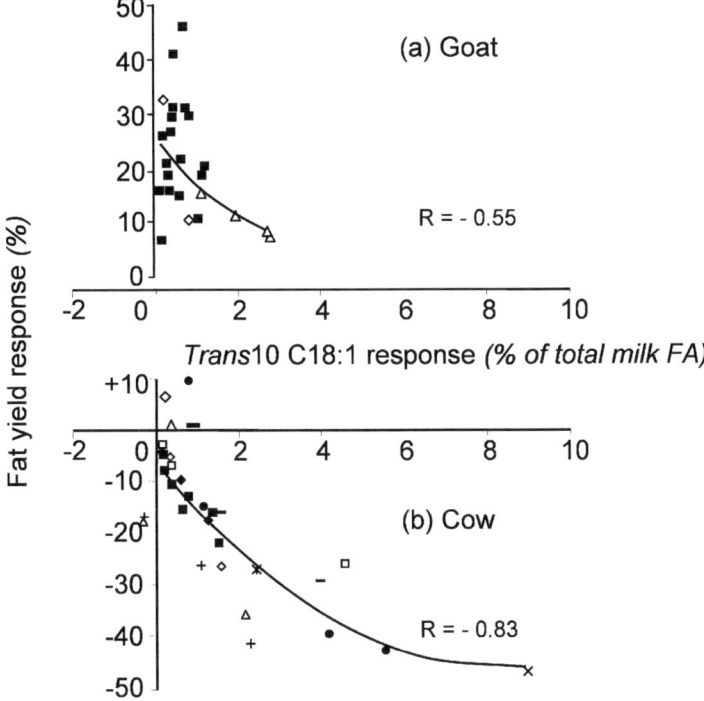

Fig. 6 Relationships between milk *trans*-10 C18:1 and fat yield responses in goats and cows. (a) Goat studies: data from 24 lipid-supplemented groups compared to 12 control groups (353 goats). The forages were either hay (■), fresh grass (◇), or corn silage (△). The lipids supplements were either sunflower oil, oleic sunflower oil, linseed oil, extruded linseeds, or rapeseeds (4–6% lipids in diet DM) and were given for 3 to 5 weeks (adapted from Chilliard et al., 2006a, b). (b) Cow studies: data from 31 lipid- or concentrate-supplemented groups, compared to control groups (13 studies) (adapted from Loor et al., 2005a)

The *trans*-10 C18:1 isomer synthesis in the rumen is probably due to the reduction of *trans*-10, *cis*-12 CLA resulting from an alternative pathway for biohydrogenation of linoleic acid, which increases when rumen pH decreases. Indeed, this "*trans*-10 pathway" seems to increase with diets rich in concentrate and/or corn silage and to occur after a one- to two-week period of latency after the start of dietary PUFA supplementation, following an earlier but transient increase in the "*trans*-11 pathway" (Chilliard & Ferlay, 2004; Roy et al., 2006a; Shingfield et al., 2006).

We must emphasize that in goats as in cows, a negative and curvilinear relationship between the responses of milk *trans*-10 C18:1 percentage and milk fat yield is observed (Chilliard et al., 2006a; Fig. 6), despite the fact that the fat yield response was always positive in goats, but always negative or null in cows. Then, in goats, the highest increases in *trans*-10 C18:1 were observed with either corn silage or fresh grass diets (Fig. 6) and matched with the lowest increases in milk fat yield. In addition, the maximum observed value of milk *trans*-10 C18:1 in goats was much lower than in cows, and no significant increases in goat milk *trans*-10, *cis*-12 CLA occurred in the 24 diet comparisons in Fig. 6. These observed similarities and differences among ruminants suggest a species specificity of FA ruminal and/or mammary metabolism. The positive effect of lipid supplementation on goat milk fat yield could be due in part to the mammary sensitivity to increased availability of stearic acid arising from dietary PUFA biohydrogenation in the rumen (Bernard et al., 2006).

Regarding the impact of PUFA on lipogenic gene expression and, in particular, on SCD, still little is known in ruminants. Conversely, in rodents the downregulation of *SCD* gene expression by (*n*-6) and (*n*-3) PUFA has been largely described in liver and AT (Ntambi, 1999), whereas little is known in the lactating mammary gland (Lin et al., 2004; Singh et al., 2004). Indeed, Singh et al. (2004) observed negative effects from olive oil or safflower oil fed to lactating mice on both *SCD* mRNA and activity in the mammary gland and *SCD* mRNA in the liver. Similarly, in the mouse mammary gland, Lin et al. (2004) reported a decrease in *SCD, ACACA,* and *FASN* mRNA abundance and SCD activity by the dietary addition of either *trans*-11 C18:1, *cis*-9, *trans*-11 CLA, or *trans*-10, *cis*-12 CLA, whereas these treatments had no effect on these transcripts in liver. These results on lactating mice indicate a tissue-specific regulation of lipogenic gene expression by *trans*-FA and outline the possibility to manipulate mammary *SCD* gene expression by nutrition.

In goats, the observed negative effect of the addition of oleic sunflower oil, sunflower oil, and linseed oil to hay diet (trials 6 and 7) on SCD activity (Fig. 5b) and that of formaldehyde-treated linseed on *SCD* mRNA might be partly attributed to dietary *cis*-9 C18:1, C18:2 (*n*-6), and/or C18:3 (*n*-3) escaping from the rumen and/or to *trans*-isomers formed during ruminal metabolism of these three FA (Chilliard et al., 2003c; Ferlay et al., 2003; Rouel et al., 2004). In addition, besides PUFA biohydrogenation processes, oleic acid could be isomerized in several *trans*-C18:1 isomers, including *trans*-10, as shown in microbial cultures from bovine rumen (Mosley et al., 2002). This finding is in agreement with the observed increase of several *trans*-C18:1 isomers in milk from goats fed oleic sunflower oil (Bernard et al., 2005c; Ferlay et al., 2003; Chilliard et al., 2006a). Nevertheless, data in Fig. 5b suggest that α-linolenic acid from formaldehyde-treated linseeds should be more efficient than dietary oleic acid-rich oil to decrease SCD activity.

Duodenal or Intravenous Infusion of Specific Fatty Acids

Duodenal infusion trials of pure-CLA isomers demonstrated that *trans*-10, *cis*-12 CLA inhibits milk fat synthesis in dairy cows, whereas the *cis*-9, *trans*-11 CLA isomer has no effect (Baumgard et al., 2000; Loor & Herbein, 2003). The severe reduction (48%) in milk fat yield due to the infusion of a high dose (13.6 g/day) of *trans*-10, *cis*-12 CLA (Baumgard et al., 2000) was accompanied by a dramatic reduction (> 35%) of mRNA abundance of enzymes involved in mammary uptake and intracellular trafficking of FA (*LPL* and *FABP*), de novo FA synthesis (*ACACA* and *FASN*), desaturation (*SCD*), and esterification (*GPAT* and *AGPAT*). Similarly, intravenous administration of *trans*-10, *cis*-12 CLA, either from 2 to up to 6 g/day (Viswanadha et al., 2003) or 10 g/day (Harvatine & Bauman, 2006), depressed milk fat yield, with, for the latter study, a joint decrease in the expression of genes involved in mammary uptake (*LPL*), de novo FA synthesis (*FASN*), and the regulation of lipid metabolism (*SREBP1, S14, INSIG-1*).

However, in the absence of duodenal infusion, the levels of *trans*-10, *cis*-12 CLA in the rumen, duodenal fluid, or milk always remained very low compared to the levels used in infusion studies (see above) or reached by the *trans*-10 C18:1 isomer (ratio between *trans*-10, *cis*-12 CLA, and *trans*-10 C18:1 of ∼0.01; Bauman & Griinari, 2003; Loor et al., 2004a, b, 2005a, d; Piperova et al., 2000). In addition, in cows fed with marine oil for which an MFD was observed, little or only traces of milk *trans*-10, *cis*-12 CLA were detected, whereas substantial increases in *trans*-10 C18:1 were observed (Loor et al., 2005b; Offer et al., 2001). Furthermore, species-specific responses to *trans*-10, *cis*-12 CLA duodenal infusion have been noted, with no effect on milk fat secretion in goats (Andrade & Schmidely, 2005), contrary to what is observed in cows (see above), whereas in lactating sheep, lipid-encapsulated CLA supplement containing *trans*-10, *cis*-12 CLA significantly reduced milk fat synthesis (Lock et al., 2006).

Due to the lack of pure material, it was not possible until recently to investigate a direct effect of *trans*-10 C18:1 on milk fat synthesis, whereas the potent inhibitory effect of *trans*-10, *cis*-12 CLA was clearly established by postruminal infusion trials in dairy cows (Bauman & Griinari, 2003). However, a recent study (Lock et al., 2007) using chemically synthesized *trans*-10 C18:1 infused postruminally over 4 days at 42.6 g/day/cow showed that despite the fact that this isomer was absorbed, taken up by the mammary gland, and transferred to milk fat, it had no effect on milk fat synthesis. As suggested by the authors (Lock et al., 2007), it is likely that the formation of *trans*-10 C18:1 and of *trans*-10, *cis*-12-CLA due to alterations in the rumen environment is accompanied by the formation of other biohydrogenation intermediates that could as well inhibit (or co-inhibit with *trans*-10, *cis*-12-CLA) milk fat synthesis. Thus, from nutritional studies, several other rumen-derived FA were proposed recently as potential inhibitors of cow milk fat synthesis due to high negative correlations between their milk fat concentrations and milk fat content and secretion. These proposed FA were several *cis*- or *trans*-C18:1 isomers, and *cis*-9, *trans*-13 C18:2, *cis*-9, *trans*-12 C18:2, *trans*-11, *cis*-13 CLA, *trans* 11, *cis*-15 C18:2 (Loor et al., 2005a), and *trans*-9, *cis*-11 CLA (Roy et al., 2006a; Shingfield et al., 2005, 2006). To confirm the potential role of these FA as well as of other minor CLA isomers found in milk, when available, few of them were postruminally infused. Whereas *trans*-8, *cis*-10 CLA, *cis*-11, *trans*-13 CLA, and *trans*-10, *trans*-12 CLA did not inhibit milk fat yield (Perfield et al., 2004, 2006; Sæbo et al., 2005), it was demonstrated that *cis*-10, *trans*-12 CLA (Sæbo et al., 2005) and *trans*-9, *cis*-11 CLA (Perfield et al., 2005) did reduce milk fat synthesis. However, the efficiency of the *trans*-9, *cis*-11 CLA was much lower than that of *trans*-10, *cis*-12 CLA (Perfield et al., 2005). This last result is in agreement with feeding trials on cows with or without MFD, with a greater slope of the equation between milk fat yield and *trans*-10, *cis*-12 CLA than *trans*-9, *cis*-11 CLA (Roy et al., 2006a).

Again, some of these isomers (*trans*-7, *cis*-9 CLA, *trans*-11, *cis*-13 CLA, *cis*-9, *trans*-13 C18:2, *trans*-11, *cis*-15 C18:2, *cis*-9, *trans*-13 C18:2) increased in goats supplemented with dietary PUFA, whereas no MFD was observed (Chilliard & Ferlay, 2004; Chilliard et al., 2003c; Ferlay et al., 2003).

In Vitro Studies on the Effect of Fatty Acids on Lipogenesis in Mammary Epithelial Cells

Studies several years ago on dispersed bovine MEC (Hansen et al., 1986; Hansen & Knudsen, 1987) demonstrated that the addition of C18:0, *cis*-9 C18:1, or C18:2 (*n*-6) inhibited *de novo* synthesis of FA with 16 carbons or less, except C4. More recently, looking at the effects of specific FA on the bovine mammary cell line (MAC-T cells), Jayan and Herbein (2000) showed that, compared to stearic acid, *trans*-11 C18:1 and, to a lesser extent, *cis*-9 C18:1 reduced ACC and FAS activities.

Furthermore, 35 years ago, Bickerstaffe and Annison (1970) observed negative effects of oleic, linoleic, and linolenic acids on goat mammary SCD activity measured *in vitro*, which have been partly confirmed by our *in vivo* studies on goats (see earlier section). The addition of *trans*-11 C18:1 increased *SCD* mRNA abundance in bovine MEC (Matitashvili & Bauman, 2000) and SCD activity in the bovine MAC-T cell line (Jayan & Herbein, 2000). Recently, Keating et al. (2006) demonstrated that treatment of bovine MAC-T cells with *trans*-10, *cis*-12 CLA (and, to a lesser extent, *cis*-9, *trans*-11 CLA) caused a significant reduction in *SCD* transcriptional activity, with this effect mediated through the stearoyl-CoA desaturase transcriptional enhancer element region (STE; see earlier section). The same study also showed that bovine *SCD* promoter was upregulated by insulin and downregulated by oleic acid, whereas linoleic, linolenic, stearic, and vaccenic acids had no effect. However, the *in vitro* effects of other specific *trans*-C18:1 and C18:2 isomers on mammary *SCD* gene expression are still unknown.

Elsewhere, in bovine mammary epithelial cell (BME-UV) cultures, the addition of *trans*-10, *cis*-12 CLA inhibited the stimulatory effect of prolactin on the cytosolic $NADP^+$-dependent isocitrate dehydrogenase (*IDH1*) gene expression, involved in the generation of NADPH required for *de novo* fatty acid synthesis, whereas the *cis*-9, *trans*-11 CLA isomer had no effect (Liu et al., 2006).

Further research is necessary in ruminants to identify the more important inhibitors of fat synthesis either *in vivo* (i.e., postruminal infusion of exogenous

FA) or *in vitro*, which is hampered by the lack of pure *trans*-C18:1 and C18:2 isomers and by the difficulty in obtaining an *in vitro* functional model for lipid synthesis and secretion (Barber et al., 1997).

Signaling Pathways Mediating Nutritional Regulation of Gene Expression

Whereas the signaling mechanisms involved in the regulation of lipogenic gene expression in rodent liver and adipose tissue have been comprehensively described (Clarke, 2001), little is known about these mechanisms in ruminants, particularly in the mammary gland. However, it was suggested that several genes involved in milk fat synthesis in the bovine mammary gland may share a common regulatory mechanism because of their coordinated downregulation observed in response to a postruminal infusion of *trans*-10, *cis*-12 CLA (Baumgard et al., 2002) and to an MFD diet (Peterson et al., 2003). Clarke (2001) reviewed rodent data and proposed that PUFA control the main metabolic pathways of lipid metabolism by governing the DNA-binding activity and nuclear abundance of selected transcription factors regulating the expression of key genes. The major transcription factors involved are sterol regulatory binding protein-1 (SREBP-1) and peroxisome proliferator-activated receptors (PPARs), with the FA or cholesterol acting by binding to the nuclear receptors PPAR (and LXR, HNF-4α), whereas the FA induce changes in the nuclear abundance of SREBP. As only few data are available in ruminants on these transcription factors, the following sections describe the state of knowledge on transcription factors mediating nutritional regulation of gene expression in rodents as well.

SREBP-1

SREBPs are basic-helix-loop-helix-leucine zipper (bHLH-LZ) transcription factors that belong to a family of transcription factors that regulate lipid homeostasis by controlling the expression of several enzymes required for endogenous cholesterol, FA, triacylglycerol, and phospholipid synthesis. Three SREBP forms have been characterized in rodents—SREBP-1a, -1c, and -2—differing in their roles in lipid synthesis. The SREBP-1a form is mainly expressed in cultured cells and tissues with a high cell proliferation capacity. The SREBP-1c form is expressed in many organs (mainly adipose tissue, brain, muscle, etc.) (Shimomura et al., 1997). These two forms derive from a single gene through the use of alternate promoters that give rise to different first exons (Brown & Goldstein, 1997). SREBP-2 derives from a different gene and is involved in the transcription of cholesterogenic enzymes.

SREBPs are synthesized as ~1,150 amino acid inactive precursors bound to the membrane of the endoplasmic reticulum (ER) through a tight association with SREBP cleavage activating protein (SCAP) (Miller et al., 2001). Upon sterol deprivation, the SREBP-SCAP complex moves to the Golgi apparatus, where two functionally active distinct proteases, site-1 and site-2 proteases (S1P and S2P), sequentially cleave the precursor protein SREBP to release the NH_2-terminal active domain (Sakai et al., 1996). It has been shown that this sterol-dependent trafficking requires an intact sterol-sensing domain located in the SCAP protein (Brown & Goldstein, 1997), demonstrating the dual role of SCAP as escorter and sensor. Recently, the insulin-induced gene (*INSIG-1*) protein that binds SCAP and thus facilitates retention of the SCAP/SREBP complex in the ER was identified (Yang et al., 2002). Upon appropriate conditions (low sterol concentrations or possibly insulin action; Eberlé et al., 2004), the interaction between INSIG and SCAP decreases and allows the SCAP to escort SREBPs to the Golgi apparatus for the cleavage activation process.

Finally, the mature SREBP, i.e., the NH_2-terminal portion (domain containing the bHLH-LZ), is translocated to the nucleus, where it binds (as a homodimeric form) its target genes on sterol binding elements or on palindromic sequences called E-boxes within their promoter regions (Wang et al., 1993). These target genes are implicated in cholesterol, FA, and lipid synthesis, including LPL, ACC, FAS, and SCD (Shimomura et al., 1998).

In rodents, FA downregulate the nuclear abundance of SREBP-1 by two described mechanisms: either an inhibition of the proteolytic activation process of SREBP-1 protein (under cholesterol dependence) or an inhibition of the *SREBP-1* gene transcription. Recently, in the ovine lactating mammary gland, Barber et al. (2003) identified SREBP-1 as a major regulator of *de novo* lipid synthesis through the activation of ACCα PIII, achieved together with NF-Y, USF-1, and USF-2 transcription factors. In addition, SREBP-1 binding motifs were also identified in the proximal promoter of ACCα PII, which is upregulated during lactation, indicating that SREBP-1 could play an important role in the joint regulation of PII and PIII in mammary tissue (Barber et al., 2003). Elsewhere it was shown that the addition of *trans*-10, *cis*-12 CLA to the bovine MEC line (MAC-T) had no effect on *SREBP-1* mRNA or SREBP-1 precursor protein content but reduced the abundance of the activated nuclear fragment of the protein (Peterson et al., 2004). This was accompanied by a reduction in transcriptional activation of the lipogenic genes *ACACA, FASN,* and *SCD.* These findings suggest that the inhibitory effect of this CLA isomer on lipid synthesis could be due to an inhibition of the proteolytic activation of SREBP1. Similarly, the use of microarray tools to characterize mammary gene profiling in cows fed an MFD diet (composed of 70% forage, 25% concentrate, and 5% soybean oil) showed a downregulation of several genes associated with fatty acid metabolism (see previous section) and of eight transcription factors without modification of *SREBP1* gene expression (Loor et al., 2005c). In addition, Harvatine and Bauman (2006) demonstrated the existence of a joint decrease in the mammary expression of *SREBP1, Spot 14 (S14), INSIG-1, FASN,* and *LPL* that could explain MFD in lactating cows either fed a low-forage/high-oil diet or infused with *trans*-10, *cis*-12 CLA. Again, the decreased expression of SREBP-1 and proteins associated with SREBP-1 activation, together with SREBP-1 responsive lipogenic enzymes, provides strong support for the central role of SREBP-1 in the regulation of milk fat synthesis. Furthermore, this study outlined a possible involvement of S14 in the regulation of FA synthesis in the bovine mammary gland, as shown in rodent liver and adipose tissue (Cunningham et al., 1998).

PPARs

PPARs belong to a superfamily of hormone receptors with, as for all of the members of this family, a DNA-binding domain, a gene-activating domain, and a ligand-binding domain. They regulate the transcription of genes involved in different lipid metabolism pathways including the transport of plasma triglycerides, the cellular FA uptake, and the peroxisomal and mitochondrial β-oxidation (Schoonjans et al., 1996a). The activating ligands of PPARs are peroxisome proliferators, including chemical molecules as fibrates, thiazolidinedione as well as molecules as FA, including PUFA, and their metabolites. PPARs heterodimerize with the *cis*-9 retinoic acid receptor (RXR) to bind to specific response elements located in the promoter region of the target genes.

Three PPAR subtypes have been identified: PPARα, expressed mainly in liver as well as in heart, kidney, intestinal mucosa, and brown adipose tissue, involved in FA transport and β- and ω-oxidation; PPARβ abundantly and ubiquitously expressed but mainly found in heart, lung, and kidney; PPARγ, most abundant in adipose tissue, stimulating adipocyte differentiation and lipogenesis of the mature adipocyte (Schoonjans et al., 1996b). PPARγ is also expressed in a number of epithelial tissues (breast, prostate, and colon), in which it seems to favor less malignant phenotype cells in human cancer (Sarraf et al., 1999). The *PPARγ* gene generates two transcripts, designated *PPARγ1* and γ2, resulting from differential mRNA splicing and promoter usage (Yeldandi et al., 2000), and leading to two protein isoforms, with PPARγ2 having 30 additional amino acid residues at the N terminal extremity. Among the activating compounds of the *PPAR* genes, (*n*-3) and (*n*-6) PUFA and mainly their metabolites (eicosanoids and oxidized FA) are the major natural activators of PPARα (Clarke, 2001), while 15-deoxy-$\Delta^{12, 14}$-prostaglandin J2 is the activator of the PPARγ subtype (Rosen & Spiegelman, 2001). In addition, *in vitro* studies on mature adipocytes revealed that *trans*-10, *cis*-12 CLA downregulates *PPARγ* gene expression, that could be a mechanism by which this CLA isomer prevents lipid accumulation in adipocytes (Granlund et al., 2003).

The few data available in bovine show similarities with those from rodent species. Thus, in bovine subcutaneous adipose tissue, the observed joint upregulation of PPARγ and FAS and *ACACA* and *LPL* gene expression by propionate infusion (Lee & Hossner, 2002) suggests an implication of PPARγ in the nutritional or insulin activation of lipogenesis. Elsewhere, bovine *PPARγ1* and *PPARγ2* cDNAs have been characterized (Sundvold et al., 1997) with expression of the two isoforms in adipose tissue, whereas only PPARγ2 was expressed in the mammary gland. Recently, in primary cultured bovine MEC, the expression of PPARγ2 in response to the addition of acetate and octanoate was increased while ACC activity decreased (Yonezawa et al., 2004a), conversely to previous observations in adipose tissue (Lee & Hossner, 2002). Elsewhere, an upregulation of mammary *PPARγ* gene expression was shown in dairy cows between -14 and $+14$ days relative to parturition, using a bovine cDNA microarray (Loor et al., 2004c).

Other Transcription Factors

The molecular mechanisms that control milk protein and lipogenic gene expression are not fully understood and probably involve undiscovered proteins within the mammary gland. Thus, in the nuclear extract of bovine mammary gland, the transcription factors Sp1 and NF-1 were identified (Wheeler et al., 1997), which are already known in rodents to act in conjunction with other proteins such as SREBP1, as well as six other proteins whose abundance was positively related with lactation or pregnancy status. Four of these proteins were identified as lactoferrin, annexin II, vimentin, and heavy-chain immunoglobulin. The presence of lactoferrin in the nuclear extracts is consistent with a study demonstrating that lactoferrin binds to DNA in a sequence-specific manner and activates transcription (He & Furmanski, 1995). Nevertheless, the function of lactoferrin as a transcription factor has not yet been confirmed.

Elsewhere, over the past few years, response elements to lactogenic hormones have been mapped within the promoters of milk protein genes; in some cases, the proteins that mediate the lactational signals are known. Signal transducer(s) and activator(s) of transcription (STAT) form a family of cytoplasmic proteins that are activated in response to a large number of cytokines, growth factors, and hormones (Hennighausen, 1997). The STAT proteins are activated via a cascade of phosphorylation events in which Janus protein tyrosine kinases (Jak2) are first phosphorylated. Then the activated Jak2 phosphorylate STAT proteins. In turn, STAT detach from the receptor complex, form homo- or heterodimers, and translocate from the cytoplasm to the nucleus, where they interact with specific promoter regions and regulate gene expression (Hennighausen, 1997). Until now, seven bovine *STAT* genes have been identified, *STAT1-6, 5a,* and *5b,* the latter two of which have already been sequenced (Seyfert et al., 2000). STAT5 was originally identified as a "mammary gland factor" mediating the prolactin signal to establish galactopoiesis (Rosen et al., 1999). Recently, Mao et al. (2002) demonstrated that STAT5 binding (at position -797) contributes to the lactational stimulation of promoter III (PIII) of the *ACACA* gene in the mammary gland. Hence, prolactin, acting through STAT5, contributes to the activation of ACC expression in milk-producing cells. Similarly, in BME-UV bovine mammary epithelial cell, Liu et al. (2006) reported that prolactin enhances *IDH1* mRNA and protein expression, but the molecular mechanisms of this regulation were not investigated. Elsewhere, in bovine mammary gland explant culture, Yang et al. (2000a) showed a rapid stimulation of STAT5 DNA binding activity by prolactin, growth hormone, and IGF1. In addition, the same authors demonstrated *in vivo* that STAT5 protein level and DNA binding activity are modulated by several physiological signals, including GH infusion and milking frequency (Yang et al., 2000b). Altogether these findings suggest that STAT5 might be important in determining the milk composition by coordinating FA and protein synthesis during lactation and that STAT5 transcription factor may represent part of the common route by which different extracellular signals (linked to hormonal status as well as to milking frequency) could converge and be transduced intracellularly to coordinate cell functions in the mammary gland. Recently, a study reported an association between *STAT1* variants and milk fat and protein yield and percentages in Holstein dairy cattle, implicating the *STAT1* gene in the regulation of milk protein and fat synthesis (Cobanoglu et al., 2006).

Despite the recent increased knowledge in ruminants on the characterization of transcription factors in the mammary gland, many questions still remain unanswered, in particular the role of STAT and PPARs in the regulation of lipid metabolism.

Conclusions

Over the last several years, the biochemical pathways of lipid synthesis in the mammary gland have been elucidated, and many of the enzymatic proteins and their cDNAs have been characterized. This has allowed the development of studies on the nutritional regulation of a few "candidate", genes involved in mammary FA uptake (*LPL*), *de novo* synthesis (*ACACA* and *FASN*), and desaturation (*SCD*). These studies showed that the responses of mammary "candidate" gene expression to nutritional factors do not always match milk FA secretion responses. In goats and cows, data suggest that the availability of substrates rather than the LPL activity is the limiting factor in the uptake of long-chain FA, except with extreme MFD diets fed to cows, in which both mammary LPL mRNA and activity decreased. In cows and goats, data converged to demonstrate that *ACACA* and *FASN* gene expressions are key factors of short- and medium-chain FA synthesis, even though they are not always repressed by the addition of PUFA to the diet, in goats at least. In this species, *ACACA* and *FASN* gene expressions are regulated by dietary factors at a transcriptional level at least, and *SCD* is regulated at a transcriptional and/or posttranscriptional level, depending on the lipid supplements. Conversely, in cows, the level of *SCD* mRNA varied little with the nutritional factors studied so far, except for a decrease when "protected" fish oil was fed. A fine balance between the exogenous unsaturated FA and the SCD desaturation products must be maintained within the mammary gland in order to preserve the fluidity of cellular membranes and milk fat (Chilliard et al., 2000; Parodi, 1982). In addition to its role on milk nutritional quality via the synthesis in *cis*-9, *trans*-11 CLA and *cis*-9-monounsaturated fatty acids, the impact of SCD on membrane fluidity underlines the importance of this enzyme.

The regulatory systems governing the nutritional response of mammary gene expression, in particular the intracellular signaling systems involved in these regulations, need to be further investigated in the future. The basis of the effects of nutrients and particularly the identification of specific *trans*-FA controlling lipogenic gene expression are obvious targets. Milk *trans*-10, *cis*-12 CLA is sometimes correlated with milk fat depression in cows (but not in goats) and, when infused postruminally at high doses, acts as a potent inhibitor of the expression of all the lipogenic genes. Conversely, *trans*-10 C18:1 does not directly control milk fat synthesis in dairy cows, although it largely responds to dietary factors, with its concentration being negatively related to milk fat response in cows and, to a lesser extent, in goats. Nevertheless, marked differences are observed between the milk fat yield responses of these two species, with few differences in milk FA profile responses, in particular lower increases in *trans*-10 C18:1 in goats. Elsewhere, more information on the promoter of the lipogenic genes should be acquired, which would help to clarify the roles and mechanisms of the action of PUFA and/or *trans*-FA, in order to better understand the molecular mechanisms involved in dietary- and/or species-related responses. Few data on transcription factors are available, and a central role for SREBP-1 has been outlined as mediator of FA effects, and STAT5 for hormonal and physiological effects at least, whereas the roles of PPARs need to be determined. It is expected that the development of *in vitro* functional systems for lipid synthesis and secretion would allow future progress in the identification of the inhibitors and activators of fat synthesis and in understanding differences between ruminant species.

This chapter reviewed studies focusing on the nutritional regulation of the expression of a few candidate genes controlling lipid synthesis. Nevertheless, the expression of specific milk fat globule membrane proteins (Mather, 2000) such as butyrophilin, xanthine oxidoreductase, and CD36, which intervene in milk lipid secretion, is also likely to have consequences on milk fat yield and composition (Ogg et al., 2004). The recent development of tools for studying the mammary transcriptome (macro- and microarrays; e.g., Bernard et al., 2005a; Leroux et al., 2003; Loor et al., 2004c, 2005c; Suchyta et al., 2003) and proteome (Daniels et al., 2006) will allow us to study the effect of nutritional changes on the expression (mRNA and protein) of a large number of genes putatively involved in mammary gland function, including lipid synthesis and secretion, in relationship to milk FA profile (Ollier et al., 2007). Such tools will allow us to identify new "candidate" genes or proteins whose expression is regulated by nutrition, and to understand their regulation pathways. In addition, further investigations on ruminal digestion and body metabolism of nutrients (absorption, partitioning between tissues) and mammary metabolic flows will also contribute to highlight the mechanisms underlying the *in vivo* responses to dietary factors and the differences among ruminant species.

Acknowledgment The authors thank S. Ollier and M. Bonnet for helpful discussions during the preparation of the manuscript and P. Béraud for secretarial assistance.

References

Abu-Elheiga, L., Almarza-Ortega, D. B., Baldini, A., & Wakil, S. J. (1997). Human acetyl-CoA carboxylase 2. Molecular cloning, characterization, chromosomal mapping, and evidence for two isoforms. *Journal of Biological Chemistry, 272*, 10669–10677.

Abu-Elheiga, L., Brinkley, W. R., Zhong, L., Chirala, S. S., Woldegiorgis, G., & Wakil, S. J. (2000). The subcellular localization of acetyl-CoA carboxylase 2. *Proceedings of the National Academy of Sciences USA, 97*, 1444–1449.

Abumrad, N. A., el-Maghrabi, M. R., Amri, E. Z., Lopez, E., & Grimaldi, P. A. (1993). Cloning of a rat adipocyte membrane protein implicated in binding or transport of long-chain fatty acids that is induced during preadipocyte differentiation. Homology with human CD36. *Journal of Biological Chemistry, 268*, 17665–17668.

Ahnadi, C. E., Beswick, N., Delbecchi, L., Kennelly, J. J., & Lacasse, P. (2002). Addition of fish oil to diets for dairy cows. II. Effects on milk fat and gene expression of mammary lipogenic enzymes. *Journal of Dairy Research, 69*, 521–531.

Andrade, P. V. D., & Schmidely, P. (2005). Effect of duodenal infusion of *trans*10, *cis*12-CLA on milk performance and milk fatty acid profile in dairy goats fed high or low concentrate diet in combination with rolled canola seed. *Reproduction Nutrition Development, 46*, 31–48.

Angiolillo, A., Amills, M., Urrutia, B., Domenech, A., Sastre, Y., Badaoui, B., & Jordana, J. (2006). Identification of a single nucleotide polymorphism at intron 16 of the caprine acyl-coenzyme A: diacylglycerol acyltransferase 1 (DGAT1) gene. *Journal of Dairy Research*, in press.

Annison, E. F., Linzell, J. L., & West, C. E. (1968). Mammary and whole animal metabolism of glucose and fatty acids in fasting lactating goats. *Journal of Physiology, 197*, 445–459.

Aoki, N., Ishii, T., Ohira, S., Yamaguchi, Y., Negi, M., Adachi, T., Nakamura, R., & Matsuda, T. (1997). Stage specific expression of milk fat globule membrane glycoproteins in mouse mammary gland: Comparison of MFG-E8, butyrophilin, and CD36 with a major milk protein, beta-casein. *Biochimica et Biophysica Acta, 1334*, 182–190.

Assumpcao, R. P., Santos, F. D., Setta, C. L., Barreto, G. F., Matta, I. E., Estadella, D., Azeredo, V. B., & Tavares do Carmo, M. G. (2002). *Trans* fatty acids in maternal diet may impair lipid biosynthesis in mammary gland of lactating rats. *Annals of Nutrition and Metabolism, 46*, 169–175.

Barber, M. C., & Travers, M. T. (1995). Cloning and characterisation of multiple acetyl-CoA carboxylase transcripts in ovine adipose tissue. *Gene, 154*, 271–275.

Barber, M. C., & Travers, M. T. (1998). Elucidation of a promoter activity that directs the expression of acetyl-CoA carboxylase alpha with an alternative N-terminus in a tissue-restricted fashion. *Biochemical Journal, 333 (Pt 1)*, 17–25.

Barber, M. C., Clegg, R. A., Travers, M. T., & Vernon, R. G. (1997). Lipid metabolism in the lactating mammary gland. *Biochimica et Biophysica Acta, 1347*, 101–126.

Barber, M. C., Vallance, A. J., Kennedy, H. T., & Travers, M. T. (2003). Induction of transcripts derived from promoter III of the acetyl-CoA carboxylase-alpha gene in mammary gland is associated with recruitment of SREBP-1 to a region of the proximal promoter defined by a DNase I hypersensitive site. *Biochemical Journal, 375*, 489–501.

Barber, M. C., Price, N. T., & Travers, M. T. (2005). Structure and regulation of acetyl-CoA carboxylase genes of metazoa. *Biochimica et Biophysica Acta, 1733*, 1–28.

Barillet, F., Arranz, J. J., & Carta, A. (2005). Mapping quantitative trait loci for milk production and genetic polymorphisms of milk proteins in dairy sheep. *Genetics Selection Evolution, 37 (Suppl 1)*, S109–123.

Bauman, D. E., & Griinari, J. M. (2001). Regulation and nutritional manipulation of milk fat: Low-fat milk syndrome. *Livestock Production Science, 70*, 15–29.

Bauman, D. E., & Griinari, J. M. (2003). Nutritional regulation of milk fat synthesis. *Annual Review of Nutrition, 23*, 203–227.

Bauman D. E., Corl, B. A., Baumgard, L. H., & Griinari, J. M. (2001). Conjugated linoleic acid (CLA) and the dairy cow. In J.Wisman and P.C. Garnsworthy (Eds.), *Recent Advances in Animal Nutrition* (pp. 221–250). Nottingham: Nottingham University Press.

Baumgard, L. H., Corl, B. A., Dwyer, D. A., Sæbo, A., & Bauman, D. E. (2000). Identification of the conjugated linoleic acid isomer that inhibits milk fat synthesis. *American Journal of Physiology—Regulatory, Integrative and Comparative Physiology, 278*, R179–184.

Baumgard, L. H., Corl, B. A., Dwyer, D. A., & Bauman, D. E. (2002). Effects of conjugated linoleic acids (CLA) on tissue response to homeostatic signals and plasma variables associated with lipid metabolism in lactating dairy cows. *Journal of Animal Science, 80*, 1285–1293.

Bernard, L., Leroux, C., Hayes, H., Gautier, M., Chilliard, Y., & Martin, P. (2001). Characterization of the caprine stearoyl-CoA desaturase gene and its mRNA showing an unusually long 3'-UTR sequence arising from a single exon. *Gene, 281*, 53–61.

Bernard, C., Degrelle, S., Ollier, S., Campion, E., Cassar-Malek, I., Charpigny, G., Dhorne-Pollet, S., Hue, I., Hocquette, J. F., Le Provost, F., Leroux, C., Piump, F., Rolland, G., Uzbekova, S., Zalachas, E., & Martin, P. (2005a). A cDNA macro-array resource for gene expression profiling in ruminant tissues involved in reproduction and production (milk and beef) traits. *Journal of Physiology and Pharmacology, 56 (Suppl 3)*, 215–224.

Bernard, L., Leroux, C., Bonnet, M., Rouel, J., Martin P., & Chilliard Y. (2005b). Expression and nutritional regulation of lipogenic genes in mammary gland and adipose tissues of lactating goats. *Journal of Dairy Research, 72*, 250–255.

Bernard, L., Rouel, J., Leroux, C., Ferlay, A., Faulconnier, Y., Legrand P., & Chilliard, Y. (2005c). Mammary lipid metabolism and fatty acid secretion in Alpine goats fed vegetable lipids. *Journal of Dairy Science, 88*,1478–1489.

Bernard, L., Leroux, C., & Chilliard, Y. (2006). Characterisation and nutritional regulation of the main lipogenic genes in the ruminant lactating mammary gland. In K. Sejrsen, T. Hvelplund, M.O. Nielsen (Eds.), Ruminant Physiology: Digestion, metabolism and impact of nutrition on gene expression, immunology and stress (pp. 295–362). Wageningen, The Netherlands: Wageningen Academic Publishers.

Beswick, N. S., & Kennelly, J. J. (1998). The influence of bovine growth hormone and growth hormone releasing factor on acetyl-CoA carboxylase and fatty acid synthase in primiparous Holstein cows. *Comparative Biochemistry and Physiology. Part C, Pharmacology, Toxicology and Endocrinology, 120*, 241–249.

Bickerstaffe, R., & Annison, E. F. (1970). Lipid metabolism in the perfused chicken liver. The uptake and metabolism of oleic acid, elaidic acid, *cis*-vaccenic acid, *trans*-vaccenic acid and stearic acid. *Biochemical Journal, 118*, 433–442.

Bonnet, M., Faulconnier, Y., Fléchet, J., Hocquette, J. F., Leroux, C., Langin, D., Martin, P., & Chilliard, Y. (1998). Messenger RNAs encoding lipoprotein lipase, fatty acid synthase and hormone-sensitive lipase in the adipose tissue of underfed-refed ewes and cows. *Reproduction, Nutrition, Development, 38*, 297–307.

Bonnet, M., Leroux, C., Chilliard, Y., & Martin, P. (2000a). Rapid communication: Nucleotide sequence of the ovine lipoprotein lipase cDNA. *Journal of Animal Science, 78*, 2994–2995.

Bonnet, M., Leroux, C., Faulconnier, Y., Hocquette, J. F., Bocquier, F., Martin, P., & Chilliard, Y. (2000b). Lipoprotein lipase activity and mRNA are up-regulated by refeeding in adipose tissue and cardiac muscle of sheep. *Journal of Nutrition, 130*, 749–756.

Brown, M. S., & Goldstein, J. L. (1997). The SREBP pathway: Regulation of cholesterol metabolism by proteolysis of a membrane-bound transcription factor. *Cell, 89*, 331–340.

Camps, L., Reina, M., Llobera, M., Vilaro, S., & Olivecrona, T. (1990). Lipoprotein lipase: Cellular origin and functional distribution. *American Journal of Physiology, 258*, C673–681.

Chilliard, Y., & Ferlay, A. (2004). Dietary lipids and forages interactions on cow and goat milk fatty acid composition and sensory properties. *Reproduction, Nutrition, Development, 44*, 467–492.

Chilliard, Y., Gagliostro, G., Fléchet, J., Lefaivre, J., & Sebastian, I. (1991). Duodenal rapeseed oil infusion in early and midlactation cows. 5. Milk fatty acids and adipose tissue lipogenic activities. *Journal of Dairy Science, 74*, 1844–1854.

Chilliard, Y., Ferlay, A., Mansbridge, R. M., & Doreau, M. (2000). Ruminant milk fat plasticity: Nutritional control of saturated, polyunsaturated, *trans* and conjugated fatty acids. *Annales de Zootechnie, 49*, 181–205.

Chilliard, Y., Ferlay, A., & Doreau, M. (2001). Effect of different types of forages, animal fat or marine oils in cow's diet on milk fat secretion and composition, especially conjugated linoleic acid (Cla) and polyunsaturated fatty acids. *Livestock Production Science, 70*, 31–48.

Chilliard, Y., Ferlay, A., Rouel, J., & Lamberet, G. (2003a). A review of nutritional and physiological factors affecting goat milk lipid synthesis and lipolysis. *Journal of Dairy Science, 86*, 1751–1770.

Chilliard, Y., Rouel, J., Capitan, P., Chabosseau, J. M., Raynal-Ljutovac, K., & Ferlay, A. (2003b). Correlations between milk fat content and fatty acid composition in goats receiving different combinations of forages and lipid supplements. In Y. van der Honing (Ed.), *Book of Abstracts of the 54th Annual Meeting of European Association for Animal Production* (p. 343), Rome, Italy, August 31–September 3, 2003. Wageningen, The Netherlands: Wageningen Academic Publishers.

Chilliard, Y., Rouel, J., Chabosseau, J. M., Capitan, P., Gaborit, P., & Ferlay, A. (2003c). Interactions between raygrass preservation and linseed oil supplementation on goat milk yield and composition, including *trans* and conjugated fatty acids. In Y. van der Honing (Ed.), *Book of Abstracts of the 54th Annual Meeting of European Association for Animal Production* (p. 343), Rome, Italy, August 31–September 3, 2003. Wageningen, The Netherlands: Wageningen Academic Publishers.

Chilliard, Y., Rouel, J., Ferlay, A., Bernard, L., Gaborit, P., Raynal-Ljutovac, K., Lauret, A., & Leroux, C. (2006a). Optimising goat's milk and cheese fatty acid composition. In C. Williams and J. Buttriss (Eds.), *Improving the Fat Content of Foods* (pp. 281–312). Cambridge: Woodhead Publishing Ltd.

Chilliard, Y., Ollier, S., Rouel, J., Bernard, L., & Leroux, C. (2006b). Milk fatty acid profile in goats receiving high forage or high concentrate diets supplemented, or not, with either whole rapeseeds or sunflower oil. In Y. van der Honing (Ed.), *Book of Abstracts of the 57th Annual Meeting of European Association for Animal Production* (p. 296), Antalya, Turkey, September 17–20, 2006. Wageningen, The Netherlands: Wageningen Academic Publishers.

Christie, W. W. (1979). The effects of diet and other factors on the lipid composition of ruminant tissues and milk. *Progress in Lipid Research, 17*, 245–277.

Chung, M., Ha, S., Jeong, S., Bok, J., Cho, K., Baik, M., & Choi, Y. (2000). Cloning and characterization of bovine stearoyl CoA desaturase cDNA from adipose tissues. *Bioscience, Biotechnology, and Biochemistry, 64*, 1526–1530.

Clarke, S. D. (2001). Nonalcoholic steatosis and steatohepatitis. I. Molecular mechanism for polyunsaturated fatty acid regulation of gene transcription. *American Journal of Physiology—Gastrointestinal and Liver Physiology, 281*, G865–869.

Cobanoglu, O., Zaitoun, I., Chang, Y. M., Shook, G. E., & Khatib, H. (2006). Effects of the signal transducer and activator of transcription 1 (STAT1) gene on milk production traits in Holstein dairy cattle. *Journal of Dairy Science, 89*, 4433–4437.

Coleman, R. A., & Lee, D. P. (2004). Enzymes of triacylglycerol synthesis and their regulation. *Progress in Lipid Research, 43*, 134–176.

Coleman, R. A., Lewin, T. M., & Muoio, D. M. (2000). Physiological and nutritional regulation of enzymes of triacylglycerol synthesis. *Annual Review of Nutrition, 20*, 77–103.

Corl, B. A., Baumgard, L. H., Dwyer, D. A., Griinari, J. M., Phillips, B. S., & Bauman, D. E. (2001). The role of Delta(9)-desaturase in the production of *cis*-9, *trans*-11 CLA. *Journal of Nutritional Biochemistry, 12*, 622–630.

Corl, B. A., Baumgard, L. H., Griinari, J. M., Delmonte, P., Morehouse, K. M., Yurawecz, M. P., & Bauman, D. E. (2002). *Trans*-7,*cis*-9 CLA is synthesized endogenously by delta9-desaturase in dairy cows. *Lipids, 37*, 681–688.

Coulon, J. B., Dupont, D., Pochet, S., Pradel, P., & Duployer, H. (2001). Effect of genetic potential and level of feeding on milk protein composition. *Journal of Dairy Research, 68*, 569–577.

Cunningham, B. A., Moncur, J. T., Huntington, J. T., & Kinlaw, W. B. (1998). ?Spot 14? protein: A metabolic integrator in normal and neoplastic cells. *Thyroid, 8*, 815–825.

Daniels, K. M., Webb, K. E., Jr., McGilliard, M. L., Meyer, M. J., Van Amburgh, M. E., & Akers, R. M. (2006). Effects of body weight and nutrition on mammary protein expression profiles in Holstein heifers. *Journal of Dairy Science, 89*, 4276–4288.

Davis, C. L., & Brown, R. E. (1970). Low-fat milk syndrome. In A.T. Phillipson (Ed.) *Physiology of Digestion and Metabolism in Ruminants* (pp. 545–565). Newcastle Upon Tyne: Oriel.

Delbecchi, L., Ahnadi, C. E., Kennelly J. J., & Lacasse, P. (2001). Milk fatty acid composition and mammary lipid metabolism in Holstein cows fed protected or unprotected canola seeds. *Journal of Dairy Science, 84*, 1375–1381.

Del Prado, M., Villalpando, S., Gordillo, J., & Hernandez-Montes, H. (1999). A high dietary lipid intake during pregnancy and lactation enhances mammary gland lipid uptake and lipoprotein lipase activity in rats. *Journal of Nutrition, 129*, 1574–1578.

Ding, S. T., Schinckel, A. P., Weber, T. E., & Mersmann, H. J. (2000). Expression of porcine transcription factors and genes related to fatty acid metabolism in different tissues and genetic populations. *Journal of Animal Science, 78*, 2127–2134.

Eberlé, D., Hegarty, B., Bossard, P., Ferré, P., & Foufelle, F. (2004). SREBP transcription factors: Master regulators of lipid homeostasis. *Biochimie, 86*, 839–848.

Enoch, H. G., Catala, A., & Strittmatter, P. (1976). Mechanism of rat liver microsomal stearyl-CoA desaturase. Studies of the substrate specificity, enzyme-substrate interactions, and the function of lipid. *Journal of Biological Chemistry, 251*, 5095–5103.

Ferlay, A., Rouel, J., Chabosseau, J. M., Capitan, P., Raynal-Ljutovac, K., & Chilliard, Y. (2003). Interactions between raygrass preservation and high-oleic sunflower oil supplementation on goat milk composition, including *trans* and conjugated fatty acids. In Y. van der Honing (Ed.), *Book of Abstracts of the 55th Annual Meeting of European Association for Animal Production* (p. 350), Bled, Slovenia, September 5–9, 2004. Wageningen, The Netherlands: Wageningen Academic Publishers.

Gagliostro, G., Chilliard, Y., & Davicco, M. J. (1991). Duodenal rapeseed oil infusion in early and midlactation cows. 3. Plasma hormones and mammary apparent uptake of metabolites. *Journal of Dairy Science, 74*, 1893–1903.

German, J. B., Morand, C., Dillard, C. J., & Xu, R. (1997). Milk fat composition: Targets for alteration of function and nutrition. In R. A. S. Welch, D. J. W. Burns, S. R. Davis, A. I. Popay, and C. G. Prosser (Eds.), *Milk Composition, Production and Biotechnology* (pp. 39–72). New York: CAB International.

Girard, J., Perdereau, D., Foufelle, F., Prip-Buus, C., & Ferré, P. (1994). Regulation of lipogenic enzyme gene expression by nutrients and hormones. *FASEB Journal, 8*, 36–42.

Girard, J., Ferré, P., & Foufelle, F. (1997). Mechanisms by which carbohydrates regulate expression of genes for glycolytic and lipogenic enzymes. *Annual Review of Nutrition, 17*, 325–352.

Granlund, L., Juvet, L. K., Pedersen, J. I., & Nebb, H. I. (2003). *Trans*10, *cis*12-conjugated linoleic acid prevents triacylglycerol accumulation in adipocytes by acting as a PPARγ modulator. *Journal of Lipid Research, 44*, 1441–1452.

Griinari, J. M., Dwyer, D. A., McGuire, M. A., Bauman, D. E., Palmquist, D. L., & Nurmela, K. V. (1998). *Trans*-octadecenoic acids and milk fat depression in lactating dairy cows. *Journal of Dairy Science, 81*, 1251–1261.

Griinari, J. M., Corl, B. A., Lacy, S. H., Chouinard, P. Y., Nurmela, K. V., & Bauman, D. E. (2000). Conjugated linoleic acid is synthesized endogenously in lactating dairy cows by delta(9)-desaturase. *Journal of Nutrition, 130*, 2285–2291.

Grisart, B., Coppieters, W., Farnir, F., Karim, L., Ford, C., Berzi, P., Cambisano, N., Mni, M., Reid, S., Simon, P., Spelman, R., Georges, M., & Snell, R. (2002). Positional candidate cloning of a QTL in dairy cattle: Identification of a missense mutation in the bovine DGAT1 gene with major effect on milk yield and composition. *Genome Research, 12*, 222–231.

Grisart, B., Farnir, F., Karim, L., Cambisano, N., Kim, J. J., Kvasz, A., Mni, M., Simon, P., Frere, J. M., Coppieters, W., & Georges, M. (2004). Genetic and functional confirmation of the causality of the DGAT1 K232A quantitative trait nucleotide in affecting milk yield and composition. *Proceedings of the National Academy of Sciences USA, 101*, 2398–2403.

Guichard, C., Dugail, I., Le Liepvre, X., & Lavau, M. (1992). Genetic regulation of fatty acid synthetase expression in adipose tissue: Overtranscription of the gene in genetically obese rats. *Journal of Lipid Research, 33*, 679–687.

Hansen, H. O., & Knudsen, J. (1987). Effect of exogenous long-chain fatty acids on lipid biosynthesis in dispersed ruminant mammary gland epithelial cells: Esterification of long-chain exogenous fatty acids. *Journal of Dairy Science, 70*, 1344–1349.

Hansen, H. O., Tornehave, D., & Knudsen, J. (1986). Synthesis of milk specific fatty acids and proteins by dispersed goat mammary-gland epithelial cells. *Biochemical Journal, 238*, 167–172.

Harvatine, K. J., & Bauman, D. E. (2006). SREBP1 and thyroid hormone responsive spot 14 (S14) are involved in the regulation of bovine mammary lipid synthesis during diet-induced milk fat depression and treatment with CLA. *Journal of Nutrition, 136*, 2468–2474.

He, J., & Furmanski, P. (1995). Sequence specificity and transcriptional activation in the binding of lactoferrin to DNA. *Nature, 373*, 721–724.

Hennighausen, L. (1997). Molecular mechanisms of hormone controlled gene expression in the breast. *Molecular Biology Reports, 24*, 169–174.

Jayakumar, A., Tai, M. H., Huang, W. Y., al-Feel, W., Hsu, M., Abu-Elheiga, L., Chirala, S. S., & Wakil, S. J. (1995). Human fatty acid synthase: Properties and molecular cloning. *Proceedings of the National Academy of Sciences USA, 92*, 8695–8699.

Jayan, G. C., & Herbein, J. H. (2000). "Healthier" dairy fat using *trans*-vaccenic acid. *Nutrition and food Science, 30*, 304–309.

Jensen, D. R., Bessesen, D. H., Etienne, J., Eckel, R. H., & Neville, M. C. (1991). Distribution and source of lipoprotein lipase in mouse mammary gland. *Journal of Lipid Research, 32*, 733–742.

Jensen, R. G. (2002). The composition of bovine milk lipids: January 1995 to December 2000. *Journal of Dairy Science, 85*, 295–350.

Keating, A. F., Stanton, C., Murphy, J. J., Smith, T. J., Ross, R. P., & Cairns, M. T. (2005). Isolation and characterization of the bovine stearoyl-CoA desaturase promoter and analysis of polymorphisms in the promoter region in dairy cows. *Mammalian Genome, 16*, 184–193.

Keating, A. F., Kennelly, J. J., & Zhao, F. Q. (2006). Characterization and regulation of the bovine stearoyl-CoA desaturase gene promoter. *Biochemical and Biophysical Research Communications, 344*, 233–240.

Kim, K. H. (1997). Regulation of mammalian acetyl-coenzyme A carboxylase. *Annual Review of Nutrition 17*, 77–99.

Kinsella, J. E. (1970). Stearic acid metabolism by mammary cells. *Journal of Dairy Science, 53*, 1757–1765.

Klein, I., Sarkadi, B., & Varadi, A. (1999). An inventory of the human ABC proteins. *Biochimica et Biophysica Acta, 1461*, 237–262.

Knudsen, J., & Grunnet, I. (1982). Transacylation as a chain-termination mechanism in fatty acid synthesis by mammalian fatty acid synthetase. Synthesis of medium-chain-length (C8-C12) acyl-CoA esters by goat mammary-gland fatty acid synthetase. *Biochemical Journal, 202*, 139–143.

Knudsen, J., Neergaard, T. B., Gaigg, B., Jensen, M. V., & Hansen, J. K. (2000). Role of acyl-CoA binding protein in acyl-CoA metabolism and acyl-CoA-mediated cell signaling. *Journal of Nutrition, 130*, 294S–298S.

Kühn, C., Thaller, G., Winter, A., Bininda-Emonds, O. R., Kaupe, B., Erhardt, G., Bennewitz, J., Schwerin, M., & Fries, R. (2004). Evidence for multiple alleles at the DGAT1 locus better explains a quantitative trait locus with major effect on milk fat content in cattle. *Genetics, 167*, 1873–1881.

Lee, S. H., & Hossner, K. L. (2002). Coordinate regulation of ovine adipose tissue gene expression by propionate. *Journal of Animal Science, 80*, 2840–2849.

Legrand, P., Catheline, D., Fichot, M. C., & Lemarchal, P. (1997). Inhibiting delta9-desaturase activity impairs triacylglycerol secretion in cultured chicken hepatocytes. *Journal of Nutrition, 127*, 249–256.

Lehner, R., & Kuksis, A. (1996). Biosynthesis of triacylglycerols. *Progress in Lipid Research, 35*, 169–201.

Leroux, C., Le Provost, F., Petit, E., Bernard, L., Chilliard, Y., & Martin, P. (2003). Real-time RT-PCR and cDNA macroarray to study the impact of the genetic polymorphism at the alphas1-casein locus on the expression of genes in the goat mammary gland during lactation. *Reproduction, Nutrition, Development, 43*, 459–469.

Leroux, C., Laubier, J., Giraud-Delville, C., Le Provost, F., Chadi, S., Bernard, L., Bonnet, M., Chilliard, Y., & Martin, P. (2007). Sharp increase of goat fatty acid synthase gene expression in mammary gland during the last part of pregnancy. *Journal of Dairy Research*, submitted.

Lin, X., Loor, J. J., & Herbein, J. H. (2004). *Trans*10,*cis*12-18:2 is a more potent inhibitor of *de novo* fatty acid synthesis and desaturation than *cis*9,*trans*11-18:2 in the mammary gland of lactating mice. *Journal of Nutrition, 134*, 1362–1368.

Liu, W., Degner, S. C., & Romagnolo, D. F. (2006). *Trans*-10, *cis*-12 conjugated linoleic acid inhibits prolactin-induced cytosolic NADP+-dependent isocitrate dehydrogenase expression in bovine mammary epithelial cells. *Journal of Nutrition, 136*, 2743–2747.

Lock, A. L., Teles, B. M., Perfield, J. W. II, Bauman, D. E., & Sinclair, L. A. (2006). A conjugated linoleic acid supplement containing *trans*-10, *cis*-12 reduces milk fat synthesis in lactating sheep. *Journal of Dairy Science, 89*, 1525–1532.

Lock, A. L., Tyburczy, C., Dwyer, D. A., Harvatine, K. J., Destaillats, F., Mouloungui, Z., Candy, L., & Bauman, D. E. (2007). *Trans*-10 octadecenoic acid does not reduce milk fat synthesis in dairy cows. *Journal of Nutrition, 137*, 71–76.

Loor, J. J., & Herbein, J. H. (2003). Reduced fatty acid synthesis and desaturation due to exogenous *trans*10,*cis*12-CLA in cows fed oleic or linoleic oil. *Journal of Dairy Science, 86*, 1354–1369.

Loor, J. J., Ueda, K., Ferlay, A., Chilliard, Y., & Doreau, M. (2004a). Biohydrogenation, duodenal flow, and intestinal digestibility of *trans* fatty acids and conjugated linoleic acids in response to dietary forage:concentrate ratio and linseed oil in dairy cows. *Journal of Dairy Science, 87*, 2472–2485.

Loor, J. J., Ueda, K., Ferlay, A., Chilliard, Y., & Doreau, M. (2004b). Short communication: Diurnal profiles of conjugated linoleic acids and *trans* fatty acids in ruminal fluid from cows fed a high concentrate diet supplemented with fish oil, linseed oil, or sunflower oil. *Journal of Dairy Science, 87*, 2468–2471.

Loor, J. J., Dann, H. M., Everts, R. E., Rodriguez-Zas, S. L., Lewin, H. A., &. Drackley, J. K. (2004c). Mammary and hepatic gene expression analysis in peripartal dairy cows using a bovine cDNA microarray. *Journal of Animal Science, 82 (Suppl 1)*, 196.

Loor, J. J., Ferlay, A., Ollier, A., Doreau, M., & Chilliard, Y. (2005a). Relationship among *trans* and conjugated fatty acids and bovine milk fat yield due to dietary concentrate and linseed oil. *Journal of Dairy Science, 88*, 726–740.

Loor, J. J., Doreau, M., Chardigny, J. M., Ollier, A., Sébédio, J. L., & Chilliard, Y. (2005b). Effects of ruminal or duodenal supply of fish oil on milk fat secretion and profiles of *trans*-fatty acids and conjugated linoleic acid isomers in dairy cows fed maize silage. *Animal Feed Science and Technology, 119*, 227–246.

Loor, J. J., Piperova, L. S., Everts, R. E., Rodriguez-Zas, S. L., Drackley, J. K., Erdman, R. A., & Lewin, H. A. (2005c). Mammary gene expression profiling in cows fed a milk fat-depressing diet using a bovine 13,000 oligonucleotide microarray. *Journal of Animal Science, 83 (Suppl 1)*, 120.

Loor, J. J., Ferlay, A., Ollier, A., Ueda, K., Doreau, M., & Chilliard, Y. (2005d). High-concentrate diets and polyunsaturated oils alter *trans* and conjugated isomers in bovine rumen, blood, and milk. *Journal of Dairy Science, 88*, 3986–3999.

Lopez-Casillas, F., Ponce-Castaneda, M. V., & Kim, K. H. (1991). In vivo regulation of the activity of the two promoters of the rat acetyl coenzyme-A carboxylase gene. *Endocrinology, 129*, 1049–1058.

Mao, J., Marcos, S., Davis, S. K., Burzlaff, J., & Seyfert, H. M. (2001). Genomic distribution of three promoters of the bovine gene encoding acetyl-CoA carboxylase alpha and evidence that the nutritionally regulated promoter I contains a repressive element different from that in rat. *Biochemical Journal, 358*, 127–135.

Mao, J., Molenaar, A. J., Wheeler, T. T., & Seyfert, H. M. (2002). STAT5 binding contributes to lactational stimulation of promoter III expressing the bovine acetyl-CoA carboxylase alpha-encoding gene in the mammary gland. *Journal of Molecular Endocrinology, 29*, 73–88.

Marshall, M. O., & Knudsen, J. (1977). The specificity of 1-acyl-sn-glycerol 3-phosphate acyltransferase in microsomal fractions from lactating cow mammary gland towards short, medium and long chain acyl-CoA esters. *Biochimica et Biophysica Acta, 489*, 236–241.

Massaro, M., Carluccio, M. A., & De Caterina, R. (1999). Direct vascular antiatherogenic effects of oleic acid: A clue to the cardioprotective effects of the Mediterranean diet. *Cardiologia, 44*, 507–513.

Mather, I. H. (2000). A review and proposed nomenclature for major proteins of the milk-fat globule membrane. *Journal of Dairy Science, 83*, 203–247.

Matitashvili, E., & Bauman, D. E. (2000). Effect of different isomers of C18:1 and C18:2 fatty acids on lipogenesis in bovine mammary epithelial cells. *Journal of Animal Science, 78*, 165.

Mayorek, N., Grinstein, I., & Bar-Tana, J. (1989). Triacylglycerol synthesis in cultured rat hepatocytes. The rate-limiting role of diacylglycerol acyltransferase. *European Journal of Biochemistry, 182*, 395–400.

Mihara, K. (1990). Structure and regulation of rat liver microsomal stearoyl-CoA desaturase gene. *Journal of Biochemistry, 108*, 1022–1029.

Mikkelsen, J., & Knudsen, J. (1987). Acyl-CoA-binding protein from cow. Binding characteristics and cellular and tissue distribution. *Biochemical Journal, 248*, 709–714.

Miller, R. T., Scappino, L. A., Long, S. M., & Corton, J. C. (2001). Role of thyroid hormones in hepatic effects of peroxisome proliferators. *Toxicologic Pathology, 29*, 149–155.

Mistry, D. H., & Medrano, J. F. (2002). Cloning and localization of the bovine and ovine lysophosphatidic acid acyltransferase (LPAAT) genes that codes for an enzyme involved in triglyceride biosynthesis. *Journal of Dairy Science, 85*, 28–35.

Molenaar, A., Mao, J., Oden, K., & Seyfert, H. M. (2003). All three promoters of the acetyl-coenzyme A-carboxylase alpha-encoding gene are expressed in mammary epithelial cells of ruminants. *Journal of Histochemistry and Cytochemistry, 51,* 1073–1081.

Mosley, E. E., Powell, G. L., Riley, M. B., & Jenkins, T. C. (2002). Microbial biohydrogenation of oleic acid to *trans* isomers *in vitro*. *Journal of Lipid Research, 43,* 290–296.

Murrieta, C. M., Hess, B. W., Scholljegerdes, E. J., Engle, T. E., Hossner, K. L., Moss, G. E., & Rule, D. C. (2006). Evaluation of milk somatic cells as a source of mRNA for study of lipogenesis in the mammary gland of lactating beef cows supplemented with dietary high-linoleate safflower seeds. *Journal of Animal Science, 84,* 2399–2405.

Mutch, D. M., Anderle, P., Fiaux, M., Mansourian, R., Vidal, K., Wahli, W., Williamson, G., & Roberts, M. A. (2004). Regional variations in ABC transporter expression along the mouse intestinal tract. *Physiological Genomics, 17,* 11–20.

Neville, M. C., & Picciano, M. F. (1997). Regulation of milk lipid secretion and composition. *Annual Review of Nutrition, 17,* 159–183.

Nielsen, M. O., & Jakobsen, K. (1994). Changes in mammary uptake of free fatty acids, triglyceride, cholesterol and phospholipid in relation to milk synthesis during lactation in goats. *Comparative Biochemistry and Physiology. Part A, Physiology, 109,* 857–867.

Ntambi, J. M. (1999). Regulation of stearoyl-CoA desaturase by polyunsaturated fatty acids and cholesterol. *Journal of Lipid Research, 40,* 1549–1558.

Ntambi, J. M., Buhrow, S. A., Kaestner, K. H., Christy, R. J., Sibley, E., Kelly, T. J., Jr., & Lane, M. D. (1988). Differentiation-induced gene expression in 3T3-L1 preadipocytes. Characterization of a differentially expressed gene encoding stearoyl-CoA desaturase. *Journal of Biological Chemistry, 263,* 17291–17300.

Offer, N. W., Speake, B. K., Dixon, J., & Marsden, M. (2001). Effect of fish-oil supplementation on levels of (N-3) poly-unsaturated fatty acids in the lipoprotein fractions of bovine plasma. *Animal Science, 73,* 523–531.

Ogg, S. L., Weldon, A. K., Dobbie, L., Smith, A. J., & Mather, I. H. (2004). Expression of butyrophilin (Btn1a1) in lactating mammary gland is essential for the regulated secretion of milk-lipid droplets. *Proceedings of the National Academy of Sciences USA, 101,* 10084–10089.

Ollier, S., Robert-Granié, C., Bes, S., Goutte, M., Faulconnier, Y., Chilliard, Y., & Leroux, C. (2006). Impact of nutrition on mammary transcriptome and its interaction with the CSN1S1 genotype in lactating goats. In Y. van der Honing (Ed.) *Book of Abstracts, 57th Annual Meeting of European Association for Animal Production* (p. 49), Antalya, Turkey, September 17–20, 2006. Wageningen, The Netherlands: Wageningen Academic Publishers.

Ollier, S., Robert-Granié, C., Bernard, L., Chilliard, Y., & Leroux, C. (2007). Mammary transcriptome analysis of food-deprived lactating goats highlights genes involved in milk secretion and programmed cell death. *Journal of Nutrition, 137,* 560–567.

Palmquist, D. L., Beaulieu, A. D., & Barbano, D. M. (1993). Feed and animal factors influencing milk fat composition. *Journal of Dairy Science, 76,* 1753–1771.

Pariza, M. W., Park, Y., & Cook, M. E. (2001). The biologically active isomers of conjugated linoleic acid. *Progress in Lipid Research, 40,* 283–298.

Parodi, P. W. (1982). Positional distribution of fatty acids in triglycerides from milk of several species of mammals. *Lipids, 17,* 437–442.

Perfield, J. W. II, Sæbo, A., & Bauman, D. E. (2004). Use of conjugated linoleic acid (CLA) enrichments to examine the effects of *trans*-8, *cis*-10 CLA, and *cis*-11, *trans*-13 CLA on milk-fat synthesis. *Journal of Dairy Science, 87,* 1196–1202.

Perfield, J. W. II, Lock, A. L., Sæbo, A., Griinari, J. M., & Bauman, D. E. (2005). *Trans*-9,*cis*11 conjugated linoleic acid (CLA) reduces milk fat synthesis in lactating dairy cows. *Journal of Dairy Science, 88 (Suppl 1),* 211.

Perfield, J. W. II, Delmonte, P., Lock, A. L., Yurawecz, M. P., & Bauman, D. E. (2006). Trans-10, trans-12 conjugated linoleic acid does not affect milk fat yield but reduces delta9-desaturase index in dairy cows. *Journal of Dairy Science, 89*, 2559–2566.

Peterson, D. G., Matitashvili, E. A., & Bauman, D. E. (2003). Diet-induced milk fat depression in dairy cows results in increased trans-10, cis-12 CLA in milk fat and coordinate suppression of mRNA abundance for mammary enzymes involved in milk fat synthesis. *Journal of Nutrition, 133*, 3098–3102.

Peterson, D. G., Matitashvili, E. A., & Bauman, D. E. (2004). The inhibitory effect of trans-10, cis-12 CLA on lipid synthesis in bovine mammary epithelial cells involves reduced proteolytic activation of the transcription factor SREBP-1. *Journal of Nutrition, 134*, 2523–2527.

Piperova, L. S., Teter, B. B., Bruckental, I., Sampugna, J., Mills, S. E., Yurawecz, M. P., Fritsche, J., Ku, K., & Erdman, R. A. (2000). Mammary lipogenic enzyme activity, trans fatty acids and conjugated linoleic acids are altered in lactating dairy cows fed a milk fat-depressing diet. *Journal of Nutrition, 130*, 2568–2574.

Rosen, E. D., & Spiegelman, B. M. (2001). PPARγ: A nuclear regulator of metabolism, differentiation, and cell growth. *Journal of Biological Chemistry, 276*, 37731–37734.

Rosen, E. D., Sarraf, P., Troy, A. E., Bradwin, G., Moore, K., Milstone, D. S., Spiegelman, B. M., & Mortensen, R. M. (1999). PPARγ is required for the differentiation of adipose tissue in vivo and in vitro. *Molecular Cell, 4*, 611–617.

Rouel, J., Ferlay, A., Bruneteau, E., Capitan, P., Raynal-Ljutovac, K., & Chilliard, Y. (2004). Interactions between starchy concentrate and linseed oil supplementation on goat milk yield and composition, including trans and conjugated fatty acids (FA). In Y. van der Honing (Ed.), *Book of Abstracts of the 55th Annual Meeting of European Association for Animal Production* (p. 124), Bled, Slovenia, September 5–9, 2004. Wageningen, The Netherlands: Wageningen Academic Publishers.

Roy, R., Taourit, S., Zaragoza, P., Eggen, A., & Rodellar, C. (2005). Genomic structure and alternative transcript of bovine fatty acid synthase gene (FASN): Comparative analysis of the FASN gene between monogastric and ruminant species. *Cytogenetic and Genome Research, 111*, 65–73.

Roy, A., Ferlay, A., Shingfield, K. J., & Chilliard, Y. (2006a). Examination of the persistency of milk fatty acid composition responses to plant oils in cows fed different basal diets, with particular emphasis on trans-$C_{18:1}$ fatty acids and isomers of conjugated linoleic acid. *Animal Science, 82*, 479–492.

Roy, R., Ordovas, L., Taourit, S., Zaragoza, P., Eggen, A., & Rodellar, C. (2006b). Genomic structure and an alternative transcript of bovine mitochondrial glycerol-3-phosphate acyltransferase gene (GPAM). *Cytogenetics and Genome Research, 112*, 82–89.

Roy, R., Ordovas, L., Zaragoza, P., Romero, A., Moreno, C., Altarriba, J., & Rodellar, C. (2006c). Association of polymorphisms in the bovine FASN gene with milk-fat content. *Animal Genetics, 37*, 215–218.

Sæbo, A., Sæbo, P. C., Griinari, J. M., & Shingfield, K. J. (2005). Effect of abomasal infusions of geometric isomers of 10,12 conjugated linoleic acid on milk fat synthesis in dairy cows. *Lipids, 40*, 823–832.

Sakai, J., Duncan, E. A., Rawson, R. B., Hua, X., Brown, M. S., & Goldstein, J. L. (1996). Sterol-regulated release of SREBP-2 from cell membranes requires two sequential cleavages, one within a transmembrane segment. *Cell, 85*, 1037–1046.

Sarraf, P., Mueller, E., Smith, W. M., Wright, H. M., Kum, J. B., Aaltonen, L. A., de la Chapelle, A., Spiegelman, B. M., & Eng, C. (1999). Loss-of-function mutations in PPAR gamma associated with human colon cancer. *Molecular Cell, 3*, 799–804.

Schoonjans, K., Peinado-Onsurbe, J., Lefebvre, A. M., Heyman, R. A., Briggs, M., Deeb, S., Staels, B., & Auwerx, J. (1996a). PPARα and PPARγ activators direct a distinct tissue-specific transcriptional response via a PPRE in the lipoprotein lipase gene. *EMBO Journal, 15*, 5336–5348.

Schoonjans, K., Staels, B., & Auwerx, J. (1996b). Role of the peroxisome proliferator-activated receptor (PPAR) in mediating the effects of fibrates and fatty acids on gene expression. *Journal of Lipid Research, 37*, 907–925.

Schweizer, M., Takabayashi, K., Laux, T., Beck, K. F., & Schreglmann, R. (1989). Rat mammary gland fatty acid synthase: Localization of the constituent domains and two functional polyadenylation/termination signals in the cDNA. *Nucleic Acids Research, 17*, 567–586.

Senda, M., Oka, K., Brown, W. V., Qasba, P. K., & Furuichi, Y. (1987). Molecular cloning and sequence of a cDNA coding for bovine lipoprotein lipase. *Proceedings of the National Academy of Sciences USA, 84*, 4369–4373.

Seyfert, H. M., Pitra, C., Meyer, L., Brunner, R. M., Wheeler, T. T., Molenaar, A., McCracken, J. Y., Herrmann, J., Thiesen, H. J., & Schwerin, M. (2000). Molecular characterization of STAT5A- and STAT5B-encoding genes reveals extended intragenic sequence homogeneity in cattle and mouse and different degrees of divergent evolution of various domains. *Journal of Molecular Evolution, 50*, 550–561.

Shimomura, I., Shimano, H., Horton, J. D., Goldstein, J. L., & Brown, M. S. (1997). Differential expression of exons 1a and 1c in mRNAs for sterol regulatory element binding protein-1 in human and mouse organs and cultured cells. *Journal of Clinical Investigation, 99*, 838–845.

Shimomura, I., Hammer, R. E., Richardson, J. A., Ikemoto, S., Bashmakov, Y., Goldstein, J. L., & Brown, M. S. (1998). Insulin resistance and diabetes mellitus in transgenic mice expressing nuclear SREBP-1c in adipose tissue: Model for congenital generalized lipodystrophy. *Genes and Development, 12*, 3182–3194.

Shingfield, K. J., Ahvenjarvi, S., Toivonen, V., Ärölä, A., Nurmela, K. V., Huhtanen, P., & Griinari, J. M. (2003). Effect of dietary fish oil on biohydrogenation of fatty acids and milk fatty acid content in cows. *Animal Science, 77*, 165–179.

Shingfield, K. J., Reynolds, C. K., Lupoli, B., Toivonen, V., Yurawecz, M. P., Delmonte, P., Griinari, J. M., Grandison, A. S., & Beever, D. E. (2005). Effect of forage type and proportion of concentrate in diet on milk fatty acid composition in cows fed sunflower oil and fish oil. *Animal Science, 80*, 225–238.

Shingfield, K. J., Reynolds, C. K., Hervas, G., Griinari, J. M., Grandison, A. S., & Beever, D. E. (2006). Examination of the persistency of milk fatty acid composition responses to fish oil and sunflower oil in the diet of dairy cows. *Journal of Dairy Science, 89*, 714–732.

Shirley, J. E., Emery, R. S., Convey, E. M., & Oxender, W. D. (1973). Enzymic changes in bovine adipose and mammary tissue, serum and mammary tissue hormonal changes with initiation of lactation. *Journal of Dairy Science, 56*, 569–574.

Singh, K., Hartley, D. G., McFadden, T. B., & Mackenzie, D. D. (2004). Dietary fat regulates mammary stearoyl coA desaturase expression and activity in lactating mice. *Journal of Dairy Research, 71*, 1–6.

Small, D. M. (1991). The effects of glyceride structure on absorption and metabolism. *Annual Review of Nutrition, 11*, 413–434.

Specht, B., Bartetzko, N., Hohoff, C., Kuhl, H., Franke, R., Borchers, T., & Spener, F. (1996). Mammary derived growth inhibitor is not a distinct protein but a mix of heart-type and adipocyte-type fatty acid-binding protein. *Journal of Biological Chemistry, 271*, 19943–19949.

Spitsberg, V. L., Matitashvili, E., & Gorewit, R. C. (1995). Association and coexpression of fatty-acid-binding protein and glycoprotein CD36 in the bovine mammary gland. *European Journal of Biochemistry, 230*, 872–878.

Suchyta, S. P., Sipkovsky, S., Halgren, R. G., Kruska, R., Elftman, M., Weber-Nielsen, M., Vandehaar, M. J., Xiao, L., Tempelman, R. J., & Coussens, P. M. (2003). Bovine mammary gene expression profiling using a cDNA microarray enhanced for mammary-specific transcripts. *Physiological Genomics, 16*, 8–18.

Sundvold, H., Brzozowska, A., & Lien, S. (1997). Characterisation of bovine peroxisome proliferator-activated receptors gamma 1 and gamma 2: Genetic mapping and differential expression of the two isoforms. *Biochemical and Biophysical Research Communications, 239*, 857–861.

Viswanadha, S., Giesy, J. G., Hanson, T. W., & McGuire, M. A. (2003). Dose response of milk fat to intravenous administration of the *trans*-10, *cis*-12 isomer of conjugated linoleic acid. *Journal of Dairy Science, 86*, 3229–3236.

Viturro, E., Farke, C., Meyer, H. H., & Albrecht, C. (2006). Identification, sequence analysis and mRNA tissue distribution of the bovine sterol transporters ABCG5 and ABCG8. *Journal of Dairy Science, 89*, 553–561.

Wakil, S. J. (1989). Fatty acid synthase, a proficient multifunctional enzyme. *Biochemistry, 28*, 4523–4530.

Wang, X., Briggs, M. R., Hua, X., Yokoyama, C., Goldstein, J. L., & Brown, M. S. (1993). Nuclear protein that binds sterol regulatory element of low density lipoprotein receptor promoter. II. Purification and characterization. *Journal of Biological Chemistry, 268*, 14497–14504.

Ward, R. J., Travers, M. T., Richards, S. E., Vernon, R. G., Salter, A. M., Buttery, P. J., & Barber, M. C. (1998). Stearoyl-CoA desaturase mRNA is transcribed from a single gene in the ovine genome. *Biochimica et Biophysica Acta, 1391*, 145–156.

West, C. E., Bickerstaffe, R., Annison, E. F., & Linzell, J. L. (1972). Studies on the mode of uptake of blood triglycerides by the mammary gland of the lactating goat. The uptake and incorporation into milk fat and mammary lymph of labelled glycerol, fatty acids and triglycerides. *Biochemical Journal, 126*, 477–490.

Wheeler, T. T., Broadhurst, M. K., Rajan, G. H., & Wilkins, R. J. (1997). Differences in the abundance of nuclear proteins in the bovine mammary gland throughout the lactation and gestation cycles. *Journal of Dairy Science, 80*, 2011–2019.

Williams, C. M. (2000). Dietary fatty acids and human health. *Annales de Zootechnie, 49*, 165–180.

Winter, A., Kramer, W., Werner, F. A., Kollers, S., Kata, S., Durstewitz, G., Buitkamp, J., Womack, J. E., Thaller, G., & Fries, R. (2002). Association of a lysine-232/alanine polymorphism in a bovine gene encoding acyl-CoA:diacylglycerol acyltransferase (DGAT1) with variation at a quantitative trait locus for milk fat content. *Proceedings of the National Academy of Sciences USA, 99*, 9300–9305.

Yang, J., Kennelly, J. J., & Baracos, V. E. (2000a). The activity of transcription factor Stat5 responds to prolactin, growth hormone, and IGF-I in rat and bovine mammary explant culture. *Journal of Animal Science, 78*, 3114–3125.

Yang, J., Kennelly, J. J., & Baracos, V. E. (2000b). Physiological levels of Stat5 DNA binding activity and protein in bovine mammary gland. *Journal of Animal Science, 78*, 3126–3134.

Yang, T., Espenshade, P. J., Wright, M. E., Yabe, D., Gong, Y., Aebersold, R., Goldstein, J. L., & Brown, M. S. (2002). Crucial step in cholesterol homeostasis: Sterols promote binding of SCAP to INSIG-1, a membrane protein that facilitates retention of SREBPs in ER. *Cell, 110*, 489–500.

Yeldandi, A. V., Rao, M. S., & Reddy, J. K. (2000). Hydrogen peroxide generation in peroxisome proliferator-induced oncogenesis. *Mutation Research, 448*, 159–177.

Yonezawa, T., Yonekura, S., Kobayashi, Y., Hagino, A., Katoh, K., & Obara, Y. (2004a). Effects of long-chain fatty acids on cytosolic triacylglycerol accumulation and lipid droplet formation in primary cultured bovine mammary epithelial cells. *Journal of Dairy Science, 87*, 2527–2534.

Yonezawa, T., Yonekura, S., Sanosaka, M., Hagino, A., Katoh, K., & Obara, Y. (2004b). Octanoate stimulates cytosolic triacylglycerol accumulation and CD36 mRNA expression but inhibits acetyl coenzyme A carboxylase activity in primary cultured bovine mammary epithelial cells. *Journal of Dairy Research, 71*, 398–404.

Yu, L., Li-Hawkins, J., Hammer, R. E., Berge, K. E., Horton, J. D., Cohen, J. C., & Hobbs, H. H. (2002). Overexpression of ABCG5 and ABCG8 promotes biliary cholesterol secretion and reduces fractional absorption of dietary cholesterol. *Journal of Clinical Investigation, 110*, 671–680.

Zhang, L., Ge, L., Parimoo, S., Stenn, K., & Prouty, S. M. (1999). Human stearoyl-CoA desaturase: Alternative transcripts generated from a single gene by usage of tandem polyadenylation sites. *Biochemical Journal, 340 (Pt 1)*, 255–264.

Lipophilic Microconstituents of Milk

Antonella Baldi and Luciano Pinotti

Abstract Milk has long been recognized as a source of macro- and micronutrients, immunological components, and biologically active substances, which not only allow growth but also promote health in mammalian newborns. Many milk lipids, lipid-soluble substances, and their digested products are bioactive, including vitamins and vitamin-like substances. Vitamins A, E, D, and K and carotenoids are known as highly lipophilic food microconstituents (HLFMs), and all occur in milk. HLFMs also include phytosterols, which, although they are not vitamins, are nevertheless biologically active and present in milk. Fat-soluble micronutrients, including fat-soluble vitamins, are embedded in the milk fat fraction, and this has important implications for their bioaccessibility and bioavailability from milk. In fact, the fat component of milk is an effective delivery system for highly lipophilic microconstituents. The vitamin content of animal products can be enhanced by increasing the feed content of synthetic or natural vitamins or precursors. An advantage of augmenting milk microconstituents by animal nutrition rather than milk fortification is that it helps safeguard animal health, which is a primary factor in determining the quality, safety, and wholesomeness of animal-origin foods for human consumption. The milk fat delivery system offers numerous possibilities for exploitation by nutritionists. For example, the payload could consist of enhanced levels of several micronutrients, opening possibilities for synergic effects that are as yet incompletely understood.

A. Baldi
Department of Veterinary Sciences and Technology for Food Safety, University of Milan, Via Trentacoste 2-20134 Milan, Italy
e-mail: antonella.baldi@unimi.it

Introduction

In recent years consumers have begun to look at foods not just for basic nutritional requirements, but also for health benefits. As a result, the concepts of "functional foods" and "nutraceuticals" have been developed, which focus on foods or the bioactive components of foods that promote health and well-being.

Milk is a remarkable source of macro- and micronutrients, immunological components, and biologically active substances, which not only allow growth but also promote health in mammalian newborns. Many milk lipids, lipid-soluble substances, and their digested products are bioactive, including triacylglycerides, diacylglycerides, saturated and polyunsaturated fatty acids, phospholipids, vitamins, and vitamin-like substances. Vitamins A, E, D, and K and carotenoids are known as highly lipophilic food microconstituents (HLFMs), meaning that their octanol–water partition coefficients (log pc) are greater than 8. HLFMs also include phytosterols (Borel, 2003), which, although they are not vitamins, are nevertheless biologically active. Small quantities of phytosterols are present in milk (Brewington et al., 1970; Goudjil et al., 2003). In what follows, we consider the roles of this broader category of milk microconstituents.

Milk as a Source of Highly Lipophilic Microconstituents

It has long been known that milk fat is a major source of lipid-soluble vitamins. The lipid-soluble vitamin composition of various milks and some dairy products is shown in Table 1. In bovine milk, the fat-soluble vitamin content is known to vary with breed, parity, physiological state (e.g., pregnancy, lactation), production level, and health status (Baldi, 2005; McDowell, 1989; Nozière et al., 2006a, b). Other factors, such as nutritional state and amount and type of forage, can affect the vitamin E and A (and β-carotene) content, while the vitamin D and K content of cow's milk is influenced by the animal's exposure to direct sunlight, the quantity of sun-cured forage in the diet, and the functional state of the rumen (McDowell, 1989, 2006).

The vitamin A (retinol) concentration of bovine milk ranges from 0.28 to 0.92 mg/L (Lindmark-Månsson & Åkesson, 2000). Vitamin A is mainly present in milk in esterified form. The mammary gland takes up retinol derived from the liver, esterifies it, and outputs it to milk (Tomlinson et al., 1974). Other sources of milk vitamin A are retinol esters derived from dietary β-carotene and dietary retinol.

Milk also contains carotenoids, mainly β-carotene (see below), which yield vitamin A by cleavage of the centrally located double bond. Because there are losses during β-carotene absorption and conversion, normally 6 μg of β-carotene are required to yield 1 μg of retinol equivalent. However, absorption of β-carotene from milk is particularly efficient; only 2 μg of β-carotene

Table 1 Fat-Soluble Vitamin Content of Some Dairy Products

	Carotene[a] (mg/100 g)	Vitamin A (mg/100 g)	Vitamin D (µg/100 g)	Vitamin E (mg/100 g)	Vitamin K (µg/100 g)
Bovine milk, raw	0.018	0.030	0.06	0.13	0.3
Human milk	0.003	0.054	0.07	0.28	0.5
Butter	0.380	0.590	1.20	2.20	7
Cheese					
Emmental	0.120	0.270	1.10	0.53	3
Camembert (60% fat)	0.290	0.500	–	0.77	–
Camembert (30% fat)	0.100	0.200	0.17	0.30	–

Adapted from Belitz et al. (2004).
[a]All carotenoids with pro-vitamin A activity.

are required to yield 1 µg of retinol equivalent [Gurr, 1995; Institute of Medicine (IOM), 2001].

Vitamin A plays a central role in many essential biological processes including vision, growth and development, immunity, and reproduction (Debier & Larondelle, 2005; McDowell, 1989). Although vitamin A does not have major chain-breaking activity, it can function as a scavenger of singlet oxygen and may also react with other reactive oxygen species (Baldi et al., 2006). Vitamin A in milk is essential for the newborn, and any deficiency during gestation and lactation may adversely affect health, growth, and development. Vitamin A is known to protect epithelia by acting as a cross-linking agent between lipid and proteins within the lipid bilayer.

Vitamin A has also been suggested to play a role in the morphogenesis, differentiation, and proliferation of the mammary gland, probably via interactions with growth factors (Meyer et al., 2005). *In vitro*, retinol acid and retinoic acid have been reported as potent inhibitors of bovine mammary epithelial cell proliferation (Cheli et al., 2003). Retinol is a strong inhibitor of xenobiotic oxidations catalyzed by various isoenzymes of cytochrome P450. At the cellular level, ochratoxin cytotoxicity in mammalian cells is related to increasing cell damage caused by reactive oxygen species but can be limited by medium supplementation with all-*trans* retinol (Baldi et al., 2004). Although the modes of action of retinol at the cellular level remain incompletely understood, the main protective effects of retinol may be due to regulatory effects on the growth of normal cells mediated by control of the gene expression of growth factors (Blomhoff & Blomhoff, 2006; Cheli et al., 2003).

Carotenoids in cow's milk consist mainly of all-*trans*-β-carotene (75–90% of total milk carotenoids), with lutein, zeaxanthin, and β-cryptoxanthin as minor constituents (Havemose et al., 2004; HulShof et al., 2006). The range of β-carotene concentration in cow's milk is 0.05–0.20 mg/L (Lindmark-Månsson

& Åkesson, 2000; Nozière et al., 2006a), which is higher than that reported for human milk (0.01 mg/L), probably due to differences in diet and digestive physiology. In spite of this, the total carotenoids content of human milk (around 0.063 mg/L according to Macias & Schweigert, 2001) is similar to that of bovine milk, because five carotenoids (lutein, cryptoxanthin, α-carotene, β-carotene, lycopene, and their isomers) contribute in more equitable measure to the total in human milk (Lindmark-Månsson & Åkesson, 2000; Macias & Schweigert, 2001).

HulShof et al. (2006) reported that raw bovine milk (4.4% fat) contains on average 0.40 mg/L of retinol and 0.20 mg/L of carotenoids. Full-fat milk (standardized to 3.5% fat) and semiskimmed milk (standardized to 1.5% fat) contain 0.34 and 0.14 mg/L of retinol and 0.18 and 0.9 mg/L of carotenoids, respectively. Based on this study (HulShof et al., 2006), the vitamin A activity of full-fat cow's milk, calculated as retinol equivalents (RE) per gram of fat, is 12.3 µg/g, assuming a β-carotene-to-retinol bioconversion ratio of 1:0.5 (IOM, 2001). This value is 20% higher than RE values published in several national nutrient databases (HulShof et al., 2006).

Carotenoids are important as precursors of vitamin A and as components of the antioxidant network; however, they are also involved in cell communication, immune function, and fertility (Chew & Park, 2004; Michel et al., 1994; Nozière et al., 2006a; Stahl & Sies, 2005). It has also been shown that dietary carotenoids have biological effects on signaling pathways and that they or their metabolites influence the expression of certain genes (Stahl et al., 2002). They also seem to inhibit certain regulatory enzymes involved in carcinogenesis, which would explain their reported cancer-preventative properties (Stahl & Sies, 2005). Carotenoids function in milk in combination with vitamin E, mainly as antioxidants, although other protective activities cannot be excluded.

The vitamin E content in cow's milk has been reported to vary between 0.2 and 1.0 mg/L of α-tocopherol (Jensen, 1995) depending on dietary regimen and other factors as discussed below. α-Tocopherol is the main form of vitamin E present in cow's milk, representing 84–92% of the total, while γ-tocopherol and α-tocotrienol contribute roughly 5% each. Kaushik and co-workers (2001) noted that the α-tocotrienol content of raw and commercial milks ranged from 17.6 µg/L for whole milk to 1.4 µg/L for nonfat milk.

Vitamin E's role as an antioxidant able to prevent free radical-mediated tissue damage, and hence prevent or delay the development of degenerative and inflammatory conditions, has been extensively investigated (Baldi et al., 2006; McDowell, 1989; Weiss & Spears, 2006). As an antioxidant, tocopherol helps maintain the integrity of fat globule membranes in milk (Atwal et al., 1990; Charmley et al., 1993; Nicholson & St-Laurent, 1991). It has been reported that bovine milk is susceptible to auto-oxidation when the level of vitamin E falls below 20 µg/g of fat (Atwal et al., 1990). The vitamin E family includes tocotrienols, which possess powerful neuroprotective, antioxidant, anticancer, and cholesterol-lowering properties in their own right and in fact

Table 2 Vitamins A and E and β-Carotene Content of Colostrum, Transition Milk, and Mature Milk

	Colostrum (First Milking)	Transition Milk (Fifth Milking)	Mature Milk
Vitamin A, IU/kg	9,834	2,455	1,126
Relative value	*100%*	*25%*	*11%*
β-Carotene, mg/kg	88.18	38.58	0.066
Relative value	*100%*	*44%*	*0.07%*
Vitamin E, IU/kg	9.03	3.75	1.10
Relative value	*100%*	*41%*	*12%*

Adapted from Seymour (2002).

are now thought to have more potent antioxidant properties than α-tocopherol (Sen et al., 2006).

Bovine colostrum contains higher amounts of vitamins A and E and β-carotene than milk: levels decline over about four days to those typical of mature milk (Table 2; Debier et al., 2005; Seymour, 2002; Zanker et al., 2000).

The vitamin D content in cow's milk is quite low (Table 1); as a consequence, milk and dairy products are fortified with vitamin D in some countries, including Canada (Calvo et al., 2004; Lamberg-Allardt, 2006). Liquid (fluid) milk in Canada is labeled as providing 44% of the recommended daily intake of vitamin D (400 IU) per 250-mL serving. Other Canadian milk products that require vitamin D fortification are evaporated milk, powdered milk, and goat's milk. Fortified milk may be used in food manufacturing (e.g., yogurt), but industrial milk used for baking and products such as soft and hard cheeses is not usually fortified. In the United States, vitamin D fortification is generally optional, the exception being "fortified" milk (Calvo et al., 2004). The risks and prevalence of vitamin D deficiency in European countries are well documented. Policies on food fortification and vitamin D supplementation have recently been revised by several European countries and the European Union (Tylavsky et al., 2006). Vitamin D has been added to liquid milk products, margarines, and butters in Finland since 2003 (Lamberg-Allardt, 2006). There are indications that good vitamin D status may be associated with benefits not directly related to bone mineralization (Hendy et al., 2006).

As is the case for vitamin D, bovine milk is not a particularly good source of vitamin K (3 µg/L; Jensen, 1995). Vitamin K was first identified as an essential factor in blood coagulation, but it has emerged recently that this substance may have protective actions against osteoporosis, atherosclerosis, and hepatocarcinoma (Kaneki et al., 2006). Accumulated evidence indicates that subclinical nonhemostatic vitamin K deficiency in extrahepatic tissues, particularly bone and the vasculature, is widely prevalent in adult populations (Kaneki et al., 2006). When co-administered with vitamin D, vitamin K may have a favorable effect on bone density (Weber, 2001). These findings, and the fact that dietary reference intakes have recently been increased by 50% in the United States (Weber, 2001), have renewed interest in vitamin K. Since milk and dairy products constitute an effective delivery system (see below), it is expected that they may be fortified with vitamin K in the near future.

The fat-soluble vitamins and β-carotene from milk contribute variously to total intakes in adults. While dairy products provide more than 20% of the total daily intake of vitamin A in Western diets (Gurr, 1995; Beitz, 2005; HulShof et al., 2006), they provide only about 2–2.5% of the total daily intake of vitamin E and β-carotene (Beitz, 2005); no similar information appears to be available for vitamins D and K. Although fruit and vegetables are the known major sources of vitamins A and E and carotenoids in human diets, increasing the levels of those compounds in milk appears an attractive way of increasing milk value and quality. Furthermore, because of the wide variety of available milk products and their high consumption, these products appear as an excellent matrix for new and functional products whose consumption may have a significant impact on public health (Herrero et al., 2002, 2006).

As noted, HLFMs include plant-derived sterols or phytosterols (Borel, 2003), of which there are two main structural types: sterols (principally sitosterol, campesterol, and stigmasterol) and stanols (principally sitostanol and campestanol). Phytosterols occur in milk as part of the sterol fraction. The main milk sterol is cholesterol (3 mg/g fat, equivalent to 100 mg/L cow's milk), while small quantities of other sterols (7-dehydrocholesterol, 22-dehydrocholesterol, ergosterol, fucosterol, lanosterol, lathosterol, 24-methylenecholesterol) as well as several phytosterols are present (Brewington et al., 1970; Walstra & Jennes, 1984). The International Dairy Federation (1992) noted that these phytosterols constitute less than <1% of milk fat. The main plant-derived sterol in ruminant milk is β-sitosterol, also called sitosterol (Brewington et al., 1970; Goudjil et al., 2003).

Phytosterols have received much attention because of their cholesterol-lowering properties, although the exact mechanism by which they decrease serum cholesterol is not well understood. They may promote cholesterol precipitation in the gut so it is not absorbed, or they may compete with cholesterol for micellar solubilization, limiting cholesterol absorption (Moreau et al., 2002). With regard to milk, recent studies (Noakes et al., 2005; Ortega et al., 2006) have indicated that novel cholesterol-lowering, low-fat dairy products containing phytosterols can be developed, expanding the food product alternatives for consumers (Ortega et al., 2006). These foods are perceived as nutritious and healthy and can easily be integrated into a heart-healthy diet, helping to maintain desirable cholesterol levels or providing an additional dietary option to help lower cholesterol levels (Noakes et al., 2005).

Milk as a Vehicle for Highly Lipophilic Food Microconstituents

In general, the fat-soluble vitamin content in milk depends on the milk fat content (Debier et al., 2005; Kaushik et al., 2001). Modifying the fat content of dairy products by reducing total fat or certain types of fat alters the fat-soluble vitamin because these compounds are mainly associated with the milk fat globule (Jensen & Nielsen, 1996; Zahar & Smith, 1995). In fact, vitamin A and carotenoids are almost entirely confined to the core and membrane of the fat globule, with negligible quantities in the serum (Mulder & Walstra, 1974; Walstra & Jenness, 1984). This distribution is affected by milk fat globule dimensions so that the retinol content of milk and dairy products depends on the quantity of fat globule membrane per gram of fat, which is inversely proportional to the globule diameter (Zahar & Smith, 1995). Bovine milk fat globules acquire carotene during formation in the mammary cells, and a minor fraction may also be extracted from the enveloping secretory membrane (Patton et al., 1980).

With regard to vitamin E, greater fat content of milk products is also associated with higher levels of the vitamin: For every 1-gram increase in total lipids, the α-tocopherol content increases by 17 µg. In this case, however, the vitamin is mainly present in the milk fat globule membrane and not the core (Jensen & Nielsen, 1996). The α-tocopherol content of milk products also varies with cholesterol content (Kaushik et al., 2001): every 1-mg increase in cholesterol is associated with a 1-µg increase in α-tocopherol (Kaushik et al., 2001).

The fact that fat-soluble micronutrients—and fat-soluble vitamins in particular—are embedded in the milk fat fraction has important implications for their bioaccessibility and bioavailability from milk. German and Dillard (2006) introduced the concept of milk fat as an excellent *nutrient delivery medium,* particularly for fat-soluble vitamins. This concept is consistent with Hayes and co-workers' study (2001), which found that absorption of vitamin E into the human bloodstream, when microdispersed in milk, is considerably more efficient (two- to threefold) than from orange juice or vitamin E capsules. They suggested that the "inherent chemistry of milk" was important in increasing vitamin E bioavailability from milk (Hayes et al., 2001) and that proteins or peptides produced during milk digestion and absorption promote α-tocopherol uptake.

The concept of milk as a nutrient delivery system seems particularly relevant when milk is the unique food for newborns. It is known that colostrum intake within the first 24 hours of life is essential for supplying the carotene, retinol, and α-tocopherol necessary for the first week of life of calves. Vitamin absorption by the neonate from colostrum is, in fact, highly efficient to help ensure that a sufficient amount of these micronutrients meets the needs of rapidly growing cells, tissues, and developing organ systems (Blum et al., 1997). This efficient transfer presumably compensates for rather limited placental transfer during gestation.

Analogous considerations also seem to apply to phytosterols. It has been suggested, for example, that milk fat globule membranes altered by acid or microbial action in yogurts may adsorb sterols differently to a native membrane and thus become an effective delivery system for phytosterols (Mensink et al., 2002; Volpe et al., 2001). Clifton and co-workers (2004) demonstrated that phytosterols are almost three times more effective when added to low-fat milk than when added to bread or cereal. These studies tested milk as a delivery vehicle for phytosterols added to milk postharvest. Naturally occurring phytosterol effects on milk and their transfer from feed to milk have not been extensively investigated. At least one study (Gulati et al., 1978) investigated the effects of protected β-sitosterol supplementation in lactating dairy goats and cows. This study noted that β-sitosterol did not cause the drop in milk fat content that occurred when protected cholesterol was added to the diet.

Fat-Soluble Vitamins in Animal Nutrition and Milk

The main way to enhance the vitamin content of animal products is to increase the feed content of synthetic or natural vitamins or precursors (Sahlin & House, 2006). Much research has been done on vitamin E supplementation to dairy cows mainly to sustain animal health and production rather than to increase the vitamin content of milk products. In addition to its role in preventing free radical-mediated tissue damage (Allison & Laven, 2001; Baldi, 2005; Van Metre & Callan, 2001; Weiss & Spears, 2006), vitamin E is involved in immune system function, and supplementation with supranutritional levels of the vitamin can improve immune responses in some cases (Baldi, 2005; Hogan et al., 1992, 1993; Politis et al., 1995, 2001; Weiss & Spears, 2006). Thus, deficiencies in vitamin E or selenium have been associated with high somatic cell counts (SCC) in cows' milk and also with increased incidence and severity of intramammary infections and mastitis. The positive effects of vitamin E on SCC depend on adequate dietary levels of selenium. Supplementation with vitamin E when dietary selenium is adequate significantly reduces the incidence of intramammary infection and clinical mastitis (Smith et al., 1984).

These benefits of vitamin E supplementation, particularly in the context of the much-reduced use of fresh (vitamin E-rich) forage in dairy cow nutrition, have led to a substantial increase in recommended intake levels for this animal. In 1989, the NRC (National Research Council, 1989) recommendation was 15 IU of vitamin E (as racemic tocopheryl acetate) per kg of dry matter intake (DMI). At present, the recommendation is 80 IU/kg DMI in the dry and immediate postpartum periods, and about 20 IU/kg DMI during lactation, in view of the higher total DMI intake at that time (National Research Council, 2001). Vitamin E intake is generally considered adequate when α-tocopherol plasma levels are in the range of 3–3.5 mg/L. No further benefits are observed above these levels.

Vitamin E supplementation may not always be effective, however, particularly during the peripartum period, in which it is established that plasma vitamin E levels fall significantly in the dairy cow and it is difficult to maintain levels that are considered adequate. The liver plays a central role in the release of α-tocopherol into the circulation and in transfer to peripheral tissues. This function requires hepatic α-tocopherol transfer protein (TTP), which incorporates the vitamin into nascent very low-density lipoproteins. TTP discriminates against tocopherol homologues and also prefers the natural stereoisomer. The greater biological activity of the natural isomer compared to synthetic isomers is probably due to TTP (Burton, 1994; Burton et al., 1998; Lauridsen et al., 2002).

The form (natural vs. synthetic) of administration can affect vitamin E bioavailability and may also influence transfer to milk (Baldi et al., 1997; Bontempo et al., 2000; Hidiroglou, 1996) (Figure 1). A recent study investigated the distribution of α-tocopherol stereoisomers in milk in relationship to the supplementation of various forms of vitamin E (natural and synthetic) to periparturient dairy cows (Meglia et al., 2006). Supplementation with RRR-α-tocopheryl acetate resulted in better plasma vitamin E status compared to supplementation with all-rac-α-tocopheryl acetate, RRR-α-tocopherol (free alcohol), and no supplementation. Moreover, irrespective of the form of supplementation, the bioavailability of the RRR stereoisomer was greatest, and this form was enriched in milk (over 86% of the total). Data from this study (Meglia et al., 2006) indicated that vitamin E transfer from feed to milk was about 1.6–2.2%, which is consistent with the transfer rate reported in other supplementation studies (Atwal et al., 1990; Allison & Laven, 2001; Baldi et al., 2000; Weiss & Wyatt, 2003; Weiss 2005).

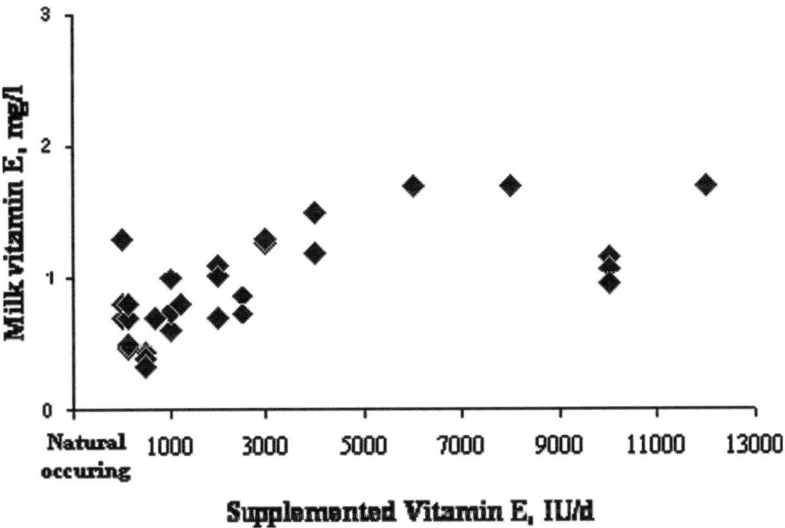

Fig. 1 Relationship between level of vitamin E supplementation in dairy cows and the vitamin E content of cow's milk. [Data from various studies; adapted from Allison and Laven (2001), Baldi et al. (2000), Bell et al. (2006), Havemose et al. (2006), and Weiss and Wyatt (2003).]

Variations in levels of other fat-soluble micronutrients have been reported in cow's plasma and milk following changes in forage and feeding level (Havemose et al., 2004, 2006; Nozière et al., 2006a, b) (Figure 2). Nozière et al. (2006b) determined the kinetics of the decrease in carotenoids in plasma, milk, and adipose tissue after switching from a high- to a low-carotenoid diet. The data indicated that uptake of plasma β-carotene by the mammary gland was dependent on lipoprotein lipase (LPL) activity, as also reported for α-tocopherol in rats (Martinez et al., 2002).

However, dietary manipulation can change the vitamin content of the milk only within certain limits. Jensen and co-workers (1999) measured the maximum secretory capacity to milk (V_{max}) of α-tocopherol and β-carotene. Mean V_{max} values were 32.4 and 2.5 mg/day for α-tocopherol and β-carotene, respectively. Thus, the daily secretion of α-tocopherol and β-carotene is limited and also found to be independent of milk yield and milk fat content.

Vitamin A partitioning and trafficking differ from those of vitamin E and carotenoids. Vitamin A in milk (present mainly as esters) originates by esterification of retinol (from the liver) in the mammary gland, and also from uptake of vitamin A esters from the diet. Conversion of β-carotene to retinol may also occur in the mammary gland of lactating dairy cows (Nozière et al., 2006b). Thus, transfer from plasma to milk will depend on the contributions of these sources to the total plasma pool (Nozière et al., 2006a). The uptake of dietary vitamin A ester packed in chylomicrons acts directly on the vitamin A content of the milk, as they are taken up by the mammary gland. Lipoprotein lipases (LPL) in lactating mammary tissue may be responsible for the hydrolysis of chylomicron-derived retinol ester, allowing retinol uptake by the gland. This process would explain why milk vitamin A concentrations can vary when

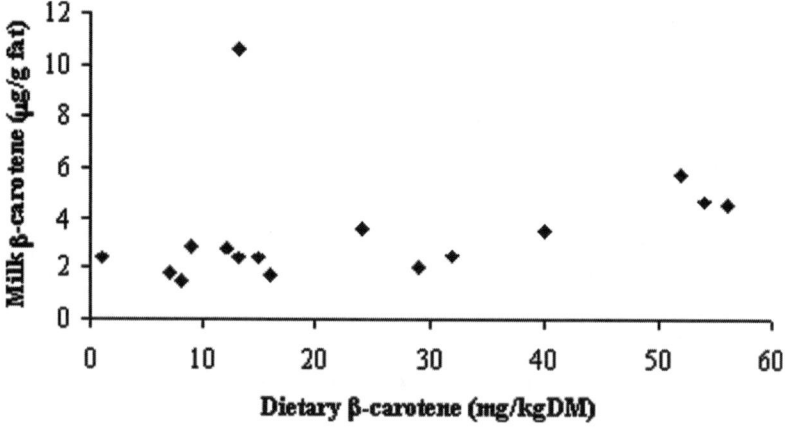

Fig. 2 Relationship between β-carotene of feed and β-carotene content of the milk fat fraction of cow's milk. [Data from various studies; adapted from Nozière et al. (2006a) and Havemose et al. (2006).]

plasma levels in the lactating animal do not change (see Debier & Larondelle, 2005, for references). By contrast, transfer from blood to milk of liver retinol (alcohol), which is secreted into the circulation bound to its specific transport protein, the retinol-binding protein (RBP), seems unaffected by vitamin A ingestion, explaining why uptake by the mammary gland does not increase with increasing dietary intake of non-esterified vitamin A. Esterification of retinol is necessary for its uptake by the mammary gland; this appears to be regulated mainly by acyl CoA-retinol acyltransferase. Considered together, these factors contribute to explaining not only why vitamin A does not increase with milk lipids, which increase as lactation progress (Debier & Larondelle, 2005), but also why its level in milk is so difficult to change by dietary manipulation.

Effects of Lipophilic Microconstituents on Milk Lipids

Vitamin E and β-carotene are highly lipophilic and, as such, not only are associated within the milk fat fraction but can also have notable effects on those fats. One such effect is that of inhibiting lipid oxidation. Originally, vitamin E supplementation to lactating dairy cows was increased in order to maintain animal health and reduce the SCC (Baldi et al., 2006). However, studies in the 1990s (Atwal et al., 1990; Charmley et al., 1993; Charmley & Nicholson, 1994; Nicholson & St-Laurent, 1991) indicated that vitamin E supplementation to the animal could help slow lipid peroxidation in milk. Bovine milk is susceptible to auto-oxidation when the level of vitamin E falls below 20 µg/g of fat (Atwal et al., 1990), leading to "oxidized" flavors variously described as cardboard-like, metallic, or tallow-like. Usually, such flavors develop after some time and in association with improper storage, but sometimes they can develop soon after milking (Weiss, 2005). Although vitamin E transfer from diet to milk is low, supra-nutritional supplementation can increase the vitamin content of milk. Thus, vitamin E supplementation at 2,000 IU/day to transition cows raises the α-tocopherol milk content by about 40% compared to supplementation at 1,000 IU/day (Baldi et al., 2000).

As discussed in depth by Chilliard in this volume, nutritionists are attempting to increase the content of bioactive lipids in milk in order to obtain "functional milks," which may benefit human health (Bauman et al. 2006; Chilliard et al., 2001). The best-known milk fatty acids thought to benefit human health are butyric acid, oleic acid, and C18 to C22 polyunsaturated fatty acids (particularly conjugated linoleic acids). Highly polar lipids, in particular sphingiolipids and their derivatives, have also come under scrutiny for their supposed functional qualities (Rombaut & Dewettinck, 2006). Specific animal feeding regimes can increase the polyunsaturated lipid content of milk with the aim of making it healthier, but at the same time the milk becomes more vulnerable to oxidation (Havemose et al., 2004). It is therefore important to protect milks high in polyunsaturates during processing so that their nutritional and organoleptic qualities are not impaired, and they do not become a source of pro-oxidants when consumed. In this connection, it has also been shown that the uptake of plasma vitamin E by the mammary gland of dairy cows increases when diets enriched in polyunsaturated fatty acids (oxidative stress-inducing) are fed (Durand et al., 2005). These findings have stimulated much interest in milk antioxidants and their transfer from dietary components.

Available data on the amount of dietary vitamin E required to prevent oxidized flavors and ensure the oxidative stability of milk, whose polyunsaturated fat content has been increased by dietary intervention, are inconclusive. Weiss (2005) suggested supplementing by at least 3,000 IU of vitamin E per day, when oxidized flavor is a problem. By contrast, Havemose and colleagues' data (2004, 2006) indicate that tocopherols and carotenoids in milk do not prevent oxidation of polyunsaturated lipids, although they can delay protein oxidation. Note, however, that these latter studies used dietary manipulation (e.g., grass silage vs. corn silage) to vary the amount of natural antioxidant in milk of lactating cows (i.e., there was no dietary supplementation).

Regardless of the effects of α-tocopherol and β-carotene on the oxidative stability of the milk when it contains increased levels of unsaturated fatty acids, high levels of these antioxidants seem important for maintaining milk quality and safety in general. The main advantage of increasing these microconstituents of milk by animal nutrition rather than by milk fortification is that they also safeguard the health of the animal, a primary factor in determining the quality, safety, and wholesomeness of foods of animal origin for human consumption. Furthermore, this "feed-to-food" approach makes it possible to reposition animal products as key foods for the delivery of important nutrients into the human diet.

Conclusions

Milk has a longstanding tradition of safety and is widely accepted as a food that promotes normal growth and development. As this chapter has shown, the fat component of milk is an effective delivery system for highly lipophilic microconstituents such as fat-soluble vitamins and phytosterols. This delivery system offers numerous possibilities that can be exploited by nutritionists. For example, the payload could consist of enhanced levels of several micronutrients opening possibilities for synergic effects that are as yet incompletely understood. Although vitamin E is best known as an antioxidant, other properties are emerging such as modulating effects on various signaling cascades at the cellular level by inhibition of protein kinase C (Azzi et al., 2000; Brigelius-Flohé et al., 2002). Carotenoids are also implicated in cell signaling (Stahl & Sies, 2005), while all-*trans*-retinoic acid is known to regulate the expression of several hundred genes through binding to nuclear transcription factors (Blomhoff & Blomhoff, 2006). An attractive way of enhancing levels of these health-promoting microconstituents in milk is by interventions at the level of animal nutrition to thereby obtain "natural fortification" of milk and dairy products that enhances the public perception of their value.

References

Allison, R. D., & Laven, R. A. (2001). Vitamin E for milk production in dairy cows: A review. *Nutrition Abstracts and Reviews, Series B: Livestock Feeds and Feeding, 71*, 43R–51R.

Atwal, A. S., Hidiroglou, M., Kramer, J. K. G., & Binns, M. R. (1990). Effects of feeding α-tocopherol and calcium salts of fatty acids on vitamin E and fatty acid composition of cow's milk. *Journal of Dairy Science, 73*, 2832–2841.

Azzi, A., Breyer, I., Feher, M., Pastori, M., Ricciarelli, R., Spycher, S., Staffieri, M., Stocker, A., Zimmer, S., & Zingg, J.-M. (2000). Specific cellular responses to α-tocopherol. *Journal of Nutrition, 130*, 1649–1652.

Baldi, A. (2005). Vitamin E in dairy cows. *Livestock Production Science, 98*, 117–122.

Baldi, A., Bontempo, V., Cheli, F., Carli, S., Sgoifo Rossi, C., & Dell'Orto, V. (1997). Relative bioavailability of vitamin E in dairy cows following intraruminal administration of three different preparations of DL-α-tocopheryl acetate. *Veterinary Research, 28*, 512–524.

Baldi, A., Savoini, G., Pinotti, L., Monfardini, E., Cheli, F., & Dell'Orto, V. (2000). Effects of vitamin E and different energy sources on vitamin E status, milk quality and reproduction in transition cows. *Journal of Veterinary Medicine Series A, 47*, 599–608.

Baldi, A., Losio, M. N., Cheli, F., Rebucci, R., Sangalli, L., Fusi, E., Bertasi, B., Pavoni, E., Carli, S., & Politis, I. (2004). Evaluation of the protective effects of α-tocopherol and retinol against ochratoxin A cytotoxicity. *British Journal of Nutrition, 91*, 507–512.

Baldi, A., Pinotti, L., & Fusi, E. (2006). Influence of antioxidants on ruminant health. *Feed Compounder, 26*, 19–25.

Bauman, D. E., Lock, A. L., Corl, B. A., Ip, C., Salter, A. M., & Parodi, P. W. (2006). Milk fatty acids and human health: Potential role of conjugated linoleic acid and *trans* fatty acids. In K. Sejrsen, T. Hvelplund, & M. O. Nielsen (Eds.), *Ruminant Physiology. Digestion, Metabolism and Impact of Nutrition on Gene Expression, Immunology and Stress* (pp. 529–561). Wageningen, The Netherlands: Wageningen Academic Publishers.

Beitz, D. C. (2005). Contributions of animal products to healthy diets. In *Proceedings 2005 Cornell Nutrition Conference for Feed Manufacture* (pp. 117–126). Ithaca, NY: Cornell University Press.

Belitz, H. D., Grosch, W., & Schieberle, P. (2004). Vitamins. In M. M. Burghagen (Ed.), *Food Chemistry* (pp. 409–426). Berlin: Springer.

Bell, J. A., Griinari, J. M., & Kennelly, J. J. (2006). Effect of safflower oil, flaxseed oil, monensin, and vitamin E on concentration of conjugated linoleic acid in bovine milk fat. *Journal of Dairy Science, 89*, 733–748.

Blomhoff, R., & Blomhoff, H. K. (2006). Overview of retinoid metabolism and function. *Journal of Neurobiology, 66*, 606–630.

Blum, J. W., Hadorn, U., Sallmann, H. P., & Schuep, W. (1997). Delaying colostrum intake by one day impairs plasma lipid, essential fatty acid, carotene, retinol and α-tocopherol status in neonatal calves. *Journal of Nutrition, 127*, 2024–2029.

Bontempo, V., Baldi, A., Cheli, F., Fantuz, F., Politis, I., Carli, S., & Dell'Orto, V. (2000). Kinetic behavior of three preparations of α-tocopherol after oral administration to postpubertal heifers. *American Journal of Veterinary Research, 61*, 589–593.

Borel, P. (2003). Factors affecting absorption of highly lipophilic food microconstituents (fat-soluble vitamins, carotenoids and phytosterols). *Clinical Chemistry Laboratory Medicine, 41*, 979–994.

Brewington, C. R., Caress, E. A., & Schwartz, D. (1970). Isolation and identification of new constituents in milk fat. *Journal of Lipid Research, 11*, 355–361.

Brigelius-Flohé, R., Kelly, F. J., Salonen, J. T., Neuzil, J., Zingg, J. M., & Azzi, A. (2002). The European perspective on vitamin E: Current knowledge and future research. *American Journal of Clinical Nutrition, 76*, 703–716.

Burton, G. W. (1994). Vitamin E: Molecular and biological function. *Proceedings of the Nutrition Society, 53*, 251–262.

Burton, G. W., Traber, M. G., Acuff, R. V., Walters, D. N., Kayden, H., Hughes, L., & Ingold, K. U. (1998). Human plasma and tissue α-tocopherol concentrations in response to supplementation with deuterated natural and synthetic vitamin E. *American Journal of Clinical Nutrition, 67*, 669–684.

Calvo, M. S., Whiting, S. J., & Barton, C. N. (2004). Vitamin D fortification in the United States and Canada: Current status and data needs. *American Journal of Clinical Nutrition, 80*, 1710S–1716S.

Charmley, E., & Nicholson, J. W. G. (1994). Influence of dietary fat source on oxidative stability and fatty acid composition of milk from cows receiving a low or high level of dietary vitamin E. *Canadian Journal of Animal Science, 74*, 657–664.

Charmley, E., Nicholson, J. W. G., & Zee, J. A. (1993). Effect of supplemental vitamin E and selenium in the diet on vitamin E and selenium levels and control of oxidized flavor in milk from Holstein cows. *Canadian Journal of Animal Science, 73*, 453–457.

Cheli, F., Politis, I., Rossi, L., Fusi, E., & Baldi, A. (2003). Effects of retinoids on proliferation and plasminogen activator expression in a bovine mammary epithelial cell line. *Journal of Dairy Research, 70*, 367–372.

Chew, B. P., & Park, J. S. (2004). Carotenoid action on immune system. *Journal of Nutrition, 134*, 257–261.

Chilliard, Y., Ferlay, A., & Doreau, M. (2001). Effect of different types of forages, animal fat or marine oils in cow's diet on milk secretion and composition, especially conjugated linoleic acid (CLA) and polyunsaturated fatty acids. *Livestock Production Science, 70*, 31–48.

Clifton, P. M., Noakes, M., Sullivan, D., Erichsen, N., Ross, D., Annison, G., Fassoulakis, A., Cehun, M., & Nestel, P. (2004). Cholesterol-lowering effects of plant sterol esters differ in milk, yoghurt, bread and cereal. *European Journal of Clinical Nutrition, 58*, 503–509.

Debier, C., & Larondelle, Y. (2005). Vitamins A and E: Metabolism, roles and transfer to offspring. *British Journal of Nutrition, 93*, 153–174.

Debier, C., Pottier, J., Goffe, C., & Larondelle, Y. (2005). Present knowledge and unexpected behaviours of vitamins A and E in colostrum and milk. *Livestock Production Science, 98*, 135–147.

Durand, D., Scislowski, V., Chilliard, Y., Gruffat, D., & Bauchart, D. (2005). High fat rations and lipid peroxidation in ruminants; consequences on animal health and quality of products. In J. F. Hocquette & S. Gigli (Eds.), *Indicators of Milk and Beef Quality* (pp. 137–150). Wageningen, The Netherlands: Wageningen Academic Publishers.

German, J. B., & Dillard, C. J. (2006). Composition, structure and absorption of milk lipids: A source of energy, fat-soluble nutrients and bioactive molecules. *Critical Reviews in Food Science and Nutrition, 46*, 57–92.

Goudjil, H., Torrado, S., Fontecha, J., Martínez-Castro, I., Fraga, J. M., & Juárez, M. (2003). Composition of cholesterol and its precursors in ovine milk. *Lait, 83*,153–160.

Gulati, S. K., Cook, L. J., Ashes, J. R., & Scott, T. W. (1978). Effect of feeding protected cholesterol on ruminant milk fat secretion. *Lipids, 13*, 814–819.

Gurr, M. I. (1995). The nutritional significance of lipids. In P. F. Fox (Ed.), *Lipids* (pp. 349–402). London: Chapman & Hall.

Havemose, M. S., Weisbjerg, M. R., Bredie, W. L. P., & Nielsen, J. H. (2004). Influence of feeding different types of roughage on the oxidative stability of milk. *International Dairy Journal, 14*, 563–570.

Havemose, M. S., Weisbjerg, M. R., Bredie, W. L. P., Poulsen, H. D., & Nielsen, J. H. (2006). Oxidative stability of milk influenced by fatty acids, antioxidants, and copper derived from feed. *Journal of Dairy Science, 89*, 1970–1980.

Hayes, K. C., Pronczuk, A., & Perlman, D. (2001). Vitamin E in fortified cow milk uniquely enriches human plasma lipoproteins. *American Journal of Clinical Nutrition, 74*, 211–218.

Hendy, G. N., Hruska, K. A., Mathew, S., & Goltzman, D. (2006). New insights into mineral and skeletal regulation by active forms of vitamin D. *Kidney International, 69*, 218–223.

Herrero, C., Granado, F., Blanco, I., & Olmedilla, B. (2002). Vitamin A and E content in dairy products: Their contribution to the recommended dietary allowances (RDA) for elderly people. *Journal of Nutrition, Health & Aging, 6*, 57–59.

Herrero, C., Olmedilla, B., Granado, F., & Blanco, I. (2006). Bioavailability of vitamins A and E from whole and vitamin-fortified milks in control subjects. *European Journal of Nutrition, 45*, 391–398.

Hidiroglou, M. (1996). Pharmacokinetic profile of plasma tocopherol following intramuscular administration of acetylated alpha-tocopherol to sheep. *Journal of Dairy Science, 79*, 1027–1030.

Hogan, J. S., Weiss, W. P., Todhunter, D. A., Smith, K. L., & Schoenberger, P. S. (1992). Bovine neutrophil responses to parenteral vitamin E. *Journal of Dairy Science, 75*, 340–399.

Hogan, J. S., Weiss, W. P., & Smith, K. L. (1993). Role of vitamin E and selenium in host defence against mastitis. *Journal of Dairy Science, 76*, 2795–2908.

HulShof, P. J. M., van Roekel-Jansen, T., van de Bovenkamp, P., & West, C. E. (2006). Variation in retinol and carotenoid content of milk and milk products in The Netherlands. *Journal of Food Composition and Analysis, 19*, 67–75.

Institute of Medicine (2001). *Dietary Reference Intakes for Vitamin A, Vitamin K, Arsenic, Boron, Chromium, Copper, Iodine, Iron, Manganese, Molybdenum, Nickel, Silicon, Vanadium, and Zinc.* Washington, DC: National Academy Press.

International Dairy Federation (1992). Milk fat and milk fat products. Determination of cholesterol content. Brussels: IDF (FIL-IDF standard no. 159).

Jensen, R. J. (1995). Fat-soluble vitamins in bovine milk. In R. G. Jensen (Ed.), *Handbook of Milk Composition* (pp. 718–726). San Diego: Academic Press.

Jensen, S. K., & Nielsen, K. N. (1996). Tocopherols, retinol, β-carotene and fatty acids in fat globule membrane and fat globule core in cows' milk. *Journal of Dairy Research, 63*, 565–574.

Jensen, S. K., Bjørnbak Johannsen, A. K., & Hermansen, J. E. (1999). Quantitative secretion and maximal secretion capacity of retinol, β-carotene and α-tocopherol into cow's milk. *Journal of Dairy Research, 66*, 511–522.

Kaneki, M., Hosoi, T., Ouchi, Y., & Orimo, H. (2006). Pleiotropic actions of vitamin K: Protector of bone health and beyond? *Nutrition, 22*, 845–852.

Kaushik, S., Wander, R., Leonard, S., German, B., & Traber, M. G. (2001). Removal of fat from cow's milk decreases the vitamin E contents of the resulting dairy products. *Lipids, 36*, 73–78.

Lamberg-Allardt, C. (2006). Vitamin D in foods and as supplements. *Progress in Biophysics and Molecular Biology, 92*, 33–38.

Lauridsen, C., Engel, H., Jensen, S. K., Craig, A. M., & Traber, M. G. (2002). Lactating sows and suckling piglets preferentially incorporate RRR- over all-raca-tocopherol into milk, plasma and tissue. *Journal of Nutrition, 132*, 1258–1264.

Lindmark-Månsson, H., & Åkesson, B. (2000). Antioxidative factors in milk. *British Journal of Nutrition, 84*, S103–S110.

Macias, C., & Schweigert, F. J. (2001). Changes in the concentration of carotenoids, vitamin A, α-tocopherol and total lipids in human milk throughout early lactation. *Annals of Nutrition & Metabolism, 45*, 82–85.

Martinez, S., Barbs, C., & Herrera, E. (2002). Uptake of α-tocopherol by the mammary gland but not by white adipose tissue is dependent on lipoprotein lipase activity around parturition and during lactation in the rat. *Metabolism, 51*, 1444–1451.

McDowell, L. R. (1989). *Vitamins in Animal Nutrition: Comparative Aspects to Human Nutrition*, 1st ed. San Diego: Academic Press.

McDowell, L. R. (2006). Vitamin nutrition of livestock animals: Overview from vitamin discovery to today. *Canadian Journal of Animal Science, 86*, 171–179.

Meglia, G. E., Jensen, S. K., Lauridsen, C., & Persson, W. K. (2006). α-Tocopherol concentration and stereoisomer composition in plasma and milk from dairy cows fed natural or synthetic vitamin E around calving. *Journal of Dairy Research, 73*, 227–234.

Mensink, R. P., Ebbing, S., Lindhout, M., Plat, J., & van Heugten, M. M. (2002). Effects of plant stanol esters supplied in low-fat yoghurt on serum lipids and lipoproteins, non-cholesterol sterols and fat soluble antioxidant concentrations. *Atherosclerosis, 160*, 205–213.

Meyer, E., Lamote, I., & Burvenich, C. (2005). Retinoids and steroids in bovine mammary gland immunobiology. *Livestock Production Science, 98*, 33–46.

Michel, J. J., Chew, B. P., Wong, T. S., Heirman, L. R., & Standaert, F. E. (1994). Modulatory effects of dietary β-carotene on blood and mammary leukocyte function in peripartum dairy cows. *Journal of Dairy Science, 77*, 1408–1422.

Moreau, R. A., Whitaker, B. D., & Hicks, K. B. (2002). Phytosterols, phytostanols, and their conjugates in foods: Structural diversity, quantitative analysis, and health-promoting uses. *Progress in Lipid Research, 41*, 457–500.

Mulder, H., & Walstra, P. (1974). *The Milk Fat Globule. Emulsion Science as Applied to Milk Products and Comparable Foods*. Pudoc, Wageningen, and Commonwealth Agricultural Bureaux, Farnham Royal, The Netherlands.

National Research Council (1989). *Nutrient Requirements of Dairy Cattle*, 6th ed. Washington, DC: National Academy Press.

National Research Council (2001). *Nutrient Requirements in Dairy Cattle*, 7th ed. Washington, DC: National Academy Press.

Nicholson, J. W. G., & St-Laurent, A. M. (1991). Effect of forage type and supplemental dietary vitamin E on milk oxidative stability. *Canadian Journal of Animal Science, 71*, 1181–1186.

Noakes, M., Clifton, P. M., Doornbos, A. M. E., & Trautwein, E. A. (2005). Plant sterol ester-enriched milk and yoghurt effectively reduce serum cholesterol in modestly hypercholesterolemic subjects. *European Journal of Nutrition, 44*, 214–222.

Nozière, P., Graulet, B., Lucas, A., Martin, B., Grolier, P., & Doreau, M. (2006a). Carotenoids for ruminants: From forages to dairy products. *Animal Feed Science and Technology, 131*, 418–450.

Nozière, P., Grolier, P., Durand, D., Ferlay, A., Pradel, P., & Martin B. (2006b). Variations in carotenoids, fat-soluble micronutrients, and color in cows' plasma and milk following changes in forage and feeding level. *Journal of Dairy Science, 89*, 2634–2648.

Ortega, R. M., Palencia, A., & López-Sobaler, A. M. (2006). Improvement of cholesterol levels and reduction of cardiovascular risk via the consumption of phytosterols. *British Journal of Nutrition, 96 (Suppl 1)*, S89–S93.

Patton, S., Kelly, J. J., & Keenan, T. W. (1980). Carotene in bovine milk fat globules: Observations on origin and high content in tissue mitochondria. *Lipids, 15*, 33–38.

Politis, I., Hidiroglou, N., Batra, T. R., Gilmore, J. A., Gorewit, R. C., & Scherf, H. (1995). Effects of vitamin E on immune function of dairy cows. *American Journal of Veterinary Research, 56*, 179–184.

Politis, I., Hidiroglou, N., Cheli, F., & Baldi A. (2001). Effects of vitamin E on urokinase-plasminogen activator receptor expression by bovine neutrophils. *American Journal of Veterinary Research, 62*, 1934–1938.

Rombaut, R., & Dewettinck, K. (2006). Properties, analysis and purification of milk polar lipids. *International Dairy Journal, 16*, 1362–1373.

Sahlin, A., & House, J. D. (2006). Enhancing the vitamin content of meat and eggs: Implications for the human diet. *Canadian Journal of Animal Science, 86*, 181–195.

Sen, C. K., Khanna, S., & Roy, S. (2006). Tocotrienols: Vitamin E beyond tocopherols. *Life Sciences, 78*, 2088–2098.

Seymour, W. (2002). Vitamin nutrition of dairy cattle. In D. E. Pritchard (Ed.), *North Carolina Dairy Nutrition Management Conference Proceedings*(pp. 81–102). Raleigh: North Carolina State University.

Smith, K. L., Harrison, J. H., Hancock, D. D., Todhunter, D. A., & Conrad, H. R. (1984). Effect of vitamin E and selenium supplementation on incidence of clinical mastitis and duration of clinical symptoms. *Journal of Dairy Science, 67*, 1293–1300.

Stahl, W., & Sies, H. (2005). Bioactivity and protective effects of natural carotenoids. *Biochimica et Biophysica Acta, Molecular Basis of Disease, 1740*, 101–107.

Stahl, W., Ale-Agha, N., & Polidori, M. C. (2002). Non-antioxidant properties of carotenoids. *Journal of Biological Chemistry, 383*, 553–558.

Tomlinson, J. E., Mitchell, G. E., Jr., Bradley, N. W., Tucker, R. E., Boling, J. A., & Schelling, G. T. (1974). Transfer of vitamin A from bovine liver to milk. *Journal of Animal Science, 39*, 813–817.

Tylavsky, F. A., Cheng, S., Lyytikäinen, A., Viljakainen, H., & Lamberg-Allardt, C. (2006). Strategies to improve vitamin D status in Northern European children: Exploring the merits of vitamin D fortification and supplementation. *Journal of Nutrition, 136*, 1130–1134.

Van Metre, D. C., & Callan, R. J. (2001). Selenium and vitamin E. *The Veterinary Clinics of North America. Food Animal Practice, 7*, 373–402.

Volpe, R., Niittynen, L., Korpela, R., Sirtori, C., Bucci, A., Fraone, N., & Pazzucconi, F. (2001). Effects of yogurt enriched with plant sterols on serum lipids in patients with moderate hypercholesterolaemia. *British Journal of Nutrition, 86*, 233–239.

Walstra, P., & Jenness, R. (1984). *Dairy Chemistry and Physics*. New York: John Wiley & Sons.

Weber, P. (2001). Vitamin K and bone health. *Nutrition, 17*, 880–887.

Weiss, W. P. (2005). Antioxidant nutrients, cow health, and milk quality. In *2005 Penn State Dairy Cattle Nutrition Workshop* (pp. 11–18). Grantville, PA: Pennsylvania State University.

Weiss, W. P., & Spears, J. W. (2006). Vitamin and trace mineral effects on immune function of ruminants. In K. Sejrsen, T. Hvelplund, & M. O. Nielsen (Eds.), *Ruminant Physiology. Digestion, Metabolism and Impact of Nutrition on Gene Expression, Immunology and Stress* (pp. 473–496). Wageningen, The Netherlands: Wageningen Academic Publishers.

Weiss, W. P., & Wyatt, D. J. (2003). Effect of dietary fat and vitamin E on α-tocopherol in milk from dairy cows. *Journal of Dairy Science, 86*, 3582–3591.

Wolpowitz, D., & Gilchrest, B. (2006). The vitamin D questions: How much do you need and how should you get it? *Journal of the American Academy of Dermatology, 54*, 301–317.

Zahar, M., & Smith, D. E. (1995). Vitamin A distribution among fat globule core, fat globule membrane, and serum fraction in milk. *Journal of Dairy Science, 78*, 498–505.

Zanker, I. A., Hammon, H. M., & Blum, J. W. (2000). Beta-carotene, retinol and alpha-tocopherol status in calves fed the first colostrum at 0–2, 6–7, 12–13 or 24–25 hours after birth. *International Journal for Vitamin and Nutrition Research, 70*, 305–310.

II
Biological Activity of Native Milk Proteins: Species-Specific Effects

Milk Fat Globule Membrane Components—A Proteomic Approach

Maria Cavaletto, Maria Gabriella Giuffrida, and Amedeo Conti

Abstract The milk fat globule membrane (MFGM) is the membrane surrounding lipid droplets during their secretion in the alveolar lumen of the lactating mammary gland. MFGM proteins represent only 1–4% of total milk protein content; nevertheless, the MFGM consists of a complex system of integral and peripheral proteins, enzymes, and lipids. Despite their low classical nutritional value, MFGM proteins have been reported to play an important role in various cellular processes and defense mechanisms in the newborn.

Using a proteomic approach, such as high-resolution, two-dimensional electrophoresis followed by direct protein identification by mass spectrometry, it has been possible to comprehensively characterize the subcellular organization of MFGM.

This chapter covers the description of MFGM proteomics from the first studies about 10 years ago through the most recent papers. Most of the investigations deal with MFGMs from human and cow milk.

Milk Fat Globule Membrane

The principal lipids of milk are triacylglycerols secreted in the alveolar lumina in the form of droplets, coated with a cellular membrane, called the milk fat globule membrane (MFGM) (Mather & Keenan, 1998). MFGM is a tripartite structure, consisting of the typical bilayer membrane as the outer coat, with an electron-dense material on the inner membrane face, and finally, the monolayer of proteins and polar lipid that covers the triacylglycerol droplet core.

M. Cavaletto
Biochemistry and Proteomics Section, DISAV Dipartimento Scienze dell' Ambiente e della Vita, Università del Piemonte Orientale, via Bellini 25/G, 15100, Alessandria, Italy
e-mail: maria.cavaletto@unipmn.it

Lactating mammary cells assemble and release lipid droplets by a unique mechanism; microlipid droplets (<0.5-μm diameter) originate in or on the surfaces of rough endoplasmic reticulum membranes. These droplets are released from the endoplasmic reticulum into the cytosol with a surface coat of proteins and polar lipids. Microlipid droplets grow by fusion with each other and form larger cytoplasmic lipid droplets (>1-μm diameter). These droplets migrate unidirectionally from their sites of origin, mostly in basal and lateral cell regions, to the apical region, probably with the involvement of the cytoskeletal elements. The materials on the surface of the lipid droplets appear to remain associated with the droplets when they are secreted as milk fat globules (Cavaletto et al., 2004; Heid & Keenan, 2005).

Cytoplasmic lipid droplets approach the apical surface, are gradually coated with plasma membrane, and then are released into the alveolar lumen completely surrounded by plasma membrane, as first described by Bargmann and Knoop (1959) and reviewed by Mather and Keenan (1998). In some cases, a cytoplasm inclusion is entrapped into the secreted globules and appears as "crescent" material between the outer membrane layer and the lipid globule.

An alternative mechanism of lipid globule secretion has been described by Wooding (1971), who proposed the progressive fusion of secretory vesicles on the surface of the lipid droplet, leading to the formation of an intracytoplasmic vacuole released by exocytosis; in this case the outer membrane of the lipid globule would be entirely derived from the secretory vesicle membrane. Such a mechanism may be common during the periparturient period or when milk secretion is inhibited (Mather & Keenan, 1998). See Fig. 1 for a schematic representation of the two proposed mechanisms for milk fat globule secretion.

Until now no definitive conclusion has been made on the contribution of the apical plasma membrane or the secretory vesicle membrane to the MFGM, and a combination of the two mechanisms of secretion may be possible.

New proteomic studies on the MFGM characterization will help in defining the molecular basis of the biological processes, involved in the origin and secretion of milk fat by mammary epithelial cells.

Proteomic Analysis

The proteome, or the protein complement of genome, is the full set of proteins expressed by a genome under a particular set of environmental conditions (Pandey & Mann, 2000). Proteomics is a relatively new field and one of the fastest-growing areas of biological research, thanks to its potential to unravel biological mechanisms not accessible by other technologies. Since proteins do not work in isolation, but function in large arrays that form

Milk Fat Globule Membrane Components—A Proteomic Approach

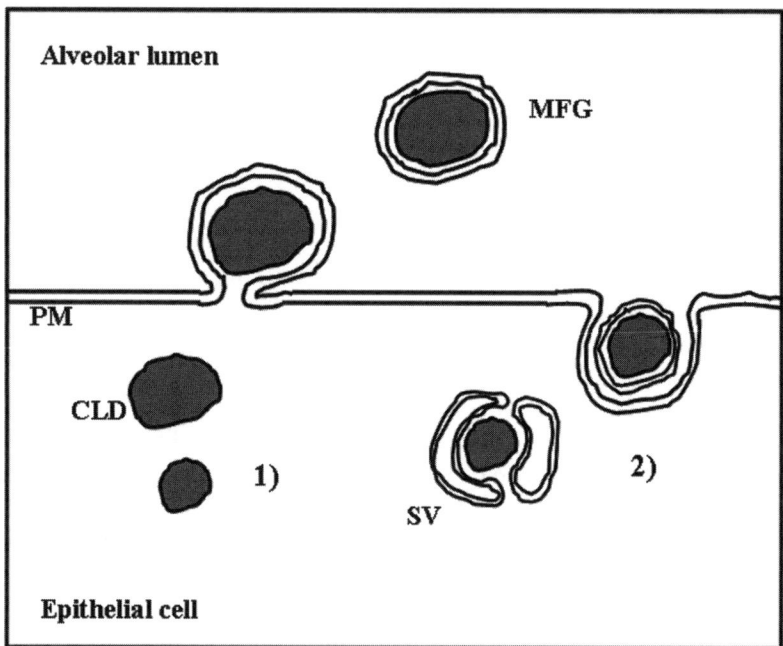

Fig. 1 Schematic representation of the two secretion mechanisms of the lipid globules in the apical region of the mammary epithelial cell. (1) Secretion by apical membrane envelopment of CLD. (2) Secretion by fusion of secretory vesicles on the surface of CLD, followed by release by exocytosis. The tripartite structure of MFGM is shown, with an intervening space between the lipid droplet surface and the surrounding outer bilayer. Size distributions between lipid droplets volume and plasma membrane bilayer are not to scale. (CLD = cytoplasmic lipid droplet; MFG = milk fat globule; PM = plasma membrane; SV = secretory vesicle.)

protein machines, proteomics is exciting because it allows one to dissect and analyze this complex machine into its component parts and to understand how it is assembled, how the proteins interact with one another, and what goes wrong in disease.

The combination of isoelectric focusing (IEF) and sodium dodecyl sulphate (SDS)-polyacrylamide gel electrophoresis (PAGE), commonly known as two-dimensional electrophoresis (2-DE), was developed in the early 1970s. It is still the method of choice for high-resolution profiling of proteins in biological samples (O'Farrel, 1975; Görg et al., 2004). With 2-DE, several thousands of proteins can be resolved on a single-slab gel, also named a bidimensional map.

Following electrophoresis, 2-DE maps may be compared between samples obtained under different physiological and/or experimental conditions; then, using image analysis software, it is possible to specifically detect up- and downregulated proteins (comparative proteomics). Recently, a number of sensitive and specific fluorescent stains have been developed that allow multiplex staining of different groups of proteins on the same gel, thus enhancing differential analysis (Patton & Beechem, 2001).

Protein identification after 2-DE separation is typically accomplished using trypsin in-gel digestion of corresponding protein spots, followed by peptide mass fingerprinting (PMF) via mass spectrometry (MS), or peptide sequencing via tandem MS (MS/MS). Proteomic MS employs soft, nondestructive ionization methods such as matrix-assisted laser desorption ionization (MALDI) and electrospray ionization (ESI). The most common analyzer platforms range from the quadrupole (Q), the ion trap (IT), to the time of flight (TOF). Several software algorithms compare the observed peptide masses and the fragmentation masses against those predicted from theoretical peptides within the sequence database (McDonald & Yates, 2000).

Although proteomic technology is advancing, some limitations become evident, such as lack of automation and insufficient dynamic range. Biological samples are characterized by large differences in the concentrations of the most and least abundant cellular proteins (approximately 5-log difference). Many proteins involved in signal transduction are present in low abundance and thus are not readily detectable in crude extracts.

Other limitations include detection of proteins with extremes in pI and molecular weight and membrane-associated proteins.

As an alternative to gel-based proteomic investigations, multidimensional liquid chromatographic methods have been combined with MS (LC MS/MS) to enable the profiling of complex protein mixtures (MudPIT technology). In general, this strategy includes a strong cation exchange in line with a reverse-phase column and allows one to directly analyze the digests of protein mixtures, yielding good results for the identification of hydrophobic proteins (Link et al., 1999).

In order to detect low-abundance proteins, a powerful strategy is prefractionation of the sample, leading to the subcellular proteome characterization. The identification of subsets of proteins at the subcellular level is therefore an initial step toward the understanding of protein translocation and cellular function (Dreger, 2003). With the fractionation of organelles and subcellular compartments, minor proteins, such as regulatory proteins or integral membrane proteins, are enriched and more easily characterized.

In this context, milk proteins can be fractionated by centrifugation into three major subsets: soluble whey proteins, the pellet of casein micelles, and the floating proteins associated with the MFGM (Cavaletto et al., 2004). The proteome of the MFGM succeeds in profiling this class of milk membrane proteins, which represent only 1–4% of total milk proteins and usually are lacking in the proteome of the whole milk, masked by the most abundant caseins.

Figure 2 summarizes the principal approaches to the proteomic analysis of the MFGM.

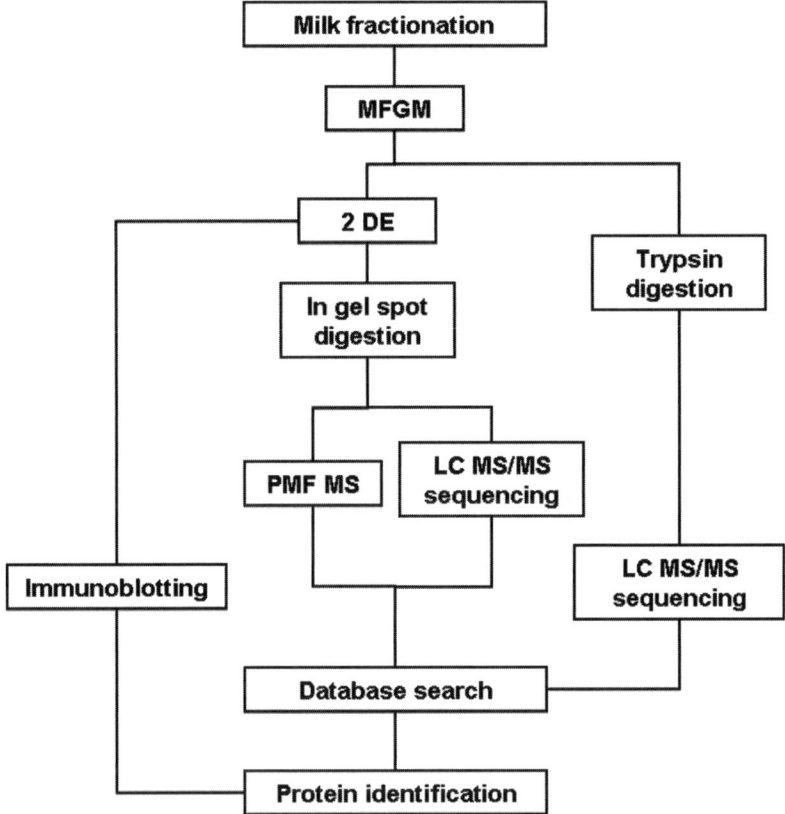

Fig. 2 Strategies applied to the proteomic analysis of the MFGM

Proteomic Approach to MFGM Characterization

Bovine MFGM

The first 2-DE separation of bovine MFGM protein was reported in the review of Mather (2000). The review describes the major proteins associated with the bovine MFGM; these corresponded to seven major bands when separated by SDS-PAGE, while in 2-DE each band was resolved into a series of related isoelectric variants. Major proteins included mucin 1, xanthine oxidase, CD36, butyrophilin, adipophilin, PAS6/7 (lactadherin), and fatty acid binding protein. Identification of the MFGM components was based largely on comparison of electrophoretic mobilities, staining characteristics, and reaction with specific antibodies. The review reported the protein characterization by means of molecular cloning, sequencing, and comparative analysis with MFGM proteins from other species.

In 2002, the effect of heat treatment on bovine MFGM proteins from early, mid-, and late season was characterized using one- and two-dimensional SDS-PAGE under reducing and nonreducing conditions (Ye et al., 2002). It was found that xanthine oxidase and butyrophilin formed aggregates via intermolecular disulfide bonds after heating.

Two papers have recently described the proteome of bovine MFGM (Fong et al., 2007; Reinhardt & Lippolis, 2006). Fong et al. (2007) used the classic proteomic approach to profile the protein and lipid composition of bovine MFGM. Protein identification was carried out using PMF and MS/MS analysis, while lipid composition was determined with a combination of capillary gas chromatography and LC-MS. The composition of MFGM resulted in 69–73% lipid and 22–24% protein; polymeric immunoglobulin receptor, apolipoprotein A and E, 71-kDa heat shock cognate protein, clusterin, lactoperoxidase, and peptidylprolyl isomerase have been identified among minor proteins.

Reinhardt and Lippolis (2006) fractionated MFGM by monodimensional electrophoresis, digested gel slices, and performed protein identification via the LC-MS/MS approach. Among the 120 identified MFGM proteins, 71% were membrane-associated, while 29% were cytoplasmic or secreted proteins; functional immune proteins such as CD14 and Toll-like receptors 2 and 4 have also been detected in the MFGM.

In another recent study, the proteome of bovine MFGM has been compared in three different conditions: from peak lactation, during the colostrum period, and during mastitis (Smolenski et al., 2007). The work is the most comprehensive characterization to date of minor proteins in bovine milk (fractionated in skim milk, whey, and MFGM); 95 distinct gene products were identified, comprising 53 proteins identified through direct LC-MS/MS and 57 through 2-DE followed by MS. The authors demonstrated that a significant fraction of minor proteins are involved in protection against infection.

Human MFGM

The first separation of human MFGM by 2-DE was described in Goldfarb (1997), in which 17 proteins were identified by immunoblotting with specific immunoprobes. The high resolution of 2-DE brought to the detection multiple spots of different pIs due to the presence of multiple isoforms. Besides the typical MFGM proteins, such as xanthine oxidase, butyrophilin, and fatty acid binding protein, other proteins were mapped, including the IgM μ chain, the IgA α chain, the HLA class I heavy chain, immunoglobulin light chains, secretory piece, J chain, actin, α acid glycoprotein, albumin, and casein, with particular attention to the pattern of apolipoproteins E, A-I, A-II, and H.

In 2001, the map of human colostral MFGM was published (Quaranta et al., 2001). This was the first report of MFGM proteome in which proteins were directly identified by PMF MS and/or N-terminal sequencing. Using a new

MFGM double-extraction method with SDS followed by urea/thiourea/ CHAPS, 23 protein spots were identified. The main spots corresponded to lactadherin, adipophilin, butyrophilin, and carbonic anhydrase; the latter had not previously been detected in association with the MFGM. Proteomic analysis revealed the presence of other minor identified MFGM components, including α-lactalbumin, casein, disulphide isomerase, and clusterin (or apolipoprotein J), the latter two as newly identified proteins in human MFGM.

Human butyrophilin expression was evaluated in a comparative proteomic approach (Cavaletto et al., 2002) between colostral and mature milk. While searching the protein complement of seven human butyrophilin transcripts, known only at the mRNA level and mapping on chromosome 6, the authors found 14 multiple forms of butyrophilin; among them, a butyrophilin at pI 6.5 was shown in mature MFGM, whereas the putative butyrophilin, named BTN2A1, was detected for the first time at a protein level.

In 2002, proteomics was applied to the characterization of N-glycosylation (glycomics) of MFGM proteins (Charlwood et al., 2002). The composition of N-linked sugars was analyzed in a hybrid mass spectrometer (MALDI-Q-TOF), after in-gel enzymatic release and subsequent derivation of glycans. Four proteins, clusterin, lactoferrin, polymeric Ig receptor, and lactadherin, were found to possess a wide range of different sugar motifs. In particular, multiple fucosylation products, probably linked to infant protection against bacterial and viral infections, were highlighted.

The first annotated database of human colostral MFGM proteins separated by 2-DE was published in 2003 (Fortunato et al., 2003) and is available in the WORLD-2DPAGE List database at http://www.expasy.org/ch2d/2d-index.html as a partially federated map (Appel et al., 1996). With PMF by MALDI-TOF MS and sequencing by nanoESI-IT MS/MS, 107 protein spots were identified, many of which were present as multiple spots due to posttranslational modifications. On the whole, they derived from 39 genes or gene families. About 60% of the identified proteins were typical MFGM or mammary gland–secreted proteins, and 10% were linked to protein folding and destination, among them cyclophilin, a peptidylprolyl isomerase involved in the response to inflammatory stimuli. Proteins involved in intracellular trafficking and/or receptorial activities were detected in 9%. The cargo selection protein or TIP47, which could interact with the lipid droplet surface, adipophilin and butyrophilin in the process of budding and secretion of the MFGM, has been identified in this group. The remaining minor proteins are correlated with signal transduction, complement complex, and glutathione metabolism.

Mouse MFGM

Wu et al. (2000) described a comparative proteomic analysis between the mouse MFGM and the cytoplasmic lipid droplets (CLDs) of the mouse liver and the mammary gland. The authors tried to dissect the complexity of the lipid secretion and provided evidence that mammary CLDs were intimately associated with membrane-like structures originating from the endoplasmic reticulum.

Since liver CLDs differed from mammary CLDs in protein composition, it was elucidated that different lipid secretion mechanisms occurred in the mammary epithelial cells and in the hepatocytes. Finally, a subset of the MFGM proteins were found also to be present in mammary CDLs, thus suggesting that the membranes and the adherent proteins associated with CDLs were involved in the secretory process.

MFGM Proteins: From Classic to Newly Identified by Proteomics

While MFGM proteins have very low classical nutritional value, they play important roles in various cell processes and in the defense mechanism for the newborn. In addition, the molecular pathways underlying the secretion of milk fat globules have not yet been elucidated, mostly due to the lack of established cell lines that secrete lipid globules.

The proteomic approach to the study of MFGM complex organization will help in defining the roles of MFGM at the level of both the mammary gland and the newborn gastrointestinal tract.

The proteomic investigations dealing with MFGM have directly confirmed, by MS identification, the presence of the classic major proteins associated to MFGM, and in some cases posttranslational modifications have been highlighted.

Thanks to its high-resolution power and high sensitivity, proteomics has resulted in the identification of numerous minor proteins that were not known to be associated with the MFGM and whose function in secreted milk still has to be elucidated. Table 1 lists the minor MFGM proteins, identified by proteomics.

Table 1 List of the Minor MFGM Proteins, Identified by Proteomic Tools

Minor Protein	Function
Actin	Cell motility
Albumin	Binding and transport
Aldehyde dehydrogenase	Metabolic enzyme
α1-Acid glycoprotein	Structural protein
Annexin 1, A2	Structural protein
Apolipoprotein A-1	Transport and lipoprotein metabolism
Apolipoprotein A-2	Transport and lipoprotein metabolism
Apolipoprotein A-4	Transport and lipoprotein metabolism
Apolipoprotein C1	Transport and lipoprotein metabolism

Table 1 (continued)

Minor Protein	Function
Apolipoprotein E	Transport and lipoprotein metabolism
Apolipoprotein H	Transport and lipoprotein metabolism
ATP synthase	Metabolic enzyme
Breast cancer suppressor 1	Mediator of metastasis suppression
α-Casein	Transport of calcium phosphate
β-Casein	Micelle stability
Cathelicidin	Protection
CD14	Immune system
CD36	Receptor and adhesion
CD59	Inhibitor
Cholesterol esterase	Triglyceride hydrolysis
Clusterin	Apoptosis
Complement C4 γ-chain	Complement activation
CRABP II	Cell differentiation
Disulfide isomerase	Protein destination
Dynein intermediate chain	Microtubule motor
Endoplasmin	Protein destination
Enolase 1	Metabolic enzyme
ERcarboxylesterase	Triglyceride synthesis
ERP29	Secretion
ERP99	Secretion
Fatty acid binding protein	Lipid transport
Fatty acid synthase	Lipid synthesis
Fibrinogen	Platelet aggregation
Gelsolin	Cytoskeletal structure
Gephyrin	Cytoskeletal interaction
Glucose regulated protein 58 kDa	Chaperone
Glutamate receptor	Signal transduction
γ-Glutamyl transferase	Glutathione metabolism
GAPDH	Metabolic enzyme
Glycerol-3-P DH	Metabolic enzyme
GTPbinding protein	Signal transduction
GTPbinding protein SAR1b	Transport
GRP 78	Protein destination
GSHH	Protection
Heat shock 27 kDa	Chaperone
Heat shock 70 kDa	Chaperone
Heme binding protein	Transport
Histone H2, H3	DNA binding
HLA class I	Immune system
Immunoglobulin A	Secretory immunity
Immunoglobulins D, G, M	Immune system
Isocitrate DH	Metabolic enzyme
J chain	Immune system
Keratin type II	Cytoskeletal structure
KIAA1586 protein	DNA binding

Table 1 (continued)

Minor Protein	Function
α-Lactalbumin	Lactose synthesis
Lactoferrin	Iron transport
Lactoperoxidase	Metabolic enzyme
Lysozyme	Protection
Macrophage protein 65 kDa	Protection
Macrophage scavenger receptor	Protection
β2-Microglobulin	Protection
Migration inhibitor factor MIF	Cytokine
Oxoprolinase	Metabolic enzyme
Peptidoglycan recognition protein	Protection
Peroxiredoxin IV	Protection
Peroxisome coactivator 1	DNA binding
Poly Ig receptor	Ig superfamily
Prohibitin	Signal transduction
14-3-3 Protein	Signal transduction
Proteose peptone 3	Structural protein
Pyruvate carboxylase	Lipogenesis
Rotamase (cyclophilin)	Folding
S100 Ca binding protein	Transport
SCY1-like2	Signal transduction
Secretory piece	Immune system
Selenium binding protein	Transport
TER ATPase	Membrane fusion
TIF32/RPG1	Cytoskeletal structure
Toll-like receptor 2, 4	Immune system
Transforming protein RhoA	Signal transduction
Tubulin	Structural protein
Villin 2	Structural protein
Vimentin	Structural protein
Voltage-dependent anion channel	Signal transduction
WNT-2B protein	Cellular development

Major MFGM Proteins

Butyrophilin, the most abundant protein in MFGM, is a type 1 membrane glycoprotein. It consists of two extracellular immunoglobulin-like domains and a large intracellular domain homologous to ret finger protein. Butyrophilin may have some receptorial function; it has been exploited to modulate the encephalitogenic T-cell response, supporting its possible involvement in autoimmune diseases (Cavaletto et al., 2002).

Mucin 1 is a highly glycosylated transmembrane protein, a fragment of which co-migrates with butyrophilin (Fong et al., 2007). Mucin is resistant to degradation in the stomach due to its high degree of glycosylation.

Additionally, O-linked sugar chains confer a protective role against the attachment of fimbriated microorganisms (Hamosh et al., 1999).

Lactadherin, the human glycoprotein homologue of bovine PAS 6/7 and mouse MFG-E8, consists of an epidermal growth factor-like domain and two C1 and C2 domains similar to those found in coagulation factors V and VIII. It does not contain the transmembrane domain but might bind to the membrane bilayer by acylation or hydrophobic interaction. Lactadherin promotes cell adhesion via integrins and inhibits rotavirus binding and infectivity (Quaranta et al., 2001).

Adipophilin and TIP47 (Fong et al., 2007; Fortunato et al., 2003; Sztalryd et al., 2006) are typical proteins associated with the surface of the cytoplasmic lipid droplets, suggesting an important structural role for lipid droplet packaging and storage. TIP47 was first identified as a cargo protein involved in the trafficking of the mannose-6-phosphate receptor. Adipophilin and TIP47 are both widely distributed among nonadipogenic tissues.

Carbonic anhydrase is a glycosylated enzyme present in many biological fluids and mostly in saliva. It has been shown to be an essential factor in the normal growth and development of the gastrointestinal tract of the newborn (Karhumaa et al., 2001; Quaranta et al., 2001).

Lactoferrin inhibits the classical pathway of complement activation and can have a bacteriostatic action by competing with bacteria for iron. Also, lactoferrin may function as an antibiotic agent for its structural properties (Cavaletto et al., 2004; Charlwood et al., 2002; Smolenski et al., 2007).

Xanthine oxidase is a cytosolic enzyme concentrated along the inner face of the MFGM. Protein–protein interactions are responsible for the formation of a supramolecular complex among xanthine oxidase, butyrophilin, and adipophilin, probably involved in lipid globule secretion (Mather, 2000). Due to its enzymatic activity, it can act as a defense protein. As reported in Aoki (2006), analysis using knockout mice has revealed that xanthine oxidase and butyrophilin are indispensable for milk fat secretion. With the accumulation of milk fat within mammary epithelial cells, larger lipid droplets with disrupted MFGMs have been described in mice where xanthine oxidase or butyrophilin expression was damaged or abolished.

Conclusions

The proteomic approach applied to the study of the MFGM components has resulted in the identification and characterization of a large set of proteins associated with this particular subcellular compartment. For the major proteins, it is now possible to depict their functional involvement in milk secretion, but for those newly identified by proteomics and for minor ones, their function is still an open and unknown field.

As a source of bioactive components, MFGMs can contribute greatly to the nutraceutical value of milk, in healthy and in pathological conditions; and milk research could explore their utility as functional food ingredients.

In the near future, the completion of the MFGM proteome and the exploitation of the progress in proteomics will elucidate the complex network of interactions between the MFGM and both the mammary gland (site of origin) and the newborn gastrointestinal environment (final destination).

References

Aoki, N. (2006). Regulation and functional relevance of milk fat globules and their components in the mammary gland. *Bioscience, Biotechnology, Biochemistry, 70*, 2019–2027.

Appel, R. D., Bairoch, A., Sanchez, J. C., Vargas, J. R., Golaz, O., Pasquali, C., & Hochstrasser, D. F. (1996). Federated 2-DE database: A simple means of publishing 2-DE data. *Electrophoresis, 17*, 540–546.

Bargmann, W., & Knoop, A. (1959). Über die morphologie der milchsekretion. Licht- und elektronenmikroskopische studien an der milchdrüse der ratte. *Z. Zellforsch, 49*, 344–388.

Cavaletto, M., Giuffrida, M. G., Fortunato, D., Gardano, L., Dellavalle, G., Napolitano, L., Giunta, C., Bertino, E., Fabris, C., & Conti, A. (2002). A proteomic approach to evaluate the butyrophilin gene family expression in human milk fat globule membrane. *Proteomics, 2*, 850–856.

Cavaletto, M., Giuffrida, M. G., & Conti, A. (2004). The proteomic approach to analysis of human milk fat globule membrane. *Clinica Chimica Acta, 347*, 41–48.

Charlwood, J., Hanrahan, S., Tyldesley, R., Langridge, J., Dwek, M., & Camilleri, P. (2002). Use of proteomic methodology for the characterization of human milk fat globular membrane proteins. *Analytical Biochemistry, 301*, 314–324.

Dreger, M. (2003). Proteome analysis at the level of subcellular structures. *European Journal of Biochemistry, 270*, 589–599.

Fong, B. Y., Norris, C. S., & MacGibbon, A. K. H. (2007). Protein and lipid composition of bovine milk-fat-globule membrane. *International Dairy Journal, 17*, 275–288.

Fortunato, D., Giuffrida, M. G., Cavaletto, M., Perono Garoffo, L., Dellavalle, G., Napolitano, L., Giunta, C., Fabris, C., Bertino, E., Coscia, A., & Conti, A. (2003). Structural proteome of human colostral fat globule membrane proteins. *Proteomics, 3*, 897–905.

Goldfarb, M. (1997). Two-dimensional electrophoretic analysis of human milk-fat-globule membrane proteins with attention to apolipoprotein E patterns. *Electrophoresis, 18*, 511–515.

Görg, A., Weiss, W., & Dunn, M. J. (2004). Current two-dimensional electrophoresis technology for proteomics. *Proteomics, 4*, 3665–3685.

Hamosh, M., Peterson, J. A., Henderson, T. R., Scallan, C. D., Kiwan, R., Ceriani, R. L., Armand, M., Mehta, N. R., & Hamosh, P. (1999). Protective function of human milk: The milk fat globule. *Seminars in Perinatology, 23*, 242–249.

Heid, H. W., & Keenan, T. W. (2005). Intracellular origin and secretion of milk fat globules. *European Journal of Cell Biology, 84*, 245–258.

Karhumaa, P., Leinonen, J., Parkkila, S., Kaunisto, K., Tapanainen, J., & Rajaniemi, H. (2001). The identification of secreted carbonic anhydrase VI as a constitutive glycoprotein of human and rat milk. *Proceedings of the National Academy of Sciences USA, 98*, 11604–11608.

Link, A. J., Eng, J., Schieltz, D. M., Carmack, E., Mize, G. J., Morris, D. R., Garvik, B. M., & Yates, J. R. III (1999). Direct analysis of protein complexes using mass spectrometry. *Nature Biotechnology, 17*, 676–682.

Mather, I. H. (2000). A review and proposed nomenclature for major proteins of the milk-fat globule membrane. *Journal of Dairy Science, 83*,203–247.

Mather, I. H., & Keenan, T. W. (1998). Origin and secretion of milk lipids. *Journal of Mammary Gland Biology Neoplasia, 3*,259–273.

McDonald, W. H., & Yates, J. R. III (2000). Proteomic tools for cell biology. *Traffic, 1*, 747–754.

O'Farrel, P. H. (1975). High resolution two-dimensional electrophoresis of proteins. *Journal of Biological Chemistry, 250*,4007–4021.

Pandey, A., & Mann, M. (2000). Proteomics to study genes and genomes. *Nature, 405*, 837–846.

Patton, W. F., & Beechem, J. M. (2001). Rainbow's end: The quest for multiplexed fluorescence quantitative analysis in proteomics. *Current Opinions in Chemical Biology, 6*, 63–69.

Quaranta, S., Giuffrida, M. G., Cavaletto, M., Giunta, C., Godovac-Zimmermann, J., Cañas, B., Fabris, C., Bertino, E., Mombrò, M., & Conti, A. (2001). Human proteome enhancement: High-recovery method and improved two-dimensional map of colostral fat globule membrane proteins. *Electrophoresis, 22*,1810–1818.

Reinhardt, T. A., & Lippolis, J. D. (2006). Bovine milk fat globule membrane proteome. *Journal of Dairy Research, 73*,406–416.

Sztalryd, C., Bell, M., Lu, X., Mertz, P., Hickenbottom, S., Chang, B. H. J., Chan, L., Kimmel, A. R., & Londos, C. (2006). Functional compensation for adipose differentiation-related protein (ADFP) by Tip47 in an ADFP null embryonic cell line. *Journal of Biological Chemistry, 281*,34341–34348.

Smolenski, G., Haines, S., Kwan, F. Y. S., Bond, J., Farr, V., Davis, S. R., Stelwagen, K., & Wheeler, T. T. (2007). Characterisation of host defence proteins in milk using a proteomic approach. *Journal of Proteome Research, 6*,207–215.

Wooding, F. B. P. (1971). The mechanism of secretion of the milk fat globule. *Journal of Cell Science, 9*,805–821.

Wu, C. C., Howell, K. E., Neville, M. C., Yates, J. R. III, & McManaman, J. L. (2000). Proteomics reveals a link between the endoplasmic reticulum and lipid secretory mechanisms in mammary epithelial cells. *Electrophoresis, 21*,3470–3482.

Ye, A., Singh, H., Taylor, M. W., & Anema, S. (2002). Characterization of protein components of natural and heat-treated milk fat globule membranes. *International Dairy Journal, 12*,393–402.

Milk Lipoprotein Membranes and Their Imperative Enzymes

Nissim Silanikove

Abstract There are two main sources of lipoprotein membranes in milk: the relatively well-defined milk fat globule membrane (MFGM) that covers the milk fat globules, and the much less attended lipoprotein source, in the form of vesicles floating in the milk serum. We challenge the common view that the milk serum lipoprotein membrane (MSLM) is secondly derived from the MFGM and present a different view suggesting that it represents Golgi-derived vesicles that are released intact to milk. The potential role of enzymes attached to the MSLM and MFGM is considered in detail for select ubiquitously expressed enzymes.

Introduction

General Introduction

The sole unique feature of mammals is the nurturing of their progeny with a complete nourishing food (milk) during infancy. For this purpose, milk contains essential nutrients such as proteins, carbohydrates, lipids, minerals, and vitamins, together with bioactive substances including immunoglobulins, peptides, antimicrobial factors, hormones, and growth factors (Clare & Swaisgood, 2000). Just about 70 indigenous enzymes have been identified so far in bovine milk (Fox, 2003), and many of them were identified in the milk of various mammalian species (Fox & Kelly, 2006a, b).

The study of indigenous enzymes started more than 50 years ago. Of the 70 identified enzymes, about 20 enzymes that are present at the highest levels in bovine milk have been isolated from milk and characterized (Fox & Kelly, 2006a, b). This is a very active and modern aspect of dairy research, which is covered in the present book by the chapters written by Zimecki, Politis

---

N. Silanikove
Agricultural Research Organization, Institute of Animal Science, P.O. Box 6, Bet Dagan, 50-250, Israel
e-mail: nsilanik@agri.huji.ac.il

& Chronopoulou, Lopez-Exposita & Reico, saito, Blum & Baumrucker and Kuntz and Ruloff. Though modern research in the 20th century significantly added to the knowledge of indigenous enzymes in milk, the vast majority of these studies were concerned with their biochemical properties and technological significance. Methodological studies aimed at identifying the physiological role of milk enzymes have been initiated recently (Silanikove et al., 2006).

It was concluded that indigenous milk enzymes have a key biological role and so far have been found to be involved in providing prodigestive support to the young, the control of milk secretion, the control of mammary gland developmental stage (involution), the gland innate immune system, and the prevention of oxidative damage to its essential nutrients. During excretion, milk enzymes constantly consume metabolites, produce free radicals, and modify their composition. In the few examples studied, milk enzyme along with other components (e.g., cytokines, enzyme inhibitors) form complex metabolic pathways. Thus, milk has proven to be an attractive and readily obtained medium for the investigation of complex biochemical networks. Only a small fraction of milk enzymes was associated with the above-described functions. In addition, the residence of indigenous milk enzymes in milk in general does not seem to be a redundant phenomenon, or a mere consequence, as their presence in milk is not due to leakage from blood (Silanikove et al., 2006). Thus, many more secrets on the topic of the biological role of milk enzymes are most likely waiting to be revealed.

The vast majority of milk enzymes in various species are associated with lipoprotein plasma serum-like membranes (LM) (Shahani et al., 1973, 1980; Fox & Kelly, 2006a, b). However, apart from the membranes covering the lipid core, (see the chapter written by Cavaletti et al. on MFGM) which is generally known as the milk fat globule membrane (MFGM), very little is known about the origin and mode of formation of other sources of LM in milk. However, the amount of LM freely floating in the milk serum as vesicles is approximately similar to the MFGM (Huang & Kuksis, 1967; Silanikove & Shapiro, 2007). Thus, the oversight research on milk serum LM (MSLM) likely represents a dearth of useful information from physiological and biotechnological contemplations.

The aim of the present chapter is to evaluate available literature on the MSLM, the MFGM, and their essential enzymes from the viewpoint of their potential biological role and comparative aspects. As the original reviewed reports mostly did not plan to evaluate the biological roles of these enzymes, this chapter consists of a critical evaluation of indirect data and inevitably slithers into some speculations. However, this review will hopefully attract researchers to this fascinating field and encourage them to put forward some promising hypotheses.

Milk Fractions with Respect to Enzyme Distribution

Milk is not a homogeneous solution of enzymes. Rather, the concentrations of given enzymes are specifically associated with one or more of the five milk phases described in Fig. 1. As Silanikove et al. (2006) discussed, the distribution of enzymes in milk most likely reflects the way in which they were secreted into the milk and their tendency to associate with particular milk constituents or phases.

Briefly, the five physical phases of milk are composed as follows:

Whey: Milk serum, commonly known as whey, is the medium in which all compartments are homogeneously dispersed.
Fat globules: The fat is dispersed in milk as small droplets that are enveloped by a plasma membrane rich in phospholipids, the MFGM. In bovine milk, these globules range in size from 1–8 μm and average 3–4 μm in diameter (Heid & Keenan, 2005).
Casein micelles: The main protein in milk is cascin, arranged as large colloidal particles commonly known as micelles.
Milk cells: The milk cells, generally referred to as somatic cells, comprise white blood cells and sloughing epithelial cells.
Milk serum lipoprotein membrane vesicles: The lost continent?

The MSLM may be considered as the fifth physical element of milk (Fig 1; Silanikove et al., 2006). In bovine milk, 40–60% of the membranous phospholipids are in the skim milk, with the remainder associated with the MFGM (Huang & Kuksis, 1967; Morton, 1954; Plantz & Patton, 1973). Most of the MSLM vesicles appear in the electronic micrograph as nanosized vesicles, but some also

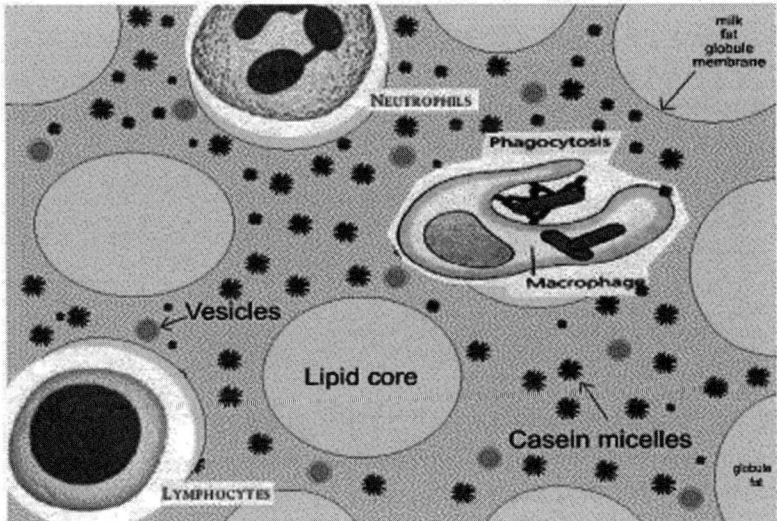

Fig. 1 Schematic representation of the physical phases of milk. The area between the milk particles represents the milk serum (whey), the phase during which all other phases are homogenously dispersed (adapted from Silanikove et al., 2006 with permission from Elsevier). Note that the number of MSLM, $\sim 10^{15}$/mL of milk (denoted in the figure as vesicles), and the number of fat globules, $\sim 10^{10}$/mL of milk, are not represented proportionally in the scheme. (*See* color plate 1)

appear as detached microvilli (Plantz & Patton, 1973; Plantz et al., 1973). Because these vesicles are in the nano range, whereas the MFGM are in the micron size, the MSLM number (10^{15}/mL of milk) is at least one order greater than that of the number of fat globules (10^{10}/mL) (Silanikove & Shapiro, 2007).

In addition, cream-derived membrane vesicles are also found in milk (Morton, 1954). However, while it has long been known that these phospholipidic membranes reside in the butter cream serum, there is no clear idea as to their origin and mode of appearance in milk. It is not clear if butter cream serum–derived vesicles are true constituents, a free-floating substance in milk (Fig. 2). As the data about them are scant and it is not clear if they are present in milk in significant amounts, we will not consider them further in this chapter.

We can consider two explanations to explain the source of MSLM:

1. These vesicles are shaded from the MFGM after it has been secreted to the milk (Kanno, 1990; Wooding, 1977).
2. As proposed in Fig. 2, the MSLM vesicles originate from some of the Golgi vesicles that travel constantly to the apical aspect of the mammary epithelial cell, and somehow and for some reasons some of them are released to the

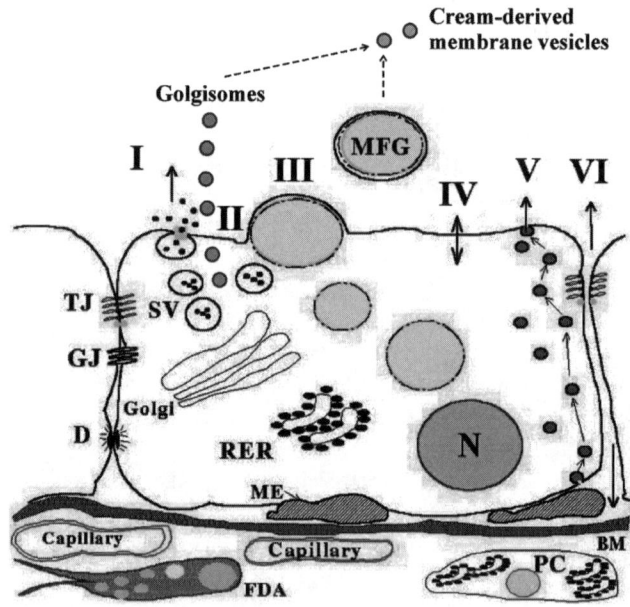

Fig. 2 A model explaining the origin of milk serum lipoprotein membranes (MSLM). According to this model, MSLM originate from a subpopulation of Golgi-derived vesicles, which constantly flow from the Golgi apparatus to the apical membrane of the alveoli and are being released to milk by unidentified mechanism. According to the proposed model, vesicles derived from the serum of milk cream are not truly components of milk, but rather reflect their coalescence-inducement from milk fat globule membranes or the trapping of the MSLM during centrifugation and other manipulations of milk. (*See* color plate 2)

milk. In addition, some of the MSLM are membrane microparticles that possibly have been shaded from the plasma membrane as detached microvilli (Plantz & Patton, 1973; Plantz et al., 1973). Kitchen (1974) proposed a somewhat similar but less defined model.

The following evidence supports the model proposed in Fig. 2:

1. Morphological and radioactive tracer data do not support the concept that MSLM vesicles arise by disintegration of the MFGM material (Patton & Keenan, 1971).
2. There is precedent from another organ (the prostate gland) that epithelial cells surrounding a lumen discharge vesicles (known as prostasomes) into their cavity. The size of these vesicles is similar to the size of the MSLM vesicles. These prostasomes are considered to play an important role in regulating sperm activity and are implicated in the disease state (cancer) of the prostate (Ekdahl et al., 2006).
3. The MFGM serves an essential function in milk in keeping milk lipid droplets homogeneously dispersed in the milk serum and in preventing their lipolyses by serum lipase and esterase. Thus, it is considered highly improbable that the MFGM can shade $\sim$50% of their mass and still maintain their essential feature.
4. Though the gross content of protein and phospholipids in the MSLM and MFGM is similar (Huang & Kuksis, 1967; Silanikove & Shapiro, 2007), there are differences between them in protein composition (Kitchen, 1974; Plantz et al., 1973) and enzyme activity (Table 1, rearranged from Kitchen, 1974; Silanikove & Shapiro, 2007).
5. Shennan (1992) found that the K^+ permeability of the MFGM is very low in both high- and low-strength ionic media, whereas the MSLM displayed a significant permeability to K^+(Shennan, 1992; Silanikove et al., 2000). The results are consistent with the MSLM being composed of either a unilamellar or bilamellar membrane, more closely resembling Golgi vesicles, whereas the MFGM is made of a trilamellar membrane, where the two of the lamellas are separated from the third lamella by a dense layer of proteins (Robenek et al., 2006a, b). The complex LM structure of the MFGM most likely explains its impermeability to K^+ions, despite the potential presence of potassium channels in the MFGM. Indeed, electronic micrographs of liposomes prepared from the MFGM show a complex multilamellar, or liposome-within-liposome, structure (Waninge et al., 2004)

Whereas the presence of enzymes in the MFGM is a certain outcome of fat secretion, at least, according to the model proposed in Fig. 2, the appearance of MSLM vesicles in milk is a regulatory phenomenon. Thus, the question that intuitively arises is, Do milk proteins including enzymes associated with the MSLM have a physiological role?

In the remainder of the chapter, we consider this question with a focus on MSLM enzymes. Some of the well-known roles of enzymes in the cell membrane in general, and in the plasma membrane in particular, serve as a guiding line.

Table 1 Ratio of the Gross and Fine Enzyme Composition of Skim Milk Lipoprotein Membranes (MSLM) and the Milk Fat Globule Membrane (MFGM)

Component	Ratio MSLM/MFGM	
Total protein	1 (S)	
Total fat	1 (S)	
Phospholipids	1 (S)	1.3 (K)
Cholesterol		3.1 (K)
Neutral hexose		1.4 (K)
Sialic acid		2.7 (K)
Alkaline phosphatase	0.9 (S)	1.6 (K)
Acid phosphatase	0.4 (S)	0.5 (K)
5′-Nucleotidase		1.2 (K)
Mg^{2+}-ATPase	3.5* (S)	3.3 (K)
Nucleotide pyrophosphatase		4.7 (K)
Inorganic phosphatase		2 (K)
Glucose-6-phosphatase		2.7 (K)
Sulphydryl phosphatase		2.7 (K)
γ-Glutamyl transpeptidase		5.7 (K)
Diaphorase		0.3 (K)
Xanthine oxidase outside	0.5 (S)	0.3 (K)
Xanthine dehydrogenase inside	0.7 (S)	

* Unpublished data.
Xanthine oxidase (XO) outside: XO attached to the external membrane side of MSLM or MFGM.
Xanthine dehydrogenase (XD) inside: XD attached to the internal membrane side of MSLM or MFGM.
Source: Modified from Kitchen (K) (1974) and Silanikove et al. (S) (2007).

In particular, the potential role of the following enzymes is more closely considered: ATP7B (ATPase, Cu^{2+}-transporting, β-polypeptide, EC 3.6.3.4), glutathione peroxidase (EC 1.11.1.9), glutathione reductase (EC 1.6.4.2), γ-glutamyl-transpeptidase (EC 2.3.2.2), alkaline phosphatase (EC 3.1.3.1), ribonucleases (EC 3.1.27.5 isolated from human milk), and 5′-nucleotidase (EC 3.1.3.5; CD73).

The enzymes in the MFGM and MSLM may be classified into four leading groups: (1) redox regulatory enzymes, (2) enzymes involved in phosphorus ion metabolism, (3) nucleotide metabolizing enzymes, and (4) a range of enzymes with miscellaneous activities, which we will not consider here.

Redox Regulating Enzymes

Redox regulating enzymes are ubiquitously abundant on plasma membranes (Goldenberg, 1998; Low & Crane, 1978), including on the MFGM and MSLM (Shahani et al., 1973, 1980). Despite evidence that endomembranes (endoplasmic reticulum, Golgi, secretory vesicles derived from the endoplasmic reticulum and from the Golgi apparatus and plasma membrane) have a joint origin, there is also evidence that differentiation occurs during their formation (Goldenberg, 1998; Low & Crane, 1978). For example, the cholesterol concentration is low in the endoplasmic reticulum and high in plasma membranes, whereas glucose-6-phosphatase is concentrated in the endoplasmic reticulum in many cells and is almost completely absent from plasma membranes. With a few notable exceptions (one of them is the residence of high activity/concentration of xanthine oxidoreductase in plasma membranes and the MFGM; Table 1), redox enzymes found in the plasma membrane fractions have been considered to be derived from contamination of the preparation with other membranes, or as possible residual material passed on during membrane flow from the endoplasmic reticulum and Golgi secretory vesicles fusing with the plasma membrane. It is possible that the high activity of xanthine oxidoreductase (XOR), mainly in the form of xanthine reductase (XD), in the MFGM derives from the fusion of XD-rich vesicles with the plasma membrane (Silanikove & Shapiro, 2007).

In the MFGM, the presence of redox enzymes likely redundantly reflects the contribution of endomembranes to the plasma membranes covering the fat globules. In addition, the activity of these enzymes toward milk components may be hindered to a large degree because some of these enzymes are buried within the complex trilamellar structure of the MFGM (Silanikove et al., 2006; Silanikove & Shapiro, 2007). To the contrary, because of the simpler unilamellar or bilamellar composition of LM in the MSLM, and because of their smaller size and greater surface area, MSLM enzymes, especially those attached to the outer surface, are potentially physiologically active in milk solution.

Potential functions of the redox enzymes in the MSLM and alveolus apical plasma membrane are discussed next.

Driving Ion and Organic Substances for the Transport or Production of Radicals

It is not clear at present whether the NADH and NADPH oxidases in plasma membranes in the apical aspect, or in the MFGM, or in the MSLM can act as free radical generators, though such a function in leukocytes was shown (e.g., Shirley et al., 1984; Takanaka & Obrien, 1975). It is well established that xanthine oxidase (XO) expressed in large amounts in the MFGM and MSLM generates superoxide during oxidation of either NADH or xanthine (Harrison, 2006). Generation of superoxide by XO located in MFGM was demonstrated *in vitro* (Harrison, 2006); however, it has not been demonstrated so far in the milk itself. Superoxide or other radicals may play an important role in generating a bactericidal environment, particularly by leukocytes. However, it may also have additional functions such as the formation and release of prostaglandins, as superoxide is required for the cyclooxygenase. In addition, the direct use of oxidation-reduction reactions to drive the transport of ions or amino acids in membranes of mitochondria and the plasma membrane of bacteria and eukaryote cells is well documented (Goldenberg, 1998; Low & Crane, 1978). In some cases, NADH oxidation in the plasma membrane facilitates amino acid transport without ATP generation (Low & Crane, 1978). In other cases, plasma membrane dehydrogenases may be involved in localized generation of ATP and NADH at sequestrated sites on the plasma membrane by bound dehydrogenase such as the glyceraldehyde-3-phosphate dehydrogenase or the lactate dehydrogenase (Low & Crane, 1978). A function like this has been proposed for ATP generated at a restricted region on the membrane by the glyceraldehyde-3-phosphate dehydrogenase to energize Na^+/K^+ transport at the site. The Na^+/K^+ ATPase pump, however, does not seem to be active on the apical aspect of the alveolus plasma membrane (Shennan, 1992), and there are no reports so far for the residence of glyceraldehyde-3-phosphate dehydrogenase in milk, MFGM, MSLM, or apical plasma membrane from the apical side.

The Toxic Milk Mouse Illnesses. Evidence for the Role of Copper P-type ATPase in Cu Translocation to Milk

The toxic milk mouse illness is regarded as an animal model for Wilson's disease in humans (Michalczyk et al., 2000). It is an autosomal recessive condition, and the mutant dams have greatly reduced transfer of copper to milk and across the placenta. Pups born to such dams are seriously copper-deficient and often die unless fostered to normal dams. The widespread copper deficiency in calves, kids, and lambs suggests that the Wilson disease-like syndrome may also be a common problem in farm animals.

Copper transporting P-type ATPases, designated ATP7A and ATP7B (the Wilson protein), play an essential role in mammalian copper balance. In humans, impairment in the intestinal transport of copper, caused by mutations in the ATP7A gene, leads to Menkes disease. Defects in a similar gene, the copper transporting ATPase, ATP7B, result in Wilson's disease. This transporter has two functions: transport of copper into the plasma protein ceruloplasmin, and elimination of copper through the bile. Variants of ATP7B can be functionally assayed to identify defects in both of these functions. Recently, the Wilson protein was found to be involved in impairments in Cu-translocation activity. The Wilson protein has been shown to be localized at the trans-Golgi network, but, unlike the wild-type protein, it is not able to undergo vesicular trafficking in response to elevated copper concentrations (Michalczyk et al., 2000).

An open question to be resolved concerns whether the copper P-type ATPase is translocated to the plasma membrane with the Golgi-derived vesicles, whereby it is fused with the plasma membrane and involved in copper transport from the cytoplasm to milk. Alternatively, the copper P-type ATPase may be involved in loading copper into transport vesicles that merge with the plasma membrane and then discharge their content into the milk.

Milk MSLM or MFGM from nursing females may nevertheless serve as a valuable source of biological specimen for simple identification of Wilson's disease.

Role of Glutathione and Related Enzymes in Controlling the Oxidative State of Milk and in Supplying the Young with the Essential Amino Acid Cysteine

Like the lungs, the mammary glands are unique in having a large epithelial surface area (~ 100 m^2) that is at risk for oxidant-mediated attack, particularly during inflammation caused by bacterial invasion to the mammary gland (Paape et al., 2003). Invasions of bacteria or inflammatory response invoke production of intracellular (within leukocytes) and extracellular formation of free radicals, which help to eliminate the infection (Paape et al., 2003). Antioxidant milk resources are therefore important in preventing oxidant injury. Silanikove et al. (2005, 2006) described and reviewed the key role of catalase in preventing nirosative oxidative injury.

Reduced glutathione (GSH) is a critical cellular antioxidant and a cofactor for enzymes that detoxify carcinogens, heavy metals, and toxic chemicals. The typical concentration of GSH in cells, including mammary gland tissue, is in the range of 1 to 2 mM (Fujikake & Ballatori, 2002). The levels of GSH in the milk of rats (100 µM) and that of humans (200 µM), though 10- to 20-fold higher than in blood plasma (10 µM), are still considerably lower than in cells or lung epithelial lining fluid (~ 1 mM; Comhair & Erzurum, 2005). Thus, although GSH may contribute to the milk antioxidant system, it may not be considered a central element.

The presence of the essential amino acid cysteine in milk appears to be particularly important to neonatal mammals because of its importance as an essential constituent in various proteins and in GSH, which is a tri-amino acid peptide that comprises cysteine (γ-L-glutamyl-L-cysteinglycine). There is evidence that the healthy characteristic of whey proteins and their contribution to prevention of carcinogenesis relate to high cysteine content and contribute to GSH formation in cells (Chuang et al., 2005).

Figure 3 depicts a scheme proposing the role of GSH in the formation of free cysteine. One of the extracellular types of glutathione peroxidase (Gp) is an LM-associated enzyme (Fujikake & Ballatori, 2002), whereas glutathione reductase and the catabolic enzyme γ-glutamyl-transpeptidase (GGT), which cleaves GSH to its amino acid components, are classical plasma membrane enzymes. In all cells, GSH is synthesized in the cytosolic space, but GSH degradation occurs extracellularly (Comhair & Erzurum, 2005). Thus, the fact that GSH concentration in milk (rats and humans) is higher than in blood plasma indicates that milk is a medium for GSH degradation and recycling and that the milk LM have a pivotal role in GSH metabolism.

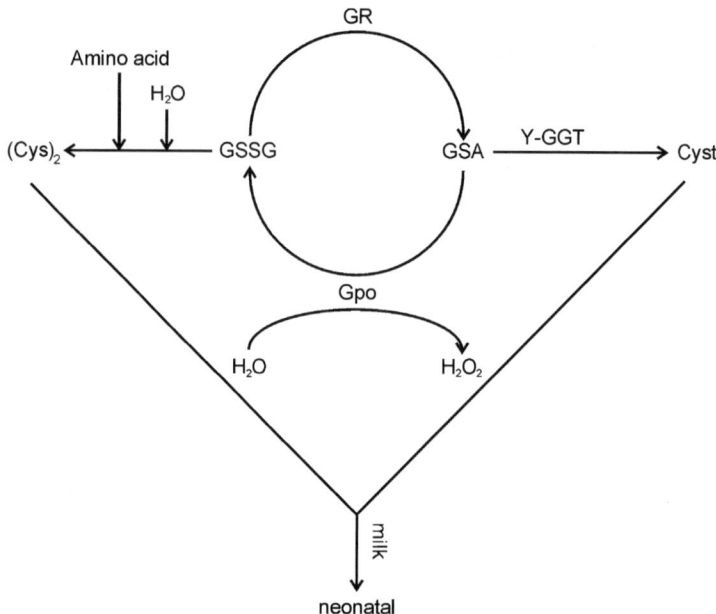

Fig. 3 A model explaining how milk provides the essential amino acid cysteine to infants. (Gpo = glutathione peroxidase; GR = glutathione reductase; GSA = reduced glutathione; GSSG = oxidized glutathione; Cys = cystein.)

Because no glutathione disulfide (GSSG) can be measured in rat's milk (Fujikake & Ballatori, 2002), it may be assumed that this molecule is rapidly breaking down to its amino acid components. This happens in the cytosol when the electron donor is shunted by competitive enzymes. In milk, such a competitive enzyme is probably GGT, which catabolizes GSH to its components, hence reducing the free GSH available for cycling with GSSG. Thus, GSSG break down and GSH catabolism may contribute to free milk cysteine, which in humans is in the range of 20 to 30 µM (Chuang et al., 2005). In support of this model, Will et al. (2002) showed that supplying mice whose GGT gene was knocked out with cysteine in the form of N-acetylcysteine was necessary to prevent growth retardation and cysteine deficiency. It is predicted that pups born to GGT knockout mice will also suffer from cysteine deficiency, even in cysteine-supplemented dams.

Phosphorus Metabolizing Enzymes

Phosphorus is a component in many metabolites (ATP, IP_3, etc.) that plays a fundamental role in the control of cell function. Phosphorylation/dephosphorylation of amino acids on proteins is a major route for propagation of signal transduction events. Thus, unsurprisingly, many cellular and extracellular enzymes that are associated with the plasma membrane metabolize phosphorus, including the MFGM and MSLM (Fox & Kelly, 2006b; Table 1).

Potential Biological Roles of Alkaline Phosphatase (AlP) in Milk

General Features of AlP

Milk contains several phosphatases, the principal ones being alkaline (AlP) and acid phosphomonoesterases (Fox & Kelly, 2006b). AlP is a membrane-bound glycoprotein that is widely distributed in animal tissues and microorganisms. The AlP activity of human and bovine milk varies considerably between individuals and herds and throughout lactation (Fox & Kelly, 2006b). The indigenous AlP in milk is similar to the enzyme in mammary tissue (Fox & Kelly, 2006b). The AlP in human milk is similar, but not identical, to human liver AlP the difference between the two AlPs stems from differences in the sialic acid content (Fox & Kelly, 2006b). Most of the AlP in the mammary gland is in the myoepithelial cells, which may suggest a role in milk secretion; there is much lower AlP activity in the epithelial secretory cells and in milk (Bingham & Malin, 1992). Bingham and Malin (1992) suggest that there are two AlPs in milk, one of which is from sloughed-off myoepithelial cells, the other originating from lipid microdroplets and acquired intracellularly. The latter is probably the AlP found in the MFGM, but unlike XOR, it is not a structural component of the MFGM (Fox & Kelly, 2006b). Most or all studies on milk AlP have been on AlP isolated from cream/MFGM. It would appear that a comparative study of AlP isolated from skimmed milk with that isolated from the MFGM is warranted. Silanikove et al. (2007) have shown that most of the skim milk AlP is associated with MSLM, thus suggesting it actually similar to MFGM AlP. AlP in milk is significant today mainly because it is used universally as an indicator of proper pasteurization. Although AlP can dephosphorylate casein under suitable conditions, as far as we know it has no direct technological significance in milk. Perhaps its optimum pH (10.5) is too far away from that of milk.

Potential Role of AlP in Deactivation of the Negative Feedback System That Regulates Milk Secretion

Silanikove et al. (2000, 2006) described and reviewed a negative-feedback system that controls milk secretion. The N-terminal product of β-casein hydrolysed by plasmin interacts with the plasma membrane in the lumen of the alveoli and downregulates milk secretion. Proteolysis of casein and liberation of this peptide occur in milk stored in the gland between sucklings and/or milking. There is evidence that phosphorylated peptides are dephosphorylated by AlP (see Fox, 2003). Hence, AlP appears to be capable of dephosphorylating peptides under physiological pH. Further work is warranted on the significance of indigenous AlP to dephosphorylate peptides. However, unpublished work from our laboratory is consistent with such a possibility.

Potential Role of AlP Located in Plasma Membranes of Polymorphonuclears in Deactivation of Lipopolysaccharide

AlPs in mammals are encoded by four distinct loci. The three tissue-specific isoenzymes (intestinal, placental, germ cell) are clustered on chromosome 2 and are 90–98% homologous in humans. The fourth AP isoenzyme, called tissue-nonspecific AP (TNAP), is expressed in a variety of tissues, including polymorphonuclear neutrophils (PMN) (Chikkappa, 1992). This enzyme is located on human chromosome 1 and is $\sim$50% identical to the other three isoenzymes (Fishman, 1990). TNAP is an extracellular universal phosphatase, possessing organic phosphatase, pyrophosphatase, and DNA phosphatase activities, and plays an important role in bone mineralization (Fishman, 1990).

Lipopolysaccharide (LPS), which constitutes the outer membrane of Gram-negative bacteria, induces acute inflammation in various tissues and organs, including the mammary gland. Administration of LPS induced the systemic upregulation of TNAP in mice (Xu et al., 2002). A significant increase in TNAP activity was observed in milk PMN during experimental *Escherichia coli* mastitis (Babaei et al., 2007).

LPS is a natural substrate of TNAP at physiological pH, as demonstrated by the capability of placental AP to dephosphorylate and to detoxify LPS (Poelstra et al., 1997a, b). It has been shown that bovine intestinal AP attenuated the inflammatory response in the polymicrobial peritonitis mouse model (Van Veen et al., 2005) and that calf intestinal AP prevented 80% of mice from lethal *E. coli* infection and attenuated LPS toxicity in piglets (Beumer et al., 2003). This evidence led to an effort to express TNAP in the milk of transgenic rabbits, as a source for therapeutic AP (Bodrogi et al., 2006). Based on the above data, it is suggested that TNAP has a physiological role in the mammary innate immune system in preventing toxicity caused by incursion of Gram-negative (in particular, *E. coli*) bacteria to the gland.

Potential Role of Phosphatase Located in MSLM and MFGM in Preventing Generation and Disposition of Insoluble Mineral Complexes Such as Inorganic Ca Pyrophosphate

Inorganic pyrophosphates (PPi) are composed of two molecules of inorganic phosphate (Pi) joined by a hydrolysable high-energy bond (Fig. 4a). PPi regulates certain intracellular functions and the extracellular crystal disposition of

Fig. 4 Formulas for pyrophosphate-related minerals. (a) The formula for inorganic pyrophosphate. (b) The formula for phosphocitrate. (c) The formula for Ca pyrophosphate

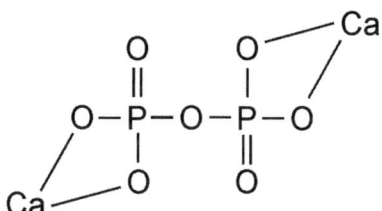

Ca phosphate. There are many direct links between extracellular PPi disposition and diseases with connective tissue matrix calcification disorder (Terkeltaub, 2001).

Natural compounds similar to the structure of PPi include phosphocitrate (Fig. 4b) and Ca-PPi (Fig. 4c). Phosphocitrate is synthesized and enriched in mitochondria by enzymes such as nucleoside triphosphate pyrophosphohydrolase (NTPPPH) and is seen to function very effectively as a crystallization inhibitor and to block hydroxyapatite-induced cell stimulation. On the other hand, Ca-PPi is only slightly soluble, and its presence in extracellular fluid is linked to the pathogenesis of crystal arthropathies such as basic Ca phosphate and Ca pyrophosphate dihydrate crystal disease (Makowski & Ramsby, 2004).

Milk is considered the best food supply of Ca in nature, which makes it particularly valuable for the development of mammals during their early growing phase. This quality of milk relates to its high content of supersaturated Ca phosphate complexes. Sequestration of Ca phosphate by caseins occurs in the Golgi apparatus of mammary epithelial cells. This event helps both to accumulate high quantities of Ca on caseins and to prevent calcification of the gland and precipitation of crystalline Ca phosphate complexes in milk. Calcium phosphate nanoclusters are formed by sequestration of amorphous Ca within a shell of casein and are generally known as the micellar Ca phosphate (Smith et al., 2004). Nevertheless, studies on the partition of salts in milk have shown that about one-third of the Ca, half of the Pi, two-thirds of the Mg, and nine-tenths of the citrate in bovine milk are diffusible through a semipermeable membrane or can be separated by ultrafiltration (Silva et al., 2001). Thus, one cannot rule out the potential formation of PPi (if the NTPPPH enzyme exists in milk) and the subsequent spontaneous formation of Ca-PPi.

Bovine milk has been shown to contain Ca phosphate-citrate complexes (McGann et al., 1983a) and Zn phosphocitrate-casein complexes (McGann et al., 1983b). Highly insoluble minerals, which resemble Ca-PPi, were isolated from processed bovine whey (Allen & Cornforth, 2007). Thus, milk phosphatases likely have a significant function in preventing the formation of insoluble PPi-like complexes in milk, which otherwise may lead to pathological inflammation.

Nucleotide Metabolizing Enzymes

Delay in the development of both the innate and acquired immune function in newborn mammals is compensated by (1) *in utero* transfer of pathogen-specific IgG to the developing fetus and (2) various protective factors found in milk (Siegrist, 2001). During the first year of life, human milk-fed infants are significantly more resistant to the development of various infectious diseases than formula-fed babies (Siegrist, 2001). Some of the various immune factors in human and other mammals' milk identified as responsible for this protection include pathogen-specific antibodies, oligosaccharides, lipids, and mucin (Siegrist, 2001, and Rosetta & Baldi, this volume). In addition, various components in milk, including ribonucleotides (Schallera et al., 2007), are assumed to facilitate immune maturation of the milk-fed offspring, and their addition to baby formulas might make them more like human milk (Schallera et al., 2007). In view of the wealth of nucleotide metabolizing enzymes in the MSLM and MFGM (Shahani et al., 1973, 1980), we now consider their potential function.

Nucleotides (Nu) are major components of RNA and DNA and participate in the mediation of energy metabolism, signal transduction, and the general regulation of cell growth (Schallera et al., 2007). They also participate in lipoprotein metabolism and enhance high-density lipoprotein (HDL) plasma concentration as well as the synthesis of apolipoprotein (Apo) A1 and Apo A1V in preterm infants, and an upregulation of long-chain polyunsaturated fatty acid synthesis in human neonates (Schallera et al., 2007).

Ribonucleotides (RNu) are considered "conditionally essential" for the proper development of the human neonate, because the supply of RNu through *de novo* synthesis and endogenous salvage pathway sources is thought to be insufficient for optimal functioning of rapidly growing intestinal and lymphoid tissues, even though their low levels might not result in an overt clinical deficiency syndrome (Schallera et al., 2007). Over the last two decades, RNu have been extensively studied as ingredients in infant formulas; several reviews for such a role have been published (Schallera et al., 2007). The main driving force for these investigations relates to the fact that RNu are present in higher amounts in human milk than in cow's milk and cow's milk-based infant formulas (Schallera et al., 2007).

The value of RNu was demonstrated also in nonhuman mammals; the fortification of animal diets with RNu derivatives improved resilience to infection (Manzano et al., 2003).

Ribonucleases (RNases) have been purified and sequenced from a number of vertebrate tissues, including pancreas, kidney, liver, brain, and milk (Dyer & Rosenberg, 2006). The ubiquitous presence of RNase in various tissues suggests that they play an important biological role, but it has yet to be more precisely assigned (Dyer & Rosenberg, 2006; Marcus et al., 2005). A protective role in retrovirus infection in milk has been envisaged for RNase, because milk RNase inhibited the reverse transcription of the RNA genome of the mouse mammary tumor virus (Fox & Kelly, 2006b). The importance of this observation relates to the transmission of mammary tumor virus through milk to pups in several strains of mice. It was also shown that RNA-dependent DNA synthesis by a retrovirus (the avian myeloblastosis virus), incubated in human milk, was inversely proportional to the amount of RNase present in the milk samples (Dyer & Rosenberg, 2006; Ye & Ng, 2000). Also, not considered so far, RNase may play a role in liberating RNu from endogenous sources of milk t-RNA for the progeny needs. The fact that RNase content is higher in colostrum than during milk secretion nevertheless also sustains the notion that this enzyme is important for offspring development (Roman et al., 1990).

Extracellular nucleotides play many biological roles, including intercellular communication and modulation of nucleotide receptor signaling, and are dependent on the phosphorylation state of the nucleotide (Kennedy et al., 2005). Regulation of nucleotide phosphorylation is necessary, and a specialized class of enzymes (nucleotide pyrophosphatases/phosphodiesterases) has been identified in various mammalian tissues. Nucleotide pyrophosphatases are an integral component of the MFGM and MSLM (Table 1), which suggests that regulation of nucleotide phosphorylation played an as-yet unidentified role in the feedback regulation of mammary function and/or in supporting the proper development of offspring.

5′-Nucleotidase, like alkaline phosphatase, is an ecto-enzyme that is ubiquitously expressed on the plasma membrane and is characterized on the MFGM and MSLM (Table1). Ecto-enzymes are catalytic membrane proteins with their active sites outside the cell (Goding, 2000). Many leukocyte antigens are ecto-enzymes (Goding, 2000). Thus, 5′-nucleotidase may play a role in milk nucleotide metabolism or in the mammary gland immune function.

Conclusions

Previously available information that clearly supports the concept that milk enzymes are there because they have a key role in regulating milk secretion, mammary gland development, and the mother–offspring interrelationship has been reviewed (Silanikove et al., 2006). In the present chapter, we have tried to critically analyze existing reports on enzymes associated with the milk membrane and with their potential physiological function. In doing so, we have tried to avoid crossing the fine line between legitimate hypotheses and wild speculations. The success of this survey would be determined in its power to stir scientific interest and initiate modern research in an area that seems fascinating and fundamentally important.

References

Allen, K., & Cornforth, D. (2007). Antioxidant mechanism of milk mineral—High-affinity iron binding. *Journal of Food Science, 72*, C78–C83.

Babaei, H., Mansouri-Najand, L., Molaei, M. M., Kheradmand, A., & Sharifan, M. (2007). Assessment of lactate dehydrogenase, alkaline phosphatase and aspartate aminotransferase activities in cow's milk as an indicator of subclinical mastitis. *Veterinary Research Communications, 31*, 419–425.

Beumer, C., Wulferink, M., Raaben, W., Fiechter, D., Brands, R., & Seinen, W. (2003). Calf intestinal alkaline phosphatase, a novel therapeutic drug for lipopolysaccharide (LPS)-mediated diseases, attenuates LPS toxicity in mice and piglets. *Journal of Pharmacology and Experimental Therapy, 307*, 737–744.

Bingham, E. W., & Malin, E. L. (1992). Alkaline phosphatase in the lactating bovine mammary gland and the milk fat globule membrane. Release by phosphatidylinositol-specific phospholipase-C. *Comparative Biochemistry and Physiology B—Biochemistry & Molcular Biology, 102*, 213–218.

Bodrogi, L., Brands, R., Roaben, W., Seinen, W., Baranyi, M., Fiechter, D., & Bosze, Z. (2006). High level experssion of tissue-nonspecific alkaline phosphatase in the milk of transgenic rabbits. *Transgenic Research, 15*, 627–636.

Clare, D. A., & Swaisgood, H. E. (2000). Bioactive milk peptides: A prospectus. *Journal of Dairy Science, 83*, 1187–1195.

Chikkappa, G. (1992). Control of neutrophils alkaline phosphatase synthesis by cytokines in health and disease. *Experimental Haematology, 20*, 388–390.

Chuang, C. K., Lin, S. P., Lee, H. C., Wang, T. J., Shih, Y. S., Huang, T. Y., & Yeung, C. Y. (2005). Free amino acids in full-term and pre-term human milk and infant formula. *Journal of Pediatric Gastroenterology and Nutrition, 40*, 496–500.

Comhair, S. A. A., & Erzurum, S. C. (2005). The regulation and role of extracellular glutathione peroxidase. *Antioxidant & Redox Signaling, 7*, 72–79.

Dyer, K. D., & Rosenberg, H. F. (2006). The RNase a superfamily: Generation of diversity and innate host defense. *Molecular Diversity, 10*, 585–597.

Ekdahl, K. N., Ronquist, G., Nilsson, B., & Babiker, A. A. (2006). Possible immunoprotective and angiogenesis-promoting roles for malignant cell-derived prostasomes: A new paradigm for prostatic cancer? *Current Topics in Complement Advances in Experimental Medicine and Biology, 586*, 107–119.

Fishman, H. (1990). Alkaline phosphatase isozymes: Recent progress. *Clinical Biochemistry, 23*, 99–104.

Fox, P. F. (2003). Indigenous enzymes in milk. In P. F. Fox & P. L. H. McSweeney (Eds.), *Advanced Dairy Chemistry*, Vol. 1, *Proteins* (pp. 447–467). New York: Kluwer Academic/Plenum Publishers.

Fox, P. F., & Kelly, A. L. (2006a). Indigenous enzymes in milk: Overview and historical aspects—Part 1. *International Dairy Journal, 16*, 500–516.

Fox, P. F., & Kelly, A. L. (2006b). Indigenous enzymes in milk: Overview and historical aspects—Part 2. *International Dairy Journal, 16*, 517–532.

Goding, J. W. (2000). Ecto-enzymes: Physiology meets pathology. *Journal of Leukocyte Biology, 67*, 285–311.

Goldenberg, H. (1998). Molecular biology of plasma membrane redox enzymes: A survey of current knowledge. *Protoplasma, 205*, 3–9.

Harrison, R. (2006). Milk xanthine oxidase: Properties and physiological roles. *International Dairy Journal, 16*, 546–554.

Fujikake, N., & Ballatori, N. (2002). Glutatthione secretion into rat milk and its subsequent γ-glutamyltranspeptidase mediated catabolisim. *Biology of the Neonate, 82*, 134–138.

Heid, H. W., & Keenan, T. W. (2005). Intracellular origin and secretion of milk fat globules. *European Journal of Cell Biology, 84*, 245–258.

Huang, T. C., & Kuksis, A. (1967). A comparative study of lipids of globule membrane and fat core and of milk serum of cows. *Lipids, 2*, 453–460.

Kanno, C. (1990). Secretory membranes of the lactating mammary gland. *Protoplasma, 159*, 184–208.

Kennedy, E. J., Pillus, L., & Ghosh, G. (2005). Pho5p and newly identified nucleotide pyrophosphatases/phosphodiesterases regulate extracellular nucleotide phosphate metabolism in *Saccharomyces cerevisiae*. *Eukaryotic Cell, 4*, 1892–1901.

Kitchen, B. J. (1974). Comparison of properties of membranes isolated from bovine skim milk and cream. *Biochimica et Biophysica Acta, 356*, 257–269.

Low, H., & Crane, F. L. (1978). Redox function in plasma-membranes. *Biochimica et Biophysica Acta, 515*, 141–161.

Makowski, G. S., & Ramsby, M. L. (2004). Differential effect of calcium phosphate and calcium pyrophosphate on binding of matrix metalloproteinases to fibrin: Comparison to a fibrin-binding protease from inflammatory joint fluids. *Clinical and Experimental Immunology, 136*, 176–187.

Manzano, M., Abadia-Molina, A. C., Olivares, E. G., Gil, A., & Rueda, R. (2003). Dietary nucleotides accelerate changes in intestinal lymphocyte maturation in weanling mice. *Journal of Pediatric Gastroenterology and Nutrition, 37*, 453–461.

Marcus, A. J., Broekman, M. J., Drosopoulos, J. H. F., Olson, K. E., Islam, N., Pinsky, D. J., & Levi, R. (2005). Role of CD39 (NTPDase-1) in thromboregulation, cerebroprotection, and cardioprotection. *Seminars in Thrombosis and Hemostasis, 31*, 234–246.

McGann, T. C., Buchheim, W., Kearney, R. D., & Richardson, T. (1983a). Composition and ultrastructure of calcium-phosphate citrate complexes in bovine-milk systems. *Biochimica et Biophysica Acta, 760,* 415–420.

McGann, T. C., Kearney, R. D., & Buchheim, W. (1983b). Zinc phosphocitrate-casein complexes in bovine-milk systems. *Kieler Milchwirtschaftliche Forschungsberichte, 35,* 409–411.

Michalczyk, A. A., Rieger, J., Allen, K. J., Mercer, J. F. B., & Ackland, M. L. (2000). Defective localization of the Wilson disease protein (ATP7B) in the mammary gland of the toxic milk mouse and the effects of copper supplementation. *Biochemical Journal, 352,* 565–571.

Morton, R. K. (1954). The lipoprotein particles in cow's milk. *Biochemical Journal, 57,* 231–237.

Paape, M. J., Bannerman, D. D., Zhao, X., & Lee J. W. (2003). The bovine neutrophil: Structure and function in blood and milk. *Veterinary Research, 34,* 597–627.

Patton, S., & Keenan, T. W. (1971). Relationship of milk phospholipids to membranes of secretory cell. *Lipids, 6,* 58–68.

Plantz, P. E., &. Patton, S. (1973). Plasma-membrane fragments in bovine and caprine skim milks. *Biochimica et Biophysica Acta, 291,*51–60.

Plantz, P. E., Keenan, T. W., & Patton, S. (1973). Further evidence of plasma-membrane material in skim milk. *Journal of Dairy Science, 56,*978–983.

Poelstra, K., Bakker, W. W., Klok, P. A., Hardonk, M. J., & Meijer, D. K. (1997a). A physiologic function for alkaline phosphatase: Endotoxin detoxification. *Laboratory Investigations, 76,* 319–327.

Poelstra, K., Bakker, W. W., Klok, P. A., Kamps, J. A., Hardonk, M. J., & Meijer, D. K. (1997b). Dephosphorylation of endotoxin by alkaline phosphatase *in vivo. American Journal of Pathology, 151,* 1163–1169.

Robenek, H., Hofnagel, O., Buers, I., Lorkowski, S., Schnoor, M., Robenek, M. J., Heid, H., Troyer, D., & Severs, N. J. (2006a). Butyrophilin controls milk fat globule secretion. *Proceedings of the National Academy of Sciences USA, 103,* 10385–10390.

Robenek, H., Hofnagel, O., Buers, I., Robenek, M. J., Troyer, D., & Severs, N. J. (2006b). Adipophilin-enriched domains in the ER membrane are sites of lipid droplet biogenesis. *Journal of Cell Science, 119,* 4215–4224.

Roman, M., Sanchez, L., & Calvo, M. (1990). Changes in ribonuclease concentration during lactation in cow's colostrum and milk. *Netherlands Milk and Dairy Journal, 44,* 207–212.

Schallera, J. P., Bucka, R. H., & Ruedab, R. (2007). Ribonucleotides: Conditionally essential nutrients shown to enhance immune function and reduce diarrheal disease in infants. *Seminars in Fetal and Neonatal Medicine, 12,* 35–44.

Shahani, K. M., Harper, W. J., Jensen, R. G., Parry, R. M., & Zittle, C. A. (1973). Enzymes in bovine milk: A review. *Journal of Dairy Science, 56,* 531–543.

Shahani, K. M., Kwan, A. J., & Friend, B. A. (1980). Role and significance of enzymes in human milk. *American Journal of Clinical Nutrition, 33,* 1861–1868.

Shennan, D. B. (1992). Is the milk-fat-globule membrane a model for mammary secretory-cell apical membrane? *Experimental Physiology, 77,*653–656.

Shirley, P. S., Bass, D. A., Lees, C. J., Parce, J. W., Waite, B. M., & Dechatelet, L. R. (1984). Co-localization of superoxide generation and NADP formation in plasma-membrane fractions from human-neutrophils. *Inflammation, 8,* 323–335.

Siegrist, C. A. (2001). Neonatal and early life vaccinology. *Vaccine, 19,* 3331–3346.

Silanikove, N., & Shapiro, F. (2007). Distribution of xanthine oxidase and xanthine dehydrogenase activity in bovine milk: Physiological and technological implications. *International Dairy Journal, 17,*1188–1194.

Silanikove, N., Shamay, A., Shinder, D., & Moran, A. (2000). Stress down regulates milk yield in cows by plasmin induced beta-casein product that blocks K+ channels on the apical membranes. *Life Sciences, 67,* 2201–2212.

Silanikove, N., Shapiro, F., Shamay, A., & Leitner, G. (2005). Role of xanthine oxidase, lactoperoxidase, and NO in the innate immune system of mammary secretion during active involution in dairy cows: Manipulation with casein hydrolysates. *Free Radical Biology Medicine, 38*, 1139–1151.

Silanikove, N., Merin, U., & Leitner, G. (2006). Physiological role of indigenous milk enzymes: An overview of an evolving picture. *International Dairy Journal, 16*, 533–545.

Silva, F. V., Lopes, G. S., Nobrega, J. A., Souza, G. B., Rita, A., & Nogueria, A. (2001). Study on the protein-bound fraction of calcium, iron, magnesium and zinc in bovine milk. *Spectrochimica Acta Part B, 56*, 1909–1916.

Smith, E., Glegg, R. A., & Holt, C. (2004). Perspective on the structure and function of caseins and casein micelles. *International Journal of Dairy Technology, 157*, 121–126.

Takanaka, K., & Obrien, P. J. (1975). Mechanisms of H_2O_2 formation by leukocytes. 1. Evidence for a plasma-membrane location. *Archives of Biochemistry and Biophysics, 169*, 428–435.

Terkeltaub, R. A. (2001). Inorganic pyrophosphate generation and disposition in pathophysiology. *American Journal of Physiology—Cell Physiology, 281*, C1–C11.

van Veen, S. Q., van Vliet, A. K., Wulferink, M., Brands, R., Boermeester, M. A., & van Gulik, T. M. (2005). Bovine intestinal alkaline phosphatase attenuates the inflammatory response in secondary peritonitis in mice. *Infection and Immunity, 73*, 4309–4314.

Waninge, R., Kalda, E., Paulsson, M., Nylander, T., & Bergenståhl, B. (2004). Cryo-TEM of isolated milk fat globule membrane structures in cream. *Physical Chemistry Chemical Physics, 6*, 1518–1523.

Will, Y., Kaetzel, R. S., Brown, M. K., Fraley, T. S., & Reed, D. J. (2002). *In vivo* reversal of glutathione deficiency and susceptibility to *in vivo* dexamethasone-induced apoptosis by N-acetylcysteine and L-2-oxothiazolidine4-carboxylic acid, but not ascorbic acid, in thymocytes from γ-glutamyltranspeptidase-deficient knockout. *Archives of Biochemistry and Biophysics, 397*, 399–406.

Wooding, F. B. P. (1977). Comparative mammary fine structure. In M. Peaker (Ed.), *Comparative Aspects of Lactation* (pp. 1–41). New York: Academic Press.

Xu, Q., Lu, Z., & Zhang, X. (2002). A novel role of alkaline phosphatase in protection from immunological liver injury in mice. *Liver, 22*, 8–14.

Ye, X. Y., & Ng, T. B. (2000). First demonstration of lactoribonuclease, a ribonuclease from bovine milk with similarity to bovine pancreatic ribonuclease. *Life Sciences, 67*, 2025–2032.

Lactoferrin Structure and Functions

Dominique Legrand, Annick Pierce, Elisabeth Elass, Mathieu Carpentier, Christophe Mariller, and Joël Mazurier

Abstract Lactoferrin (Lf) is an iron binding glycoprotein of the transferrin family that is expressed in most biological fluids and is a major component of mammals' innate immune system. Its protective effect ranges from direct antimicrobial activities against a large panel of microorganisms, including bacteria, viruses, fungi, and parasites, to anti-inflammatory and anticancer activities. This plethora of activities is made possible by mechanisms of action implementing not only the capacity of Lf to bind iron but also interactions of Lf with molecular and cellular components of both host and pathogens. This chapter summarizes our current understanding of the Lf structure-function relationships that explain the roles of Lf in host defense.

Introduction

When lactoferrin (Lf) was first discovered in human milk (Montreuil et al., 1960), it was named "lactotransferrin," suggesting a functionally related variant of transferrin. The possibility of Lf having functions other than simple iron sequestration emerged when it was reported that Lf binds to microbes, host cells, and components of the immune system. Besides its antimicrobial effects, immunomodulatory and anticancer properties were also reported. Although the many activities of Lf depend on mechanisms of action that are sometimes extremely different, most of these mechanisms are now well defined and understood. We will emphasize the structure-function relationships of this fascinating molecule.

D. Legrand
Unité de Glycobiologie Structurale et Fontionnelle, UMR n°8576 du CNRS, IFR 147
Université des Sciences et des Technologies de Lille, F-59655 Villeneuve d'Ascq Cedex, France
e-mail: dominique.legrand@univ-lille1.fr

Synthesis and Localization of Lf

Lf synthesis can be continuous (exocrine fluids), under hormonal control (genital tract, mammary gland) (Teng et al., 2002), or at well-defined stages of cell differentiation [neutrophils (PMNs)] (Masson et al., 1969). Lf is secreted in the apo-form from epithelial cells in most exocrine fluids such as saliva, bile, pancreatic and gastric fluids, tears, and milk (Montreuil et al., 1960). In milk, Lf is mainly synthesized by glandular epithelial cells; its concentration in humans may vary from 1 g/L (mature milk) to 7 g/L (colostrum). In mature bovine milk, its mean concentration is 30 mg/L. Furthermore, Lf is synthesized during the transition from promyelocytes to myelocytes and is thus a major component of the secondary granules of PMNs (Masson et al., 1969). During inflammation and in pathologies, Lf levels of biological fluids may increase greatly and constitute a marker for inflammatory diseases. This is particularly noticeable in plasma, where the Lf concentration can be as low as 0.4–2 mg/L under normal conditions but increases to 200 mg/L in septicemia. In fact, plasma Lf only represents the tip of the iceberg since (1) most neutrophil Lf is delivered by PMNs at the sites of inflammation and (2) Lf can bind to glycosaminoglycans (GAGs) of proteoglycans (Mann et al., 1994; Legrand et al., 1997), so that cells may provide high local Lf concentrations on their surfaces.

Lf Gene and Protein Structures

Lf *Gene Structure and Regulation*

The *Lf* gene appears in mammals and is highly conserved among species, with an identical organization (17 exons with 15 encoding Lf) and conserved codon interruptions at the intron-exon splice junctions (reviewed in Teng et al., 2002). *Lf* gene polymorphisms are widely distributed in the human population and may lead to modulation of Lf activity. Transcription of the *Lf* gene, mapped to human chromosome 3 at 3p21.3, leads to two products, Lf and delta-Lf (ΔLf) mRNAs, which result from the use of alternative promoters: the P1 promoter for *Lf* and P2 for ΔLf (Siebert & Huang, 1997; Liu et al., 2003).

Comparison of *Lf* gene promoters from different species showed both common and different characteristics. The *Lf* gene has been shown to be highly sensitive to nuclear receptors, and *Lf* expression is upregulated by estrogen with a magnitude of response that is cell-type-specific (mammary glands, uterus) and by retinoic acids. Transcription factors such as Sp1, Ets, PU.1, C/EBP, CDP/cut, and KLF5 also modulate *Lf* gene expression, mainly in myeloid cells (Teng, 2006). *Lf* expression is also increased following oxidative damage, in response to infection, or during early steps of embryogenesis (reviewed in Ward et al., 2005). For the P2 promoter, potential upstream regulatory elements different

Table 1 Lf Amino Acid Polymorphism

Amino acid position	5	11	29	86	100	175	390	464	561
Wild type	LAGRRRRS	A	K	E	K	A	A	V	E
Variant type	LAG - - RRS	T	R	K	R	V	T	L	D

Source: Teng and Gladwell (2006).

from those of P1, an elevated expression in lymphoid cells, and upregulation by Ets have been described (Siebert & Huang, 1997; Liu et al., 2003).

Characteristics of the Lf Sequence

The amino acid sequence of human milk Lf (hLf) was first determined by Metz-Boutigue et al. (1984) and the nucleotide sequence of human mammary gland hLf by Rey et al. (1990). Lf consists of a single polypeptide chain of about 690 amino acid residues containing an internal duplication, whereby the N-terminal half of the protein has ~40% sequence identity with the C-terminal half. The hLf sequence was found to have ~60% sequence identity with human serum transferrin (Tf). Given the close similarity between Lf and Tf, it is relevant to ask what properties define the Lf molecule. One specific feature of Lf is its highly basic character and the distribution of positive charges at the N terminus (1–7), along the outside of the first helix (13–30), and in the interlobe region (reviewed in Baker & Baker, 2005). Structurally, the feature that most readily distinguishes Lf from Tf is the peptide linker between the two lobes, which is helical in Lf but irregular in Tf, containing several proline residues.

Polymorphism

Recently, the common single-nucleotide polymorphism (SNP) in the *hLf* gene has been established (Teng & Gladwell, 2006). Sixteen SNPs have been detected, but only nine cause amino acid changes (Table 1). Two frequently occuring SNPs at positions 11 and 29 are located in the lactoferricin (Lfcin) domain, described below, and, interestingly, seminal hLf presents this polymorphism (Kumar et al., 2003). The Lf molecule presenting the K29R polymorphism has been shown to exhibit significantly greater bactericidal activity against *Streptococcus mutans* and *Streptococcus mitis* (Velliyagounder et al., 2003).

Lf Active Clusters

Lf possesses amino acid clusters or consensus sequences for which binding capacity or biological activities have been demonstrated. These clusters are presented in Table 2. Many are located in the basic domain, including the N terminus (residues 1–7) and the first helix (residues 13–30). Limited proteolysis of Lfs or

Table 2 Location and Putative Functions of Lf Peptide Clusters

Location on the hLf Polypeptide Chain	Name and Reference	Putative Function	Other Lfs
N-t: D^{60}, Y^{92}, Y^{192}, H^{253}, R^{121} C-t: D^{396}, Y^{436}, Y^{509}, H^{578}, R^{466}	Iron binding sites (Anderson et al., 1987)	Iron sequestration	Conserved in all Lfs and Tfs
N^{138}, N^{475}, N^{635}	N-glycosylation sites (Spik et al., 1988)	Protein stability, antigenic masking	Random
6–309: K^{73} and S^{259}	Peptidase S60 (Hendrixson et al., 2003)	Serine protease activity, antimicrobial activity	bLf
1–5	NLS (Penco et al., 2001)	Nuclear targeting	Not studied
442–457	NLS (Mariller et al., 2007)	Nuclear targeting	Not studied
8–25	Glycation site (Li et al., 1995)	Inhibition of antibacterial activity	Not studied
17–42	Lfcin (Bellamy et al., 1992)	Antimicrobial and anti-tumoral activities	bLf
1–47	OmpC and OmpB binding (Sallmann et al., 1999)	Antimicrobial activity	bLf
152–182	Kaliocin-1 (Viejo-Diaz et al., 2003)	Antimicrobial activity	bLf
319–324, 524–528, and 561–667	Lactoferroxin (Tani et al., 1990)	Opioid antagonist activity	Not studied
268–284	Lactoferrampin (van der Kraan et al., 2005)	Antimicrobial activity	Only found in bLf
1–6 and 28–31	Proteoglycan-binding site, LPS and sCD14-binding site (Mann et al., 1994; Elass-Rochard et al., 1995; Baveye et al., 2000)	Cell recognition, modulation of the inflammatory response	bLf
19–29	Lf 11 (Japelj et al., 2005)	LPS binding site	bLf

Table 2 (continued)

Location on the hLf Polypeptide Chain	Name and Reference	Putative Function	Other Lfs
25–30	Lipoprotein receptor-related proteins (LRPs) and lymphocyte receptor-binding site (Meilinger et al., 1995; Legrand et al., 1992)	Cell recognition, capture and internalization of Lf, immune activity	bLf
1–31	DNA binding site (van Berkel et al., 1997)	Transcription factor activity (He & Furmanski, 1995)	bLf
679–695	Procaspase 3 domain (Katunuma et al., 2006)	Stimulation of apoptosis cascade	Not studied

synthesis of Lf motifs leads to the production of active peptides that reproduce the biological activities of Lf, sometimes with a higher potency: N-terminal peptide (residues 1–5) for nuclear targeting (Penco et al., 2001), Lfcin for antimicrobial and antitumor activities and binding to proteoglycans (see below), and Lfl1 (residues 19–29) (Japelj et al., 2005) for lipopolysaccharide (LPS) binding. In contrast, amino acid mutations or deletions in these clusters abolish the activities of Lf: Mutations to $D^{60/396}$, $Y^{92/436}$, and $R^{121/466}$ inhibit the iron binding (Baker & Baker, 2005); mutations to K^{73} and S^{259} abolish the protease activity (Hendrixson et al., 2003); deletion of residues 1–4 prevents binding to proteoglycans (Legrand et al., 1997); and mutations $RR^{417-418}AA$ and $KK^{431-432}AA$ in the ΔLf isoform abolish the short bipartite nuclear localization signal (Mariller et al., 2007).

Lfcin Structure

Limited proteolysis leads to the release of Lf fragments: N-t and C-t lobes, the N-2 domain (Legrand et al, 1984), and Lfcin (Bellamy et al., 1992). Lfcin (Lfcin-B from bLf and Lfcin-H from hLf) is a 25 amino acid peptide (residues 17–42) including two Cys residues linked by a disulfide bridge and containing many hydrophobic and positively charged residues. The secondary structure of Lfcin is markedly different from the same sequence in intact Lf (reviewed in Gifford et al., 2005). The long α-helix observed in the Lf structure is replaced by a single β-sheet strand. This structure seems to be better suited for making contact with bacterial membranes. In biological fluids, Lf exists in an iron-free form that is very susceptible to proteolysis. It cannot be overlooked that a posttranslational process of maturation by proteolysis leads to the release of Lf-derived active peptides in biological fluids (Goldman et al., 1990).

Glycosylation of Lf

All Lfs contain biantennary N-acetyllactosamine-type glycans, $\alpha,1$-6 fucosylated on the N-acetylglucosamine residue linked to the polypeptide chain (Spik et al., 1988). Human Lf may also possess additional poly-N-acetyllactosamine antennae that may be $\alpha,1$-3-fucosylated on N-acetylglucosamine residues, whereas the Lf of other species contains additional high-mannose-type glycans (Coddeville et al., 1992). Both the number and location of the glycosylation sites vary among species. Furthermore, heterogeneity in the number of glycosylated sites is observed in individuals. The role of the glycan moiety seems to be restricted to a decrease in the immunogenicity of the protein and its protection from proteolysis (Spik et al., 1988; van Veen et al., 2004).

Lf *Structure*

The 3D structure of hLf has been determined by Anderson et al. (1987) (Fig. 1). The polypeptide is folded into two globular lobes representing the N- and C-terminal halves (residues 1–333 and 345–691 in hLf) linked by a short α-helix (residues 334–344 in hLf). Noncovalent interactions, mostly hydrophobic, provide a cushion between the two lobes and allow interactions with the C-terminal helix (residues 678–691 in hLf). Both lobes are folded similarly: Two α/β

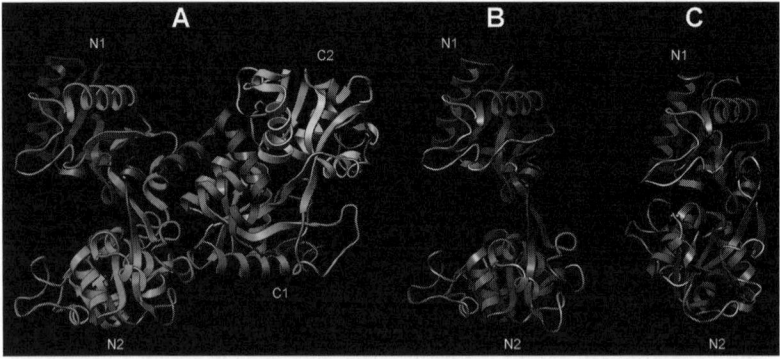

Fig. 1 Structure of hLf. (a) Ribbon diagram showing the polypeptide folding of iron-saturated hLf (Jameson et al., 1998). The N-t lobe is on the left; the polypeptide chain is colored from the N- to the C-terminal end according to a red-shift. (b) Open and (c) closed structures of the N-terminal lobe of hLf (α-helices are colored in magenta and β-sheets in blue). Domains N1, N2, C1, and C2 are indicated. (*See* color plate 3)

domains, referred to as N1 and N2, or C1 and C2, delineate a deep cleft within which the iron binding site is located. The α/β fold of each domain consists of a central, mostly parallel β-sheet, with a helix packed against it. The helical N-terminus faces the interdomain cleft, making it somewhat positively charged, and one of the helices, H5 from the N2 (or C2) domain, serves as the binding site for the essential carbonate anion at the metal binding site.

Two structures have been observed for Lfs: an open conformation, originally described for the iron-free Lf, and a closed conformation, mainly observed with the iron-saturated molecule. The conformational transition could be involved in basic functions such as transportation and catalysis. According to crystallogaphic data, the domains move essentially as rigid bodies (Fig. 1) that close over the bound metal or open to release it.

Iron Binding and Release

The iron binding site has the same composition and geometry in both lobes of Lfs and Tfs (Anderson et al., 1987; Baker & Baker, 2005). It comprises four protein ligands (2 Tyr, 1 Asp, and 1 His) that provide three negative charges to balance the 3+ charge of Fe^{3+}, together with the side chain of an Arg residue whose positive charge balances the negative charge of a CO_3^{2-} anion.

Iron release depends on the destabilization of the closed form (Baker & Baker, 2005). In the absence of receptor binding (as is the case for Tf), release is triggered by lowering the pH (Mazurier & Spik, 1980). As suggested by kinetic studies, protonation of the CO_3^{2-} ion, and then the Tyr and/or His ligands, should progressively weaken the iron coordination to the point where it no longer holds the two domains together. The property of hLf of retaining iron down to pH 2–3, while studies on isolated half-molecules of Lf indicate iron release pH profiles similar to Tf, suggests that interactions between the two Lf lobes play a key role in ensuring iron retention to low pH. The orientation of hLf lobes differ by 8.2° from those of horse Lf, 11.3° from bLf, and 14.8° from buffalo Lf. This difference may explain why bLf loses iron at pH 4 and camel Lf at pH 6 and might be functionally relevant. Contrary to the iron cargo Tf, Lf is not involved in the redistribution of iron between storage compartments and cells Lf acts instead as an iron chelator, and sequestration of iron confers to Lf several of its functions in bacteriostasis, or modulation of inflammation or other processes.

Lf Receptors in Microorganisms and Mammals

The search for specific Lf receptors, comparable with that of Tf, has consistently mobilized the energy of researchers. Surprisingly, Lf receptors with the highest specificity were discovered on bacteria (reviewed in Ling & Schryvers, 2006), whereas specific mammalian receptors were only encountered on enterocytes (reviewed in Suzuki et al., 2005). In fact, most molecular targets on the host cells are multiligand receptors and, interestingly, as reviewed hereinafter, many of them were reported as signaling, endocytosis, and nuclear targeting molecules.

Lf *Binding Molecules on Microorganisms*

Lf binding molecules have been characterized in many types of microorganisms. In *Toxoplasma gondii*, two Lf binding proteins were recently identified as the ROP4 and ROP2 antigens (Dziadek et al., 2007). In viruses, interactions of Lf with the V3-loop of gp120 and proteins E1-2 of the human immunodeficiency virus (HIV) and the hepatitis C virus (HCV), respectively, have been proposed (Swart et al., 1996; Yi et al., 1997). Strong interactions of bLf with adenovirus polypeptides III and IIIa that bind to integrins of host cells have also been demonstrated (Pietrantoni et al., 2003).

Concerning bacteria, many studies reported Lf binding to cells and its subsequent bactericidal effect. On Gram-positive (Gram+) *Staphylococcus aureus* and *Streptococcus uberis*, both glycosidic and proteic Lf binding sites were evidenced but not further characterized (Naidu et al., 1992; Moshynskyy et al., 2003). In the case of Gram-negative (Gram-) bacteria, although evidence was provided that Lf binds to the lipid-A moiety and/or the negative charges in the inner core of LPS with a high affinity (Appelmelk et al., 1994), it is unlikely that Lf/LPS interactions occur when LPS is integrated in the cell wall of bacteria. Interestingly, it has been hypothesized that Lf may use porins as anchoring sites on the surface of bacteria (Erdei et al., 1994; Sallmann et al., 1999).

In the last decade, specific Lf receptors promoting the growth of pathogenic bacteria have been thoroughly characterized. Such an iron acquisition system was extensively characterized among the members of the *Neisseriaceae* family that include not only some of the most important human pathogens: *Neisseria meningitidis, Neisseria gonorrhoeae*, and *Moraxella catarrhalis*, but also animal pathogens such as *Moraxella bovis* (reviewed in Ling & Schryvers, 2006). Its role was clearly demonstrated through experiments with mutant strains defective in the expression of Lf binding proteins. These strains were indeed unable to grow with Lf provided as the unique source of iron. The Lf binding molecules are two proteins named LbpA and LbpB that can be compared to the Tf binding proteins TbpA and TbpB, respectively, also present in all these strains. LbpA is highly homologous to the TonB-dependent receptors and consists of a trans-membrane C-terminal beta-barrel with large external loops (Prinz et al., 1999) and plugged with the N-terminal domain. Interestingly, LbpB has clusters of negative charges in its sequence that could bind the cationic Lf whose C-terminal domains are both recognized by LbpA (Wong & Schryvers, 2003). Unlike the mammalian receptors, the bacterial receptors are species-specific (Ling & Schryvers, 2006).

Mammalian Lf *Binding Proteins*

The cationic nature of Lf accounts for its propensity to bind to anionic molecules (Baker & Baker, 2005). This results in massive binding to mammalian cells and makes identification of receptors contributing specifically to the biological roles of Lf difficult. However, several Lf binding sites were clearly evidenced on target cells as well as the corresponding interaction regions on the Lf molecule.

Lf is a glycoprotein whose glycan moiety has rarely been reported to participate in cell binding. Nevertheless, Lf can bind with high affinity to rat hepatic lectin 1, the major subunit of the asialoglycoprotein receptor (McAbee et al., 2000). Furthermore, in the case of bLf, which possesses N-glycans of the oligomannosidic type, the adjuvant effects of the molecule could be due to bLf binding to the mannose receptor of immature antigen-presenting skin cells (Zimecki et al., 2002). Such binding was recently confirmed in a study showing that the binding of bLf to the DC-SIGN receptor on dendritic cells blocks its interaction with HIV glycoprotein 120 and subsequent virus transmission (Groot et al., 2005).

At the surface of cells, the sulfated chains of proteoglycans represent the major Lf binding sites (80% of total Lf binding) (Legrand et al., 1997; Damiens et al., 1998a). Although the low affinity (Kd ~ 1 μM) and ionic nature of the interactions call their physiological relevance into question, it is widely accepted that proteoglycans are responsible for the high-density binding of Lf at the surface of cells. We recently demonstrated that heparan sulfate (HS) proteoglycans are required for surface nucleolin-mediated endocytosis of Lf into cells (Legrand et al., 2004).

In addition to proteoglycans, other Lf binding sites have been evidenced. A 105-kDa receptor was demonstrated on activated lymphocytes, platelets, and mammary gland cells that could permit signaling in cells as well as endocytosis of Lf (Mazurier et al., 1989; Dhennin-Duthille et al., 2000). Recently, binding of Lf to surface nucleolin on dividing cells, endocytosis, and partial nuclear targeting of the nucleolin/Lf complex were evidenced (Legrand et al., 2004). Nucleolin is a major ubiquitous, 105-kDa nucleolar protein of exponentially growing eukaryotic cells, involved in the regulation of cell proliferation and growth, and described as a cell surface receptor for several ligands (reviewed in Srivastava & Pollard, 1999). Nucleolin is also an attachment target for viruses such as HIV (Callebaut et al., 1998). It has been hypothesized that nucleolin could be the 105-kDa lymphocyte receptor, but this is still controversial (Legrand et al., 2004).

Other important Lf receptors, namely the low-density lipoprotein receptor-related proteins (LRPs), were found on hepatocytes, fibroblasts, osteoblasts, and brain endothelial cells (Willnow et al., 1992; Meilinger et al., 1995; Fillebeen et al., 1999; Takayama et al., 2003; Grey et al., 2004). These molecules are members of a family of large receptors widely expressed on several cell types including hepatocytes, macrophages, smooth muscle cells, and neurons. LRPs are frequently referred to as scavenger receptors, but they have also been implicated in signaling pathways (Herz & Strickland, 2001).

Lastly, a specific receptor visualized as a 34–37-kDa protein under reducing conditions and responsible for Lf endocytosis was evidenced and characterized at the surface of intestinal cells (reviewed in Suzuki et al., 2005). Although the exact role of the intestinal receptor was not reported, it has been hypothesized that the intestinal Lf receptor may transduce some signals to synthesize IL-18 (Suzuki et al., 2005).

Antimicrobial Activities of Lf

Lf is a key element of the innate host defense system and, as such, it has crucial antimicrobial activities against a broad range of pathogens. In the case of bacteria, Lf affects many Gram+ and Gram- pathogens including *E. coli* spp., *Haemophilus influenzae, Salmonella typhimurium, Shigella dysenteriae, Listeria monocytogenes, Streptococcus* spp., *Vibrio cholerae, Legionella pneumophila, Enterococcus* spp., *Staphylococcus* spp., *Bacillus stearothermophilus,* and *Bacillus subtilis* (reviewed in Valenti & Antonini, 2005). In contrast, it seems to promote the growth of beneficial bacteria like *Lactobacillus* and Bifidobacteria (Sherman et al., 2004). Lf has beneficial effects against many viruses such as the Friend virus complex, cytomegalovirus (CMV), polyomavirus, herpes simplex virus (HSV), HIV, hepatitis B and C (HBV and HCV) viruses, rotaviruses, and adenovirus. Finally, Lf exhibits antifungal activity against many *Candida* spp., more particularly *Candida albicans,* and antiparasitic activities against *Entamoeba histolytica, Tritrichomonas fœtus, Trypanosoma cruzi, T. brucei, Plasmodium falciparum, T. gondii,* and *Eimeria stiedai* (reviewed in Valenti & Antonini, 2005). It is not perfectly clear whether these antimicrobial properties are related to a direct action on microbes or to the activation of the immune system, but several lines of evidence now indicate that both forms of action come into play. The difficulty of getting a clear picture of Lf antimicrobial activities stems from the fact that the mechanisms of action of Lf, as well as the ecological niches of microbes, often differ from one organism to another. We report hereafter the different mechanisms by which Lf exerts its antimicrobial properties and the microorganisms that are affected.

Antimicrobial Activities Related to Metal and Ion Chelation by Lf

Because Lf is a strong and stable iron chelator (Mazurier & Spik, 1980), it may compete with the iron acquisition systems of pathogens. Indeed, Lf competes efficiently with most of the bacterial siderophores and can limit the growth of a broad range of bacteria. However, this iron limitation effect was not observed with fungi such as *Candida* spp. In terms of viruses, the iron binding capability of Lf is obviously ineffective, but in some cases such as the rotaviruses, inhibition of replication was higher with apo Lf than with iron-saturated Lf (Superti et al., 1997). In this case, Lf could modulate the hemagglutination of rotaviruses and their binding to host cells. However, the impact of iron saturation on Lf is not well defined. With HIV, viral replication and the formation of syncytium are inhibited by Lf regardless of its degree of iron saturation or the nature of the metal bound to the protein (Puddu et al., 1998). Singh (2004) proposed a role for the iron binding ability of Lf on the adhesive properties of bacteria and the formation of biofilms during respiratory and oral infections. During cystic fibrosis (CF), apo Lf was indeed shown to chelate iron released in large quantities during the disease, to increase the motility of *Pseudomonas aeruginosa,* and to inhibit their aggregation and the formation of biofilms (Rogan et al., 2004; Singh, 2004). It is interesting to note, however, that *Neisseriaceae*, possibly *Helicobacter pylori*, and parasites like *T. fœtus* may use Lf as a source of iron for their growth (reviewed in Ling & Schryvers, 2006).

Finally, it has been reported that the Ca^{2+} binding ability of Lf influences the release of LPS from bacteria (Rossi et al., 2002). This phenomenon could explain, at least in part, the destabilization of the cell wall of Gram- bacteria and its modulation by Ca^{2+} and Mg^{2+} (Ellison et al., 1990).

Antimicrobial Activities Related to Direct Interactions of Lf *with Microbes*

Although iron chelation was initially identified as the major mechanism for Lf bactericidal activity, it is now well established that this activity is mainly due to a direct binding of Lf and Lf-derived peptides to microbes. This interaction may have several consequences, the first and major one being the destabilization of membranes of a broad spectrum of Gram- and Gram+ bacteria (reviewed in Ling & Schryvers, 2006) and fungi (*Candida*) (Xu et al., 1999), but such an effect has never been reported on viruses. The phenomenon can be compared to the effect of many other cationic antimicrobial peptides such as polymyxin B. Lfcin also possesses this activity with an even greater potency (reviewed in Gifford et al., 2005). Recently, lactoferrampin (Table 2) was also reported as an effective molecule, especially on *Candida,* but also on Gram- and Gram+ bacteria (van der Kraan et al., 2005). It is unlikely that Lf acts in a similar way to its peptides, but it has been demonstrated that the Lfcin domain is involved in Lf binding to negative charges on bacteria and fungi (Elass-Rochard et al., 1995; Nibbering et al., 2001). In Gram- bacteria, these negative charges may consist of the LPS themselves; Sallmann et al. (1999) showed that the negatively charged external loops of porins, such as *E. coli* OmpC and PhoE, were able to bind Lf. With parasites such as *T. gondii* and *E. stiedai,* preincubation of sporozoites with Lfcin significantly inhibited their infectivity in animals (Omata et al., 2001).

Another important effect of Lf binding to microbes is the modification of the interactions of microbes between themselves, with the host cells, or with the extracellular matrix. This results in modified motility and aggregation of microbes and decreased endocytosis into host cells. Experiments on *Shigella flexneri* (Gomez et al., 2003) and *Helicobacter felis* (Dial & Lichtenberger, 2002) have shown that the glycans of Lf may bind to bacterial adhesins. It is now admitted that Lf may significantly limit the formation of biofilms and thus ensure the protection of mucosa, as Kawasaki et al. (2000) showed with several *E. coli* strains. Furthermore, the inhibition of adhesion of *S. mutans* (Visca et al., 1989) and *P. aeruginosa* (Williams et al., 2003) on abiotic surfaces such as hydroxyapatite or contact lenses, respectively, has been demonstrated. These findings show possibilities for very promising applications. In addition, direct binding by Lf to HSV (Marchetti et al., 1996), HIV (Swart et al., 1996), HCV (Yi et al., 1997), adenovirus (Pietrantoni et al., 2003), and rotavirus (Superti et al., 1997) particles was evidenced and connected, at least in part, with decreased infection of host cells by Lf. Bovine Lf could also inhibit the integrin-mediated internalization of adenovirus into host cells through its binding to viral polypeptides III and IIIa (Pietrantoni et al., 2003).

Antimicrobial Activities Related to Proteolysis by Lf

Recently, a serine-type endopeptidase activity, although much lower than that of trypsin, was evidenced for Lf (Table 2). This proteolytic activity could be used to degrade virulence factors such as the IgA1 protease and the adhesin Hap from *H. influenzae* (Qiu et al., 1998), but also the invasive plasmid antigens B and C (IpaB and IpaC) from *S. flexneri* (Gomez et al., 2003) and the secreted proteins A, B, and D (EspABD) from enteropathogenic *E. coli*. Aae, an autotransporter involved in the adhesion of *Actinobacillus actinomycetemcomitans* to epithelial cells, could also be degraded by Lf (Ochoa et al., 2003).

Antimicrobial Activities Related to Interactions of Lf *with Host Cells*

Lf is present in biological fluids and binds to most cells. This property confers protection to the host against infections, particularly against viral infections. Viruses indeed use common co-receptors at the surface of host cells, namely the GAGs and integrins. Because Lf can bind to GAGs (Damiens et al., 1998a), it would be able to interfere with viruses such as HSV, HBV, CMV, adenoviruses, HIV, and other microbes and to prevent their internalization into cells. Such interference could account for the decreased infectivity of HBV on cells preincubated with Lf (Hara et al., 2002), a result that designates Lf as a possible candidate for the treatment of chronic hepatitis. Furthermore, it was demonstrated that Lf interferes with the binding of HSV glycoprotein C to HS and/or chondroitin sulphate (CS) (Hasegawa et al., 1994). Lf may also compete with adenovirus that binds to HS-GAGs (Di Biase et al., 2003). Although the cationic N-terminus of Lf was reported as the major GAG binding region of Lf (Mann et al., 1994; Legrand et al., 1997) and is important for its anti-adenovirus activity, both lobes seem important for the anti-HSV activity (Siciliano et al., 1999), suggesting other mechanisms than a simple competition between Lf and the particles for GAG binding. Interestingly, Lf also binds to three of the many co-receptors of HIV, namely GAGs, surface nucleolin (Legrand et al., 2004), and the DC-SIGN receptor (Groot et al., 2005). The interaction of Lf with surface nucleolin was shown to block the attachment and entry of HIV particles into HeLa P4 cells (Legrand et al., 2004). This interaction, together with Lf binding to proteoglycans and to the V3-loop of gp120 on viral particles (Swart et al., 1996), probably contributes to the anti-HIV activity of Lf. In the case of HSV, the cationic N-terminus does not seem essential for Lf activity, and negative charges bound to Lf increase its antiviral activity (Swart et al., 1996).

In terms of bacterial infections, Lf bound to the surface of epithelial cells would be able to inhibit the adherence of bacteria and the formation of biofilms. While many experiments showed no beneficial effect of Lf incubated with host cells prior to infection, some demonstrated the inhibition of enteroinvasive *E. coli* HB101 entry into cultured cells by bLf (Longhi et al., 1993). Such an inhibitory effect was also observed for Gram- *Y. enterocolitica* and *Y. pseudotuberculosis* with bovine Lfcin (Di Biase et al., 2004) and for Gram+ *S. aureus* with Lf (Diarra et al., 2003) and was connected to Lf's ability to bind to bacterial adhesins, host cell integrins, and GAGs.

In the case of fungi, the effect of Lf on *Candida* spp. has been thoroughly studied and often related to Lf adsorption onto fungi and cell wall destabilization (Xu et al., 1999). It was suggested, however, that Lf could bind to host cells and work by some host-mediated mechanism of action. Concerning parasites, Lf adsorption on host cells has been related to the enhancement of *T. cruzi* phagocytosis and lysis by macrophages (Lima & Kierszenbaum, 1987) and also to the inhibition of the intracellular growth of *P. falciparum* in erythrocytes (Fritsch et al., 1987). Shakibaei & Frevert (1996) demonstrated that Lf might limit the infection of fibroblasts by *Plasmodium*, probably through its binding to HS-GAGs and to the LRP, which is recognized by the CS protein of *Plasmodium*.

Finally, adsorption of Lf on host cells might involve much more active processes than a simple competition for binding to microbial receptors. Thus, Lf may be able to activate immune cells through nuclear targeting and/or activation pathways. For example, in the case of CMV infection, an increased activity of NK cells has been reported (Shimizu et al., 1996). Furthermore, a recent study suggests that the antimycotic activity of Lf is related to an activation of the immune system (Yamaguchi et al., 2004).

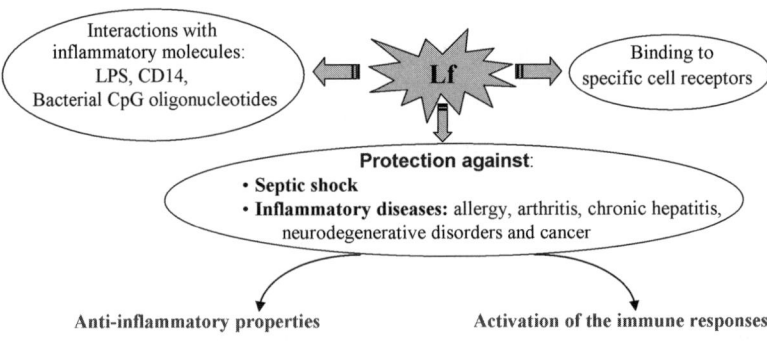

Fig. 2 Modulation of inflammation by Lf

Lf Is a Modulator of Inflammation

Lf is a potent modulator of inflammatory and immune responses, revealing host-protective effects not only against microbial infections but also in inflammatory disorders such as allergies, arthritis, and cancer (Legrand et al., 2005; Ward et al., 2005; Yamauchi et al., 2006). The up- or downregulating effects of Lf are related to its ability to interact with pro-inflammatory bacterial components, mainly LPS (Appelmelk et al., 1994; Elass-Rochard et al., 1995), and specific cellular receptors on a wide range of epithelial and immune cells (Suzuki et al., 2005) (Fig. 2). This results in the modulation of the production of various cytokines and of the recruitment of immune cells at the infected sites.

Anti-inflammatory Properties of Lf

Downregulation of Pro-inflammatory Cytokine Production

Lf may limit the inflammation associated with various bacterial infections, thus preventing septic shock. In particular, orally administred bLf has a beneficial effect on infections and protects animals against a lethal dose of LPS (Zagulski et al., 1989; Dial et al., 2005). Additionally, Lf plays anti-inflammatory roles in noninfectious pathologies such as rheumatoid arthritis, inflammatory bowel disorders, neurodegenerative diseases, and skin allergies. It has been shown that administration of Lf protects against chemically induced cutaneous inflammation (Cumberbatch et al., 2003) and nonsteroidal anti-inflammatory drug-induced intestinal injury (Dial et al., 2005). In collagen-induced and septic arthritis mouse models, peri-articular injection of hLf reduced inflammation (Guillen et al., 2000).

The protection against bacterial infections is well understood and is explained by Lf's ability to enhance the secretion of anti-inflammatory cytokines such as IL-10 and to inhibit the production of several pro-inflammatory cytokines, mainly tumor necrosis factor TNF-α, IL-1β, IL-6, and IL-8 (reviewed in Legrand et al., 2005). It is now known that these effects are mostly mediated by Lf's neutralization of pro-inflammatory microbial molecules such as LPS and bacterial unmethylated CpG-containing oligonucleotides (Britigan et al., 2001). Through its Lfcin domain, Lf binds to the lipid A of LPS (Appelmelk et al., 1994). The sequestration of LPS by Lf prevents the interaction of endotoxin with the LPS binding protein (LBP), therefore preventing LPS binding to membrane CD14 (mCD14) (Elass-Rochard et al., 1998) and its further presentation to the cell-signaling Toll-like receptor 4 (TLR4). Furthermore, other mechanisms may account for the inhibition of LPS-induced cytokine release by Lf. High-affinity interactions were indeed evidenced between the cationic N-terminal lobe of Lf and soluble CD14 (sCD14), either in a free form or complexed with LPS (Baveye et al., 2000). Through this mechanism, Lf inhibits the secretion of IL-8 by endothelial cells (Elass et al., 2002). Interestingly, it should be observed that following pretreatment with the Lf-LPS complex, cells are rendered tolerant to LPS challenge (Na et al., 2004). The presence of specific receptors on immune cells suggests that the modulation of host responses by Lf may be related to a direct effect via receptor-mediated signaling pathways. Indeed, Lf downregulates IL-6 secretion in monocytes through a mechanism involving Lf translocation to the nucleus and inhibition of NF-κB activation (Haversen et al., 2002).

The protective roles of Lf in noninfectious pathologies have not been clearly elucidated. However, as demonstrated in skin allergies (Cumberbatch et al., 2003) and in adjuvant-stimulated arthritis in rats (Hayashida et al., 2004), a decrease in pro-inflammatory cytokines, particularly TNF-α and IL1-β, and an increase in anti-inflammatory cytokines, including IL-10, were observed. In skin and lung allergies, Lf could be internalized into mast cells and interact with tryptase, chymase, and cathepsin G, three potent inflammatory proteases (He et al., 2003). Moreover, Lf can inhibit anti-IgE–induced histamine release from colon mast cells.

Inhibition of Reactive Oxygen Species Overproduction

During inflammation, reactive oxygen species (ROS) are overproduced by activated granulocytes, and their synthesis can be catalyzed in the presence of free iron. Thus, the chelation of iron by apoLf may prevent the formation of hydroxyl radicals and subsequent lipid peroxidation, particularly in patients with chronic hepatitis (Konishi et al., 2006). Lastly, during neurodegenerative diseases, transcytosis of plasma Lf through the blood-brain barrier may contribute to limit iron deposits and oxidative stress in the brain (Fillebeen et al., 1999).

Regulation of the Recruitment of Immune Cells at Inflammatory Sites

Cell migration is critical for a variety of biological processes. Recently, during influenza virus infection (pneumonia), bLf was shown to reduce the number of infiltrating leukocytes in bronchoalveolar lavage fluid, thus suppressing the hyperreaction of the host (Yamauchi et al., 2006). Although the exact mechanism of Lf's action in pneumonia is not defined, Baveye et al. (2000) demonstrated *in vitro* that Lf may modulate the activation of human umbilical endothelial cells by LPS and the expression of the adhesion molecules E-selectin and ICAM-1. Lf also decreases the recruitment of eosinophils, reduces pollen antigen-induced allergic airway inflammation in a murine model of asthma (Kruzel et al., 2006), and reduces migration of Langherans cells in cutaneous inflammation (Griffiths et al., 2001). Finally, Lf can modulate fibroblast motility by regulating MMP-1 gene expression, a matrix metalloproteinase involved in the extracellular matrix turnover and the promotion of cell migration (Oh et al., 2001).

Stimulation of the Host Immune Responses by **Lf**

Lf displays immunological properties influencing both innate and acquired immunities. In particular, oral administration of bLf seems to influence mucosal and systemic immune responses in mice (Sfeir et al., 2004). It was also recently demonstrated that a complex of Lf with monophosphoryl lipid A is an efficient adjuvant of the humoral and cellular immune responses (Chodaczek et al., 2006). Its stimulating effect on the immune system concerns mainly the maturation and differentiation of T-lymphocytes, the Th1/Th2 cytokine balance (Fischer et al., 2006), and the activation of phagocytes.

Effects of Lf on T Lymphocytes

Under nonpathogenic conditions, Lf is able to stimulate the differentiation of T cells from their immature precursors through the induction of CD4 antigen (Dhennin-Duthille et al., 2000). A similar effect was described on isolated thymocytes incubated overnight with Lf. Furthermore, oral delivery of Lf significantly increased the number of CD4+ cells in lymphoid tissues (Kuhara et al., 2000). Lf induces a Th1 polarization in diseases in which the ability to control infection or tumor relies on a strong response, but may also reduce Th1 cytokines to limit excessive inflammatory responses, as previously described. Thus, Lf enhances both the ability of Th cells to assist the fungicidal actions of macrophages (Wakabayashi et al., 2003) and BCG vaccine efficacity against challenge with *Mycobacterium tuberculosis* (Hwang et al., 2005). An upregulation of the Th1 response was associated with *S. aureus* clearance in Lf transgenic mice (Guillen et al., 2002). Oral administration of Lf increased the splenocyte production of IFN-γ and Il-12 in response to herpes simplex virus type 1 infection (Wakabayashi et al., 2004). Furthermore, the eradication of chronic hepatitis C virus by IFN therapy is optimized by administration of bLf, which induces a Th-1 cytokine-dominant environment in peripheral blood (Ishii et al., 2003). The proposed mechanism is that oral Lf induces IL-18 production in the small intestine, therefore leading to an increase in the level of Th1 cells. Furthermore, bLf may promote systemic host immunity by activating the transcription in the small intestine of important genes such as *IL-12p40*, *IFN-β*, and *NOD2* (Yamauchi et al., 2006).

Monocyte/Macrophage and PMN Activation by Lf

During infection, Lf secreted from neutrophil granules can bind to PMNs and monocytes/macrophages (Gahr et al., 1991) and promotes the secretion of inflammatory molecules such as TNF-α (Sorimachi et al., 1997). A recent study has shown that Lf may activate macrophages via TLR4-dependent and -independent signaling pathways (Curran et al., 2006). This activation induces CD40 expression and IL-6 secretion. The binding of Lf to macrophages also enhances the phagocytosis of pathogens, as demonstrated during the infection of bovine mammary gland by *S. aureus* (Kai et al., 2002). Lf may also influence myelopoiesis by modulating granulocyte-macrophage colony stimulating factor (GM-CSF) production (Sawatzki & Rich, 1989).

Lf Protects Against Tumorigenesis and Metastasis

Although the current views on the physiological role of Lf are dominated by its activity as an antimicrobial agent and immune modulator, Lf also directly affects cell growth and acts as a regulatory element in the defense against tumorigenesis and metastasis. Its antitumoral activities have long been suggested since numerous *in vivo* studies showed that bLf significantly reduces chemically induced tumorigenesis when administered orally to rodents (reviewed in Tsuda et al., 2002, 2004). Both bLf and recombinant hLf (rhLf) also possess antimetastatic effects, and injection or ingestion of bLf inhibits the growth of transplanted tumors and prevents experimental metastasis in rodents (Varadhachary et al., 2004; Bezault et al., 1994; Iigo et al., 1999). The spliced isoform of Lf (ΔLf) and Lf-derived peptides such as Lfcin also possess antitumor activities (Breton et al., 2004; Yoo et al., 1998).

Mechanisms of Action of Lf in Cancer

Inhibition of Carcinogenesis-Promoting Enzymes

The concomitant administration of bLf and carcinogens to rodents inhibits the induction of activating enzymes for carcinogenic heterocyclic amines, modulates lipid peroxidation, and activates antioxidant and carcinogen detoxification enzyme activities, blocking cancer development (Tsuda et al., 2002; Chandra Mohan et al., 2006b).

Activation of Immune Cells

Lf promotes NK cell cytotoxicity, as shown by *in vitro* and *in vivo* studies (Damiens et al., 1998b; Bezault et al., 1994; Sekine et al., 1997). In animal models of carcinogenesis, a marked increase in the number of NK and $CD8^+$, $CD4^+$, and $IFN\gamma^+$ cells was observed in the mucosal layer of the small intestine as well as in the peripheral cell population of bLf-treated rodents (Iigo et al., 1999). Oral administration of rhLf to tumor-bearing mice also led to an enhancement of both local mucosal and systemic immune responses (Varadhachary et al., 2004).

Production of IL-18 and Inhibition of Angiogenesis

Recently, a mechanism involving the enhancement by Lf of caspase-1 activity followed by a transient increase in the active form of IL-18 has been reported to elevate both mucosal and systemic immune responses via the regulation of cytokine production, the mechanism of which remains to be elucidated (Varadhachary et al., 2004; Kuhara et al., 2006; Iigo et al., 2004). Moreover, IL-18 as an anti-angiogenic compound might also be partially responsible for bLf inhibition of angiogenesis (Shimamura et al., 2004). Curiously, bLf and hLf exert opposite effects on angiogenesis. Whereas orally administered bLf inhibits angiogenesis in rats (Norrby et al., 2001) and tumor-induced angiogenesis in mice (Shimamura et al., 2004), hLf exerts a specific pro-angiogenic effect (Norrby, 2004). Lf was recently shown to stimulate *in vivo* angiogenesis via the upregulation of the VEGF receptor KDR/Flk-1, subsequently promoting VEGF-induced proliferation and migration of endothelial cells (Kim et al., 2006). Since tumor growth is angiogenesis-dependent, the extensive therapeutic potential warrants further studies to elucidate the contradictory effects of Lf on angiogenesis. In contrast to IL-18, some cytokines such as IL-8, IL-6, and IL-1β are potent angiogenic stimulators; therefore, Lf action *in vivo* might partially involve suppression of the production of angiogenic cytokines (Haversen et al., 2002).

Promotion of Apoptosis

Lf is able to promote apoptosis in cancerous cells. Bovine Lf changed apoptosis-related gene expression in colon mucosa (Fujita et al., 2004a) and in chemically induced lung proliferative lesions (Matsuda et al., 2007). Thus, activation of caspases (3 and 8), death-inducing receptor (Fas), and pro-apoptotic Bcl-2 members (Bid, Bax) has been reported (Fujita et al., 2004b; Chandra Mohan et al., 2006a). Caspase-3 plays a central role in various apoptosis cascades, and Lf has recently been shown to enhance procaspase-3 maturation *in vitro* (Katunuma et al., 2004). The Y^{679}-K^{695} domain of Lf, which binds procaspase-3, is able as a synthetic peptide to compete with Lf and block procaspase-3 processing (Katunuma et al., 2006). This result is further strengthened by *in vivo* data since in D-galactosamine–induced apoptotic rat liver, Lf was shown to translocate from lysosomes into the cytoplasm of hepatocytes by an unknown mechanism and to lead to procaspase-3 activation (Katunuma et al., 2006). Other reports indicate that Lf could either promote or inhibit apoptosis in a dose-dependent manner. *In vitro* studies on PC12 cells showed that high doses of Lf lead to activation of caspases-3 and -8 and to a decrease in the protein expression of phosphorylated ERK1/2 and Bcl-2, whereas low doses upregulate phosphorylated ERK1/2 and Bcl-2 expression and protect cells from FasL-induced apoptosis (Lin et al., 2005). Lf may also indirectly interfere with apoptosis by inhibiting retinoid signaling pathways since translocation of exogeneous Lf to the nucleus in the presence of retinoids leads to the alteration of retinoid-induced gene transactivation in mammary cells *in vitro* (Baumrucker et al., 2005). Relationships between Lf and retinoids are complex since the *Lf* gene promoter contains a retinoid response element (Teng, 2006) and require further investigation.

Natural and synthetic bLf peptides also showed cytotoxic effects for cancer cells *in vitro* and *in vivo*. Bovine Lfcin induces apoptosis in different cancer cell lines, and cellular exposure to it caused caspase-3 activation, G1 arrest, and/or triggering of the mitochondrial pathway of apoptosis through the production of ROS (Yoo et al., 1997; Mader et al., 2005). A strong correlation between antitumor activity and the net positive charge close to +7 of a shorter peptide, analogous to bovine Lfcin (residues 14–31) (Yang et al., 2004), might partially explain bovine Lfcin activity. Pepsin-digested bLf (Lfn-p) induced apoptosis via the activation of both caspase-3 and JNK/SAP kinase in squamous carcinoma cells (Sakai et al., 2005), but these authors could not clearly elucidate whether a relationship exists between these two pathways.

Regulation of Cell Growth

Lf also limits the growth of tumor cells by inducing cell cycle arrest at the G1/S transition. In the MDA-MB-231 breast cancer cell line, the molecular mechanism underlying this G1 arrest associates both inhibition of cdk2 and cdk4 activities and an increase in cdk inhibitor (CKI) p21 expression, involving the MAPK pathway (Damiens et al., 1999). In head and neck cancer cells, Lf downregulation of cell growth is mediated through the p27/cyclin E-dependent pathway involving changes in the phosphorylation status of AKT (Xiao et al., 2004). The NF-κB signaling pathway has been implicated in HeLa cervical carcinoma cells, and Lf was shown to upregulate the tumor suppressor protein p53 and its target genes *mdm2* and *p21* (Oh et al., 2004). Recently, retinoblastoma protein (Rb)-mediated growth arrest has been demonstrated. Thus, Lf induces overexpression of Rb in tumor cell lines. A majority of Rb then remains as a hypophosphorylated isoform, and this form binds to E2F1, downregulating its transcriptional activity and therefore leading to cell cycle arrest (Son et al., 2006). In addition, Lf-induced overexpression of the CKI protein p21 was also observed and appears to occur independently of p53, whereas the expression of p27, another member of the Kip/Cip group of CKI, was not altered by Lf (Son et al., 2006).

Expression of cytoplasmic ΔLf, the cytoplasmic Lf isoform, also led to cell cycle arrest (Breton et al., 2004) and upregulation of both *Rb* and *Skp1* (S-phase kinase-associated protein) genes (Mariller et al., 2007). Since Skp1 belongs to the SCF (Skp1/Cullin-1/F-box ubiquitin ligase) complex responsible for the ubiquitination of proteins leading to their degradation by the proteasome at the G1/S transition, ΔLf may therefore indirectly modulate the half-life of cell cycle actors. ΔLf acts as a transactivating factor via a direct *in situ* interaction with two Lf response elements detected in both *Skp1* and *Rb* promoters (Mariller et al., 2007). The GGCACTTGC sequence (He & Furmanski, 1995) was also found to be functional in the IL-1β promoter and shown to be responsible for IL-1β transactivation by Lf *in vitro* (Son et al., 2002). Further studies will be necessary to confirm whether Lf induces *in situ* transcription of genes and if so, to find out whether both Lf isoforms target the same genes.

Lf Expression Is Downregulated in Certain Types of Cancer

Since Lf isoforms exert cancer-suppressive effects, they may act as tumor suppressor proteins, whose expression would be suppressed upon carcinogenesis and their gene downregulated in cancerous cells. Lf expression has been found to be decreased or absent in numerous cancers (reviewed in Ward et al., 2005). Downregulation is due to structural alterations such as mutations, allelic loss of part of chromosome 3, and modification of the degree, as well as the pattern, of methylation (Teng et al., 2004; Iijima et al., 2006). These genetic and epigenetic inactivations of the *Lf* gene in cancer may therefore provide the tumor cells with a selective growth advantage.

ΔLf expression was observed in normal tissues but was downregulated in their malignant counterparts (Siebert & Huang, 1997; Liu et al., 2003; Benaïssa et al., 2005). Interestingly, it was shown that the expression level of either Lf or ΔLf mRNAs was of good prognosis value in human breast cancer, with high concentrations associated with longer relapse-free and overall survival, suggesting the usefulness of the detection of either Lf or ΔLf transcripts as markers for the follow-up of breast cancer patients (Benaïssa et al., 2005). Lf, which was also identified as a cancer-specific marker of endocervical adenocarcinomas, may also be useful for the early detection of disease and for prognosis (Farley et al., 1997).

Lf May Also Display Mitogenic Effects

Lf has been shown to stimulate osteoblast proliferation *in vitro* and to exert an anabolic effect promoting bone formation *in vivo* (Cornish et al., 2006). It also limits bone resorption by inhibiting osteoclastogenesis (Lorget et al., 2002; Cornish et al., 2006). Taken together, these data suggest that Lf has a physiological role in bone growth and might therefore act as a potential therapeutic agent in osteoporosis.

Commercial and Clinical Applications of Lf

The large-scale preparation of bLf from cheese whey or skim milk (up to 100 metric tonnes per year) and of recombinant hLf produced in microorganisms and plants makes Lf available for human and animal (fish farming) health purposes and commercial applications. The first major application of bLf was the supplementation of infant formulas, but it is now added to cosmetics, pet care supplements, and immune system–enhancing nutraceuticals, including drinks, fermented milks, and chewing gums. In all these media, Lf is expected to exert its natural antimicrobial, antioxidative, anti-inflammatory, anticancerous, and immunomodulatory properties. Furthermore, clinical trials demonstrated the efficiency of Lf against infections and in inflammatory diseases. For example, a recent clinical study concluded that the combination of Lf and fluconazole at the threshold minimal inhibitory concentrations elicited potent synergism, leading to total fungistasis of *C. albicans* and *C. glabrata* vaginal pathogens (Naidu et al., 2004). Lf was also reported as a potent molecule in the treatment of common inflammatory diseases (reviewed in Legrand et al., 2005). In addition, extensive clinical trials are underway in Japan to further explore its preventive potential against colon carcinogenesis (Tsuda et al., 2002).

Lf also offers applications in food preservation and safety, either by retarding lipid oxidation (Medina et al., 2002) or by limiting the growth of microbes. For example, incorporation of Lf into edible films has a great potential to enhance the safety of foods since the film can function as a physical barrier as well as an antimicrobial agent. Lf can be also directly used as a spray applied to beef carcasses (Taylor et al., 2004).

Lastly, Lf can be used as a clinical marker of inflammatory diseases since Lf levels in blood and biological fluids may greatly increase in septicemia or during Severe Acute Respiratory Syndrome (Reghunathan et al., 2005). In the same way, fecal Lf levels quickly increase with the influx of leukocytes into the intestinal lumen during inflammation. Fecal Lf is thus used as a noninvasive diagnostic tool to evaluate the severity of intestinal inflammation in patients presenting with abdominal pain and diarrhea (Greenberg et al., 2002). This biomarker has been shown to be a sensitive and specific marker of disease activity in chronic inflammatory bowel disease (Kane et al., 2003) and in Crohn's disease (Buderus et al., 2004).

Conclusion

When Lf was discovered in milk in the early 1960s, it would have been difficult to imagine that it could exert so many biological activities. Its protective roles against pathogens, inflammation, and cancer make the molecule a centerpiece of the nonspecific immune system. The time has now come for the development and application of promising health-enhancing nutraceuticals for food and pharmaceutical applications of this intriguing molecule.

References

Anderson, B. F., Baker, H. M., Dodson, E. J., Norris, G. E., Rumball, S. V., Waters, J. M., & Baker, E. N. (1987). Structure of human lactoferrin at 3.2-Å resolution. *Proceedings of the National Academy of Sciences USA, 84*, 1769–1773.

Appelmelk, B. J., An, Y. Q., Geerts, M., Thijs, B. G., de Boer, H. A., MacLaren, D. M., de Graaff, J., & Nuijens, J. H. (1994). Lactoferrin is a lipid A-binding protein. *Infection and Immunity, 62*, 2628–2632.

Baker, E. N., & Baker, H. M. (2005). Molecular structure, binding properties and dynamics of lactoferrin. *Cellular and Molecular Life Sciences, 62*, 2531–2539.

Baumrucker, C. R., Schanbacher, F., Shang, Y., & Green, M. H. (2006). Lactoferrin interaction with retinoid signaling: Cell growth and apoptosis in mammary cells. *Domestic Animal Endocrinology, 30*, 289–303.

Baveye, S., Elass, E., Fernig, D. G., Blanquart, C., Mazurier, J., & Legrand, D. (2000). Human lactoferrin interacts with soluble CD14 and inhibits expression of endothelial adhesion molecules, E-selectin and ICAM-1, induced by the CD14-lipopolysaccharide complex. *Infection and Immunity, 68*, 6519–6525.

Bellamy, W., Takase, M., Yamauchi, K., Wakabayashi, H., Kawase, K., & Tomita, M. (1992). Identification of the bactericidal domain of lactoferrin. *Biochimica et Biophysica Acta, 1121*, 130–136.

Benaïssa, M., Peyrat, J. P., Hornez, L., Mariller, C., Mazurier, J., & Pierce, A. (2005). Expression and prognostic value of lactoferrin mRNA isoforms in human breast cancer. *International Journal of Cancer, 114*, 299–306.

Bezault, J., Bhimani, R., Wiprovnick, J., & Furmanski, P. (1994). Human lactoferrin inhibits growth of solid tumors and development of experimental metastases in mice. *Cancer Research, 54*, 2310–2312.

Breton, M., Mariller, C., Benaïssa, M., Caillaux, K., Browaeys, E., Masson, M., Vilain, J. P., Mazurier, J., & Pierce, A. (2004). Expression of delta-lactoferrin induces cell cycle arrest. *Biometals, 17*, 325–329.

Britigan, B. E., Lewis, T. S., Waldschmidt, M., McCormick, M. L., & Krieg, A. M. (2001). Lactoferrin binds CpG-containing oligonucleotides and inhibits their immunostimulatory effects on human B cells. *Journal of Immunology, 167*, 2921–2928.

Buderus, S., Boone, J., Lyerly, D., & Lentze, M. J. (2004). Fecal lactoferrin: A new parameter to monitor infliximab therapy. *Digestive Diseases and Sciences, 49*, 1036–1039.

Callebaut, C., Blanco, J., Benkirane, N., Krust, B., Jacotot, E., Guichard, G., Seddiki, N., Svab, J., Dam, E., Muller, S., Briand, J. P., & Hovanessian, A. G. (1998). Identification of V3 loop-binding proteins as potential receptors implicated in the binding of HIV particles to CD4(+) cells. *Journal of Biological Chemistry, 273*, 21988–21997.

Chandra Mohan, K. V., Devaraj, H., Prathiba, D., Hara, Y., & Nagini, S. (2006a). Antiproliferative and apoptosis inducing effect of lactoferrin and black tea polyphenol combination on hamster buccal pouch carcinogenesis. *Biochimica et Biophysica Acta, 1760*, 1536–1544.

Chandra Mohan, K. V., Kumaraguruparan, R., Prathiba, D., & Nagini, S. (2006b). Modulation of xenobiotic-metabolizing enzymes and redox status during chemoprevention of hamster buccal carcinogenesis by bovine lactoferrin. *Nutrition, 22*, 940–946.

Chodaczek, G., Zimecki, M., Lukasiewicz, J., & Lugowski, C. (2006). A complex of lactoferrin with monophosphoryl lipid A is an efficient adjuvant of the humoral and cellular immune response in mice. *Medical Microbiology and Immunology, 195*, 207–216.

Coddeville, B., Strecker, G., Wieruszeski, J. M., Vliegenthart, J. F., van Halbeek, H., Peter-Katalinic, J., Egge, H., & Spik, G. (1992). Heterogeneity of bovine lactotransferrin glycans. Characterization of α-D-Galp-(1–>3)-β-D-Gal- and α-NeuAc-(2–>6)-β-D-GalpNAc-(1–>4)-β-D-GlcNAc-substituted N-linked glycans. *Carbohydrate Research, 236*, 145–164.

Cornish, J., Palmano, K., Callon, K. E., Watson, M., Lin, J. M., Valenti, P., Naot, D., Grey, A. B., & Reid, I. R. (2006). Lactoferrin and bone; structure-activity relationships. *Biochemistry and Cell Biology, 84*, 297–302.

Cumberbatch, M., Bhushan, M., Dearman, R. J., Kimber, I., & Griffiths, C. E. (2003). IL-1β-induced Langerhans' cell migration and TNF-α production in human skin: Regulation by lactoferrin. *Clinical and Experimental Immunology, 132*, 352–359.

Curran, C. S., Demick, K. P., & Mansfield, J. M. (2006). Lactoferrin activates macrophages via TLR4-dependent and -independent signaling pathways. *Cellular Immunology, 242*, 23–30.

Damiens, E., El Yazidi, I., Mazurier, J., Elass-Rochard, E., Duthille, I., Spik, G., & Boilly-Marer, Y. (1998a). Role of heparan sulphate proteoglycans in the regulation of human lactoferrin binding and activity in the MDA-MB-231 breast cancer cell line. *European Journal of Cell Biology, 77*, 344–351.

Damiens, E., Mazurier, J., El Yazidi, I., Masson, M., Duthille, I., Spik, G., & Boilly-Marer, Y. (1998b). Effects of human lactoferrin on NK cell cytotoxicity against haematopoietic and epithelial tumour cells. *Biochimica et Biophysica Acta, 1402*, 277–287.

Damiens, E., El Yazidi, I., Mazurier, J., Duthille, I., Spik, G., & Boilly-Marer, Y. (1999). Lactoferrin inhibits G1 cyclin-dependent kinases during growth arrest of human breast carcinoma cells. *Journal of Cellular Biochemistry, 74*, 486–498.

Dhennin-Duthille, I., Masson, M., Damiens, E., Fillebeen, C., Spik, G., & Mazurier, J. (2000). Lactoferrin upregulates the expression of CD4 antigen through the stimulation of the mitogen-activated protein kinase in the human lymphoblastic T Jurkat cell line. *Journal of Cellular Biochemistry, 79*, 583–593.

Di Biase, A. M., Pietrantoni, A., Tinari, A., Siciliano, R., Valenti, P., Antonini, G., Seganti, L., & Superti, F. (2003). Heparin-interacting sites of bovine lactoferrin are involved in anti-adenovirus activity. *Journal of Medical Virology, 69*, 495–502.

Di Biase, A. M., Tinari, A., Pietrantoni, A., Antonini, G., Valenti, P., Conte, M. P., & Superti, F. (2004). Effect of bovine lactoferricin on enteropathogenic Yersinia adhesion and invasion in HEp-2 cells. *Journal of Medical Microbiology, 53*, 407–412.

Dial, E. J., & Lichtenberger, L. M. (2002). Effect of lactoferrin on Helicobacter felis induced gastritis. *Biochemistry and Cell Biology, 80*, 113–117.

Dial, E. J., Dohrman, A. J., Romero, J. J., & Lichtenberger, L. M. (2005). Recombinant human lactoferrin prevents NSAID-induced intestinal bleeding in rodents. *Journal of Pharmacy and Pharmacology, 57*, 93–99.

Diarra, M. S., Petitclerc, D., Deschenes, E., Lessard, N., Grondin, G., Talbot, B. G., & Lacasse, P. (2003). Lactoferrin against *Staphylococcus aureus* mastitis. Lactoferrin alone or in combination with penicillin G on bovine polymorphonuclear function and mammary epithelial cells colonisation by *Staphylococcus aureus*. *Veterinary Immunology and Immunopathology, 95*, 33–42.

Dziadek, B., Dziadek, J., & Dlugonska, H. (2007). Identification of *Toxoplasma gondii* proteins binding human lactoferrin: A new aspect of rhoptry proteins function. *Experimental Parasitology, 115*, 277–282.

Elass, E., Masson, M., Mazurier, J., & Legrand, D. (2002). Lactoferrin inhibits the lipopolysaccharide-induced expression and proteoglycan-binding ability of interleukin-8 in human endothelial cells. *Infection and Immunity, 70*, 1860–1866.

Elass-Rochard, E., Roseanu, A., Legrand, D., Trif, M., Salmon, V., Motas, C., Montreuil, J., & Spik, G. (1995). Lactoferrin-lipopolysaccharide interaction: Involvement of the 28-34 loop region of human lactoferrin in the high-affinity binding to *Escherichia coli* 055B5 lipopolysaccharide. *Biochemical Journal, 312 (Pt 3)*, 839–845.

Elass-Rochard, E., Legrand, D., Salmon, V., Roseanu, A., Trif, M., Tobias, P. S., Mazurier, J., & Spik, G. (1998). Lactoferrin inhibits the endotoxin interaction with CD14 by competition with the lipopolysaccharide-binding protein. *Infection and Immunity, 66*, 486–491.

Ellison, R. T., III, LaForce, F. M., Giehl, T. J., Boose, D. S., & Dunn, B. E. (1990). Lactoferrin and transferrin damage of the Gram-negative outer membrane is modulated by Ca^{2+} and Mg^{2+}. *Journal of General Microbiology, 136*, 1437–1446.

Erdei, J., Forsgren, A., & Naidu, A. S. (1994). Lactoferrin binds to porins OmpF and OmpC in *Escherichia coli*. *Infection and Immunity, 62*, 1236–1240.

Farley, J., Loup, D., Nelson, M., Mitchell, A., Esplund, G., Macri, C., Harrison, C., & Gray, K. (1997). Neoplastic transformation of the endocervix associated with downregulation of lactoferrin expression. *Molecular Carcinogenesis, 20*, 240–250.

Fillebeen, C., Descamps, L., Dehouck, M. P., Fenart, L., Benaïssa, M., Spik, G., Cecchelli, R., & Pierce, A. (1999). Receptor-mediated transcytosis of lactoferrin through the blood-brain barrier. *Journal of Biological Chemistry, 274*, 7011–7017.

Fischer, R., Debbabi, H., Dubarry, M., Boyaka, P., & Tomé, D. (2006). Regulation of physiological and pathological Th1 and Th2 responses by lactoferrin. *Biochemistry and Cell Biology, 84*, 303–311.

Fritsch, G., Sawatzki, G., Treumer, J., Jung, A., & Spira, D. T. (1987). Plasmodium falciparum: Inhibition *in vitro* with lactoferrin, desferriferrithiocin, and desferricrocin. *Experimental Parasitology, 63*, 1–9.

Fujita, K., Matsuda, E., Sekine, K., Iigo, M., & Tsuda, H. (2004a). Lactoferrin enhances Fas expression and apoptosis in the colon mucosa of azoxymethane-treated rats. *Carcinogenesis, 25*, 1961–1966.

Fujita, K., Matsuda, E., Sekine, K., Iigo, M., & Tsuda, H. (2004b). Lactoferrin modifies apoptosis-related gene expression in the colon of the azoxymethane-treated rat. *Cancer Letters, 213*, 21–29.

Gahr, M., Speer, C. P., Damerau, B., & Sawatzki, G. (1991). Influence of lactoferrin on the function of human polymorphonuclear leukocytes and monocytes. *Journal of Leukocyte Biology, 49*, 427–433.

Gifford, J. L., Hunter, H. N., & Vogel, H. J. (2005). Lactoferricin: A lactoferrin-derived peptide with antimicrobial, antiviral, antitumor and immunological properties. *Cellular and Molecular Life Sciences, 62*, 2588–2598.

Goldman, A. S., Garza, C., Schanler, R. J., & Goldblum, R. M. (1990). Molecular forms of lactoferrin in stool and urine from infants fed human milk. *Pediatric Research, 27*, 252–255.

Gomez, H. F., Ochoa, T. J., Carlin, L. G., & Cleary, T. G. (2003). Human lactoferrin impairs virulence of *Shigella flexneri*. *Journal of Infectious Diseases, 187*, 87–95.

Greenberg, D. E., Jiang, Z. D., Steffen, R., Verenker, M. P., & DuPont, H. L. (2002). Markers of inflammation in bacterial diarrhea among travelers, with a focus on enteroaggregative *Escherichia coli* pathogenicity. *Journal of Infectious Diseases, 185*, 944–949.

Grey, A., Banovic, T., Zhu, Q., Watson, M., Callon, K., Palmano, K., Ross, J., Naot, D., Reid, I. R., & Cornish, J. (2004). The low-density lipoprotein receptor-related protein 1 is a mitogenic receptor for lactoferrin in osteoblastic cells. *Molecular Endocrinology, 18*, 2268–2278.

Griffiths, C. E., Cumberbatch, M., Tucker, S. C., Dearman, R. J., Andrew, S., Headon, D. R., & Kimber, I. (2001). Exogenous topical lactoferrin inhibits allergen-induced Langerhans cell migration and cutaneous inflammation in humans. *British Journal of Dermatology, 144*, 715–725.

Groot, F., Geijtenbeek, T. B., Sanders, R. W., Baldwin, C. E., Sanchez-Hernandez, M., Floris, R., van Kooyk, Y., de Jong, E. C., & Berkhout, B. (2005). Lactoferrin prevents dendritic cell-mediated human immunodeficiency virus type 1 transmission by blocking the DC-SIGN–gp120 interaction. *Journal of Virology, 79*, 3009–3015.

Guillen, C., McInnes, I. B., Vaughan, D., Speekenbrink, A. B., & Brock, J. H. (2000). The effects of local administration of lactoferrin on inflammation in murine autoimmune and infectious arthritis. *Arthritis and Rheumatism, 43*, 2073–2080.

Guillen, C., McInnes, I. B., Vaughan, D. M., Kommajosyula, S., van Berkel, P. H., Leung, B. P., Aguila, A., & Brock, J. H. (2002). Enhanced Th1 response to *Staphylococcus aureus* infection in human lactoferrin-transgenic mice. *Journal of Immunology, 168*, 3950–3957.

Hara, K., Ikeda, M., Saito, S., Matsumoto, S., Numata, K., Kato, N., Tanaka, K., & Sekihara, H. (2002). Lactoferrin inhibits hepatitis B virus infection in cultured human hepatocytes. *Hepatology Research, 24*, 228.

Hasegawa, K., Motsuchi, W., Tanaka, S., & Dosako, S. (1994). Inhibition with lactoferrin of *in vitro* infection with human herpes virus. *Japanese Journal of Medical Science and Biology, 47*, 73–85.

Haversen, L., Ohlsson, B. G., Hahn-Zoric, M., Hanson, L. A., & Mattsby-Baltzer, I. (2002). Lactoferrin down-regulates the LPS-induced cytokine production in monocytic cells via NF-κB. *Cellular Immunology, 220*, 83–95.

Hayashida, K., Kaneko, T., Takeuchi, T., Shimizu, H., Ando, K., & Harada, E. (2004). Oral administration of lactoferrin inhibits inflammation and nociception in rat adjuvant-induced arthritis. *Journal of Veterinary Medical Science, 66*, 149–154.

He, J., & Furmanski, P. (1995). Sequence specificity and transcriptional activation in the binding of lactoferrin to DNA. *Nature, 373*, 721–724.

He, S., McEuen, A. R., Blewett, S. A., Li, P., Buckley, M. G., Leufkens, P., & Walls, A. F. (2003). The inhibition of mast cell activation by neutrophil lactoferrin: Uptake by mast cells and interaction with tryptase, chymase and cathepsin G. *Biochemical Pharmacology, 65*, 1007–1015.

Hendrixson, D. R., Qiu, J., Shewry, S. C., Fink, D. L., Petty, S., Baker, E. N., Plaut, A. G., & St Geme, J. W., III (2003). Human milk lactoferrin is a serine protease that cleaves Haemophilus surface proteins at arginine-rich sites. *Molecular Microbiology, 47*, 607–617.

Herz, J., & Strickland, D. K. (2001). LRP: A multifunctional scavenger and signaling receptor. *Journal of Clinical Investigation*, *108*, 779–784.

Hwang, S. A., Kruzel, M. L., & Actor, J. K. (2005). Lactoferrin augments BCG vaccine efficacy to generate T helper response and subsequent protection against challenge with virulent *Mycobacterium tuberculosis*. *International Immunopharmacology*, *5*, 591–599.

Iigo, M., Kuhara, T., Ushida, Y., Sekine, K., Moore, M. A., & Tsuda, H. (1999). Inhibitory effects of bovine lactoferrin on colon carcinoma 26 lung metastasis in mice. *Clinical and Experimental Metastasis*, *17*, 35-40.

Iigo, M., Shimamura, M., Matsuda, E., Fujita, K., Nomoto, H., Satoh, J., Kojima, S., Alexander, D. B., Moore, M. A., & Tsuda, H. (2004). Orally administered bovine lactoferrin induces caspase-1 and interleukin-18 in the mouse intestinal mucosa: A possible explanation for inhibition of carcinogenesis and metastasis. *Cytokine*, *25*, 36–44.

Iijima, H., Tomizawa, Y., Iwasaki, Y., Sato, K., Sunaga, N., Dobashi, K., Saito, R., Nakajima, T., Minna, J. D., & Mori, M. (2006). Genetic and epigenetic inactivation of LTF gene at 3p21.3 in lung cancers. *International Journal of Cancer*, *118*, 797–801.

Ishii, K., Takamura, N., Shinohara, M., Wakui, N., Shin, H., Sumino, Y., Ohmoto, Y., Teraguchi, S., & Yamauchi, K. (2003). Long-term follow-up of chronic hepatitis C patients treated with oral lactoferrin for 12 months. *Hepatology Research*, *25*, 226–233.

Jameson, G. B., Anderson, B. F., Norris, G. E., Thomas, D. H., & Baker, E. N. (1998). Structure of human apolactoferrin at 2.0 Å resolution. Refinement and analysis of ligand-induced conformational change. *Acta Crystallographica D*, *54*, 1319–1335.

Japelj, B., Pristovsek, P., Majerle, A., & Jerala, R. (2005). Structural origin of endotoxin neutralization and antimicrobial activity of a lactoferrin-based peptide. *Journal of Biological Chemistry*, *280*, 16955–16961.

Kai, K., Komine, K., Komine, Y., Kuroishi, T., Kozutsumi, T., Kobayashi, J., Ohta, M., Kitamura, H., & Kumagai, K. (2002). Lactoferrin stimulates *Staphylococcus aureus* killing activity of bovine phagocytes in the mammary gland. *Microbiology and Immunology*, *46*, 187–194.

Kane, S. V., Sandborn, W. J., Rufo, P. A., Zholudev, A., Boone, J., Lyerly, D., Camilleri, M., & Hanauer, S. B. (2003). Fecal lactoferrin is a sensitive and specific marker in identifying intestinal inflammation. *American Journal of Gastroenterology*, *98*, 1309–1314.

Katunuma, N., Le, Q. T., Murata, E., Matsui, A., Majima, E., Ishimaru, N., Hayashi, Y., & Ohashi, A. (2006). A novel apoptosis cascade mediated by lysosomal lactoferrin and its participation in hepatocyte apoptosis induced by D-galactosamine. *FEBS Letters*, *580*, 3699–3705.

Katunuma, N., Murata, E., Le, Q.T., Hayashi, Y., & Ohashi, A. (2004). New apoptosis cascade mediated by lysosomal enzyme and its protection by epigallo-catechin gallate. *Advances in Enzyme Regulation*, *44*, 1–10.

Kawasaki, Y., Tazume, S., Shimizu, K., Matsuzawa, H., Dosako, S., Isoda, H., Tsukiji, M., Fujimura, R., Muranaka, Y., & Isihida, H. (2000). Inhibitory effects of bovine lactoferrin on the adherence of enterotoxigenic *Escherichia coli* to host cells. *Bioscience, Biotechnology, and Biochemistry*, *64*, 348–354.

Kim, C. W., Son, K. N., Choi, S. Y., & Kim, J. (2006). Human lactoferrin upregulates expression of KDR/Flk-1 and stimulates VEGF-A-mediated endothelial cell proliferation and migration. *FEBS Letters*, *580*, 4332–4336.

Konishi, M., Iwasa, M., Yamauchi, K., Sugimoto, R., Fujita, N., Kobayashi, Y., Watanabe, S., Teraguchi, S., Adachi, Y., & Kaito, M. (2006). Lactoferrin inhibits lipid peroxidation in patients with chronic hepatitis C. *Hepatology Research*, *36*, 27–32.

Kruzel, M. L., Bacsi, A., Choudhury, B., Sur, S., & Boldogh, I. (2006). Lactoferrin decreases pollen antigen-induced allergic airway inflammation in a murine model of asthma. *Immunology*, *119*, 159–166.

Kuhara, T., Iigo, M., Itoh, T., Ushida, Y., Sekine, K., Terada, N., Okamura, H., & Tsuda, H. (2000). Orally administered lactoferrin exerts an antimetastatic effect and enhances production of IL-18 in the intestinal epithelium. *Nutrition and Cancer, 38,* 192–199.

Kuhara, T., Yamauchi, K., Tamura, Y., & Okamura, H. (2006). Oral administration of lactoferrin increases NK cell activity in mice via increased production of IL-18 and type I IFN in the small intestine. *Journal of Interferon Cytokine Research, 26,* 489–499.

Kumar, J., Weber, W., Münchau, S., Yadav, S., Bhaskar Singh, S., Saravanan, K., Paramasivam, M., Sharma, S., Kaur, P., Bhushan, A., Srinivasan, A., Betzel, C., & Singh, T. P. (2003). Crystal structure of human seminal diferric lactoferrin at 3.4 Å resolution. *Indian Journal of Biochemistry and Biophysics, 40,* 14–21.

Legrand, D., Mazurier, J., Metz-Boutigue, M. H., Jollès, J., Jollès, P., Montreuil, J., & Spik, G. (1984). Characterization and localization of an iron-binding 18-kDa glycopeptide isolated from the N-terminal half of human lactotransferrin. *Biochimica et Biophysica Acta, 787,* 90–96.

Legrand, D., van Berkel, P. H., Salmon, V., van Veen, H. A., Slomianny, M. C., Nuijens, J. H., & Spik, G. (1997). The N-terminal Arg2, Arg3 and Arg4 of human lactoferrin interact with sulphated molecules but not with the receptor present on Jurkat human lymphoblastic T-cells. *Biochemical Journal, 327 (Pt 3),* 841–846.

Legrand, D., Vigié, K., Said, E. A., Elass, E., Masson, M., Slomianny, M. C., Carpentier, M., Briand, J. P., Mazurier, J., & Hovanessian, A. G. (2004). Surface nucleolin participates in both the binding and endocytosis of lactoferrin in target cells. *European Journal of Biochemistry, 271,* 303–317.

Legrand, D., Elass, E., Carpentier, M., & Mazurier, J. (2005). Lactoferrin: A modulator of immune and inflammatory responses. *Cellular and Molecular Life Sciences, 62,* 2549–2559.

Li, Y. M., Tan, A. X., & Vlassara, H. (1995). Antibacterial activity of lysozyme and lactoferrin is inhibited by binding of advanced glycation-modified proteins to a conserved motif. *Nature Medicine, 1,* 1057–1061.

Lima, M. F., & Kierszenbaum, F. (1987). Lactoferrin effects on the interaction of blood forms of *Trypanosoma cruzi* with mononuclear phagocytes. *International Journal for Parasitology, 17,* 1205–1208.

Lin, T. Y., Chiou, S. H., Chen, M., & Kuo, C. D. (2005). Human lactoferrin exerts bi-directional actions on PC12 cell survival via ERK1/2 pathway. *Biochemical and Biophysical Research Communications, 337,* 330–336.

Ling, J. M., & Schryvers, A. B. (2006). Perspectives on interactions between lactoferrin and bacteria. *Biochemistry and Cell Biology, 84,* 275–281.

Liu, D., Wang, X., Zhang, Z., & Teng, C. T. (2003). An intronic alternative promoter of the human lactoferrin gene is activated by Ets. *Biochemical and Biophysical Research Communications, 301,* 472–479.

Longhi, C., Conte, M. P., Seganti, L., Polidoro, M., Alfsen, A., & Valenti, P. (1993). Influence of lactoferrin on the entry process of *Escherichia coli* HB101 (pRI203) in HeLa cells. *Medical Microbiology and Immunology, 182,* 25–35.

Lorget, F., Clough, J., Oliveira, M., Daury, M. C., Sabokbar, A., & Offord, E. (2002). Lactoferrin reduces *in vitro* osteoclast differentiation and resorbing activity. *Biochemical and Biophysical Research Communications, 296,* 261–266.

Mader, J. S., Salsman, J., Conrad, D. M., & Hoskin, D. W. (2005). Bovine lactoferricin selectively induces apoptosis in human leukemia and carcinoma cell lines. *Molecular Cancer Therapeutics, 4,* 612–624.

Mann, D. M., Romm, E., & Migliorini, M. (1994). Delineation of the glycosaminoglycan-binding site in the human inflammatory response protein lactoferrin. *Journal of Biological Chemistry, 269,* 23661–23667.

Marchetti, M., Longhi, C., Conte, M. P., Pisani, S., Valenti, P., & Seganti, L. (1996). Lactoferrin inhibits herpes simplex virus type 1 adsorption to Vero cells. *Antiviral Research, 29,* 221–231.

Mariller, C., Benaïssa, M., Hardivillé, S., Breton, M., Pradelle, G., Mazurier, J., & Pierce, A. (2007). Human delta-lactoferrin is a transcription factor which enhances Skp1 (S-phase kinase associated protein) gene expression. *FEBS Journal, 274*, 2038–2053.

Masson, P. L., Heremans, J. F., & Schonne, E. (1969). Lactoferrin, an iron-binding protein in neutrophilic leukocytes. *Journal of Experimental Medicine, 130*, 643–658.

Matsuda, Y., Saoo, K., Hosokawa, K., Yamakawa, K., Yokohira, M., Zeng, Y., Takeuchi, H., & Imaida, K. (2007). Post-initiation chemopreventive effects of dietary bovine lactoferrin on 4-(methylnitrosamino)-1-(3-pyridyl)-1-butanone-induced lung tumorigenesis in female A/J mice. *Cancer Letters, 246*, 41–46.

Mazurier, J., & Spik, G. (1980). Comparative study of the iron-binding properties of human transferrins. I. Complete and sequential iron saturation and desaturation of the lactotransferrin. *Biochimica et Biophysica Acta, 629*, 399–408.

Mazurier, J., Legrand, D., Hu, W. L., Montreuil, J., & Spik, G. (1989). Expression of human lactotransferrin receptors in phytohemagglutinin-stimulated human peripheral blood lymphocytes. Isolation of the receptors by antiligand-affinity chromatography. *European Journal of Biochemistry, 179*, 481–487.

McAbee, D. D., Jiang, X., & Walsh, K. B. (2000). Lactoferrin binding to the rat asialoglycoprotein receptor requires the receptor's lectin properties. *Biochemical Journal, 348 (Pt 1)*, 113–117.

Medina, I., Tombo, I., Satue-Gracia, M. T., German, J. B., & Frankel, E. N. (2002). Effects of natural phenolic compounds on the antioxidant activity of lactoferrin in liposomes and oil-in-water emulsions. *Journal of Agricultural and Food Chemistry, 50*, 2392–2399.

Meilinger, M., Haumer, M., Szakmary, K. A., Steinbock, F., Scheiber, B., Goldenberg, H., & Huettinger, M. (1995). Removal of lactoferrin from plasma is mediated by binding to low density lipoprotein receptor-related protein/alpha 2-macroglobulin receptor and transport to endosomes. *FEBS Letters, 360*, 70–74.

Metz-Boutigue, M. H., Jollès, J., Mazurier, J., Schoentgen, F., Legrand, D., Spik, G., Montreuil, J., & Jollès, P. (1984). Human lactotransferrin: Amino acid sequence and structural comparisons with other transferrins. *European Journal of Biochemistry, 145*, 659–676.

Montreuil, J., Tonnelat, J., & Mullet, S. (1960). Preparation and properties of lactosiderophilin (lactotransferrin) of human milk. [in French] *Biochimica et Biophysica Acta, 45*, 413–421.

Moshynskyy, I., Jiang, M., Fontaine, M. C., Perez-Casal, J., Babiuk, L. A., & Potter, A. A. (2003). Characterization of a bovine lactoferrin binding protein of *Streptococcus uberis*. *Microbial Pathogenesis, 35*, 203–215.

Na, Y. J., Han, S. B., Kang, J. S., Yoon, Y. D., Park, S. K., Kim, H. M., Yang, K. H., & Joe, C. O. (2004). Lactoferrin works as a new LPS-binding protein in inflammatory activation of macrophages. *International Immunopharmacology, 4*, 1187–1199.

Naidu, A. S., Andersson, M., & Forsgren, A. (1992). Identification of a human lactoferrin-binding protein in *Staphylococcus aureus*. *Journal of Medical Microbiology, 36*, 177–183.

Naidu, A. S., Chen, J., Martinez, C., Tulpinski, J., Pal, B. K., & Fowler, R. S. (2004). Activated lactoferrin's ability to inhibit *Candida* growth and block yeast adhesion to the vaginal epithelial monolayer. *Journal of Reproductive Medicine, 49*, 859–866.

Nibbering, P. H., Ravensbergen, E., Welling, M. M., van Berkel, L. A., van Berkel, P. H., Pauwels, E. K., & Nuijens, J. H. (2001). Human lactoferrin and peptides derived from its N terminus are highly effective against infections with antibiotic-resistant bacteria. *Infection and Immunity, 69*, 1469–1476.

Norrby, K. (2004). Human apo-lactoferrin enhances angiogenesis mediated by vascular endothelial growth factor A *in vivo*. *Journal of Vascular Research, 41*, 293–304.

Norrby, K., Mattsby-Baltzer, I., Innocenti, M., & Tuneberg, S. (2001). Orally administered bovine lactoferrin systemically inhibits VEGF(165)-mediated angiogenesis in the rat. *International Journal of Cancer, 91*, 236–240.

Ochoa, T. J., Noguera-Obenza, M., Ebel, F., Guzman, C. A., Gomez, H. F., & Cleary, T. G. (2003). Lactoferrin impairs type III secretory system function in enteropathogenic *Escherichia coli*. *Infection and Immunity, 71*, 5149–5155.

Oh, S. M., Hahm, D. H., Kim, I. H., & Choi, S. Y. (2001). Human neutrophil lactoferrin trans-activates the matrix metalloproteinase 1 gene through stress-activated MAPK signaling modules. *Journal of Biological Chemistry, 276*, 42575–42579.

Oh, S. M., Pyo, C. W., Kim, Y., & Choi, S. Y. (2004). Neutrophil lactoferrin upregulates the human p53 gene through induction of NF-κB activation cascade. *Oncogene, 23*, 8282–8291.

Omata, Y., Satake, M., Maeda, R., Saito, A., Shimazaki, K., Yamauchi, K., Uzuka, Y., Tanabe, S., Sarashina, T., & Mikami, T. (2001). Reduction of the infectivity of *Toxoplasma gondii* and *Eimeria stiedai* sporozoites by treatment with bovine lactoferricin. *Journal of Veterinary Medical Science, 63*, 187–190.

Penco, S., Scarfi, S., Giovine, M., Damonte, G., Millo, E., Villaggio, B., Passalacqua, M., Pozzolini, M., Garre, C., & Benatti, U. (2001). Identification of an import signal for, and the nuclear localization of, human lactoferrin. *Biotechnology and Applied Biochemistry, 34*, 151–159.

Pietrantoni, A., Di Biase, A. M., Tinari, A., Marchetti, M., Valenti, P., Seganti, L., & Superti, F. (2003). Bovine lactoferrin inhibits adenovirus infection by interacting with viral structural polypeptides. *Antimicrobial Agents and Chemotherapy, 47*, 2688–2691.

Prinz, T., Meyer, M., Pettersson, A., & Tommassen, J. (1999). Structural characterization of the lactoferrin receptor from *Neisseria meningitidis*. *Journal of Bacteriology, 181*, 4417–4419.

Puddu, P., Borghi, P., Gessani, S., Valenti, P., Belardelli, F., & Seganti, L. (1998). Antiviral effect of bovine lactoferrin saturated with metal ions on early steps of human immunodeficiency virus type 1 infection. *International Journal of Biochemistry and Cell Biology, 30*, 1055–1062.

Qiu, J., Hendrixson, D. R., Baker, E. N., Murphy, T. F., St Geme, J. W., III, & Plaut, A. G. (1998). Human milk lactoferrin inactivates two putative colonization factors expressed by *Haemophilus influenzae*. *Proceedings of the National Academy of Sciences USA, 95*, 12641–12646.

Reghunathan, R., Jayapal, M., Hsu, L. Y., Chng, H. H., Tai, D., Leung, B. P., & Melendez, A. J. (2005). Expression profile of immune response genes in patients with Severe Acute Respiratory Syndrome. *BMC Immunology, 6*, 2.

Rey, M. W., Woloshuk, S. L., deBoer, H. A., & Pieper, F. R. (1990). Complete nucleotide sequence of human mammary gland lactoferrin. *Nucleic Acids Research, 18*, 5288.

Rogan, M. P., Taggart, C. C., Greene, C. M., Murphy, P. G., O'Neill, S. J., & McElvaney, N. G. (2004). Loss of microbicidal activity and increased formation of biofilm due to decreased lactoferrin activity in patients with cystic fibrosis. *Journal of Infectious Diseases, 190*, 1245–1253.

Rossi, P., Giansanti, F., Boffi, A., Ajello, M., Valenti, P., Chiancone, E., & Antonini, G. (2002). Ca^{2+} binding to bovine lactoferrin enhances protein stability and influences the release of bacterial lipopolysaccharide. *Biochemistry and Cell Biology, 80*, 41–48.

Sakai, T., Banno, Y., Kato, Y., Nozawa, Y., & Kawaguchi, M. (2005). Pepsin-digested bovine lactoferrin induces apoptotic cell death with JNK/SAPK activation in oral cancer cells. *Journal of Pharmacological Sciences, 98*, 41–48.

Sallmann, F. R., Baveye-Descamps, S., Pattus, F., Salmon, V., Branza, N., Spik, G., & Legrand, D. (1999). Porins OmpC and PhoE of *Escherichia coli* as specific cell-surface targets of human lactoferrin. Binding characteristics and biological effects. *Journal of Biological Chemistry, 274*, 16107–16114.

Sawatzki, G., & Rich, I. N. (1989). Lactoferrin stimulates colony stimulating factor production *in vitro* and *in vivo*. *Blood Cells, 15*, 371–385.

Sekine, K., Ushida, Y., Kuhara, T., Iigo, M., Baba-Toriyama, H., Moore, M. A., Murakoshi, M., Satomi, Y., Nishino, H., Kakizoe, T., & Tsuda, H. (1997). Inhibition of initiation and early stage development of aberrant crypt foci and enhanced natural killer activity in male rats administered bovine lactoferrin concomitantly with azoxymethane. *Cancer Letters, 121,* 211–216.

Sfeir, R. M., Dubarry, M., Boyaka, P. N., Rautureau, M., & Tomé, D. (2004). The mode of oral bovine lactoferrin administration influences mucosal and systemic immune responses in mice. *Journal of Nutrition, 134,* 403–409.

Shakibaei, M., & Frevert, U. (1996). Dual interaction of the malaria circumsporozoite protein with the low density lipoprotein receptor-related protein (LRP) and heparan sulfate proteoglycans. *Journal of Experimental Medicine, 184,* 1699–1711.

Sherman, M. P., Bennett, S. H., Hwang, F. F., & Yu, C. (2004). Neonatal small bowel epithelia: Enhancing anti-bacterial defense with lactoferrin and Lactobacillus GG. *Biometals, 17,* 285–289.

Shimamura, M., Yamamoto, Y., Ashino, H., Oikawa, T., Hazato, T., Tsuda, H., & Iigo, M. (2004). Bovine lactoferrin inhibits tumor-induced angiogenesis. *International Journal of Cancer, 111,* 111–116.

Shimizu, K., Matsuzawa, H., Okada, K., Tazume, S., Dosako, S., Kawasaki, Y., Hashimoto, K., & Koga, Y. (1996). Lactoferrin-mediated protection of the host from murine cytomegalovirus infection by a T-cell-dependent augmentation of natural killer cell activity. *Archives of Virology, 141,* 1875–1889.

Siciliano, R., Rega, B., Marchetti, M., Seganti, L., Antonini, G., & Valenti, P. (1999). Bovine lactoferrin peptidic fragments involved in inhibition of herpes simplex virus type 1 infection. *Biochemical and Biophysical Research Communications, 264,* 19–23.

Siebert, P. D., & Huang, B. C. (1997). Identification of an alternative form of human lactoferrin mRNA that is expressed differentially in normal tissues and tumor-derived cell lines. *Proceedings of the National Academy of Sciences USA, 94,* 2198–2203.

Singh, P. K. (2004). Iron sequestration by human lactoferrin stimulates *P. aeruginosa* surface motility and blocks biofilm formation. *Biometals, 17,* 267–270.

Son, H. J., Lee, S. H., & Choi, S. Y. (2006). Human lactoferrin controls the level of retinoblastoma protein and its activity. *Biochemistry and Cell Biology, 84,* 345–350.

Son, K. N., Park, J., Chung, C. K., Chung, D. K., Yu, D. Y., Lee, K. K., & Kim, J. (2002). Human lactoferrin activates transcription of IL-1β gene in mammalian cells. *Biochemical and Biophysical Research Communications, 290,* 236–241.

Sorimachi, K., Akimoto, K., Hattori, Y., Ieiri, T., & Niwa, A. (1997). Activation of macrophages by lactoferrin: Secretion of TNF-α, IL-8 and NO. *Biochemistry and Molecular Biology International, 43,* 79–87.

Spik, G., Coddeville, B., & Montreuil, J. (1988). Comparative study of the primary structures of sero-, lacto- and ovotransferrin glycans from different species. *Biochimie, 70,* 1459–1469.

Srivastava, M., & Pollard, H. B. (1999). Molecular dissection of nucleolin's role in growth and cell proliferation: New insights. *FASEB Journal, 13,* 1911–1922.

Superti, F., Ammendolia, M. G., Valenti, P., & Seganti, L. (1997). Antirotaviral activity of milk proteins: Lactoferrin prevents rotavirus infection in the enterocyte-like cell line HT-29. *Medical Microbiology and Immunology, 186,* 83–91.

Suzuki, Y. A., Lopez, V., & Lönnerdal, B. (2005). Mammalian lactoferrin receptors: Structure and function. *Cellular and Molecular Life Sciences, 62,* 2560–2575.

Swart, P. J., Kuipers, M. E., Smit, C., Pauwels, R., deBethune, M. P., de Clercq, E., Meijer, D. K., & Huisman, J. G. (1996). Antiviral effects of milk proteins: Acylation results in polyanionic compounds with potent activity against human immunodeficiency virus types 1 and 2 *in vitro. AIDS Research and Human Retroviruses, 12,* 769–775.

Takayama, Y., Takahashi, H., Mizumachi, K., & Takezawa, T. (2003). Low density lipoprotein receptor-related protein (LRP) is required for lactoferrin-enhanced collagen gel contractile activity of human fibroblasts. *Journal of Biological Chemistry, 278,* 22112–22118.

Tani, F., Iio, K., Chiba, H., & Yoshikawa, M. (1990). Isolation and characterization of opioid antagonist peptides derived from human lactoferrin. *Agricultural and Biological Chemistry, 54*, 1803–1810.

Taylor, S., Brock, J., Kruger, C., Berner, T., & Murphy, M. (2004). Safety determination for the use of bovine milk-derived lactoferrin as a component of an antimicrobial beef carcass spray. *Regulatory Toxicology and Pharmacology, 39*, 12–24.

Teng, C., Gladwell, W., Raphiou, I., & Liu, E. (2004). Methylation and expression of the lactoferrin gene in human tissues and cancer cells. *Biometals, 17*, 317–323.

Teng, C. T. (2006). Factors regulating lactoferrin gene expression. *Biochemistry and Cell Biology, 84*, 263–267.

Teng, C. T., & Gladwell, W. (2006). Single nucleotide polymorphisms (SNPs) in human lactoferrin gene. *Biochemistry and Cell Biology, 84*, 381–384.

Teng, C. T., Beard, C., & Gladwell, W. (2002). Differential expression and estrogen response of lactoferrin gene in the female reproductive tract of mouse, rat, and hamster. *Biology of Reproduction, 67*, 1439–1449.

Tsuda, H., Sekine, K., Fujita, K., & Ligo, M. (2002). Cancer prevention by bovine lactoferrin and underlying mechanisms—A review of experimental and clinical studies. *Biochemistry and Cell Biology, 80*, 131–136.

Tsuda, H., Ohshima, Y., Nomoto, H., Fujita, K., Matsuda, E., Iigo, M., Takasuka, N., & Moore, M. A. (2004). Cancer prevention by natural compounds. *Drug Metabolism and Pharmacokinetics, 19*, 245–263.

Valenti, P., & Antonini, G. (2005). Lactoferrin: An important host defence against microbial and viral attack. *Cellular and Molecular Life Sciences, 62*, 2576–2587.

van Berkel, P. H., Geerts, M. E., van Veen, H. A., Mericskay, M., de Boer, H. A., & Nuijens, J. H. (1997). N-terminal stretch Arg2, Arg3, Arg4 and Arg5 of human lactoferrin is essential for binding to heparin, bacterial lipopolysaccharide, human lysozyme and DNA. *Biochemical Journal, 328 (Pt 1)*, 145–151.

van der Kraan, M. I., van Marle, J., Nazmi, K., Groenink, J., van't Hof, W., Veerman, E. C., Bolscher, J. G., & Nieuw Amerongen, A. V. (2005). Ultrastructural effects of antimicrobial peptides from bovine lactoferrin on the membranes of *Candida albicans* and *Escherichia coli*. *Peptides, 26*, 1537–1542.

van Veen, H. A., Geerts, M. E., van Berkel, P. H., & Nuijens, J. H. (2004). The role of N-linked glycosylation in the protection of human and bovine lactoferrin against tryptic proteolysis. *European Journal of Biochemistry, 271*, 678–684.

Varadhachary, A., Wolf, J. S., Petrak, K., O'Malley, B. W., Jr., Spadaro, M., Curcio, C., Forni, G., & Pericle, F. (2004). Oral lactoferrin inhibits growth of established tumors and potentiates conventional chemotherapy. *International Journal of Cancer, 111*, 398–403.

Velliyagounder, K., Kaplan, J. B., Furgang, D., Legarda, D., Diamond, G., Parkin, R. E., & Fine, D. H. (2003). One of two human lactoferrin variants exhibits increased antibacterial and transcriptional activation activities and is associated with localized juvenile periodontitis. *Infection and Immunity, 71*, 6141–6147.

Viejo Diaz, M., Andres, M. T., Perez-Gil, J., Sanchez, M., & Fierro, J. F. (2003). Potassium efflux induced by a new lactoferrin-derived peptide mimicking the effect of native human lactoferrin on the bacterial cytoplasmic membrane. *Biochemistry, 68*, 217–227.

Visca, P., Berlutti, F., Vittorioso, P., Dalmastri, C., Thaller, M. C., & Valenti, P. (1989). Growth and adsorption of *Streptococcus mutans* 6715-13 to hydroxyapatite in the presence of lactoferrin. *Medical Microbiology and Immunology, 178*, 69–79.

Wakabayashi, H., Takakura, N., Teraguchi, S., & Tamura, Y. (2003). Lactoferrin feeding augments peritoneal macrophage activities in mice intraperitoneally injected with inactivated *Candida albicans*. *Microbiology and Immunology, 47*, 37–43.

Wakabayashi, H., Kurokawa, M., Shin, K., Teraguchi, S., Tamura, Y., & Shiraki, K. (2004). Oral lactoferrin prevents body weight loss and increases cytokine responses during herpes simplex virus type 1 infection of mice. *Bioscience, Biotechnology, and Biochemistry, 68*, 537–544.

Ward, P. P., Paz, E., & Conneely, O. M. (2005). Multifunctional roles of lactoferrin: A critical overview. *Cellular and Molecular Life Sciences, 62*, 2540–2548.

Williams, T. J., Schneider, R. P., & Willcox, M. D. (2003). The effect of protein-coated contact lenses on the adhesion and viability of Gram negative bacteria. *Current Eye Research, 27*, 227–235.

Willnow, T. E., Goldstein, J. L., Orth, K., Brown, M. S., & Herz, J. (1992). Low density lipoprotein receptor-related protein and gp330 bind similar ligands, including plasminogen activator-inhibitor complexes and lactoferrin, an inhibitor of chylomicron remnant clearance. *Journal of Biological Chemistry, 267*, 26172–26180.

Wong, H., & Schryvers, A. B. (2003). Bacterial lactoferrin-binding protein A binds to both domains of the human lactoferrin C-lobe. *Microbiology, 149*, 1729–1737.

Xiao, Y., Monitto, C. L., Minhas, K. M., & Sidransky, D. (2004). Lactoferrin down-regulates G1 cyclin-dependent kinases during growth arrest of head and neck cancer cells. *Clinical Cancer Research, 10*, 8683–8686.

Xu, Y. Y., Samaranayake, Y. H., Samaranayake, L. P., & Nikawa, H. (1999). In vitro susceptibility of *Candida* species to lactoferrin. *Medical Mycology, 37*, 35–41.

Yamaguchi, H., Abe, S., & Takakura, N. (2004). Potential usefulness of bovine lactoferrrin for adjunctive immunotherapy for mucosal *Candida* infections. *Biometals, 17*, 245–248.

Yamauchi, K., Wakabayashi, H., Shin, K., & Takase, M. (2006). Bovine lactoferrin: Benefits and mechanism of action against infections. *Biochemistry and Cell Biology, 84*, 291–296.

Yang, N., Strom, M. B., Mekonnen, S. M., Svendsen, J. S., & Rekdal, O. (2004). The effects of shortening lactoferrin derived peptides against tumour cells, bacteria and normal human cells. *Journal of Peptide Science, 10*, 37–46.

Yi, M., Kaneko, S., Yu, D. Y., & Murakami, S. (1997). Hepatitis C virus envelope proteins bind lactoferrin. *Journal of Virology, 71*, 5997–6002.

Yoo, Y. C., Watanabe, R., Koike, Y., Mitobe, M., Shimazaki, K., Watanabe, S., & Azuma, I. (1997). Apoptosis in human leukemic cells induced by lactoferricin, a bovine milk protein-derived peptide: Involvement of reactive oxygen species. *Biochemical and Biophysical Research Communications, 237*, 624–628.

Yoo, Y. C., Watanabe, S., Watanabe, R., Hata, K., Shimazaki, K., & Azuma, I. (1998). Bovine lactoferrin and Lactoferricin inhibit tumor metastasis in mice. *Advances in Experimental Medicine and Biology, 443*, 285–291.

Zagulski, T., Lipinski, P., Zagulska, A., Broniek, S., & Jarzabek, Z. (1989). Lactoferrin can protect mice against a lethal dose of *Escherichia coli* in experimental infection *in vivo*. *British Journal of Experimental Pathology, 70*, 697–704.

Zimecki, M., Kocieba, M., & Kruzel, M. (2002). Immunoregulatory activities of lactoferrin in the delayed type hypersensitivity in mice are mediated by a receptor with affinity to mannose. *Immunobiology, 205*, 120–131.

CD14: A Soluble Pattern Recognition Receptor in Milk

Karine Vidal and Anne Donnet-Hughes

Abstract An innate immune system capable of distinguishing among self, non-self, and danger is a prerequisite for health. Upon antigenic challenge, pattern recognition receptors (PRRs), such as the Toll-like receptor (TLR) family of proteins, enable this system to recognize and interact with a number of microbial components and endogenous host proteins. In the healthy host, such interactions culminate in tolerance to self-antigen, dietary antigen, and commensal microorganisms but in protection against pathogenic attack. This duality implies tightly regulated control mechanisms that are not expected of the inexperienced neonatal immune system. Indeed, the increased susceptibility of newborn infants to infection and to certain allergens suggests that the capacity to handle certain antigenic challenges is not inherent. The observation that breast-fed infants experience a lower incidence of infections, inflammation, and allergies than formula-fed infants suggests that exogenous factors in milk may play a regulatory role.

There is increasing evidence to suggest that upon exposure to antigen, breast milk educates the neonatal immune system in the decision-making processes underlying the immune response to microbes. Breast milk contains a multitude of factors such as immunoglobulins, glycoproteins, glycolipids, and antimicrobial peptides that, qualitatively or quantitatively, may modulate how neonatal cells perceive and respond to microbial components. The specific role of several of these factors is highlighted in other chapters in this book. However, an emerging concept is that breast milk influences the neonatal immune system's perception of "danger." Here we discuss how CD14, a soluble PRR in milk, may contribute to this education.

K. Vidal
Nestlé Research Center, Nestec Ltd, Vers-Chez-Les-Blanc, P.O. Box 44, CH-1000 Lausanne 26, Switzerland
e-mail: karine.vidal@rdls.nestle.com

The Discovery of CD14

CD14, first described in 1990 as a receptor for the bacterial endotoxin lipopolysaccharide (LPS) (Wright et al., 1990), is the first documented PRR (Pugin et al., 1994). Originally characterized as a monocyte/macrophage differentiation antigen, it is constitutively expressed by a variety of other cell types, including polymorphonuclear neutrophils, chondrocytes (Goyert & Ferrero, 1987; Matsuura et al., 1994; Tobias & Ulevitch, 1993), B cells (Schumann et al., 1994), dendritic cells (Verhasselt et al., 1997), gingival fibroblasts (Watanabe et al., 1996; Sugawara et al., 1998), keratinocytes (Song et al., 2002), hepatocytes (Liu et al., 1998), tracheal epithelial cells (Diamond et al., 1996), and human intestinal epithelial cell lines (Funda et al., 2001). However, a soluble form of CD14 (sCD14) also exists. First discovered in cell culture supernatants and in normal serum that blocked monocyte staining with anti-CD14 monoclonal antibodies (Maliszewski et al., 1985), sCD14 has been described in substantial concentrations in serum (Landmann et al., 1996; Bazil et al., 1986), cerebrospinal fluid (Cauwels et al., 1999), urine (Bazil et al., 1989), seminal plasma (Harris et al., 2001), saliva (Uehara et al., 2003), tears (Blais et al., 2005), and breast milk (Labeta et al., 2000; Vidal et al., 2001).

Molecular Characteristics of CD14

Genomic DNA encoding human CD14 was cloned in 1988 (Ferrero & Goyert, 1988). The CD14 gene has been mapped to region 5q 23-31 of chromosome 5, a region that codes for growth factors and receptors (Goyert et al., 1988). It consists of only two exons, starting with an ATG sequence directly followed by an 88-bp intron. A single mRNA species is translated.

During processing in the endoplasmic reticulum, a 19 amino acid-signal sequence is removed (Haziot et al., 1988) to yield the mature human membrane CD14 (mCD14) protein, which is composed of 356 amino acids with multiple leucine-rich repeats (Ferrero et al., 1990) and has sites for N-linked and O-linked glycosylation (Stelter et al., 1996). The sequence ends with a 22 amino acid-hydrophobic domain that lacks the characteristic basic residues of a stop transfer domain. CD14 is attached to the cell surface by a glycosylphosphatidylinositol (GPI) anchor that is added to the C terminus in the endoplasmic reticulum (Haziot et al., 1988; Simmons et al., 1989). CD14 has a pI of 4.5–5.1 (Nasu et al., 1991) and a carbohydrate content that accounts for about 20% of the total molecular mass of the mature 53-kDa glycoprotein (Bazil et al., 1986).

The primary sequence of human CD14 is highly homologous (61–73%) to the deduced amino acid sequence of its mouse (Setoguchi et al., 1989), rat (Takai et al., 1997), rabbit (Tobias et al., 1992), and bovine (Ikeda et al., 1997) counterparts. The bovine CD14 cDNA, isolated from a genomic library

(Ikeda et al., 1997), encodes a protein of 373 amino acids whose coding sequence is separated by a 90-nucleotide intron.

Soluble Forms of CD14

In normal human plasma, sCD14 occurs at concentrations of about 3 µg/mL (Bazil et al., 1989; Grunwald et al., 1992; Durieux et al., 1994), but in human milk, its concentrations are ~10-fold higher (Table 1) (Labeta et al., 2000; Vidal et al., 2001). Of note, sCD14 has also been detected in bovine colostrum (Labeta et al., 2000; Filipp et al., 2001) and milk (Filipp et al., 2001). Quantification of sCD14 in bovine milk has been difficult due to the lack of a specific assay. Nevertheless, it is estimated to be present in quantities that are slightly inferior to those in human milk (Lee et al., 2003a; Vangroenweghe et al., 2004).

The origin of sCD14 is uncertain. There are many potential sources for sCD14, including cleavage of the receptor from monocytes by proteases or phospholipases (Bazil et al., 1989; Bazil & Strominger, 1991), and direct secretion of full-length molecules that have bypassed the GPI-linking mechanism (Labeta et al., 1993; Durieux et al., 1994; Bufler et al., 1995). Cells treated with phosphoinositol-phospholipase C release CD14, which migrates in SDS-PAGE as a doublet in the 50-kDa range. In contrast, CD14 released via the action of cellular proteases migrates as a single molecular-weight moiety in SDS-PAGE (Bazil & Strominger, 1991).

At least two soluble forms of CD14 are constitutively generated in serum (Bazil et al., 1989; Durieux et al., 1994; Stelter et al., 1996), one form is a ~50-kDa molecule that is released after shedding of the cell surface form (Bazil et al., 1986; Haziot et al., 1988, 1993b). A second form, with higher molecular weight (~56 kDa), is released from cells before addition of the GPI anchor (Labeta et al., 1993; Landmann et al., 1995; Bufler et al., 1995). The sCD14 found in serum and urine appears to be a mixture of the two forms. More recently, a new isoform of sCD14, named sCD14 subtype, has been identified in the serum of patients with sepsis (Yaegashi et al., 2005).

The sCD14 present in human milk during most of the lactating period is a singe molecular form of ~48 kDa (Labeta et al., 2000; Vidal et al., 2001). This is identical to that produced by differentiated mammary epithelial cells *in vitro* (Labeta et al., 2000; Vidal et al., 2001) and suggests that the mammary gland

Table 1 Levels of sCD14 in Human Breast Milk During Lactation

	Days Postpartum		
	≤6 days	≥8 days	0–71 days
sCD14 (µg/mL)	20.10 ± 8.74 ($n = 10$)	13.09 ± 4.31 ($n = 30$)	14.84 ± 6.39 ($n = 40$)

Data represent mean ± standard deviation of samples tested by enzyme-linked immunosorbent assay.

epithelium is the main source of milk sCD14. Interestingly, the early milk samples have a complex sCD14 pattern with three polypeptide bands (~48, 50, and 56 kDa). It is not known whether these different forms reflect distinct glycosylation patterns and/or different sizes of the core proteins. However, since milk macrophages express the typical serum sCD14 isoforms (i.e., 50 and 56 kDa), their production of sCD14 may contribute to the total pool of sCD14 within the first week postpartum. It is also possible that there is leakage of serum sCD14 into the milk via mammary gland alveolar cell tight junctions or by transient active transport.

Regulation of sCD14 Expression

In vitro studies show that while treatment of blood monocytes with LPS and TNF-α causes an increase in serum sCD14 levels, exposure to IFN-γ and IL-4 decreases the levels (Schutt et al., 1992; Landmann et al., 1992). Little is known about the regulation of sCD14 expression *in vivo*. However, increased circulating sCD14 levels correlate with infectious, autoimmune, and inflammatory diseases (Kruger et al., 1991; Nockher et al., 1994; Yu et al., 1998). In addition, it has been demonstrated that sCD14 can be considered as a type 2 acute-phase protein (APP) (Bas et al., 2004). Liver levels of CD14 mRNA increase in IL-6-/- mice injected with turpentine, an experimental model of acute-phase response (Bas et al., 2004). Furthermore, serum sCD14 levels in patients with various arthropathies correlate with those of C-reactive protein, a classical APP, and with IL-6, a cytokine known to regulate the synthesis of APP in the liver. Serum levels of sCD14 also correlate with disease activity in rheumatoid arthritis and reactive arthritis patients.

Interestingly, human milk sCD14 levels correlate with those of specific fatty acids (Dunstan et al., 2004; Laitinen et al., 2006) and are influenced by fish oil supplementation in the maternal diet (Dunstan et al., 2004). In bovine milk, sCD14 levels are elevated during mastitis or following intramammary challenge with LPs or *Escherichia coli* (Bannerman et al., 2003; Lee et al., 2003a, b; Vangroenweghe et al., 2004) and may be due to infiltration of neutrophils to the inflamed mammary gland.

Although highly conserved across a wide range of species, genes involved in innate immunity demonstrate considerable interethnic variability predominantly as single-nucleotide polymorphisms (SNPs). It has been recently demonstrated that genetic variation in the promoter region of the cDNA encoding CD14 affects sCD14 levels (LeVan et al., 2006; Baldini et al., 1999). No correlation has been observed between CD14 promoter polymorphisms at positions -4190, -2838, -1720, and -260 and the levels of sCD14 at birth. However, an association between genotypes and sCD14 is evident at three months of age, and longitudinal analyses suggest that CD14 polymorphisms modulate sCD14 levels through the first year of life in healthy infants (LeVan et al., 2006).

Interestingly, it has been reported that sCD14 levels in plasma and milk are differentially regulated by the same genetic variants (Guerra et al., 2004a). More specifically, plasma and milk sCD14 levels differ significantly both by CD14/-1619 and CD14/-550 genotypes and by haplotypes. Moreover, the CD14/-550T allele and the corresponding ATC haplotype are associated with high levels of sCD14 in milk but low levels of sCD14 in plasma. Taken together, these suggest the existence of cell-specific regulation of CD14 gene expression in the two compartments (Guerra et al., 2004a).

Role of CD14 in the Host Response to Bacterial Ligands

Host–Microbe Interactions

Our understanding of microbiota–host immune system interactions has made great progress in recent years, and there is increasing evidence to suggest that such interactions benefit both the bacteria and the host. Intestinal epithelial cells (IEC) are at the interface between the luminal environment and the mucosal immune system. For immune homeostasis and effective immune defense, these cells, together with sentinel dendritic cells, sense the bacterial load in the intestinal lumen and determine the outcome of the primary, innate response.

In healthy individuals, transient local immune responses but not systemic responses are initiated against the intestinal flora (Macpherson & Harris, 2004; Haller et al., 2000), while active and more aggressive responses are generated against the "danger signals" from pathogens (Matzinger, 2002). It is now clear that the responses to commensals are essential for the development and maturation of the intestinal immune system (Cebra, 1999; Macpherson & Harris, 2004), the integrity of the intestinal epithelium, the maintenance of immune homeostasis, as well as for the production of antimicrobial peptides and tissue repair (Rakoff-Nahoum et al., 2005). Indeed, the hygiene hypothesis suggests that in early life, a modified interaction between microbial antigens and the innate immune system underlies the increased incidence of allergic and autoimmune diseases in developed countries (Bloomfield et al., 2006). The original hypothesis arose from epidemiological studies that reported an inverse association between family size and the development of atopy (Strachan, 1989). Later studies have found a similar relationship using other measures of microbial exposure such as farm living, bed sharing, and attending nursery school as well as more direct forms of exposure such as infection, exposure to endotoxins, food-borne microbes, or the gut microbiota (Bloomfield et al., 2006). In this context, it is interesting that lower circulating levels of sCD14 levels in children are associated with the development of atopy (Zdolsek & Jenmalm, 2004) and wheezing (Guerra et al., 2004b) and that blood cells from farmers' children have higher amounts of CD14 mRNA than those from nonfarmers' children (Lauener et al., 2002).

CD14 as a PRR

Wright et al. (1990) were the first to identify the monocyte differentiation antigen CD14 as a key monocyte receptor for bacterial LPS. The binding stoichiometry of LPS to CD14 has been reported to be one to one (Kitchens & Munford, 1995), with a K_d value of 27 nM (Kirkland et al., 1993). However, since LPS tends to aggregate in solution, stoichiometric data are conflicting. It is through the formation of large ternary complexes consisting of LPS, CD14, and lipopolysaccharide binding protein (LBP) (Gegner et al., 1995) that monocytes detect the presence of LPS (Thomas et al., 2002). Cellular activation begins when the acute-phase protein LBP binds to LPS and catalyzes its binding to CD14 on the cell surface (Tobias et al., 1986; Schumann et al., 1990; Wright et al., 1990; Mathison et al., 1992; Martin et al., 1992; Hailman et al., 1994). Binding and phagocytosis of whole Gram-negative bacteria is also mediated by membrane CD14 in an LBP-dependent manner (Grunwald et al., 1996). It is noteworthy that serum from LBP knockout mice is unable to mediate LPS-induced oxidative burst responses in mouse peritoneal exudate cells (Jack et al., 1997).

Since CD14 lacks a cytoplasmic domain, it cannot signal the presence of LPS. Rather, the transmembrane TLR family of proteins, which discriminate conserved motifs present in pathogens and commensals, transduce intracellular signals (Aderem & Ulevitch, 2000). To date, 12 mouse and 10 human TLRs capable of recognizing single or multiple motifs from bacteria, viruses, and fungi have been identified (Kaisho & Akira, 2006). Of these, the most studied TLR for bacterial recognition are TLR2, which recognizes the lipoteichoic acid (LTA) and peptidoglycan (PGN) from Gram-positive bacteria; TLR4 and TLR5, which bind, respectively, to the lipopolysaccharide (LPS) and the flagellin from Gram-negative bacteria; and TLR9, the receptor for unmethylated CpG DNA. The discoveries that TLR4 mediates LPS-induced signal transduction (Chow et al., 1999; Takeuchi et al., 1999) and that for some bacterial ligands, TLR cooperate with other soluble and cell-surface proteins led to CD14's reclassification as a co-receptor for LPS.

It is now known that for intracellular signal transduction, LPS is transferred from the CD14-LBP complex to TLR4 that is bound to myeloid differentiation protein (MD)-2 on the cell surface (da Silva et al., 2001). TLR belong to the TLR/IL-1R superfamily whose members have a Toll/IL-1 receptor (TIR) domain (O'Neill, 2002). In response to ligand binding, this cytoplasmic TIR domain then sequentially recruits a series of adapter molecules including myeloid differentiation marker-88 (MyD88), IL-1 receptor-associated kinase (IRAK), and tumor necrosis factor receptor-associated factor (TRAF)-6. Ultimately, this recruitment activates nuclear factor (NF)-κB and mitogen-activated proteins (MAP) kinases (O'Neill et al., 2003) and induces the production of pro-inflammatory cytokines such as IL-8, TNF-α, and IL-1 (Dentener et al., 1993), as well as anti-inflammatory cytokines (e.g., IL-10, TGF-β).

LPS is not the only bacterial ligand recognized by CD14. In 1994, Pugin et al. (1994) also identified CD14 as a PRR for PGN, the major cell wall component of Gram-positive bacteria and for lipoarabinomannan, a glycolipid from *Mycobacterium tuberculosis*. Indeed, it is now known that several other microbial ligands from bacteria, yeast, and spirochetes are able to interact with CD14 (Table 2) (Heumann et al., 1998). In addition, CD14 may function as a receptor for several endogenous human proteins such as heat shock protein 60 (Kol et al., 2000), ceramide, anionic phospholipids, modified lipoproteins, and opsonized particles (Schmitz & Orso, 2002).

Like that of LPS, cellular activation by the LTA and PGN of Gram-positive bacteria also involves a receptor complex comprising CD14, but instead of TLR4, signaling is transduced by TLR-2 (Schwandner et al., 1999; Schroder et al., 2003). Although both TLR4 and TLR2 ligands activate NF-κB via MyD88, IRAK, and TRAF-6 (O'Neill et al., 2003), they elicit different biological responses. For example, TLR4 agonists promote dendritic cell production of the Th1-inducing

Table 2 Interactions of CD14 with Ligands Derived from Different Microbial Sources

Microbial Sources	Ligands
Gram-negative bacteria	LPS from *E. coli*[a]
	Whole Gram-negative *E. coli*[b]
	Polymannuronic acid from *Pseudomonas*[c]
Gram-positive bacteria	Lipoarabinomannan from *Mycobacteria tuberculosis*[d]
	Cell wall constituents from *Bacillus subtilis*[e] and *Staphylococcus aureus*[f]
	Soluble peptidoglycan from *Staphylococcus aureus*[g]
	Rhamnose glucose polymers of *Streptococcus mutans*[h]
	Lipoteichoic acid (LTA) from *Staphylococcus aureus* and *Streptococcus pyrogenes*[i]
	Lipoproteins and lipopeptides from *Treponema palladium* and *Borrelia burdorferi*[j]
Yeast	Peptide derived from the WI-1 protein of *Blastomyces dermatidis*[k]
Spirochetes	Outer surface protein (Osp) from *Borrelia burgdorferi*[l]

Sources: [a]Gallay et al. (1993); [b]Jack et al. (1995); [c]Espevik et al. (1993); [d]Pugin et al. (1994), Savedra et al. (1996), Zhang et al. (1993); [e]Pugin et al. (1994); [f]Kusunoki et al. (1995), Kusunoki and Wright (1996); [g]Weidemann et al. (1994), Gupta et al. (1996), Weidemann et al. (1997), Dziarski et al. (1998); [h]Soell et al. (1995); [i]Cleveland et al. (1996), Hattor et al. (1997); [j]Sellati et al. (1998), Wooten et al. (1998); [k]Newman et al. (1995); [l]Wooten et al. (1998)

cytokine IL-12p70 and the chemokine, interferon-inducible protein (IP)-10. In contrast, TLR2 agonists fail to stimulate the production of these proteins but rather induce the production of the IL-12 inhibitory p40 homodimer that favors Th2 development (Re & Strominger, 2001).

Roles of sCD14

Like membrane CD14 (mCD14), sCD14 also forms a complex with LPS (Vita et al., 1997) that is capable of activating both CD14-positive cells and CD14-negative cells, such as epithelial, endothelial, and smooth muscle cells (Frey et al., 1992; Pugin et al., 1993; Haziot et al., 1993a; Arditi et al., 1993; Goldblum et al., 1994; Loppnow et al., 1995; Read et al., 1993). This complex can stimulate cells even in the absence of LBP (Hailman et al., 1994). However, the nature of the cellular response depends on the concentration of LPS, sCD14, and LBP present (Kitchens & Thompson, 2005). Indeed, the range of sCD14 concentrations found in normal and septic humans can significantly decrease monocyte responses to LPS. By competing with mCD14, the soluble form limits the amount of LPS binding to the cells (Jacque et al., 2006) and thereby inhibits LPS-induced cellular activation (Grunwald et al., 1993; Haziot et al., 1994; Schutt et al., 1992). However, in addition to binding bacterial motifs, sCD14 also binds phospholipids, such as phosphatidylinositol and phosphatidylethanolamine, and transfers them to high-density lipoproteins (Yu et al., 1997). It also mediates the influx of phospholipids into cells as well as their efflux out of cells and into plasma (Sugiyama & Wright, 2001). By shuttling LPS from mCD14 to plasma lipoproteins (Yu et al., 1997; Kitchens et al., 1999, 2001), sCD14 may retain LPS in the circulation and prevent LPS-mediated lethality (Jacque et al., 2006). The suggestion that LPS moves between the membrane and soluble forms of CD14 until an equilibrium is reached and is progressively removed from both forms when plasma lipoproteins are present (Kitchens & Thompson, 2005) may explain the apparent ambiguity in sCD14 function.

Relevance of sCD14 for Infant Health

The fetal intestine is sterile, and the events governing colonization and subsequent assembly of the intestinal microbiota remain elusive. At parturition, the immature intestinal immune system is immediately challenged by a massive bacterial insult in the birth canal and the external environment. In the early postnatal period, the composition of the microbiota constantly changes as a large number of organisms compete for an intestinal niche. However, some stability is achieved around weaning (Edwards & Parrett, 2002) and maintained throughout adulthood (Zoetendal et al., 1998). It is remarkable that the inexperienced neonatal immume system accommodates all these changes in the absence of adverse immune responses and finally permits an estimated 10^{14} bacteria from at least 400 different species (Berg, 1996) to establish themselves in the intestinal lumen. Clearly, in the full-term, healthy infant, basic protective measures to ensure immune tolerance to commensals, even in the face of pathogenic attack, are already in place at birth. However, interaction with a high density of microbial antigens from a wide spectrum of species is a potentially hazardous process, particularly for preterm infants. Indeed, an inappropriate inflammatory response to intestinal microorganisms may contribute to the development of pathological conditions such as necrotizing enterocolitis (NEC) (Caplan & MacKendrick, 1993; Hoy et al., 1990). Furthermore, perturbations in the microbiota can lead to septicemia (van Saene et al., 2003) or allergic disease (Kirjavainen et al., 2002) in infants.

In the adult intestine, low expression of TLR2 and TLR4 (Otte et al., 2004; Abreu et al., 2001) and the absence of MD2 (Abreu et al., 2001) on IEC could explain, at least in part, the lack of response to the microbiota. On the other hand, fetal IEC express both TLR2 and TLR4 and are hyperresponsive to LPS (Claud et al., 2004; Fusunyan et al., 2001). Indeed, an animal model suggests that an interaction between intestinal bacteria and neonatal IEC expressing high levels of TLR4 underlies the development of NEC (Jilling et al., 2006). Thus, increased expression of TLR on fetal IEC could explain the increased susceptibility of preterm infants to NEC. Continuous activation of IEC in the immediate postnatal period may cause downregulated TLR expression on IEC (Abreu et al., 2001) and tolerance to endotoxins (Lotz et al., 2006). It is not known if soluble forms of CD14 influence the outcome of such activation. However, cellular activation by LPS can lead to shedding of CD14 in a soluble form and subsequent downregulation of inflammatory cytokine production (Sohn et al., 2007; Kitchens & Thompson, 2005).

Amniotic fluid contains sCD14, the concentration of which is increased with intrauterine infection and preterm labor (Espinoza et al., 2002). It is possible that sCD14 in the amniotic fluid still bathes mucosal surfaces during the birthing process and determines the extent of the initial microbial interaction with the associated lymphoid tissue. It is also possible that breast milk sCD14 continues this in the postnatal period. To date, few studies have examined the effect of sCD14 in amniotic fluid or breast milk on host–microbe interactions, the immune status of the infant, or the development of infection and disease. Nevertheless, the levels of sCD14 in amniotic fluid and breast milk are associated with subsequent development of atopy, eczema, and asthma (Jones et al., 2002; Rothenbacher et al., 2005).

Breastfeeding certainly influences the composition of intestinal microbiota (Falk et al., 1998) and protects against infection (Lonnerdal, 2003), bacterial translocation in neonatal animals, and septicemia and NEC in premature human infants (Steinwender et al., 1996; Ronnestad et al., 2005). It also reduces the risk of developing asthma, atopic dermatitis, eczema, and allergy (Kull et al., 2004; Lawrence, 2005) and pathological diseases later in life

(Jackson & Nazar, 2006). It is plausible that regulatory factors in breast milk educate the neonatal immune system to recognize and respond appropriately to bacterial components and that milk sCD14 may contribute to such an education. Certainly, human breast milk contains high levels of sCD14 (Labeta et al., 2000; Filipp et al., 2001), which may participate in the transient activation of the innate immune response to bacterial components and its subsequent downregulation. For example, milk sCD14 mediates the LPS-induced production of the pro-inflammatory cytokines IL-8 and TNF-α and of the chemokine ENA-78 by IEC (Labeta et al., 2000) as well as the production of IL-8 by monocytes and dendritic cells (Labeta et al., 2000; Lebouder et al., 2006). An excessive, unresolved inflammatory response may be avoided by the limited amount of LBP in milk (Vidal et al., 2001) and/or by the presence of other milk proteins such as lactoferrin, which has been shown to inhibit the LPS-induced production of pro-inflammatory mediators via NF-κB (Haversen et al., 2002) as well as the sCD14-LPS–induced expression of IL-8 and adhesion molecules by endothelial cells (Elass et al., 2002; Baveye et al., 2000). Interestingly, milk sCD14 does not mediate the production of pro-inflammatory cytokines by IEC exposed to Gram-positive bacteria or their LTA *in vitro* (Vidal et al., 2002). This differential response may be due to the expression of TLR4 and absence of TLR2 on the cell line used. However, LTA inhibits the LPS-milk sCD14–induced response, most probably by competitive binding to the sCD14 molecule (Vidal et al., 2002). It is also noteworthy that sCD14 mediates LPS-induced expression of IL-6, IL-8, IL-12, and co-stimulatory molecules by dendritic cells, which, like IEC, do not express mCD14 (Verhasselt et al., 1997).

Direct immunoregulatory effects of sCD14 on activated T and B cells have also been reported. More specifically, sCD14 inhibits the proliferation of activated T cells and their production of the Th1 cytokines IL-2 and IFN-γ and the Th2 cytokine IL-4 (Rey Nores et al., 1999). It also induces a progressive accumulation of IκBα, an inhibitor of NF-κB. In B cells, sCD14 interferes with the CD40 signaling pathway and the production of IgE (Arias et al., 2000) and, when administered to neonatal mice, bovine milk-derived sCD14 induces immunoglobulin secretion (Filipp et al., 2001). A correlation between sCD14 concentrations in colostrum and the numbers of IgA and IgM secreting cells in human neonates lends further support to this observation (Rinne et al., 2005).

It is tempting to speculate that sCD14 in milk instigates a beneficial, innate immune response to specific microbes in sentinel IEC and dendritic cells that lack mCD14 and thwarts an exaggerated response to microbial antigens through the production of protective immunoglobulins and the regulation of effector T cells. The observation that breast milk sCD14 survives intact in conditions that mimic the upper digestive tract but is digested by pancreatin (Blais et al., 2006) suggests that such innate responses are initiated in an environment with low bacterial density but are avoided in the densely populated distal intestine.

Relevance of sCD14 to Pathological Diseases

The expression and activation of TLRs in the gastrointestinal tract must be tightly regulated to prevent unremitting inflammation in the face of microbial exposure. Ideally, molecules coordinating TLR effects should limit but not completely eliminate microbial interaction with the mucosal immune system. The previous section suggests that sCD14 possesses this quality and contributes to immune homeostasis in the healthy individual. However, such an attribute may secure immune defense and prevent a dysregulated inflammatory response during infection or aberrant TLR signaling. Admittedly, increased levels of sCD14 are associated with the severity of infection in Gram-positive sepsis (Burgmann et al., 1996) and, notably, with a high mortality in Gram-negative septic shock (Landmann et al., 1995), but to date, there is no clear indication whether these relationships are the cause or effect. Nevertheless, some *in vivo* animal studies suggest that increased levels of sCD14 can counteract the detrimental effects of LPS. For example, levels of sCD14 increase in milk following mammary gland infection or the injection of LPS (Lee et al., 2003a; Vangroenweghe et al., 2004), and the administration of recombinant sCD14 has been shown to reduce the infection in mice (Lee et al., 2003c) and cows (Lee et al., 2003b; Nemchinov et al., 2006). Recombinant sCD14 also protects against mortality in mice treated with LPS (Stelter et al., 1998; Haziot et al., 1995). Furthermore, in transgenic mice expressing different copy numbers of the human CD14 transgene on a murine CD14-/- background, mice with high levels of human CD14 retain LPS in the circulation and prevent its delivery to tissues and organs (Jacque et al., 2006). In so doing, these mice are hyporesponsive to LPS and survive a lethal dose (Jacque et al., 2006).

CD14 is an APP, and several clinical studies have reported increased serum levels of sCD14 in a range of inflammatory conditions (Bas et al., 2004). Higher serum levels are associated with several insulin-resistance–related phenotypes (Fernandez-Real et al., 2003), systemic lupus erythematosus (Egerer et al., 2000), atopic dermatitis (Wuthrich et al., 1992), systemic inflammatory response syndrome (Stoiser et al., 1998), angina (Zalai et al., 2001), preterm labor even in the absence of infection (Gardella et al., 2001), multiple organ failure (Endo et al., 1994), rheumatoid arthritis (Horneff et al., 1993; Yu et al., 1998), multiple sclerosis (Lutterotti et al., 2006), Kawasaki disease (Takeshita et al., 2000), and Gaucher's disease (Hollak et al., 1997). These associations may reflect CD14's capacity to bind to nonmicrobial factors such as monosodium urate crystals (Scott et al., 2006), host heat shock proteins (Kol et al., 2000), integrins (Humphries & Humphries, 2007), surfactant proteins (Sano et al., 2000), atherogenic lipids, and lipoproteins (Schmitz & Orso, 2002), but they also suggest that sCD14 may modulate host immune responses to other "danger" signals besides those of microbial origin.

Conclusion

An emerging concept is that breast milk influences the neonatal immune system's perception of "danger." To do so, soluble PRRs in milk, such as sCD14, may actually facilitate the intestinal response to specific microbial motifs by activating intracellular signaling pathways such as that of NF-κB, a process necessary for maturation of immune tissues. There is evidence that sCD14 mediates both pro- and anti-inflammatory responses depending on the type and location of the responder cell and the nature and dose of the stimulus. It is tempting to speculate that the presence of this versatile molecule in breast milk instructs the neonatal immune system to recognize and respond appropriately to self, non-self, and different forms of danger. To date, little work has specifically addressed the biological activity of milk sCD14 or the function of sCD14 administered orally. Nevertheless, the possibility of developing sCD14-containing products using animal milk holds much promise, not only for infant nutrition but also for clinical application.

References

Abreu, M. T., Vora, P., Faure, E., Thomas, L. S., Arnold, E. T., & Arditi, M. (2001). Decreased expression of Toll-like receptor-4 and MD-2 correlates with intestinal epithelial cell protection against dysregulated proinflammatory gene expression in response to bacterial lipopolysaccharide. *Journal of Immunology, 167*, 1609–1616.

Aderem, A. & Ulevitch, R. J. (2000). Toll-like receptors in the induction of the innate immune response. *Nature, 406*, 782–787.

Arditi, M., Zhou, J., Dorio, R., Rong, G. W., Goyert, S. M., & Kim, K. S. (1993). Endotoxin-mediated endothelial cell injury and activation: Role of soluble CD14. *Infectious Immunology, 61*, 3149–3156.

Arias, M. A., Rey Nores, J. E., Vita, N., Stelter, F., Borysiewicz, L. K., Ferrara, P., et al. (2000). Cutting edge: Human B cell function is regulated by interaction with soluble CD14: Opposite effects on IgG1 and IgE production. *Journal of Immunology, 164*, 3480–3486.

Baldini, M., Lohman, I. C., Halonen, M., Erickson, R. P., Holt, P. G., & Martinez, F. D. (1999). A polymorphism* in the 5' flanking region of the CD14 gene is associated with circulating soluble CD14 levels and with total serum immunoglobulin E. *American Journal of Respiriratory Cell and Molecular Biology, 20*, 976–983.

Bannerman, D. D., Paape, M. J., Hare, W. R., & Sohn, E. J. (2003). Increased levels of LPS-binding protein in bovine blood and milk following bacterial lipopolysaccharide challenge. *Journa of Dairy Science, 86*, 3128–3137.

Bas, S., Gauthier, B. R., Spenato, U., Stingelin, S., & Gabay, C. (2004). CD14 is an acute-phase protein. *Journal of Immunology, 172*, 4470–4479.

Baveye, S., Elass, E., Fernig, D. G., Blanquart, C., Mazurier, J., & Legrand, D. (2000). Human lactoferrin interacts with soluble CD14 and inhibits expression of endothelial adhesion molecules, E-selectin and ICAM-1, induced by the CD14-lipopolysaccharide complex. *Infectious Immunology, 68*, 6519–6525.

Bazil, V., & Strominger, J. L. (1991). Shedding as a mechanism of down-modulation of CD14 on stimulated human monocytes. *Journal of Immunology, 147*, 1567–1574.

Bazil, V., Horejsi, V., Baudys, M., Kristofova, H., Strominger, J. L., Kostka, W., et al. (1986). Biochemical characterization of a soluble form of the 53-kDa monocyte surface antigen. *European Journal of Immunology, 16,* 1583–1589.

Bazil, V., Baudys, M., Hilgert, I., Stefanova, I., Low, M. G., Zbrozek, J., et al. (1989). Structural relationship between the soluble and membrane-bound forms of human monocyte surface glycoprotein CD14. *Molecular Immunology, 26,* 657–662.

Berg, R. D. (1996). The indigenous gastrointestinal microflora. *Trends in Microbiology, 4,* 430–435.

Blais, D. R., Vascotto, S. G., Griffith, M., & Altosaar, I. (2005). LBP and CD14 secreted in tears by the lacrimal glands modulate the LPS response of corneal epithelial cells. *Investigative Ophthalmology and Visual Science, 46,* 4235–4244.

Blais, D. R., Harrold, J., & Altosaar, I. (2006). Killing the messenger in the nick of time: Persistence of breast milk sCD14 in the neonatal gastrointestinal tract. *Pediatric Research, 59,* 371–376.

Bloomfield, S. F., Stanwell-Smith, R., Crevel, R. W., & Pickup, J. (2006). Too clean, or not too clean: The hygiene hypothesis and home hygiene. *Clinical and Experimental Allergy, 36,* 402–425.

Bufler, P., Stiegler, G., Schuchmann, M., Hess, S., Kruger, C., Stelter, F., et al. (1995). Soluble lipopolysaccharide receptor (CD14) is released via two different mechanisms from human monocytes and CD14 transfectants. *European Journal of Immunology, 25,* 604–610.

Burgmann, H., Winkler, S., Locker, G. J., Presterl, E., Laczika, K., Staudinger, T., et al. (1996). Increased serum concentration of soluble CD14 is a prognostic marker in Gram-positive sepsis. *Clinical Immunology and Immunopathology, 80,* 307–310.

Caplan, M. S., & MacKendrick, W. (1993). Necrotizing enterocolitis: A review of pathogenetic mechanisms and implications for prevention. *Pediatric Pathology, 13,* 357–369.

Cauwels, A., Frei, K., Sansano, S., Fearns, C., Ulevitch, R., Zimmerli, W., et al. (1999). The origin and function of soluble CD14 in experimental bacterial meningitis. *Journal of Immunology, 162,* 4762–4772.

Cebra, J. J. (1999). Influences of microbiota on intestinal immune system development. *American Journal of Clinical Nutrition, 69,* 1046S–1051S.

Chow, J. C., Young, D. W., Golenbock, D. T., Christ, W. J., & Gusovsky, F. (1999). Toll-like receptor-4 mediates lipopolysaccharide-induced signal transduction. *Journal of Biological Chemistry, 274,* 10689–10692.

Claud, E. C., Lu, L., Anton, P. M., Savidge, T., Walker, W. A., & Cherayil, B. J. (2004). Developmentally regulated IκB expression in intestinal epithelium and susceptibility to flagellin-induced inflammation. *Proceedings of the National Academy of Sciences USA, 101,* 7404–7408.

Cleveland, M. G., Gorham, J. D., Murphy, T. L., Tuomanen, E., & Murphy, K. M. (1996). Lipoteichoic acid preparations of Gram-positive bacteria induce interleukin-12 through a CD14-dependent pathway. *Infectious Immunology, 64,* 1906–1912.

da Silva, C. J., Soldau, K., Christen, U., Tobias, P. S., & Ulevitch, R. J. (2001). Lipopolysaccharide is in close proximity to each of the proteins in its membrane receptor complex. Transfer from CD14 to TLR4 and MD-2. *Journal of Biological Chemistry, 276,* 21129–21135.

Dentener, M. A., Bazil, V., Von Asmuth, E. J., Ceska, M., & Buurman, W. A. (1993). Involvement of CD14 in lipopolysaccharide-induced tumor necrosis factor-α, IL-6 and IL-8 release by human monocytes and alveolar macrophages. *Journal of Immunology, 150,* 2885–2891.

Diamond, G., Russell, J. P., & Bevins, C. L. (1996). Inducible expression of an antibiotic peptide gene in lipopolysaccharide-challenged tracheal epithelial cells. *Proceedings of the National Academy of Sciences USA, 93,* 5156–5160.

Dunstan, J. A., Roper, J., Mitoulas, L., Hartmann, P. E., Simmer, K., & Prescott, S. L. (2004). The effect of supplementation with fish oil during pregnancy on breast milk immunoglobulin A, soluble CD14, cytokine levels and fatty acid composition. *Clinical and Experimental Allergy, 34,* 1237–1242.

Durieux, J. J., Vita, N., Popescu, O., Guette, F., Calzada-Wack, J., Munker, R., et al. (1994). The two soluble forms of the lipopolysaccharide receptor, CD14: Characterization and release by normal human monocytes. *European Journal of Immunology, 24,* 2006–2012.

Dziarski, R., Tapping, R. I., & Tobias, P. S. (1998). Binding of bacterial peptidoglycan to CD14. *Journal of Biological Chemistry, 273,* 8680–8690.

Edwards, C. A., & Parrett, A. M. (2002). Intestinal flora during the first months of life: New perspectives. *British Journal of Nutrition, 88(Suppl 1),* S11–S18.

Egerer, K., Feist, E., Rohr, U., Pruss, A., Burmester, G. R., & Dorner, T. (2000). Increased serum soluble CD14, ICAM-1 and E-selectin correlate with disease activity and prognosis in systemic lupus erythematosus. *Lupus, 9,* 614–621.

Elass, E., Masson, M., Mazurier, J., & Legrand, D. (2002). Lactoferrin inhibits the lipopolysaccharide-induced expression and proteoglycan-binding ability of interleukin-8 in human endothelial cells. *Infectious Immunology, 70,* 1860–1866.

Endo, S., Inada, K., Kasai, T., Takakuwa, T., Nakae, H., Kikuchi, M., et al. (1994). Soluble CD14 (sCD14) levels in patients with multiple organ failure (MOF). *Research Communications in Chemical Pathology and Pharmacology, 84,* 17–25.

Espevik, T., Otterlei, M., Skjak-Braek, G., Ryan, L., Wright, S. D., & Sundan, A. (1993). The involvement of CD14 in stimulation of cytokine production by uronic acid polymers. *European Journal of Immunology, 23,* 255–261.

Espinoza, J., Chaiworapongsa, T., Romero, R., Gomez, R., Kim, J. C., Yoshimatsu, J., et al. (2002). Evidence of participation of soluble CD14 in the host response to microbial invasion of the amniotic cavity and intra-amniotic inflammation in term and preterm gestations. *Journal of Maternal and Fetal Neonatal Medicine, 12,* 304–312.

Falk, P. G., Hooper, L. V., Midtvedt, T., & Gordon, J. I. (1998). Creating and maintaining the gastrointestinal ecosystem: What we know and need to know from gnotobiology. *Microbiology and .Molecular Biology Reviews, 62,* 1157–1170.

Fernandez-Real, J. M., Broch, M., Richart, C., Vendrell, J., Lopez-Bermejo, A., & Ricart, W. (2003). CD14 monocyte receptor, involved in the inflammatory cascade, and insulin sensitivity. *Journal of Clinical Endocrinology and Metabolism, 88,* 1780–1784.

Ferrero, E., & Goyert, S. M. (1988). Nucleotide sequence of the gene encoding the monocyte differentiation antigen, CD14. *Nucleic Acids Research, 16,* 4173.

Ferrero, E., Hsieh, C. L., Francke, U., & Goyert, S. M. (1990). CD14 is a member of the family of leucine-rich proteins and is encoded by a gene syntenic with multiple receptor genes. *Journal of Immunology, 145,* 331–336.

Filipp, D., Alizadeh-Khiavi, K., Richardson, C., Palma, A., Paredes, N., Takeuchi, O., et al. (2001). Soluble CD14 enriched in colostrum and milk induces B cell growth and differentiation. *Proceedings of the National Academy of Sciences USA, 98,* 603–608.

Frey, E. A., Miller, D. S., Jahr, T. G., Sundan, A., Bazil, V., Espevik, T., et al. (1992). Soluble CD14 participates in the response of cells to lipopolysaccharide. *Journal of Experimental Medicine, 176,* 1665–1671.

Funda, D. P., Tuckova, L., Farre, M. A., Iwase, T., Moro, I., & Tlaskalova-Hogenova, H. (2001). CD14 is expressed and released as soluble CD14 by human intestinal epithelial cells *in vitro*: Lipopolysaccharide activation of epithelial cells revisited. *Infectious Immunology, 69,* 3772–3781.

Fusunyan, R. D., Nanthakumar, N. N., Baldeon, M. E., & Walker, W. A. (2001). Evidence for an innate immune response in the immature human intestine: Toll-like receptors on fetal enterocytes. *Pediatric Research, 49,* 589–593.

Gallay, P., Jongeneel, C. V., Barras, C., Burnier, M., Baumgartner, J. D., Glauser, M. P., et al. (1993). Short time exposure to lipopolysaccharide is sufficient to activate human monocytes. *Journal of Immunology, 150,* 5086–5093.

Gardella, C., Hitti, J., Martin, T. R., Ruzinski, J. T., & Eschenbach, D. (2001). Amniotic fluid lipopolysaccharide-binding protein and soluble CD14 as mediators of the inflammatory response in preterm labor. *American Journal of Obstetrics and Gynecology, 184*, 1241–1248.

Gegner, J. A., Ulevitch, R. J., & Tobias, P. S. (1995). Lipopolysaccharide (LPS) signal transduction and clearance. Dual roles for LPS binding protein and membrane CD14. *Journal of Biological Chemistry, 270*, 5320–5325.

Goldblum, S. E., Brann, T. W., Ding, X., Pugin, J., & Tobias, P. S. (1994). Lipopolysaccharide (LPS)-binding protein and soluble CD14 function as accessory molecules for LPS-induced changes in endothelial barrier function, *in vitro*. *Journal of Clinical Investigations, 93*, 692–702.

Goyert, S. M., & Ferrero, E. (1987). Biochemical analysis of myeloid antigens and cDNA expression of gp 55 (CD14). In A. McMichael (Ed.), *Leucocyte Typing III* (pp. 613–619). Oxford: Oxford University Press.

Goyert, S. M., Ferrero, E., Rettig, W. J., Yenamandra, A. K., Obata, F., & Le Beau, M. M. (1988). The CD14 monocyte differentiation antigen maps to a region encoding growth factors and receptors. *Science, 239*, 497–500.

Grunwald, U., Kruger, C., Westermann, J., Lukowsky, A., Ehlers, M., & Schutt, C. (1992). An enzyme-linked immunosorbent assay for the quantification of solubilized CD14 in biological fluids. *Journal of .Immunology Methods, 155*, 225–232.

Grunwald, U., Kruger, C., & Schutt, C. (1993). Endotoxin-neutralizing capacity of soluble CD14 is a highly conserved specific function. *Circulatory Shock, 39*, 220–225.

Grunwald, U., Fan, X., Jack, R. S., Workalemahu, G., Kallies, A., Stelter, F., et al. (1996). Monocytes can phagocytose Gram-negative bacteria by a CD14-dependent mechanism. *Journal of Immunology, 157*, 4119–4125.

Guerra, S., Carla, L., I, LeVan, T. D., Wright, A. L., Martinez, F. D., & Halonen, M. (2004a). The differential effect of genetic variation on soluble CD14 levels in human plasma and milk. *American Journal of Reproductive Immunology, 52*, 204–211.

Guerra, S., Lohman, I. C., Halonen, M., Martinez, F. D., & Wright, A. L. (2004b). Reduced interferon gamma production and soluble CD14 levels in early life predict recurrent wheezing by 1 year of age. *American Journal of Respiration and Critical Care Medicine, 169*, 70–76.

Gupta, D., Kirkland, T. N., Viriyakosol, S., & Dziarski, R. (1996). CD14 is a cell-activating receptor for bacterial peptidoglycan. *Journal of Biological Chemistry, 271*, 23310–23316.

Hailman, E., Lichenstein, H. S., Wurfel, M. M., Miller, D. S., Johnson, D. A., Kelley, M., et al. (1994). Lipopolysaccharide (LPS)-binding protein accelerates the binding of LPS to CD14. *Journal of Experimental Medicine, 179*, 269–277.

Haller, D., Bode, C., Hammes, W. P., Pfeifer, A. M., Schiffrin, E. J., & Blum, S. (2000). Non-pathogenic bacteria elicit a differential cytokine response by intestinal epithelial cell/leucocyte co-cultures. *Gut, 47*, 79–87.

Harris, C. L., Vigar, M. A., Rey Nores, J. E., Horejsi, V., Labeta, M. O., & Morgan, B. P. (2001). The lipopolysaccharide co-receptor CD14 is present and functional in seminal plasma and expressed on spermatozoa. *Immunology, 104*, 317–323.

Hattor, Y., Kasai, K., Akimoto, K., & Thiemermann, C. (1997). Induction of NO synthesis by lipoteichoic acid from *Staphylococcus aureus* in J774 macrophages: Involvement of a CD14-dependent pathway. *Biochemistry and Biophysics Research Community, 233*, 375–379.

Haversin, L., Ohlsson, B. G., Hahn-Zoric, M., Hanson, L. A., & Mattsby-Baltzer, I. (2002). Lactoferrin down-regulates the LPS-induced cytokine production in monocytic cells via NF-κB. *Cell Immunology, 220*, 83–95.

Haziot, A., Chen, S., Ferrero, E., Low, M. G., Silber, R., & Goyert, S. M. (1988). The monocyte differentiation antigen, CD14, is anchored to the cell membrane by a phosphatidylinositol linkage. *Journal of Immunology, 141*, 547–552.

Haziot, A., Rong, G. W., Silver, J., & Goyert, S. M. (1993a). Recombinant soluble CD14 mediates the activation of endothelial cells by lipopolysaccharide. *Journal of Immunology, 151*, 1500–1507.

Haziot, A., Tsuberi, B. Z., & Goyert, S. M. (1993b). Neutrophil CD14: Biochemical properties and role in the secretion of tumor necrosis factor-alpha in response to lipopolysaccharide. *Journal of Immunology, 150*, 5556–5565.

Haziot, A., Rong, G. W., Bazil, V., Silver, J., & Goyert, S. M. (1994). Recombinant soluble CD14 inhibits LPS-induced tumor necrosis factor-alpha production by cells in whole blood. *Journal of Immunology, 152*, 5868–5876.

Haziot, A., Rong, G. W., Lin, X. Y., Silver, J., & Goyert, S. M. (1995). Recombinant soluble CD14 prevents mortality in mice treated with endotoxin (lipopolysaccharide). *Journal of Immunology, 154*, 6529–6532.

Heumann, D., Glauser, M. P., & Calandra, T. (1998). Molecular basis of host-pathogen interaction in septic shock. *Current Opinions in Microbiology, 1*, 49–55.

Hollak, C. E., Evers, L., Aerts, J. M., & van Oers, M. H. (1997). Elevated levels of M-CSF, sCD14 and IL8 in type 1 Gaucher disease. *Blood Cells Molecular Disease, 23*, 201–212.

Horneff, G., Sack, U., Kalden, J. R., Emmrich, F., & Burmester, G. R. (1993). Reduction of monocyte-macrophage activation markers upon anti-CD4 treatment. Decreased levels of IL-1, IL-6, neopterin and soluble CD14 in patients with rheumatoid arthritis. *Clinical Experiments in Immunology, 91*, 207–213.

Hoy, C., Millar, M. R., MacKay, P., Godwin, P. G., Langdale, V., & Levene, M. I. (1990). Quantitative changes in faecal microflora preceding necrotising enterocolitis in premature neonates. *Archives of Disease in Childhood, 65*, 1057–1059.

Humphries, J. D., & Humphries, M. J. (2007). CD14 is a ligand for the integrin α4β1. *FEBS Letters, 581*, 757–763.

Ikeda, A., Takata, M., Taniguchi, T., & Sekikawa, K. (1997). Molecular cloning of bovine CD14 gene. *Journal of Veterinary Medicine and Science, 59*, 715–719.

Ismail, A. S., & Hooper, L. V. (2005). Epithelial cells and their neighbors. IV. Bacterial contributions to intestinal epithelial barrier integrity. *American Journal of Physiology: Gastrointestinal and Liver Physiology, 289*, G779–G784.

Jack, R. S., Grunwald, U., Stelter, F., Workalemahu, G., & Schutt, C. (1995). Both membrane-bound and soluble forms of CD14 bind to Gram-negative bacteria. *European Journal of Immunology, 25*, 1436–1441.

Jack, R. S., Fan, X., Bernheiden, M., Rune, G., Ehlers, M., Weber, A., et al. (1997). Lipopolysaccharide-binding protein is required to combat a murine Gram-negative bacterial infection. *Nature, 389*, 742–745.

Jackson, K. M., & Nazar, A. M. (2006). Breastfeeding, the immune response, and long-term health. *Journal of American Osteopath Association, 106*, 203–207.

Jacque, B., Stephan, K., Smirnova, I., Kim, B., Gilling, D., & Poltorak, A. (2006). Mice expressing high levels of soluble CD14 retain LPS in the circulation and are resistant to LPS-induced lethality. *European Journal of Immunology, 36*, 3007–3016.

Jilling, T., Simon, D., Lu, J., Meng, F. J., Li, D., Schy, R., et al. (2006). The roles of bacteria and TLR4 in rat and murine models of necrotizing enterocolitis. *Journal of Immunology, 177*, 3273–3282.

Jones, C. A., Holloway, J. A., Popplewell, E. J., Diaper, N. D., Holloway, J. W., Vance, G. H., et al. (2002). Reduced soluble CD14 levels in amniotic fluid and breast milk are associated with the subsequent development of atopy, eczema, or both. *Journal of Allergy and Clinical Immunology, 109*, 858–866.

Kaisho, T., & Akira, S. (2006). Toll-like receptor function and signaling. *Journal of Allergy and Clinical Immunology, 117*, 979–987.

Kirjavainen, P. V., Arvola, T., Salminen, S. J., & Isolauri, E. (2002). Aberrant composition of gut microbiota of allergic infants: A target of bifidobacterial therapy at weaning? *Gut, 51*, 51–55.

Kirkland, T. N., Finley, F., Leturcq, D., Moriarty, A., Lee, J. D., Ulevitch, R. J., et al. (1993). Analysis of lipopolysaccharide binding by CD14. *Journal of Biological Chemistry, 268,* 24818–24823.

Kitchens, R. L., & Munford, R. S. (1995). Enzymatically deacylated lipopolysaccharide (LPS) can antagonize LPS at multiple sites in the LPS recognition pathway. *Journal of Biological Chemistry, 270,* 9904–9910.

Kitchens, R. L., & Thompson, P. A. (2005). Modulatory effects of sCD14 and LBP on LPS-host cell interactions. *Journal of Endotoxin Research, 11,* 225–229.

Kitchens, R. L., Wolfbauer, G., Albers, J. J., & Munford, R. S. (1999). Plasma lipoproteins promote the release of bacterial lipopolysaccharide from the monocyte cell surface. *Journal of Biological Chemistry, 274,* 34116–34122.

Kitchens, R. L., Thompson, P. A., Viriyakosol, S., O'Keefe, G. E., & Munford, R. S. (2001). Plasma CD14 decreases monocyte responses to LPS by transferring cell-bound LPS to plasma lipoproteins. *Journal of Clinical Investigations, 108,* 485–493.

Kol, A., Lichtman, A. H., Finberg, R. W., Libby, P., & Kurt-Jones, E. A. (2000). Cutting edge: Heat shock protein (HSP) 60 activates the innate immune response: CD14 is an essential receptor for HSP60 activation of mononuclear cells. *Journal of Immunology, 164,* 13–17.

Kruger, C., Schutt, C., Obertacke, U., Joka, T., Muller, F. E., Knoller, J., et al. (1991). Serum CD14 levels in polytraumatized and severely burned patients. *Clinical Experiments in Immunology, 85,* 297–301.

Kull, I., Almqvist, C., Lilja, G., Pershagen, G., & Wickman, M. (2004). Breast-feeding reduces the risk of asthma during the first 4 years of life. *Journal of Allergy and Clinical Immunology, 114,* 755–760.

Kusunoki, T., & Wright, S. D. (1996). Chemical characteristics of *Staphylococcus aureus* molecules that have CD14-dependent cell-stimulating activity. *Journal of Immunology, 157,* 5112–5117.

Kusunoki, T., Hailman, E., Juan, T. S., Lichenstein, H. S., & Wright, S. D. (1995). Molecules from *Staphylococcus aureus* that bind CD14 and stimulate innate immune responses. *Journal of Experimental Medicine, 182,* 1673–1682.

Labeta, M. O., Durieux, J. J., Fernandez, N., Herrmann, R., & Ferrara, P. (1993). Release from a human monocyte-like cell line of two different soluble forms of the lipopolysaccharide receptor, CD14. *European Journal of Immunology, 23,* 2144–2151.

Labeta, M. O., Vidal, K., Nores, J. E., Arias, M., Vita, N., Morgan, B. P., et al. (2000). Innate recognition of bacteria in human milk is mediated by a milk-derived highly expressed pattern recognition receptor, soluble CD14. *Journal of Experimental Medicine, 191,* 1807–1812.

Laitinen, K., Hoppu, U., Hamalainen, M., Linderborg, K., Moilanen, E., & Isolauri, E. (2006). Breast milk fatty acids may link innate and adaptive immune regulation: Analysis of soluble CD14, prostaglandin E2, and fatty acids. *Pediatric Research, 59,* 723–727.

Landmann, R., Fisscher, A. E., & Obrecht, J. P. (1992). Interferon-γ and interleukin-4 down-regulate soluble CD14 release in human monocytes and macrophages. *Journal of Leukocyte Biology, 52,* 323–330.

Landmann, R., Zimmerli, W., Sansano, S., Link, S., Hahn, A., Glauser, M. P., et al. (1995). Increased circulating soluble CD14 is associated with high mortality in Gram-negative septic shock. *Journal of Infectious Diseases, 171,* 639–644.

Landmann, R., Reber, A. M., Sansano, S., & Zimmerli, W. (1996). Function of soluble CD14 in serum from patients with septic shock. *Journal of Infectious Disease, 173,* 661–668.

Lauener, R. P., Birchler, T., Adamski, J., Braun-Fahrlander, C., Bufe, A., Herz, U., et al. (2002). Expression of CD14 and Toll-like receptor 2 in farmers' and non-farmers' children. *Lancet, 360,* 465–466.

Lawrence, R. M. (2005). Host-resistance factors and immunologic significance of human milk. In R. A. Lawrence & R. M. Lawrence (Eds.), *Breastfeeding. A Guide for the Medical Profession* (pp. 171–214). Philadelphia: Elsevier Mosby.

Lebouder, E., Rey-Nores, J. E., Raby, A. C., Affolter, M., Vidal, K., Thornton, C. A., et al. (2006). Modulation of neonatal microbial recognition: TLR-mediated innate immune responses are specifically and differentially modulated by human milk. *Journal of Immunology, 176,* 3742–3752.

Lee, J. W., Paape, M. J., Elsasser, T. H., & Zhao, X. (2003a). Elevated milk soluble CD14 in bovine mammary glands challenged with *Escherichia coli* lipopolysaccharide. *Journal of Dairy Science, 86,* 2382–2389.

Lee, J. W., Paape, M. J., Elsasser, T. H., & Zhao, X. (2003b). Recombinant soluble CD14 reduces severity of intramammary infection by *Escherichia coli*. *Infectious Immunology, 71,* 4034–4039.

Lee, J. W., Paape, M. J., & Zhao, X. (2003c). Recombinant bovine soluble CD14 reduces severity of experimental *Escherichia coli* mastitis in mice. *Veterinary Research, 34,* 307–316.

LeVan, T. D., Guerra, S., Klimecki, W., Vasquez, M. M., Lohman, I. C., Martinez, F. D., et al. (2006). The impact of CD14 polymorphisms on the development of soluble CD14 levels during infancy. *Genes and Immunology, 7,* 77–80.

Liu, S., Khemlani, L. S., Shapiro, R. A., Johnson, M. L., Liu, K., Geller, D. A., et al. (1998). Expression of CD14 by hepatocytes: Upregulation by cytokines during endotoxemia. *Infectious Immunology, 66,* 5089–5098.

Lonnerdal, B. (2003). Nutritional and physiologic significance of human milk proteins. *American Journal of Clinical Nutrition, 77,* 1537S–1543S.

Loppnow, H., Stelter, F., Schonbeck, U., Schluter, C., Ernst, M., Schutt, C., et al. (1995). Endotoxin activates human vascular smooth muscle cells despite lack of expression of CD14 mRNA or endogenous membrane CD14. *Infectious Immunology, 63,* 1020–1026.

Lotz, M., Gutle, D., Walther, S., Menard, S., Bogdan, C., & Hornef, M. W. (2006). Postnatal acquisition of endotoxin tolerance in intestinal epithelial cells. *Journal of Experimental Medicine, 203,* 973–984.

Lutterotti, A., Kuenz, B., Gredler, V., Khalil, M., Ehling, R., Gneiss, C., et al. (2006). Increased serum levels of soluble CD14 indicate stable multiple sclerosis. *Journal of Neuroimmunology, 181,* 145–149.

Macpherson, A. J., & Harris, N. L. (2004). Interactions between commensal intestinal bacteria and the immune system. *Nature Reviews Immunology, 4,* 478–485.

Maliszewski, C. R., Ball, E. D., Graziano, R. F., & Fanger, M. W. (1985). Isolation and characterization of My23, a myeloid cell-derived antigen reactive with the monoclonal antibody AML-2-23. *Journal of Immunology, 135,* 1929–1936.

Martin, T. R., Mathison, J. C., Tobias, P. S., Leturcq, D. J., Moriarty, A. M., Maunder, R. J., et al. (1992). Lipopolysaccharide binding protein enhances the responsiveness of alveolar macrophages to bacterial lipopolysaccharide. Implications for cytokine production in normal and injured lungs. *Journal of Clinical Investigations, 90,* 2209–2219.

Mathison, J. C., Tobias, P. S., Wolfson, E., & Ulevitch, R. J. (1992). Plasma lipopolysaccharide (LPS)-binding protein. A key component in macrophage recognition of Gram-negative LPS. *Journal of Immunology, 149,* 200–206.

Matsuura, K., Ishida, T., Setoguchi, M., Higuchi, Y., Akizuki, S., & Yamamoto, S. (1994). Upregulation of mouse CD14 expression in Kupffer cells by lipopolysaccharide. *Journal of Experimental Medicine, 179,* 1671–1676.

Matzinger, P. (2002). The danger model: A renewed sense of self. *Science, 296,* 301–305.

Nasu, N., Yoshida, S., Akizuki, S., Higuchi, Y., Setoguchi, M., & Yamamoto, S. (1991). Molecular and physiological properties of murine CD14. *International Immunology, 3,* 205–213.

Nemchinov, L. G., Paape, M. J., Sohn, E. J., Bannerman, D. D., Zarlenga, D. S., & Hammond, R. W. (2006). Bovine CD14 receptor produced in plants reduces severity of intramammary bacterial infection. *FASEB Journal, 20,* 1345–1351.

Newman, S. L., Chaturvedi, S., & Klein, B. S. (1995). The WI-1 antigen of *Blastomyces dermatitidis* yeasts mediates binding to human macrophage CD11b/CD18 (CR3) and CD14. *Journal of Immunology, 154*, 753–761.

Nockher, W. A., Wigand, R., Schoeppe, W., & Scherberich, J. E. (1994). Elevated levels of soluble CD14 in serum of patients with systemic lupus erythematosus. *Clinical Experiments in Immunology, 96*, 15–19.

O'Neill, L. A. (2002). Signal transduction pathways activated by the IL-1 receptor/Toll-like receptor superfamily. *Current Topics in Microbiology and Immunology, 270*, 47–61.

O'Neill, L. A., Dunne, A., Edjeback, M., Gray, P., Jefferies, C., & Wietek, C. (2003). Mal and MyD88: Adapter proteins involved in signal transduction by Toll-like receptors. *Journal of Endotoxin Research, 9*, 55–59.

Otte, J. M., Cario, E., & Podolsky, D. K. (2004). Mechanisms of cross hyporesponsiveness to Toll-like receptor bacterial ligands in intestinal epithelial cells. *Gastroenterology, 126*, 1054–1070.

Pugin, J., Schurer-Maly, C. C., Leturcq, D., Moriarty, A., Ulevitch, R. J., & Tobias, P. S. (1993). Lipopolysaccharide activation of human endothelial and epithelial cells is mediated by lipopolysaccharide-binding protein and soluble CD14. *Proceedings of the National Academy of Sciences USA, 90*, 2744–2748.

Pugin, J., Heumann, I. D., Tomasz, A., Kravchenko, V. V., Akamatsu, Y., Nishijima, M., et al. (1994). CD14 is a pattern recognition receptor. *Immunity, 1*, 509–516.

Rakoff-Nahoum, S., Paglino, J., Eslami-Varzaneh, F., Edberg, S., & Medzhitov, R. (2004). Recognition of commensal microflora by Toll-like receptors is required for intestinal homeostasis. *Cell, 118*, 229–241.

Re, F., & Strominger, J. L. (2001). Toll-like receptor 2 (TLR2) and TLR4 differentially activate human dendritic cells. *Journal of Biological Chemistry, 276*, 37692–37699.

Read, M. A., Cordle, S. R., Veach, R. A., Carlisle, C. D., & Hawiger, J. (1993). Cell-free pool of CD14 mediates activation of transcription factor NF-κB by lipopolysaccharide in human endothelial cells. *Proceedings of the National Academy of Sciences USA, 90*, 9887–9891.

Rey Nores, J. E., Bensussan, A., Vita, N., Stelter, F., Arias, M. A., Jones, M., et al. (1999). Soluble CD14 acts as a negative regulator of human T cell activation and function. *European Journal of Immunology, 29*, 265–276.

Rinne, M., Kalliomaki, M., Arvilommi, H., Salminen, S., & Isolauri, E. (2005). Effect of probiotics and breastfeeding on the bifidobacterium and lactobacillus/enterococcus microbiota and humoral immune responses. *Journal of Pediatrics, 147*, 186–191.

Ronnestad, A., Abrahamsen, T. G., Medbo, S., Reigstad, H., Lossius, K., Kaaresen, P. I., et al. (2005). Septicemia in the first week of life in a Norwegian national cohort of extremely premature infants. *Pediatrics, 115*, e262–e268.

Rothenbacher, D., Weyermann, M., Beermann, C., & Brenner, H. (2005). Breastfeeding, soluble CD14 concentration in breast milk and risk of atopic dermatitis and asthma in early childhood: Birth cohort study. *Clinical and Experimental Allergy, 35*, 1014–1021.

Sano, H., Chiba, H., Iwaki, D., Sohma, H., Voelker, D. R., & Kuroki, Y. (2000). Surfactant proteins A and D bind CD14 by different mechanisms. *Journal of Biological Chemistry, 275*, 22442–22451.

Savedra, R., Jr., Delude, R. L., Ingalls, R. R., Fenton, M. J., & Golenbock, D. T. (1996). Mycobacterial lipoarabinomannan recognition requires a receptor that shares components of the endotoxin signaling system. *Journal of Immunology, 157*, 2549–2554.

Schmitz, G., & Orso, E. (2002). CD14 signalling in lipid rafts: New ligands and co-receptors. *Current Opinions in Lipidology, 13*, 513–521.

Schroder, N. W., Morath, S., Alexander, C., Hamann, L., Hartung, T., Zahringer, U., et al. (2003). Lipoteichoic acid (LTA) of *Streptococcus pneumoniae* and *Staphylococcus aureus* activates immune cells via Toll-like receptor (TLR)-2, lipopolysaccharide-binding protein

(LBP), and CD14, whereas TLR-4 and MD-2 are not involved. *Journal of Biological Chemistry, 278,* 15587–15594.

Schumann, R. R., Leong, S. R., Flaggs, G. W., Gray, P. W., Wright, S. D., Mathison, J. C., et al. (1990). Structure and function of lipopolysaccharide binding protein. *Science, 249,* 1429–1431.

Schumann, R. R., Rietschel, E. T., & Loppnow, H. (1994). The role of CD14 and lipopolysaccharide-binding protein (LBP) in the activation of different cell types by endotoxin. *Medical Microbiology and Immunology (Berlin), 183,* 279–297.

Schutt, C., Schilling, T., Grunwald, U., Schonfeld, W., & Kruger, C. (1992). Endotoxin-neutralizing capacity of soluble CD14. *Research in Immunology, 143,* 71–78.

Schwandner, R., Dziarski, R., Wesche, H., Rothe, M., & Kirschning, C. J. (1999). Peptidoglycan- and lipoteichoic acid-induced cell activation is mediated by Toll-like receptor 2. *Journal of Biological Chemistry, 274,* 17406–17409.

Scott, P., Ma, H., Viriyakosol, S., Terkeltaub, R., & Liu-Bryan, R. (2006). Engagement of CD14 mediates the inflammatory potential of monosodium urate crystals. *Journal of Immunology, 177,* 6370–6378.

Sellati, T. J., Bouis, D. A., Kitchens, R. L., Darveau, R. P., Pugin, J., Ulevitch, R. J., et al. (1998). *Treponema pallidum* and *Borrelia burgdorferi* lipoproteins and synthetic lipopeptides activate monocytic cells via a CD14-dependent pathway distinct from that used by lipopolysaccharide. *Journal of Immunology, 160,* 5455–5464.

Setoguchi, M., Nasu, N., Yoshida, S., Higuchi, Y., Akizuki, S., & Yamamoto, S. (1989). Mouse and human CD14 (myeloid cell-specific leucine-rich glycoprotein) primary structure deduced from cDNA clones. *Biochimica et Biophysica Acta, 1008,* 213–222.

Simmons, D. L., Tan, S., Tenen, D. G., Nicholson-Weller, A., & Seed, B. (1989). Monocyte antigen CD14 is a phospholipid anchored membrane protein. *Blood, 73,* 284–289.

Soell, M., Lett, E., Holveck, F., Scholler, M., Wachsmann, D., & Klein, J. P. (1995). Activation of human monocytes by streptococcal rhamnose glucose polymers is mediated by CD14 antigen, and mannan binding protein inhibits TNF-α release. *Journal of Immunology, 154,* 851–860.

Sohn, E. J., Paape, M. J., Bannerman, D. D., Connor, E. E., Fetterer, R. H., & Peters, R. R. (2007). Shedding of sCD14 by bovine neutrophils following activation with bacterial lipopolysaccharide results in down-regulation of IL-8. *Veterinary Research, 38,* 95–108.

Song, P. I., Park, Y. M., Abraham, T., Harten, B., Zivony, A., Neparidze, N., et al. (2002). Human keratinocytes express functional CD14 and Toll-like receptor 4. *Journal of Investigations in Dermatology, 119,* 424–432.

Steinwender, G., Schimpl, G., Sixl, B., Kerbler, S., Ratschek, M., Kilzer, S., et al. (1996). Effect of early nutritional deprivation and diet on translocation of bacteria from the gastrointestinal tract in the newborn rat. *Pediatric Research, 39,* 415–420.

Stelter, F., Pfister, M., Bernheiden, M., Jack, R. S., Bufler, P., Engelmann, H., et al. (1996). The myeloid differentiation antigen CD14 is N- and O-glycosylated. Contribution of N-linked glycosylation to different soluble CD14 isoforms. *European Journal of Biochemistry, 236,* 457–464.

Stelter, F., Witt, S., Furll, B., Jack, R. S., Hartung, T., & Schutt, C. (1998). Different efficacy of soluble CD14 treatment in high- and low-dose LPS models. *European Journal of Clinical Investigations, 28,* 205–213.

Stoiser, B., Knapp, S., Thalhammer, F., Locker, G. J., Kofler, J., Hollenstein, U., et al. (1998). Time course of immunological markers in patients with the systemic inflammatory response syndrome: Evaluation of sCD14, sVCAM-1, sELAM-1, MIP-1α and TGF-β2. *European Journal of Clinical Investigations, 28,* 672–678.

Strachan, D. P. (1989). Hay fever, hygiene, and household size. *British Medical Journal, 299,* 1259–1260.

Sugawara, S., Sugiyama, A., Nemoto, E., Rikiishi, H., & Takada, H. (1998). Heterogeneous expression and release of CD14 by human gingival fibroblasts: Characterization and

CD14-mediated interleukin-8 secretion in response to lipopolysaccharide. *Infectious Immunology, 66*, 3043–3049.
Sugiyama, T., & Wright, S. D. (2001). Soluble CD14 mediates efflux of phospholipids from cells. *Journal of Immunology, 166*, 826–831.
Takai, N., Kataoka, M., Higuchi, Y., Matsuura, K., & Yamamoto, S. (1997). Primary structure of rat CD14 and characteristics of rat CD14, cytokine, and NO synthase mRNA expression in mononuclear phagocyte system cells in response to LPS. *Journal of Leukocyte Biology, 61*, 736–744.
Takeshita, S., Nakatani, K., Tsujimoto, H., Kawamura, Y., Kawase, H., & Sekine, I. (2000). Increased levels of circulating soluble CD14 in Kawasaki disease. *Clinical Experiments in Immunology, 119*, 376–381.
Takeuchi, O., Hoshino, K., Kawai, T., Sanjo, H., Takada, H., Ogawa, T., et al. (1999). Differential roles of TLR2 and TLR4 in recognition of Gram-negative and Gram-positive bacterial cell wall components. *Immunity, 11*, 443–451.
Thomas, C. J., Kapoor, M., Sharma, S., Bausinger, H., Zyilan, U., Lipsker, D., et al. (2002). Evidence of a trimolecular complex involving LPS, LPS binding protein and soluble CD14 as an effector of LPS response. *FEBS Letters, 531*, 184–188.
Tobias, P. S., & Ulevitch, R. J. (1993). Lipopolysaccharide binding protein and CD14 in LPS dependent macrophage activation. *Immunobiology, 187*, 227–232.
Tobias, P. S., Soldau, K., & Ulevitch, R. J. (1986). Isolation of a lipopolysaccharide-binding acute phase reactant from rabbit serum. *Journal of Experimental Medicine, 164*, 777–793.
Tobias, P. S., Mathison, J., Mintz, D., Lee, J. D., Kravchenko, V., Kato, K., et al. (1992). Participation of lipopolysaccharide-binding protein in lipopolysaccharide-dependent macrophage activation. *American Journal of Respiratory Cell and Molecular Biology, 7*, 239–245.
Uehara, A., Sugawara, S., Watanabe, K., Echigo, S., Sato, M., Yamaguchi, T., et al. (2003). Constitutive expression of a bacterial pattern recognition receptor, CD14, in human salivary glands and secretion as a soluble form in saliva. *Clinical and Diagnostic Laboratory Immunology, 10*, 286–292.
van Saene, H. K., Taylor, N., Donnell, S. C., Glynn, J., Magnall, V. L., Okada, Y., et al. (2003). Gut overgrowth with abnormal flora: The missing link in parenteral nutrition-related sepsis in surgical neonates. *European Journal of Clinical Nutrition, 57*, 548–553.
Vangroenweghe, F., Rainard, P., Paape, M., Duchateau, L., & Burvenich, C. (2004). Increase of *Escherichia coli* inoculum doses induces faster innate immune response in primiparous cows. *Journal of Dairy Science, 87*, 4132–4144.
Verhasselt, V., Buelens, C., Willems, F., De Groote, D., Haeffner-Cavaillon, N., & Goldman, M. (1997). Bacterial lipopolysaccharide stimulates the production of cytokines and the expression of costimulatory molecules by human peripheral blood dendritic cells: Evidence for a soluble CD14-dependent pathway. *Journal of Immunology, 158*, 2919–2925.
Vidal, K., Labeta, M. O., Schiffrin, E. J., & Donnet-Hughes, A. (2001). Soluble CD14 in human breast milk and its role in innate immune responses. *Acta Odontologica Scandinavica, 59*, 330–334.
Vidal, K., Donnet-Hughes, A., & Granato, D. (2002). Lipoteichoic acids from *Lactobacillus johnsonii* strain La1 and *Lactobacillus acidophilus* strain La10 antagonize the responsiveness of human intestinal epithelial HT29 cells to lipopolysaccharide and Gram-negative bacteria. *Infectious Immunology, 70*, 2057–2064.
Vita, N., Lefort, S., Sozzani, P., Reeb, R., Richards, S., Borysiewicz, L. K., et al. (1997). Detection and biochemical characteristics of the receptor for complexes of soluble CD14 and bacterial lipopolysaccharide. *Journal of Immunology, 158*, 3457–3462.
Watanabe, A., Takeshita, A., Kitano, S., & Hanazawa, S. (1996). CD14-mediated signal pathway of *Porphyromonas gingivalis* lipopolysaccharide in human gingival fibroblasts. *Infectious Immunology, 64*, 4488–4494.

Weidemann, B., Brade, H., Rietschel, E. T., Dziarski, R., Bazil, V., Kusumoto, S., et al. (1994). Soluble peptidoglycan-induced monokine production can be blocked by anti-CD14 monoclonal antibodies and by lipid A partial structures. *Infectious Immunology, 62*, 4709–4715.

Weidemann, B., Schletter, J., Dziarski, R., Kusumoto, S., Stelter, F., Rietschel, E. T., et al. (1997). Specific binding of soluble peptidoglycan and muramyldipeptide to CD14 on human monocytes. *Infectious Immunology, 65*, 858–864.

Wooten, R. M., Morrison, T. B., Weis, J. H., Wright, S. D., Thieringer, R., & Weis, J. J. (1998). The role of CD14 in signaling mediated by outer membrane lipoproteins of *Borrelia burgdorferi*. *Journal of Immunology, 160*, 5485–5492.

Wright, S. D., Ramos, R. A., Tobias, P. S., Ulevitch, R. J., & Mathison, J. C. (1990). CD14, a receptor for complexes of lipopolysaccharide (LPS) and LPS binding protein. *Science, 249*, 1431–1433.

Wuthrich, B., Kagi, M. K., & Joller-Jemelka, H. (1992). Soluble CD14 but not interleukin-6 is a new marker for clinical activity in atopic dermatitis. *Archives in Dermatology Research, 284*, 339–342.

Yaegashi, Y., Shirakawa, K., Sato, N., Suzuki, Y., Kojika, M., Imai, S., et al. (2005). Evaluation of a newly identified soluble CD14 subtype as a marker for sepsis. *Journal of Infectious Chemotherapy, 11*, 234–238.

Yu, B., Hailman, E., & Wright, S. D. (1997). Lipopolysaccharide binding protein and soluble CD14 catalyze exchange of phospholipids. *Journal of Clinical Investigations, 99*, 315–324.

Yu, S., Nakashima, N., Xu, B. H., Matsuda, T., Izumihara, A., Sunahara, N., et al. (1998). Pathological significance of elevated soluble CD14 production in rheumatoid arthritis: In the presence of soluble CD14, lipopolysaccharides at low concentrations activate RA synovial fibroblasts. *Rheumatology International, 17*, 237–243.

Zalai, C. V., Kolodziejczyk, M. D., Pilarski, L., Christov, A., Nation, P. N., Lundstrom-Hobman, M., et al. (2001). Increased circulating monocyte activation in patients with unstable coronary syndromes. *Journal of the American College of Cardiology, 38*, 1340–1347.

Zdolsek, H. A., & Jenmalm, M. C. (2004). Reduced levels of soluble CD14 in atopic children. *Clinical Experiments in Allergy, 34*, 532–539.

Zhang, Y., Doerfler, M., Lee, T. C., Guillemin, B., & Rom, W. N. (1993). Mechanisms of stimulation of interleukin-1β and tumor necrosis factor-α by *Mycobacterium tuberculosis* components. *Journal of Clinical Investigations, 91*, 2076–2083.

Zoetendal, E. G., Akkermans, A. D., & De Vos, W. M. (1998). Temperature gradient gel electrophoresis analysis of 16S rRNA from human fecal samples reveals stable and host-specific communities of active bacteria. *Applied and Environmental Microbiology, 64*, 3854–3859.

Apoptosis and Tumor Cell Death in Response to HAMLET (Human α-Lactalbumin Made Lethal to Tumor Cells)

Oskar Hallgren, Sonja Aits, Patrick Brest, Lotta Gustafsson, Ann-Kristin Mossberg, Björn Wullt, and Catharina Svanborg

Abstract HAMLET (human α-lactalbumin made lethal to tumor cells) is a molecular complex derived from human milk that kills tumor cells by a process resembling programmed cell death. The complex consists of partially unfolded α-lactalbumin and oleic acid, and both the protein and the fatty acid are required for cell death. HAMLET has broad antitumor activity *in vitro*, and its therapeutic effect has been confirmed *in vivo* in a human glioblastoma rat xenograft model, in patients with skin papillomas and in patients with bladder cancer. The mechanisms of tumor cell death remain unclear, however. Immediately after the encounter with tumor cells, HAMLET invades the cells and causes mitochondrial membrane depolarization, cytochrome c release, phosphatidyl serine exposure, and a low caspase response. A fraction of the cells undergoes morphological changes characteristic of apoptosis, but caspase inhibition does not rescue the cells and Bcl-2 overexpression or altered *p53* status does not influence the sensitivity of tumor cells to HAMLET. HAMLET also creates a state of unfolded protein overload and activates 20S proteasomes, which contributes to cell death. In parallel, HAMLET translocates to tumor cell nuclei, where high-affinity interactions with histones cause chromatin disruption, loss of transcription, and nuclear condensation. The dying cells also show morphological changes compatible with macroautophagy, and recent studies indicate that macroautophagy is involved in the cell death response to HAMLET. The results suggest that HAMLET, like a hydra with many heads, may interact with several crucial cellular organelles, thereby activating several forms of cell death, in parallel. This complexity might underlie the rapid death response of tumor cells and the broad antitumor activity of HAMLET.

Keywords: HAMLET · lactalbumin · cancer · programmed cell death · apoptosis · macroautophagy · Bcl-2 · p53 · caspase

---

C. Svanborg
Institute of Laboratory Medicine, Section for Microbiology, Immunology and Glycobiology, Sölvegatan 23, 22362 Lund, Sweden
e-mail: catharina.svanborg@med.lu.se

HAMLET's Structure

HAMLET was discovered by serendipity when testing the effect of human breast milk fractions on bacterial attachment to alveolar type II lung carcinoma cells. Bacterial attachment was inhibited, but, in addition, the fraction killed the tumor cells to which the bacteria should adhere. The tumoricidal activity resided in the casein fraction, which had been obtained by low pH treatment of milk, and further fractionation revealed that the active molecular complex contained α-lactalbumin. The native protein had no effect on tumor cells, suggesting that the cell death–inducing variant had been structurally modified. After excluding posttranslational modifications, we examined if differences in the tertiary structure might explain the activity (Håkansson et al., 1995; Svensson, 1999). Previous studies had shown that α-lactalbumin can form relatively stable folding intermediates when calcium binding is impeded by low pH. The active form of the protein was shown to be partially unfolded, in a "molten globule"-like state. Unlike known molten globules of α-lactalbumin the active fraction did not revert to the native folded state in cell culture medium or in the presence of calcium. This suggested that α-lactalbumin in the active fraction was bound to a stabilizing co-factor. In a series of experiments, the co-factor was identified as oleic acid (C18:1, 9 *cis*). The need for unfolding and fatty acid binding was subsequently proven by deliberate conversion of native α-lactalbumin to HAMLET. It was achieved by EDTA treatment to remove the calcium ion from the protein and by binding of oleic acid presented on an ion exchange matrix. (Fig. 1).

Alpha-lactalbumin is abundant in human milk and functions as a galactosyltransferase co-enzyme in lactose synthesis. The crystal structure has been solved, revealing α-helical and β-sheet domains and four disulfide bonds. The human protein is 14 kDa in size and is a metalloprotein with a high-affinity binding site for Ca^{2+}, although other divalent ions can also interact. To study if Ca^{2+} is involved in the tumoricidal activity, Ca^{2+}-binding site mutants were constructed (Svensson et al., 2003a). The Ca^{2+}-binding site is coordinated by oxygens contributed by side-chain carboxylates of aspartate residues at positions 82, 87, and 88 and by carbonyl oxygens of lysine 79 and aspartate 84. When Ca^{2+} is released, the protein adopts the apo state, with a loss of defined tertiary structure. Mutational inactivation of Ca^{2+} binding prevents the protein

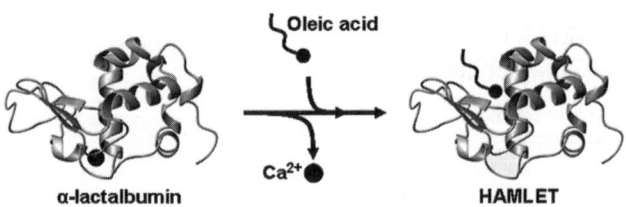

Fig. 1 HAMLET is defined as the product of partially unfolded α-lactalbumin and oleic acid

from reverting to the native state. We used the Ca^{2+}-binding site mutants to study if unfolding was sufficient to make α-lactalbumin tumoricidal and if the Ca^{2+}-free mutants, unfolded could form an active complex with oleic acid. Interestingly, the mutant protein alone was inactive in the cell death assay, but it formed a highly active complex with oleic acid. The results demonstrated that unfolding and oleic acid are required and that cell death is independent of the α-lactalbumin Ca^{2+} content.

In subsequent studies, the fatty acid specificity of α-lactalbumin was investigated (Svensson et al., 2003b). Partially unfolded α-lactalbumin was exposed on an ion-exchange matrix to fatty acids differing in carbon chain length, saturation, and orientation of the double bonds, and eluted complexes were tested for tumoricidal activity. Functional tumoricidal HAMLET complexes were only formed with oleic acid and other unsaturated C18.1 fatty acids and saturated fatty acids. However, fatty acids with shorter carbon chains and unsaturated fatty acids with the double bond in the *trans* orientation failed to form active complexes. The results suggested that there is a stereospecific fit between the fatty acid and the partially unfolded protein and that oleic acid and related fatty acids may fit the tumor cell targets better than other fatty acids.

More than 40 transformed cell lines from different origins and species have been tested for sensitivity to the original milk fraction or the defined HAMLET complex so far, all cell lines have been sensitive, but with somewhat different kinetics (Svanborg et al., 2003). Lymphoid cells died more rapidly and at lower HAMLET concentrations than carcinoma cells, while healthy differentiated cells were resistant to the effects of HAMLET unless concentrations were so high that the fatty acid became lytic. The effect was also unrelated to the *p53* status, and the Bcl-2 genotype, in the cell lines where such information was available (see below).

Apoptosis

Programmed cell death (PCD) is crucial for the development and maintenance of multicellular organisms. It is required to counteract excessive proliferation of cell populations, but also to eliminate and unwanted cells without harming surrounding tissues. Impaired regulation of PCD has been shown in a multiplicity of disease states, including cancer. PCD is an active and strictly regulated process in contrast to necrosis, which is described as a passive form of cell destruction (Leist & Jaattela, 2001; Lockshin & Zakeri, 2001). Cell death has been known to exist since the 19th century, but the term "programmed" was introduced by Lockshin and Williams in 1965 when describing the death of neural embryonic insect cells as caused by predictable "programmed" changes (Lockshin & Williams, 1965). In 1972, Kerr et al. showed that the programmed morphological changes described by Lockshin and Williams were not restricted to embryonic cells but existed in all cell types (Kerr et al., 1972). They called the phenomenon "apoptosis" from the Greek word for "leaves falling from a tree" and based it on morphological criteria. The dying cell showed shrinkage, membrane blebbing, chromatin condensation and fragmentation, detachment, and the formation of apoptotic bodies. The apoptotic cells were recognized and eliminated by macrophages without causing damage to surrounding tissues.

Horvitz et al. (1994) later showed that the morphological changes in the apoptotic cell were regulated and executed by discrete signaling pathways introduced from studies in the nematode *Caenorhabditis elegans*. Homologues of the nematode apoptotic proteins have been identified in humans even though the mammalian cell death programs are more complex. PCD has been operationally defined as an active process that is dependent on signaling events in the dying cell (Leist & Jaattela, 2001; Lockshin & Zakeri, 2001). PCD includes apoptosis, or type I PCD, and autophagy, or type II PCD. To further discriminate among different forms, researchers have proposed that the nuclear morphology of the dying cells serve as a criterion (Leist & Jaattela, 2001; Jaattela & Tschopp, 2003). Classic apoptosis involves compact chromatin condensation and fragmentation into discrete and simple geometric shapes. Apoptosis-like cell death is characterized by less compact condensation and fragmentation often marginalized to the nuclear periphery. Necrosis-like cell death proceeds with little or no chromatin condensation.

Regulation of Programmed Cell Death: Mitochondria and the Bcl-2 Family

Many cell death stimuli release proapoptotic proteins from the intermembrane space of the mitochondria, a phenomenon called mitochondrial outer membrane permeabilization (MOMP) (Green & Kroemer, 2004). The mitochondrial outer membrane integrity and the MOMP response are controlled by the Bcl-2 protein family. This family was named after its first member: B-cell CLL/lymphoma 2, which was observed in follicular lymphomas carrying the t(14;18) translocation. The cells survived longer and became resistant to treatment, which indicated that Bcl-2 was an oncogene (Bakhshi et al., 1985; Tsujimoto et al., 1985). To date, at least 23 members have been classified into three groups according to the presence of four Bcl-2 homology (BH) domains (Fig. 2). The anti-apoptotic proteins Bcl-2, Bcl-xl, Mcl-1, and Bfl-1 contain all four BH-4 domains. The multidomain pro-apoptotic subfamily includes Bax, Bak, Mtd, and Bcl-rambo and shares BH1-3. The proapoptotic "BH3-only" family only shares the BH3 domain and consists of Bik, Bad, Bid, Bim, Hrk, Noxa, Puma, Blk, Bnip3, Bnip3L, p193, Bmf, and Bcl-G (Tsujimoto, 2003). The anti-apoptotic Bcl-2 family members serve as stabilizers of the outer mitochondrial membrane, while the pro-apoptotic Bcl-2 family proteins

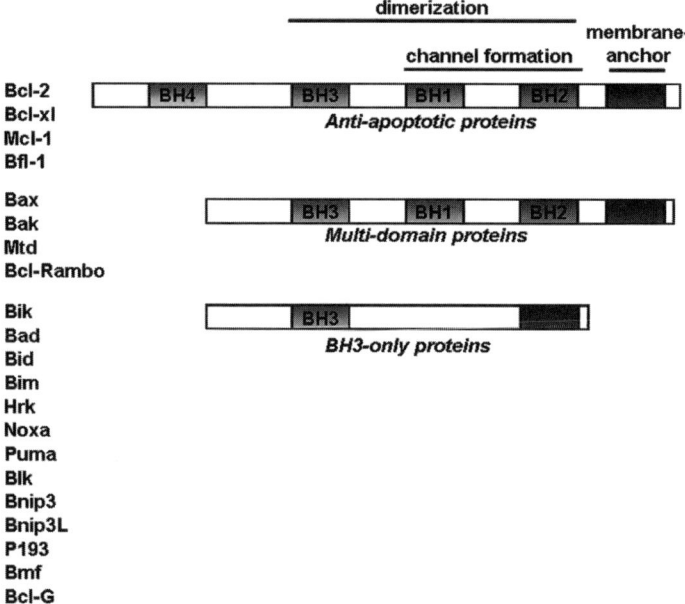

Fig. 2 The Bcl-2 family is classified into three categories based on the presence of four Bcl-2 homology (BH) domains. Anti-apoptotic members contain all four BH domains, and the pro-apoptotic multidomain members contain BH1-3. The pro-apoptotic BH3 only members only contain BH3. The BH1-3 domains functionally control dimerization, while BH1-2 domains control the channel formation events

perturb the membrane integrity. The pro-apoptotic multidomain family members Bax and Bak oligomerize and form pores in the outer membrane, which allow leakage of the intermembrane proteins (Tsujimoto, 2003). These events are antagonized by Bcl-2 and Bcl-xl. The "BH3-only" proteins may play a role as sensors of cell death signals (Bouillet & Strasser, 2002). Following an apoptotic stimulus, they translocate to the mitochondria where some members activate multidomain pro-apoptotic family members such as Bax and Bak, while others inactivate anti-apoptotic family members.

In many models the release of mitochondrial intermembrane components, such as cytochrome c, is mediated by opening of the permeability transition (PT) pore (Green & Kroemer, 2004). The pore is controlled by adenine nucleotide transporter in the inner membrane and voltage-dependent anion channel (VDAC) in the outer membrane. PT pore opening results in loss of the inner transmembrane potential ($\Delta\psi$m) and the influx of water. Bax and Bak may induce PT pore opening either by physically interacting with VDAC (Shimizu et al., 1999) or indirectly by inducing a conformational change of VDAC (Madeo et al., 1997). In contrast to Bak and Bax, Bcl-xl inhibits VDAC activity.

Execution of Programmed Cell Death: Caspases

Classic apoptosis is mediated by a family of cysteine proteases called caspases. The importance of caspases in apoptosis was first described in *C. elegans;* later it was shown that the *C. elegans* gene *ced-3* had high sequence homology with the mammalian inteleukin-1β-converting enzyme (ICE-3), which was later renamed caspase-1 (Yuan et al., 1993; Miura et al., 1995). To date, there are 12 members in the family. In mammals caspases can be divided into subgroups based on function (Garcia-Calvo et al., 1998; Thornberry & Lazebnik, 1998). The pro-inflammatory caspases (caspase-1, -4, -5, and -14) are implicated in the maturation of cytokines but may also play a role in cell death (Creagh et al., 2003). The initiator caspases (caspase-2, -8, -9, -10, and -12) serve to transduct various death signals into proteolytic activity by activating effector caspases. The effector caspases (caspase-3, -6, and -7) are responsible for the cleavage of intracellular substrates (Miura et al., 1993).

Caspases reside as inactive pro-enzymes in the cytosol and are activated by cleavage of the N-terminal pro-domain. The initiator caspases trigger a cascade of downstream caspase activity, which results in the cleavage of intracellular substrates, including inhibitors of effector molecules and inhibitors of apoptosis and molecules involved in cytoskeletal or DNA integrity, thereby causing morphological and functional changes such as cell shrinkage, chromatin condensation and fragmentation, plasma-membrane blebbing, and apoptotic body formation (Martin et al., 1995; Brown et al., 1997; Gueth-Hallonet et al., 1997). Caspase-mediated DNA fragmentation is mediated by caspase-activated DNase (CAD) (Enari et al., 1998), which is normally kept inactive in the cell nucleus by binding to its negative regulator ICAD. During apoptosis, effector caspases cleave ICAD, resulting in the release of active CAD. The dying cell externalizes surface receptors, like phosphatidyl serine (PS), to the outer membrane leaflet, allowing phagocytes to bind and engulf the dying cell. Activated caspases are controlled by the inhibitor of apoptosis (IAP) protein family (Deveraux et al., 1998). They may be important under normal cellular conditions by eliminating unwanted caspase activity. Upon an apoptotic stimulus, the IAPs are neutralized by the activity of Smac/Diablo [second activator of caspases/direct inhibitor of apoptosis (IAP)-binding protein with low pI] and Omi/HtrA2 (Du et al., 2000; Suzuki et al., 2001; Verhagen et al., 2002). The activation of caspases can be meditated through two different pathways: extrinsic and intrinsic.

The Extrinsic Pathway

Ligand binding to death receptors such as FAS (CD95/APO-1), TRAIL-RI, or TNFR1 is sufficient to cause a death signal (Tartaglia et al., 1993; Nagata, 1997). This pathway is especially important in the immune system. Association between ligands and their receptors promotes receptor trimerization and recruitment of adaptor proteins to the cytosolic death domains (DD) of the receptors. The adaptor proteins FADD (Chinnaiyan et al., 1995) (FAS-associated death domain) and TRADD (Hsu et al., 1995) (TNFR- and TRAIL-R-associated death domain) bind DDs homodimers and form the death-inducing signaling complex (DISC). Adaptor proteins contain death effector domains that recruit procaspase-8, and two procaspase-8 molecules induce proteolytic autoactivation (Muzio et al., 1996).

The apoptosis cascade then proceeds in two individual pathways depending on the cell type (Scaffidi et al., 1998). In the first pathway, activated caspase-8 directly activates effector caspases. Alternatively, in the second pathway, caspase-8 cleaves the pro-apoptotic Bcl-2 family protein Bid, which translocates and activates mitochondria and the intrinsic pathway. In addition, an alternative pathway can be activated in response to FAS ligand via recruitment of the protein Daxx to the DD cluster of the receptors, which results in activation of the apoptosis signal-regulating kinase-1 (ASK-1) and Jun N-terminal kinase (JNK) pathways (Ashe & Berry, 2003). ASK-1 mediates the death cascade by interacting with caspase-9 and the mitochondria, while JNK has been suggested to inactivate Bcl-2 and thereby stimulate Bax-mediated MOMP. An analogous pathway in TNRF1-treated cells is mediated by the kinase receptor interacting protein (RIP) and the death domain protein RAIDD/CRADD (Ahmad et al., 1997; Duan & Dixit, 1997). Caspase-2 is recruited and activated by RAIDD/CRADD, which results in MOMP. RIP can also induce necrosis-like PCD in response to both FAS ligand and TRAIL triggered by the production of reactive oxygen species (ROS) (Holler et al., 2000).

The Intrinsic Pathway

The mitochondria play an important role in the induction of apoptosis by releasing pro-apoptotic molecules. Numerous stimuli trigger MOMP directly without upstream activity of the caspases. These include hypoxia, DNA damage and cellular stress, calcium fluctuations, ROS, nitric oxide, fatty acids, and proteases that cleave constituents of the respiratory chain (Green & Kroemer, 2004) (see above). Upon stimulation, pro-apoptotic factors such as apoptosis inducing factor (AIF), Smac/Diablo, Endonuclease G (Endo G), and cytochrome c are released from the mitochondria. Cytochrome c associates with APAF-1, dATP, and procaspase-9 to form the apoptosome complex, which activates effector caspases.

Caspase-Independent Pathways

When programmed cell death is executed in the absence of caspase activity, many of the morphological changes attributed to caspases still occur, indicating that alternative pathways may have very similar endpoints. A number of proteases and nucleases have been suggested to be responsible for these events, including cathepsins, calpains, serine proteases, Endo G, and AIF (Jaattela, 2002).

Cathepsins are cysteine proteases that are associated with protein degradation in lysosomes and with the degradation of the extracellular matrix (Johnson, 2000; Turk et al., 2000). They are activated by other proteases or by autoproteolysis in acidic environments, as in the lysosomes, where they act as general proteases. In response to a variety of death stimuli, cathepsins translocate to the cytoplasm or nucleus (Roberts et al., 1997; Foghsgaard et al., 2001; Roberg et al., 2002). The neutral pH in the cytoplasm and nucleus has been suggested to alter their specificity, which then share many substrates with caspases (Gobeil et al., 2001). In addition, cathepsins have been shown to trigger the caspase cascade either by cleaving and activating caspases directly or through Bid-mediated release of cytochrome c (Stoka et al., 2001; Roberg et al., 2002).

Calpains are cysteine proteases that reside in the cytoplasm in an inactive pro-form and are activated by stimuli that trigger elevated intracellular Ca^{2+}-levels. They act either upstream or downstream of caspases. In addition, calpains can mediate apoptosis-like cell death in the absence of caspase activity (Mathiasen et al., 1999; Nakagawa & Yuan, 2000; Choi et al., 2001).

The serine proteases Granzyme A and B are located in the granules of cytotoxic T lymphocytes (CTL). When activated, CTLs release their granular contents, which are internalized in target cells, mainly through endocytosis (Browne et al., 2000). In the cytoplasm, granzymes trigger rapid caspase-dependent PCD. Granzyme B cleaves substrates after aspartate residues and can therefore directly activate caspases. However, when caspases are blocked, Granzyme B can also trigger a slower necrosis-like form of cell death. Granzyme-mediated cell death involves Granzyme A–activated DNase that triggers DNA single-stranded breaks (Beresford et al., 2001).

Omi/HtrA2 is a serine protease that normally resides inside the intermembrane space of the mitochondria. Upon death stimuli, it is released into the cytoplasm, where it triggers caspase-dependent cell death by inhibiting IAPs. In addition, Omi/HtrA2 may execute cell death independently of caspases by its serine protease activity (Suzuki et al., 2001). AIF and Endo G mediate caspase-independent DNA condensation and fragmentation (Susin et al., 1999). They are released from the mitochondria and are translocated to the nucleus in response to various death stimuli.

p53 and Resistance to Cell Death

The tumor suppressor p53 is a transcription factor, initially described in SV40-infected cells by co-precipitation with SV40 large and small T antigens (Lane & Crawford, 1979; Linzer & Levine, 1979). The *p53* gene is mutated or deleted in approximately 50% of all human malignancies, and the mutations disable the tumor suppressor functions (Chiba et al., 1990; Hollstein et al., 1991; Lowe et al., 1994). In normal cells, p53 plays a protective role by limiting the propagation of cells exposed to stress stimuli, like DNA damage, aberrant growth signals, and UV light. p53 then initiates cell cycle arrest and DNA repair, but when cells harbor irreparable DNA damage, p53 activates cell death programs and the cells undergo apoptosis (Lane, 1993). Under normal cellular conditions, p53 is present at low levels due to a tight regulation by its negative regulation partner MDM-2. The E3 ubiquitin ligase MDM-2 mediates p53 ubiquitinylation and translocation to the cytoplasm and the subsequent degradation by the proteasomal machinery (Kubbutat et al., 1997; Vogelstein et al., 2000). The *mdm-2* gene is a target for the transcriptional activity of *p53*, causing an autoregulatory loop where p53 is negatively regulated by MDM-2 and MDM-2 is positively regulated by p53 (Wu et al., 1997).

At least three independent pathways result in p53 activation (Fig. 3). The first pathway is triggered by double-stranded DNA breaks in response to ionizing radiation and is mediated by the protein kinase Chk2 and ATM, which phosphorylate p53 and reduce its affinity for MDM-2 (Carr, 2000). The second pathway is triggered by aberrant growth signals, such as expression of the oncogenes Ras or Myc, and is mediated by p14^{ARF} (Lowe & Lin, 2000;

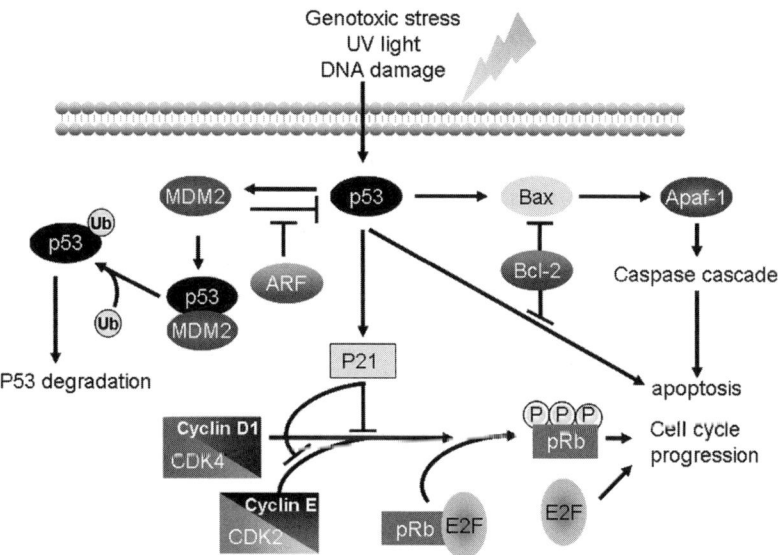

Fig. 3 p53 can trigger apoptosis and/or cell cycle arrest in response to stress signals. Under normal conditions p53 is continuously degraded by the action of MDM-2. During cellular stress, p53 is stabilized and activated by factors, like ARF, that lowers the affinity for MDM-2 binding. Activated p53 may induce cell cycle arrest by transactivation of p21 that blocks the cyclin D/Cdk4- and cyclin E/Cdk2- mediated phosphorylation (P) of Rb. Apoptosis can be induced by p53 by the activation of Bax and the intrinsic pathway

Sherr & Weber, 2000), which indirectly activates p53 by binding to MDM-2, and thereby sequestering MDM-2, which blocks p53 degradation. It may seem counterproductive for tumor progression that p53 is activated in response to oncogenic growth signals, but tumor cells that express high levels of p14^{ARF} usually have functionally inactive p53 (Lozanon et al., 1994). The third pathway is triggered by chemotherapeutic agents, ultraviolet light, and protein kinase inhibitors and is mediated by ATR and casein kinase II, which phosphorylate MDM-2 and block the subsequent degradation of p53 (Meek, 1999).

p53 induces apoptosis in response to DNA damage mainly through its transcriptional activity by activating pro-apoptotic genes, such as FAS/CD95 (Muller et al., 1998), pro-apoptotic Bcl-2 family members Noxa and Puma (Oda et al., 2000; Nakano & Vousden, 2001), or apoptosis-inducing factor-1 (APAF-1) (Meier et al., 1992; Robles et al., 2001). In addition, p53 can repress anti-apoptotic genes like *Bcl-2* and *survivin*, which encode proteins capable of inhibiting apoptosis. Mihara et al. (2003) proposed that p53 may have non-transcriptional effects by interacting directly with mitochondria and inducing the release of cytochrome c.

The loss of p53 function is usually caused by a deletion in one allele and a missense mutation in the other. Mutations in p53 may not only result in loss of wild-type activities, as is the case for other tumor suppressor genes, but may also give rise to a dominant gain of function mutants that may contribute to tumorigenesis (Greenblatt et al., 1994; Hollstein et al., 1996). Breast cancer tumors with mutations in certain domains of p53 are more aggressive than tumors with deleted p53 (Thorlacius et al., 1995; Aas et al., 1996). Moreover, when gain of function p53 mutant was introduced into p53 null, murine bladder carcinoma cells, the differentiation was inhibited and the metastatic potential was increased. Mutated p53 is incapable of transactivating the target genes of the wild-type protein including MDM-2, resulting in elevated levels of mutant protein. The difference in p53 expression between healthy cells with wild-type p53 and tumors carrying p53 mutations therefore makes p53 a desirable target for therapeutic drugs. Small molecules that restore mutant p53 activity to wild-type activity have been shown to be successful both *in vitro* and *in vivo* (Foster et al., 1999; Samuels-Lev et al., 2001; Bykov et al., 2002).

HAMLET and Apoptosis

The morphology of tumor cells changes rapidly after HAMLET exposure, with cell shrinkage, nuclear condensation, and DNA fragmentation characteristic of apoptosis (Håkansson et al., 1995). In tumor cells, HAMLET co-localizes with mitochondria and causes membrane depolarization and the release at cytochrome c (Kohler et al., 1999a, b). HAMLET-treated cells show low caspase-3 and caspase-6-like activities, with cleavage of caspase substrates such as PARP, lamin B, and α-fodrin. HAMLET also triggers DNA fragmentation, indicating that the cells might die of classical apoptosis. This is not the case, however, as the pan-caspase inhibitor zVAD-fmk does not block cell death, even though it abolishes the caspase response to HAMLET and the formation of small DNA fragments in Jurkat cells. Furthermore, antibodies blocking the FAS/CD95 receptor pathway had no effect on cell death, when the milk fraction was examined.

HAMLET, the Bcl-2 Family, and p53

When HAMLET's effect on the transcription of Bcl-2 family members was investigated, there was no change in Bcl-2 family mRNAs after HAMLET treatment, showing no *de novo* synthesis (Hallgren et al., 2006). Furthermore, Bcl-2 overexpression partially inhibited the caspase-3 activity in response to etoposide, but not in response to HAMLET, suggesting that HAMLET induces caspase activation independently of Bcl-2. Overexpression of Bcl-2 or Bcl-xl also had no impact on cell viability in response to HAMLET (Fig. 4). To elucidate if p53 is involved in HAMLET-induced cell death, we used cell lines differing in *p53* status. There were no differences in HAMLET susceptibility between tumor cells with wild-type, deleted, or mutant *p53*, suggesting that *p53* is not involved. To further examine the role of *p53*, cells with modified *p53* status were used. There was no difference in HAMLET sensitivity between colon carcinoma cells with wild-type or deleted *p53*, or between lung carcinoma cells with *p53* deletion or a gain of function *p53* mutant, confirming that *p53* is not involved in HAMLET-induced cell death (Fig. 4).

Autophagy

Autophagic or type II cell death has been suggested as a caspase-independent cell death pathway, but it is still debated whether autophagy contributes to cell death or if it only constitutes a survival mechanism. Autophagic processes are present as a normal cellular response to eliminate damaged organelles and long-lived proteins. During stress, such as starvation, cells can reuse organelles and long lived proteins as a source of nutrients. Autophagic degradation of proteins can follow several routes: (1) microautophagy, where the cytoplasm is engulfed directly by lysosomes; (2) chaperon-mediated autophagy, where proteins are targeted to lysosomal degradation aided by chaperones; and (3) macroauto-phagy, where cytosol and organelles are circumscribed by multimembrane autophagosomes, which are fused with lysosomes where the content is degraded (Gonzalez-Polo et al., 2005). The latter can be induced by cellular stress, while microautophagy is a constitutive process. Type II cell death only includes macroautophagy (Schweichel & Merker, 1973). During macroautophagy, upstream signals promote the formation of small membrane structures

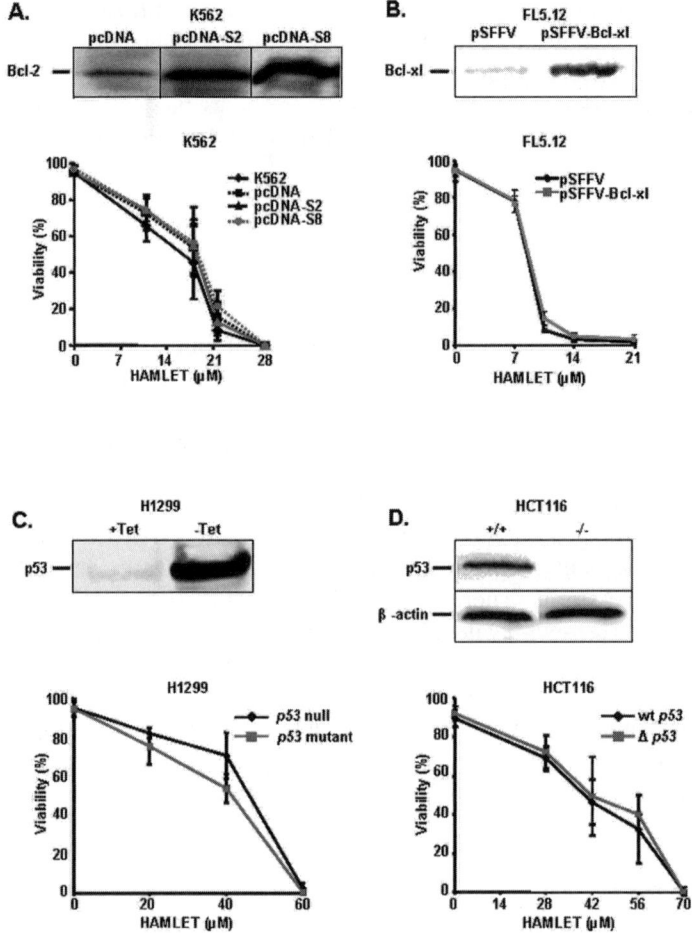

Fig. 4 Bcl-2 and Bcl-xl overexpression or p53 deletions do not protect tumor cells from HAMLET. (a) The two human chronic myelogenous leukemia K562 cell clones, pcDNA-S2 and pcDNA-S8, were stably transfected to overexpress Bcl-2 as shown by western blot, but Bcl-2 overexpression did not influence the susceptibility to HAMLET compared to the vector control. (b) Overexpression of Bcl-xl in a murine pro-B lymphocytic clone did not alter the susceptibility to HAMLET as compared to the vector control. (c) There was no difference in susceptibility to HAMLET between human lung carcinoma H1299 cells expressing a *p53* mutant or *p53* negative cells. (d) There was no difference in susceptibility to HAMLET between colon carcinoma HCT116 cells with wild-type or deleted *p53*

(Noda et al., 2002). The membranes enclose by cytoplasmic contents, are elongated and finally closed, and are then called autophagosomes (Fig. 5). Eventually, the autophagosomes are fused with lysosomes and the contents are proteolytically degraded and reutilized.

The genes and proteins involved in macroautophagy have been identified in yeast, and homologous genes have been found in higher organisms (Klionsky et al., 2003). The genes are called ATGs, for AuTophaGy genes. The most extensively studied human homologues are beclin-1 (homologue to ATG6) and MAP-LC-3 (homologue to ATG8). Under normal conditions, LC3 is present in a cytosolic form, LC3-I (Kabeya et al., 2000), but upon autophagic stimuli, a portion of the LC3-I is modified to a variant able to bind autophagosomal membranes (LC3-II) (Tanida et al., 2001, 2002). LC3 modification is essential for the formation of autophagosomes (Kabeya et al., 2000). ATG6 and beclin-1 have been shown to have a role in the class III phosphatidylinositol 3-kinase complex that is required in the early stages of autophagosome formation (Petiot et al., 2000). Class III phosphatidylinositol 3-kinase inhibitors such as 3-methyl adenine (3-MA) have been shown to inhibit autophagosome formation and macroautophagy.

Autophagy and Cell Death

The role of macroautophagy in programmed cell death has been intensely debated. Macroautophagy has been recognized as a survival mechanism during starvation conditions, when cells reutilize cytoplasmic material as a source of nutrients. Under these conditions, macroautophagy is an adaptive stress response in dying cells to prolong cell survival (Kihara et al., 2001; Klionsky et al., 2003; Levine & Klionsky, 2004). In yeast, macroautophagy is well documented as a survival mechanism in response to nutrient depletion (Tsukada & Ohsumi, 1993; Schlumpberger et al., 1997). This has also been reported in mammalian cells, where inhibition of macroautophagy can result in increased sensitivity to apoptosis during starvation conditions. Furthermore, turnover of damaged organelles such as mitochondria is accompanied by a macroautophagic response. In primary hepatocytes, depolarized mitochondria are eliminated by macroautophagy, resulting in increased resistance to

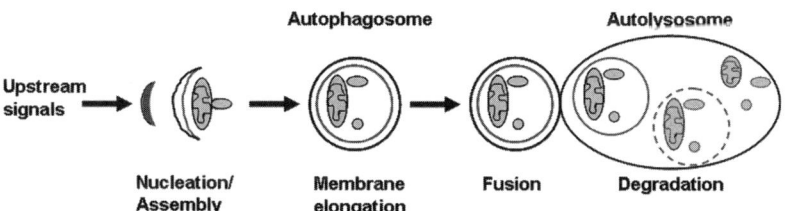

Fig. 5 During macroautophagy, upstream signals trigger formation of membrane sacs that are nucleated by organelles and cytoplasmic constituents. Membranes are elongated and finally closed, which results in the formation of double-membrane vesicles called "autophagosomes." After fusion with lysosomes, the content is degraded in autolysosomes

cell death by apoptosis (Lemasters et al., 1998). Furthermore, inhibiting macroautophagy can result in increased sensitivity to apoptosis during starvation conditions (Boya et al., 2005). These results indicate a survival role for autophagy that inhibits apoptosis. However, in other models autophagy has been described to be a cell death pathway autophagy (Edinger & Thompson, 2004; Lockshin & Zakeri, 2004). It has been suggested that the overall autophagic activity in cells undergoing autophagic cell death is far more extensive than the autophagy activity associated with organelle turnover in healthy cells (Bursch, 2004), indicating that the extent of the autophagic activity may determine if a cell is doomed to live or die. Treatment of MCF-7 breast cancer cells with the estrogen antagonist 4-hydroxytamoxifen causes cell death characterized by extensive vacuole formation. Since MCF-7 cells lack important apoptosis mediators such as caspase-3, it raises the possibility that type II cell death compensates for defects in other types of cell death pathways (Schulte-Hermann et al., 1997; Janicke et al., 1998). Moreover, embryonic fibroblasts from mice lacking the pro-apoptotic Bcl-2 family proteins Bax and Bak died with macroautophagic morphology when treated with agonists that normally induce apoptosis (Shimizu et al., 2004). Death was suppressed when macroautophagy was inhibited, indicating that apoptosis and macroautophagy may serve complementary roles to overcome blocks in death pathways. Cells dying with macroautophagic morphology have also been observed in development, during the regression of the corpus luteum (Paavola, 1978), the involution of mammary and prostate glands (Helminen & Ericsson, 1971; Sensibar et al., 1991), and the regression of Mullerian duct structure, which shows that autophagic cell death may be a physiological process (Dyche, 1979).

HAMLET and Autophagy

HAMLET-treated cells show changes characteristic of macroautophagy. Cytoplasmic vacuoles and double-membrane vesicles typical for macroautophagy were detected with electron microscopy. In addition, HAMLET changed the staining in GFP-LC3–transfected cells from a diffuse to a granular staining pattern reflecting LC3 translocation to autophagosomes. HAMLET also induced LC3-II accumulation, suggestive of macroautophagy. The response may be initiated by organelle damage, as the mitochondria were swollen with disrupted membranes and disintegrated cristae. HAMLET also caused a rapid dose-dependent decrease in ATP levels. Interestingly, cell death was reduced when macroautophagy was inhibited, indicating that macroautophagy might play an important role in HAMLET-induced cell death.

Nuclear Interactions of HAMLET with Histones and Chromatin

To formulate hypotheses about HAMLET's mechanism of action, biotinylated or Alexa fluor-stained HAMLET was used in confocal microscopy, and the interaction of the complex with different tumor cell compartments was examined. HAMLET was shown to bind to the surface of both tumor cells and healthy cells (Håkansson et al., 1999; Gustafsson et al., 2004), but a marked difference in uptake was noticed. Large quantities of HAMLET appeared to "invade" tumor cells while more moderate amounts were observed in healthy cells. The broad antitumor activity suggests that HAMLET binds to highly conserved cell surface domains and that the internalization of HAMLET must be mediated by highly active mechanisms, but no specific receptors or uptake mechanisms have been identified so far. In tumor cells, HAMLET is further translocated to the nucleus, where about 90% of HAMLET is found after 1 hour at the LD50 concentration. In healthy cells, HAMLET is retained in the cytoplasm.

The pattern of chromatin condensation has been proposed to distinguish apoptosis from other forms of PCD. The nuclei of HAMLET-treated cells undergo rapid chromatin condensation, forming patterns described for apoptotic cells. These changes were caspase-dependent, as the number of cells with these chromatin morphologies decreased when caspases were inhibited. In the presence of the caspase inhibitor, there was an increase in cells with marginalized chromatin. The total number of cells with condensed chromatin did not change when caspases were inhibited, however, suggesting that caspase activity is essential for chromatin remodeling in response to HAMLET, but not for cell death.

HAMLET targets histones in tumor cell nuclei (Duringer et al., 2003), as shown in overlay assays and affinity chromatography of nuclear extracts. HAMLET-bound histones H2B, H3, and H4 in the nuclear extracts with high affinity, as shown using purified bovine histones in a BIAcore assay. Furthermore, HAMLET prevented the assembly of core histones to DNA in mixing experiments. The high-affinity binding of HAMLET to histones may impair the nucleosome function, which affects transcription but also makes the DNA accessible for endonucleases.

HAMLET: *In Vivo* Effects

The therapeutic effects of HAMLET have been examined in several models. In a rat glioblastoma xenograft model, human glioblastoma tissue explants are grown as spheroids and cell suspensions are injected into the brain of nude rats (Fischer et al., 2004). This method establishes invasively growing human glioblastoma (GBM) tumors in nude rats and makes it possible to test different therapeutic approaches. HAMLET or α-lactalbumin was administered on day 7 when the tumor cells had been allowed to establish, and tumor progression was followed until the rats developed symptoms. In our study, α-lactalbumin– treated rats developed pressure symptoms after eight weeks, but at this time, the HAMLET-treated rats remained asymptomatic. Magnetic resonance scans revealed large tumors in α-lactalbumin–treated control rats, while HAMLET-treated rats had smaller tumors. There were no signs of toxicity when HAMLET was infused into the brain of healthy animals.

The effect of HAMLET on skin papillomas was examined in a placebo-controlled study in human patients, who had tested a variety of treatments without success (Gustafsson et al., 2004). HAMLET or placebo was applied topically once a day for three weeks. The lesion size was documented by measurements once a week during the time of treatment. A 75% decrease in the lesion volume was considered successful. Using this criterion, there was an effect from HAMLET in 100% (20/20) of the patients in the HAMLET-treated group compared to 15% (3/20) in the placebo group. The results suggested that HAMLET should be further tested as a topical agent in patients with different forms of papillomas.

HAMLET has also been tested in patients with superficial transitional cell carcinomas. The patients received instillations of HAMLET during the week before scheduled surgery. HAMLET stimulated rapid shedding of tumor cells and aggregates into the urine daily, during the five days of instillation, and most of the cells showed an apoptotic response. After five days, a reduction in tumor size or a change in tumor character was detected, and there was apoptotic cells in sections of the remaining tumors. The results suggest that topical HAMLET treatment may be used *in vivo*.

HAMLET and Breastfeeding

Human milk has many beneficial effects for the nursing child. In addition to providing a well-balanced diet, it also protects against a number of pathogens and diseases. Epidemiological data show that breastfed children have a lower incidence of gastrointestinal infections, respiratory tract infections, meningitis, and urinary tract infections (Cunningham et al., 1991; Golding et al., 1997a, b; Hanson, 1998) than formula-fed infants. The protection breastfeeding provides has been attributed to two factors: the antimicrobial and immune-modulating factors in the milk (see other chapters in this book). For example, milk contains molecules that prevent bacterial attachment, including antibodies to type 1 fimbriae on Gram-negative bacteria (Andersson et al., 1985) and oligosaccharide receptor analogs against *S. pneumoniae* and *H. pylori* (Andersson et al., 1983, 1986).

Epidemiological evidence has also suggested that breastfeeding protects against tumor development in children. The incidence of tumors in children up to 15 years was lower in those who were breastfed compared to the formula-fed controls (Davis et al., 1988). The effect was most pronounced for lymphomas. This difference is compatible with a direct effect of milk components on tumor precursor cells in the intestine of the breastfed child. As the infant acquires bacterial and viral flora, the mucosa undergoes a rapid proliferative response and rapidly proliferating cells may acquire mutations that risk converting them into tumor cells. We therefore speculate that HAMLET may exemplify a factor with a direct, local antitumor effect. Human α-lactalbumin is the most abundant protein in milk, and long-chained fatty acids such as oleic acid predominate in human milk triglycerides. The HAMLET complex is not present in newly synthesized milk, as α-lactalbumin is needed for lactose synthesis. In the stomach, at low pH, the conditions make α-lactalbumin unfold to the apo state, and a pH-sensitive lipase is activated that releases oleic acid from the milk oligosaccharides. *In vitro* mixing studies have shown that the HAMLET complex can be made from these constituents in solution, even if the efficiency is low. We therefore speculate that HAMLET might be formed in the stomach of the breastfed child and that the complex may help remove unwanted cells from gut mucosa. Due to the stabilizing fatty acid in the HAMLET complex, it is resistant to proteolysis and may survive the passage through the intestinal canal.

Conclusion

While HAMLET originally was derived from human milk, in the future it can be produced in larger amounts and tested for effects on different human tumors. HAMLET differs in spectrum and mode of action from current therapies. The complex shows broad antitumor activity likely due to parallel activation of apoptosis and macroautophagy, interference with the function of mitochondria and proteasomes, and accumulation in cell nuclei. In contrast to many conventional cancer treatments, which lack selectivity and have severe side effects, HAMLET appears to maintain tumor selectivity *in vivo* and, so far, there has been a lack of side effects. HAMLET provides a very interesting new tool in the understanding of tumor cell death and may be used to develop alternative therapeutic approaches.

Acknowledgment This study was supported by the Sharon D. Lund Foundation grant and the American Cancer Society, the Swedish Cancer Society, the Swedish Pediatric Cancer Foundation, the Medical Faculty (Lund University), the Segerfalk Foundation, the French Medical Research Foundation (FRM, Paris), the Anna-Lisa and Sven-Erik Lundgren Foundation for Medical Research, the Knut Alice Wallenberg Foundation, the Lund City Jubileumsfond, the John and Augusta Persson Foundation for Medical Research, the Maggie Stephens Foundation, the Gunnar Nilssons Cancer Foundation, the Inga-Britt och Arne Lundbergs Foundation, the Söderberg Foundation, the HJ Forssman Foundation for Medical Investigations, and the Royal Physiographic Society.

References

Aas, T., Borresen, A. L., Geisler, S., Smith-Sorensen, B., Johnsen, H., Varhaug, J. E., Akslen, L. A., & Lonning, P. E. (1996). Specific P53 mutations are associated with *de novo* resistance to doxorubicin in breast cancer patients. *Nature Medicine, 2*, 811–814.

Ahmad, M., Srinivasula, S. M., Wang, L., Talanian, R. V., Litwack, G., Fernandes-Alnemri, T., & Alnemri, E. S. (1997). CRADD, a novel human apoptotic adaptor molecule for caspase-2, and FasL/tumor necrosis factor receptor-interacting protein RIP. *Cancer Research, 57*, 615–619.

Andersson, B., Dahmén, J., Freijd, T., Leffler, H., Magnusson, G., Noori, G., & Svanborg-Edén, C. (1983). Identification of a disaccharide unit of a glycoconjugate receptor for pneumocci attaching to human pharyngeal epithelial cells. *Journal of Experimental Medicine, 158*, 559–570.

Andersson, B., Porras, O., Hanson, L. A., Svanborg-Eden, C., & Leffler, H. (1985). Non-antibody-containing fractions of breast milk inhibit epithelial attachment of *Streptococcus pneumoniae* and *Haemophilus influenzae*. *Lancet, 1*, 643.

Andersson, B., Porras, O., Hanson, L. A., Lagergard, T., & Svanborg-Eden, C. (1986). Inhibition of attachment of *Streptococcus pneumoniae* and *Haemophilus influenzae* by human milk and receptor oligosaccharides. *Journal of Infectious Diseases, 153*, 232–237.

Ashe, P. C., & Berry, M. D. (2003). Apoptotic signaling cascades. *Progress in Neuropsychopharmacology Biological Psychiatry, 27*, 199–214.

Bakhshi, A., Jensen, J. P., Goldman, P., Wright, J. J., McBride, O. W., Epstein, A. L., & Korsmeyer, S. J. (1985). Cloning the chromosomal breakpoint of t (14;18) human lymphomas: Clustering around JH on chromosome 14 and near a transcriptional unit on 18. *Cell, 41*, 899–906.

Beresford, P. J., Zhang, D., Oh, D. Y., Fan, Z., Greer, E. L., Russo, M. L., Jaju, M., & Lieberman, J. (2001). Granzyme A activates an endoplasmic reticulum-associated caspase-independent nuclease to induce single-stranded DNA nicks. *Journal of Biological Chemistry, 276*, 43285–43293.

Bouillet, P., & Strasser, A. (2002). BH3-only proteins—Evolutionarily conserved proapoptotic Bcl-2 family members essential for initiating programmed cell death. *Journal of Cell Science, 115*, 1567–1574.

Boya, P., Gonzalez-Polo, R. A., Casares, N., Perfettini, J. L., Dessen, P., Larochette, N., Metivier, D., Meley, D., Souquere, S., Yoshimori, T., Pierron, G., Codogno, P., & Kroemer, G. (2005). Inhibition of macroautophagy triggers apoptosis. *Molecular and Cell Biology, 25*, 1025–1040.

Brown, S. B., Bailey, K., & Savill, J. (1997). Actin is cleaved during constitutive apoptosis. *Biochemistry Journal, 323*, 233–237.

Browne, K. A., Johnstone, R. W., Jans, D. A., & Trapani, J. A. (2000). Filamin (280-kDa actin-binding protein) is a caspase substrate and is also cleaved directly by the cytotoxic T lymphocyte protease granzyme B during apoptosis. *Journal of Biological Chemistry, 275*, 39262–39266.

Bursch, W. (2004). Multiple cell death programs: Charon's lifts to Hades. *FEMS Yeast Research, 5*, 101–110.

Bykov, V. J., Issaeva, N., Shilov, A., Hultcrantz, M., Pugacheva, E., Chumakov, P., Bergman, J., Wiman, K. G., & Selivanova, G. (2002). Restoration of the tumor suppressor function to mutant p53 by a low-molecular-weight compound. *Nature Medicine, 8*, 282–288.

Carr, A. M. (2000). Cell cycle. Piecing together the p53 puzzle. *Science, 287*, 1765–1766.

Chiba, I., Takahashi, T., Nau, M. M., D'Amico, D., Curiel, D. T., Mitsudomi, T., Buchhagen, D. L., Carbone, D., Piantadosi, S., Koga, H., et al. (1990). Mutations in the *p53* gene are frequent in primary, resected non-small cell lung cancer. Lung Cancer Study Group. *Oncogene, 5*, 1603–1610.

Chinnaiyan, A. M., O'Rourke, K., Tewari, M., & Dixit, V. M. (1995). FADD, a novel death domain-containing protein, interacts with the death domain of Fas and initiates apoptosis. *Cell, 81,* 505–512.

Choi, W. S., Lee, E. H., Chung, C. W., Jung, Y. K., Jin, B. K., Kim, S. U., Oh, T. H., Saido, T. C., & Oh, Y. J. (2001). Cleavage of Bax is mediated by caspase-dependent or -independent calpain activation in dopaminergic neuronal cells: Protective role of Bcl-2. *Journal of Neurochemistry, 77,* 1531–1541.

Creagh, E. M., Conroy, H., & Martin, S. J. (2003). Caspase-activation pathways in apoptosis and immunity. *Immunology Reviews, 193,* 10–21.

Cunningham, A. S., Jelliffe, D. B., & Jelliffe, E. F. (1991). Breast-feeding and health in the 1980s: A global epidemiologic review [see comments]. *Journal of Pediatrics, 118,* 659–666.

Davis, M. K., Savitz, D. A., & Graubard, B. I. (1988). Infant feeding and childhood cancer. *Lancet, 2,* 365–368.

Deveraux, Q. L., Roy, N., Stennicke, H. R., Van Arsdale, T., Zhou, Q., Srinivasula, S. M., Alnemri, E. S., Salvesen, G. S., & Reed, J. C. (1998). IAPs block apoptotic events induced by caspase-8 and cytochrome c by direct inhibition of distinct caspases. *EMBO Journal, 17,* 2215–2223.

Du, C., Fang, M., Li, Y., Li, L., & Wang, X. (2000). Smac, a mitochondrial protein that promotes cytochrome c-dependent caspase activation by eliminating IAP inhibition. *Cell, 102,* 33–42.

Duan, H., & Dixit, V. M. (1997). Raidd is a new death adaptor molecule. *Nature, 385,* 86–89.

Duringer, C., Hamiche, A., Gustafsson, L., Kimura, H., & Svanborg, C. (2003). HAMLET interacts with histones and chromatin in tumor cell nuclei. *Journal of Biological Chemistry, 278,* 42131–42135.

Dyche, W. J. (1979). A comparative study of the differentiation and involution of the Mullerian duct and Wolffian duct in the male and female fetal mouse. *Journal of Morphology, 162,* 175–209.

Edinger, A. L., & Thompson, C. B. (2004). Death by design: Apoptosis, necrosis and autophagy. *Current Opinion in Cell Biology, 16,* 663–669.

Enari, M., Sakahira, H., Yokoyama, H., Okawa, K., Iwamatsu, A., & Nagata, S. (1998). A caspase-activated DNase that degrades DNA during apoptosis, and its inhibitor ICAD [see comments] [published erratum appears in *Nature,* May 28, 1998; 393 (6683): 396]. *Nature, 391,* 43–50.

Fischer, W., Gustafsson, L., Mossberg, A. K., Gronli, J., Mork, S., Bjerkvig, R., & Svanborg, C. (2004). Human α-lactalbumin made lethal to tumor cells (HAMLET) kills human glioblastoma cells in brain xenografts by an apoptosis-like mechanism and prolongs survival. *Cancer Research, 64,* 2105–2112.

Foghsgaard, L., Wissing, D., Mauch, D., Lademann, U., Bastholm, L., Boes, M., Elling, F., Leist, M., & Jaattela, M. (2001). Cathepsin B acts as a dominant execution protease in tumor cell apoptosis induced by tumor necrosis factor. *Journal of Cell Biology, 153,* 999–1010.

Foster, B. A., Coffey, H. A., Morin, M. J., & Rastinejad, F. (1999). Pharmacological rescue of mutant p53 conformation and function. *Science, 286,* 2507–2510.

Garcia-Calvo, M., Peterson, E. P., Leiting, B., Ruel, R., Nicholson, D. W., & Thornberry, N. A. (1998). Inhibition of human caspases by peptide-based and macromolecular inhibitors. *Journal of Biological Chemistry, 273,* 32608–32613.

Gobeil, S., Boucher, C. C., Nadeau, D., & Poirier, G. G. (2001). Characterization of the necrotic cleavage of poly (ADP-ribose) polymerase (PARP-1): Implication of lysosomal proteases. *Cell Death and Differentiation, 8,* 588–594.

Golding, J., Emmett, P. M., & Rogers, I. S. (1997a). Does breast feeding protect against non-gastric infections? *Early Human Development, 49 (Suppl),* S105–120.

Golding, J., Emmett, P. M., & Rogers, I. S. (1997b). Gastroenteritis, diarrhoea and breast feeding. *Early Human Development, 49 (Suppl),* S83–103.

Gonzalez-Polo, R. A., Boya, P., Pauleau, A. L., Jalil, A., Larochette, N., Souquere, S., Eskelinen, E. L., Pierron, G., Saftig, P., & Kroemer, G. (2005). The apoptosis/autophagy paradox: Autophagic vacuolization before apoptotic death. *Journal of Cell Science, 118,* 3091–3102.

Green, D. R., & Kroemer, G. (2004). The pathophysiology of mitochondrial cell death. *Science, 305,* 626–629.

Greenblatt, M. S., Bennett, W. P., Hollstein, M., & Harris, C. C. (1994). Mutations in the *p53* tumor suppressor gene: Clues to cancer etiology and molecular pathogenesis. *Cancer Research, 54,* 4855–4878.

Gueth-Hallonet, C., Weber, K., & Osborn, M. (1997). Cleavage of the nuclear matrix protein NuMA during apoptosis. *Experiments in Cell Research, 233,* 21–24.

Gustafsson, L., Leijonhufvud, I., Aronsson, A., Mossberg, A. K., & Svanborg, C. (2004). Treatment of skin papillomas with topical α-lactalbumin-oleic acid. *New England Journal of Medicine, 350,* 2663–2672.

Håkansson, A., Zhivotovsky, B., Orrenius, S., Sabharwal, H., & Svanborg, C. (1995). Apoptosis induced by a human milk protein. *Proceedings of the National Academy of Sciences USA, 92,* 8064–8068.

Håkansson, A., Andreasson, J., Zhivotovsky, B., Karpman, D., Orrenius, S., & Svanborg, C. (1999). Multimeric α-lactalbumin from human milk induces apoptosis through a direct effect on cell nuclei. *Experiments in Cell Research, 246,* 451–460.

Hallgren, O., Gustafsson, L., Irjala, H., Selivanova, G., Orrenius, S., & Svanborg, C. (2006). HAMLET triggers apoptosis but tumor cell death is independent of caspases, Bcl-2 and p53. *Apoptosis, 11,* 221–233.

Hanson, L. A. (1998). Breastfeeding provides passive and likely long-lasting active immunity. *Annals in Allergy and Asthma Immunology, 81,* 523–533; quiz 533–534, 537.

Helminen, H. J., & Ericsson, J. L. (1971). Ultrastructural studies on prostatic involution in the rat. Mechanism of autophagy in epithelial cells, with special reference to the rough-surfaced endoplasmic reticulum. *Journal of Ultrastructured Research, 36,* 708–724.

Holler, N., Zaru, R., Micheau, O., Thome, M., Attinger, A., Valitutti, S., Bodmer, J. L., Schneider, P., Seed, B., & Tschopp, J. (2000). Fas triggers an alternative, caspase-8-independent cell death pathway using the kinase RIP as effector molecule. *Nature Immunology, 1,* 489–495.

Hollstein, M., Sidransky, D., Vogelstein, B., & Harris, C. C. (1991). p53 Mutations in human cancers. *Science, 253,* 49–53.

Hollstein, M., Shomer, B., Greenblatt, M., Soussi, T., Hovig, E., Montesano, R., & Harris, C. C. (1996). Somatic point mutations in the *p53* gene of human tumors and cell lines: Updated compilation. *Nucleic Acids Research, 24,* 141–146.

Horvitz, H. R., Shaham, S., & Hengartner, M. O. (1994). The genetics of programmed cell death in the nematode *Caenorhabditis elegans. Cold Spring Harbor Symposia on Quantitative Biology, 59,* 377–385.

Hsu, H., Xiong, J., & Goeddel, D. V. (1995). The TNF receptor 1-associated protein TRADD signals cell death and NF-κB activation. *Cell, 81,* 495–504.

Jaattela, M. (2002). Programmed cell death: Many ways for cells to die decently. *Annals in Medicine, 34,* 480–488.

Jaattela, M., & Tschopp, J. (2003). Caspase-independent cell death in T lymphocytes. *Nature Immunology, 4,* 416–423.

Janicke, R. U., Ng, P., Sprengart, M. L., & Porter, A. G. (1998). Caspase-3 is required for α-fodrin cleavage but dispensable for cleavage of other death substrates in apoptosis. *Journal of Biological Chemistry, 273,* 15540–15545.

Johnson, D. E. (2000). Noncaspase proteases in apoptosis. *Leukemia, 14,* 1695–1703.

Kabeya, Y., Mizushima, N., Ueno, T., Yamamoto, A., Kirisako, T., Noda, T., Kominami, E., Ohsumi, Y., & Yoshimori, T. (2000). LC3, a mammalian homologue of yeast Apg8p, is localized in autophagosome membranes after processing. *EMBO Journal, 19,* 5720–5728.

Kerr, J. F. (1972). Shrinkage necrosis of adrenal cortical cells. *Journal of Pathology, 107,* 217–219.

Kihara, A., Kabeya, Y., Ohsumi, Y., & Yoshimori, T. (2001). Beclin-phosphatidylinositol 3-kinase complex functions at the trans-Golgi network. *EMBO Reports, 2,* 330–335.

Klionsky, D. J., Cregg, J. M., Dunn, W. A., Jr., Emr, S. D., Sakai, Y., Sandoval, I. V., Sibirny, A., Subramani, S., Thumm, M., Veenhuis, M., & Ohsumi, Y. (2003). A unified nomenclature for yeast autophagy-related genes. *Developmental Cell, 5,* 539–545.

Kohler, C., Gahm, A., Noma, T., Nakazawa, A., Orrenius, S., & Zhivotovsky, B. (1999a). Release of adenylate kinase 2 from the mitochondrial intermembrane space during apoptosis. *FEBS Letters, 447,* 10–12.

Kohler, C., Håkansson, A., Svanborg, C., Orrenius, S., & Zhivotovsky, B. (1999b). Protease activation in apoptosis induced by MAL. *Experiments in Cell Research, 249,* 260–268.

Kubbutat, M. H., Jones, S. N., & Vousden, K. H. (1997). Regulation of p53 stability by Mdm2. *Nature, 387,* 299–303.

Lane, D. P. (1993). Cancer. A death in the life of p53 [news; comment]. *Nature, 362,* 786–787.

Lane, D. P., & Crawford, L. V. (1979). T antigen is bound to a host protein in SV40-transformed cells. *Nature, 278,* 261–263.

Leist, M., & Jaattela, M. (2001). Four deaths and a funeral: From caspases to alternative mechanisms. *Nature Reviews of Molecular and Cell Biology, 2,* 589–598.

Lemasters, J. J., Qian, T., Elmore, S. P., Trost, L. C., Nishimura, Y., Herman, B., Bradham, C. A., Brenner, D. A., & Nieminen, A. L. (1998). Confocal microscopy of the mitochondrial permeability transition in necrotic cell killing, apoptosis and autophagy. *Biofactors, 8,* 283–285.

Levine, B., & Klionsky, D. J. (2004). Development by self-digestion: Molecular mechanisms and biological functions of autophagy. *Developmental Cell, 6,* 463–477.

Linzer, D. I., & Levine, A. J. (1979). Characterization of a 54K Dalton cellular SV40 tumor antigen present in SV40-transformed cells and uninfected embryonal carcinoma cells. *Cell, 17,* 43–52.

Lockshin, R. A., & Williams, C. M. (1965). *Journal of Insect Physiology, 11,* 123–133.

Lockshin, R. A., & Zakeri, Z. (2001). Programmed cell death and apoptosis: Origins of the theory. *Nature Reviews of Molecular and Cell Biology, 2,* 545–550.

Lockshin, R. A., & Zakeri, Z. (2004). Apoptosis, autophagy, and more. *International Journal of Biochemistry and Cell Biology, 36,* 2405–2419.

Lowe, S. W., & Lin, A. W. (2000). Apoptosis in cancer. *Carcinogenesis, 21,* 485–495.

Lowe, S. W., Bodis, S., McClatchey, A., Remington, L., Ruley, H. E., Fisher, D. E., Housman, D. E., & Jacks, T. (1994). p53 Status and the efficacy of cancer therapy *in vivo*. *Science, 266,* 807–810.

Lozanon, J., Berra, E. M., Diaz-Meco, M. T., Dominguez, I., Sanz, L., & Moscat, J. (1994). Protein kinase C isoform is critical for kB-dependent promoter activation by sphingomyelinase. *Biological Chemistry, 269,* 19200–19202.

Madeo, F., Frohlich, E., & Frohlich, K. U. (1997). A yeast mutant showing diagnostic markers of early and late apoptosis. *Journal of Cellular Biology, 139,* 729–734.

Martin, S. J., O'Brien, G. A., Nishioka, W. K., McGahon, A. J., Mahboubi, A., Saido, T. C., & Green, D. R. (1995). Proteolysis of fodrin (non-erythroid spectrin) during apoptosis. *Journal of Biological Chemistry, 270,* 6425–6428.

Mathiasen, I. S., Lademann, U., & Jaattela, M. (1999). Apoptosis induced by vitamin D compounds in breast cancer cells is inhibited by Bcl-2 but does not involve known caspases or p53. *Cancer Research, 59,* 4848–4856.

Meek, D. W. (1999). Mechanisms of switching on p53: A role for covalent modification? *Oncogene, 18,* 7666–7675.

Meier, T., Arni, S., Malarkannan, S., Poincelet, M., & Hoessli, D. (1992). Immunodetection of biotinylated lymphocyte-surface proteins by enhanced chemiluminescence: A nonradioactive method for cell-surface protein analysis. *Analytical Biochemistry, 204,* 220–226.

Mihara, M., Erster, S., Zaika, A., Petrenko, O., Chittenden, T., Pancoska, P., & Moll, U. M. (2003). p53 Has a direct apoptogenic role at the mitochondria. *Molecular Cell, 11,* 577–590.
Miura, M., Zhu, H., Rotello, R., Hartwieg, E. A., & Yuan, J. (1993). Induction of apoptosis in fibroblasts by IL-1 beta-converting enzyme, a mammalian homolog of the *C. elegans* cell death gene *ced-3*. *Cell, 75,* 653–660.
Miura, M., Friedlander, R. M., & Yuan, J. (1995). Tumor necrosis factor-induced apoptosis is mediated by a CrmA-sensitive cell death pathway. *Proceedings of the National Academy of Sciences USA, 92,* 8318–8322.
Muller, M., Wilder, S., Bannasch, D., Israeli, D., Lehlbach, K., Li-Weber, M., Friedman, S. L., Galle, P. R., Stremmel, W., Oren, M., & Krammer, P. H. (1998). p53 Activates the CD95 (APO-1/Fas) gene in response to DNA damage by anticancer drugs. *Journal of Experimental Medicine, 188,* 2033–2045.
Muzio, M., Chinnaiyan, A. M., Kischkel, F. C., O'Rourke, K., Shevchenko, A., Ni, J., Scaffidi, C., Bretz, J. D., Zhang, M., Gentz, R., Mann, M., Krammer, P. H., Peter, M. E., & Dixit, V. M. (1996). FLICE, a novel FADD-homologous ICE/CED-3-like protease, is recruited to the CD95 (Fas/APO-1) death-inducing signaling complex. *Cell, 85,* 817–827.
Nagata, S. (1997). Apoptosis by death factor. *Cell, 88,* 355–365.
Nakagawa, T., & Yuan, J. (2000). Cross-talk between two cysteine protease families. Activation of caspase-12 by calpain in apoptosis. *Journal of Cellular Biology, 150,* 887–894.
Nakano, K., & Vousden, K. H. (2001). PUMA, a novel proapoptotic gene, is induced by p53. *Molecular Cell, 7,* 683–694.
Noda, T., Suzuki, K., & Ohsumi, Y. (2002). Yeast autophagosomes: De novo formation of a membrane structure. *Trends in Cellular Biology, 12,* 231–235.
Oda, E., Ohki, R., Murasawa, H., Nemoto, J., Shibue, T., Yamashita, T., Tokino, T., Taniguchi, T., & Tanaka, N. (2000). Noxa, a BH3-only member of the Bcl-2 family and candidate mediator of p53-induced apoptosis. *Science, 288,* 1053–1058.
Paavola, L. G. (1978). The corpus luteum of the guinea pig. III. Cytochemical studies on the Golgi complex and GERL during normal postpartum regression of luteal cells, emphasizing the origin of lysosomes and autophagic vacuoles. *Journal of Cell Biology, 79,* 59–73.
Petiot, A., Ogier-Denis, E., Blommaart, E. F., Meijer, A. J., & Codogno, P. (2000). Distinct classes of phosphatidylinositol 3'-kinases are involved in signaling pathways that control macroautophagy in HT-29 cells. *Journal of Biological Chemistry, 275,* 992–998.
Roberg, K., Kagedal, K., & Ollinger, K. (2002). Microinjection of cathepsin D induces caspase-dependent apoptosis in fibroblasts. *American Journal of Pathology, 161,* 89–96.
Roberts, L. R., Kurosawa, H., Bronk, S. F., Fesmier, P. J., Agellon, L. B., Leung, W. Y., Mao, F., & Gores, G. J. (1997). Cathepsin B contributes to bile salt-induced apoptosis of rat hepatocytes. *Gastroenterology, 113,* 1714–1726.
Robles, A. I., Bemmels, N. A., Foraker, A. B., & Harris, C. C. (2001). APAF-1 is a transcriptional target of p53 in DNA damage-induced apoptosis. *Cancer Research, 61,* 6660–6664.
Samuels-Lev, Y., O'Connor, D. J., Bergamaschi, D., Trigiante, G., Hsieh, J. K., Zhong, S., Campargue, I., Naumovski, L., Crook, T., & Lu, X. (2001). ASPP proteins specifically stimulate the apoptotic function of p53. *Molecular Cell, 8,* 781–794.
Scaffidi, C., Fulda, S., Srinivasan, A., Friesen, C., Li, F., Tomaselli, K. J., Debatin, K. M., Krammer, P. H., & Peter, M. E. (1998). Two CD95 (APO-1/Fas) signaling pathways. *EMBO Journal, 17,* 1675–1687.
Schlumpberger, M., Schaeffeler, E., Straub, M., Bredschneider, M., Wolf, D. H., & Thumm, M. (1997). *AUT1*, a gene essential for autophagocytosis in the yeast *Saccharomyces cerevisiae*. *Journal of Bacteriology, 179,* 1068–1076.

Schulte-Hermann, R., Bursch, W., Grasl-Kraupp, B., Marian, B., Torok, L., Kahl-Rainer, P., & Ellinger, A. (1997). Concepts of cell death and application to carcinogenesis. *Toxicologic Pathology, 25,* 89–93.

Schweichel, J. U., & Merker, H. J. (1973). The morphology of various types of cell death in prenatal tissues. *Teratology, 7,* 253–266.

Sensibar, J. A., Griswold, M. D., Sylvester, S. R., Buttyan, R., Bardin, C. W., Cheng, C. Y., Dudek, S., & Lee, C. (1991). Prostatic ductal system in rats: regional variation in localization of an androgen-repressed gene product, sulfated glycoprotein-2. *Endocrinology, 128,* 2091–2102.

Sherr, C. J., & Weber, J. D. (2000). The ARF/p53 pathway. *Current Opinion in Genetics and Development, 10,* 94–99.

Shimizu, S., Narita, M., & Tsujimoto, Y. (1999). Bcl-2 family proteins regulate the release of apoptogenic cytochrome c by the mitochondrial channel VDAC [see comments]. *Nature, 399,* 483–487.

Shimizu, S., Kanaseki, T., Mizushima, N., Mizuta, T., Arakawa-Kobayashi, S., Thompson, C. B., & Tsujimoto, Y. (2004). Role of Bcl-2 family proteins in a non-apoptotic programmed cell death dependent on autophagy genes. *Nature Cell Biology, 6,* 1221–1228.

Stoka, V., Turk, B., Schendel, S. L., Kim, T. H., Cirman, T., Snipas, S. J., Ellerby, L. M., Bredesen, D., Freeze, H., Abrahamson, M., Bromme, D., Krajewski, S., Reed, J. C., Yin, X. M., Turk, V., & Salvesen, G. S. (2001). Lysosomal protease pathways to apoptosis. Cleavage of bid, not pro-caspases, is the most likely route. *Journal of Biological Chemistry, 276,* 3149–3157.

Susin, S. A., Lorenzo, H. K., Zamzami, N., Marzo, I., Snow, B. E., Brothers, G. M., Mangion, J., Jacotot, E., Costantini, P., Loeffler, M., Larochette, N., Goodlett, D. R., Aebersold, R., Siderovski, D. P., Penninger, J. M., & Kroemer, G. (1999). Molecular characterization of mitochondrial apoptosis-inducing factor. *Nature, 397,* 441–446.

Suzuki, Y., Imai, Y., Nakayama, H., Takahashi, K., Takio, K., & Takahashi, R. (2001). A serine protease, HtrA2, is released from the mitochondria and interacts with XIAP, inducing cell death. *Molecular Cell, 8,* 613–621.

Svanborg, C., Agerstam, H., Aronson, A., Bjerkvig, R., Duringer, C., Fischer, W., Gustafsson, L., Hallgren, O., Leijonhuvud, I., Linse, S., Mossberg, A. K., Nilsson, H., Pettersson, J., & Svensson, M. (2003). HAMLET kills tumor cells by an apoptosis-like mechanism—Cellular, molecular, and therapeutic aspects. *Advances in Cancer Research, 88,* 1–29.

Svensson, M. (1999). Studies on an apoptosis-inducing folding variant of human α-lactalbumin. In *Institute og Laboratory Medicine* (p. 33). Lund, Sweden: Lund University.

Svensson, M., Fast, J., Mossberg, A. K., Duringer, C., Gustafsson, L., Hallgren, O., Brooks, C. L., Berliner, L., Linse, S., & Svanborg, C. (2003a). Alpha-lactalbumin unfolding is not sufficient to cause apoptosis, but is required for the conversion to HAMLET (human α-lactalbumin made lethal to tumor cells). *Protein Science, 12,* 2794–2804.

Svensson, M., Mossberg, A. K., Pettersson, J., Linse, S., & Svanborg, C. (2003b). Lipids as cofactors in protein folding: Stereo-specific lipid-protein interactions are required to form HAMLET (human α-lactalbumin made lethal to tumor cells). *Protein Science, 12,* 2805–2814.

Tanida, I., Nishitani, T., Nemoto, T., Ueno, T., & Kominami, E. (2002). Mammalian Apg12p, but not the Apg12p.Apg5p conjugate, facilitates LC3 processing. *Biochemistry and Biophysics Research Community, 296,* 1164–1170.

Tanida, I., Tanida-Miyake, E., Ueno, T., & Kominami, E. (2001). The human homolog of Saccharomyces cerevisiae Apg7p is a protein-activating enzyme for multiple substrates including human Apg12p, GATE-16, GABARAP, and MAP-LC3. *Journal of Biological Chemistry, 276,* 1701–1706.

Tartaglia, L. A., Ayres, T. M., Wong, G. H., & Goeddel, D. V. (1993). A novel domain within the 55 kd TNF receptor signals cell death. *Cell, 74,* 845–853.

Thorlacius, S., Thorgilsson, B., Bjornsson, J., Tryggvadottir, L., Borresen, A. L., Ogmundsdottir, H. M., & Eyfjord, J. E. (1995). TP53 mutations and abnormal p53 protein staining in breast carcinomas related to prognosis. *European Journal of Cancer, 31A*, 1856–1861.

Thornberry, N. A., & Lazebnik, Y. (1998). Caspases: Enemies within. *Science, 281*, 1312–1316.

Tsujimoto, Y. (2003). Cell death regulation by the Bcl-2 protein family in the mitochondria. *Journal of Cellular Physiology, 195*, 158–167.

Tsujimoto, Y., Cossman, J., Jaffe, E., & Croce, C. M. (1985). Involvement of the *bcl-2* gene in human follicular lymphoma. *Science, 228*, 1440–1443.

Tsukada, M., & Ohsumi, Y. (1993). Isolation and characterization of autophagy-defective mutants of *Saccharomyces cerevisiae*. *FEBS Letters, 333*, 169–174.

Turk, B., Turk, D., & Turk, V. (2000). Lysosomal cysteine proteases: More than scavengers. *Biochimica et Biophysica Acta, 1477*, 98–111.

Verhagen, A. M., Silke, J., Ekert, P. G., Pakusch, M., Kaufmann, H., Connolly, L. M., Day, C. L., Tikoo, A., Burke, R., Wrobel, C., Moritz, R. L., Simpson, R. J., & Vaux, D. L. (2002). HtrA2 promotes cell death through its serine protease activity and its ability to antagonize inhibitor of apoptosis proteins. *Journal of Biological Chemistry, 277*, 445–454.

Vogelstein, B., Lane, D., & Levine, A. J. (2000). Surfing the p53 network. *Nature, 408*, 307–310.

Wu, H. Y., Nahm, M. H., Guo, Y., Russell, M. W., & Briles, D. E. (1997). Intranasal immunization of mice with PspA (pneumococcal surface protein A) can prevent intranasal carriage, pulmonary infection, and sepsis with *Streptococcus pneumoniae*. *Journal of Infectious Diseases, 175*, 839–846.

Yuan, J., Shaham, S., Ledoux, S., Ellis, H. M., & Horvitz, H. R. (1993). The *C. elegans* cell death gene ced-3 encodes a protein similar to mammalian interleukin-1 beta-converting enzyme. *Cell, 75*, 641–652.

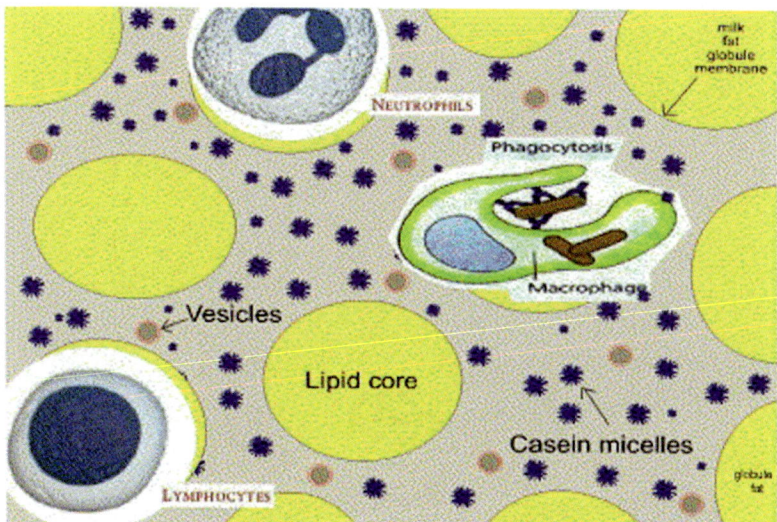

Color plate 1. Schematic representation of the physical phases of milk. The area between the milk particles represents the milk serum (whey), the phase during which all other phases are homogenously dispersed (adapted from Silanikove et al., 2006 with permission from Elsevier). Note that the number of MSLM, $\sim 10^{15}$/mL of milk (denoted in the figure as vesicles), and the number of fat globules, $\sim 10^{10}$/mL of milk, are not represented proportionally in the scheme

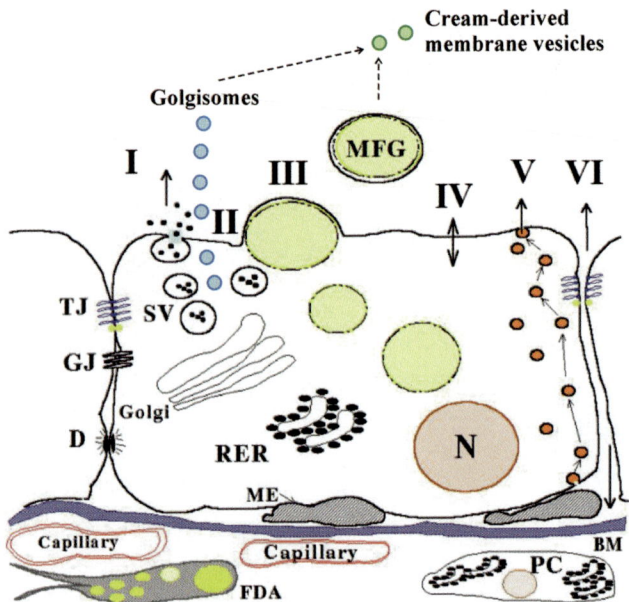

Color plate 2. A model explaining the origin of milk serum lipoprotein membranes (MSLM). According to this model, MSLM originate from a subpopulation of Golgi-derived vesicles, which constantly flow from the Golgi apparatus to the apical membrane of the alveoli and are being released to milk by unidentified mechanism. According to the proposed model, vesicles derived from the serum of milk cream are not truly components of milk, but rather reflect their coalescence-inducement from milk fat globule membranes or the trapping of the MSLM during centrifugation and other manipulations of milk

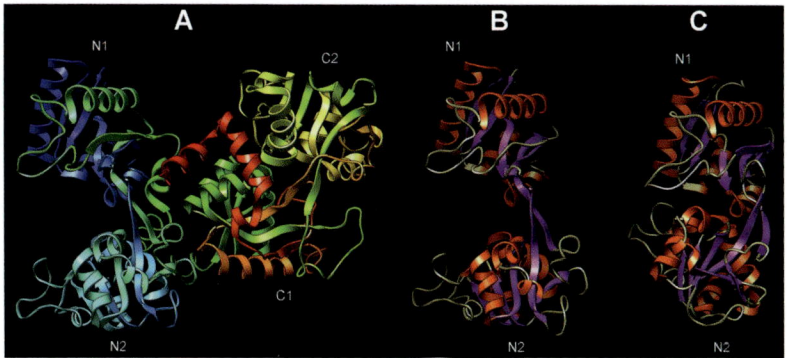

Color plate 3. Structure of hLf. (a) Ribbon diagram showing the polypeptide folding of iron-saturated hLf (Jameson et al., 1998). The N-t lobe is on the left; the polypeptide chain is colored from the N- to the C-terminal end according to a red-shift. (b) Open and (c) closed structures of the N-terminal lobe of hLf (α-helices are colored in magenta and β-sheets in blue). Domains N1, N2, C1, and C2 are indicated

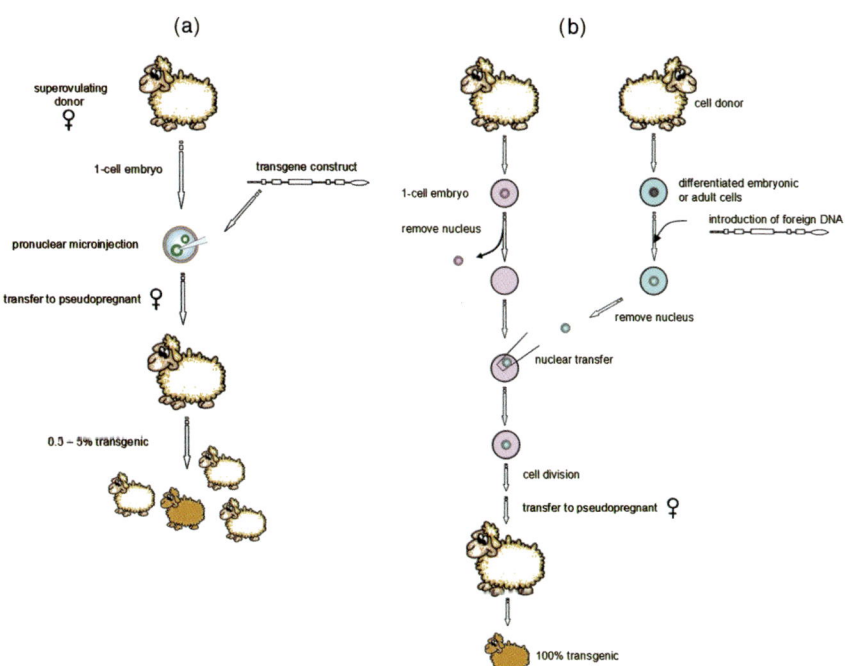

Color plate 4. Methods for producing transgenic livestock animals. (a) Pronuclear microinjection and (b) nuclear transfer

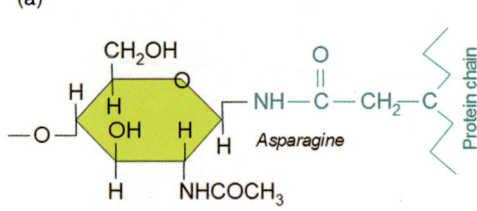

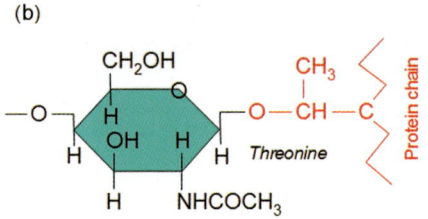

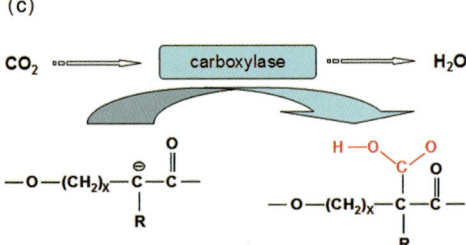

Color plate 5. Representative types of posttranslational modifications. Basic types of glycosylation: (a) N-linked glycosylation to the amide nitrogen of asparagine side chains; (b) O-linked glycosylation to the hydroxy oxygen of serine and threonine side chains. (c) Carboxylation

A Proline-Rich Polypeptide from Ovine Colostrum: Colostrinin with Immunomodulatory Activity

Michal Zimecki

Abstract A proline-rich polypeptide (PRP), later called colostrinin (CLN), was originally found as a fraction accompanying sheep colostral immunoglobulins. Extensive *in vitro* and *in vivo* studies in mice revealed its interesting T cell-tropic activities. The polypeptide promoted T cell maturation from early thymic precursors that acquired the phenotype and function of mature, helper cells; on the other hand, it also affected the phenotype and function of mature T cells. In particular, PRP was shown to recruit suppressor T cells in a model of T cell-independent humoral immune response and suppressed autoimmune hemolytic anemia in New Zealand Black mice. Subsequent *in vitro* studies in the human model revealed that CLN regulated mitogen-induced cytokine production in whole blood cultures. A discovery that CLN promoted pro-cognitive functions in experimental animal models, supported by other laboratory findings, indicating prevention of pathological processes in the central nervous system, led to application of CLN in multicenter clinical trials. The trials demonstrated the therapeutic benefit of CLN in Alzheimer's disease (AD) patients by delaying progress of the disease.

Immunological Effects of the Proline-Rich Polypeptide and Its Peptide Fragments in the *in Vivo* and *in Vitro* Experimental Models in Rodents

The proline-rich polypeptide (PRP) was isolated from ovine colostrum as a substance accompanying IgG_2 immunoglobulins (Janusz et al., 1974). This chapter overviews the *in vivo* and *in vitro* studies in rodents, in human volunteers, as well as in clinical trials revealing the potential use of PRP in delaying

---

M. Zimecki
The Institute of Immunology and Experimental Therapy, R. Weigla str. 12, 53-114 Wroclaw, Poland, Tel: 48 071 370 9953, Fax: 48 071 337 1382
e-mail: zimecki@iitd.pan.wroc.pl

progress of the neurodegenerative disease in Alzheimer patients. Initial physicochemical studies (Janusz et al., 1978) showed that PRP was soluble at 4 °C but reversibly precipitated by warming to room temperature. The molecular weight of the polypeptide, as determined by thin-layer gel filtration on Sephadex G-200, was 38 kDa. The main amino acids of PRP are proline (23%), glutamic acid (14.9%), and valine (12.9%) (Janusz et al., 1974). Subsequent physicochemical investigations revealed (Janusz et al., 1981) that the molecular weight of PRP, as determined by gel filtration on Sephadex G-100, was 17.2 kDa, but only 6 kDa in the presence of guanidinium chloride. C.d. (circular dichroism) spectra in water and in 50% (v/v) trifluoroethanol suggested the presence of block sequences of proline residues forming helices of the polyproline II type. Maximal precipitation at room temperature was observed at pH 4.6 and at ionic strength above 0.6.

Preliminary studies on the immunological activity showed that that PRP regulated the humoral immune response to sheep red blood cells (SRBC) depending on the magnitude of the response in a given experiment. The polypeptide was also shown to increase permeability of skin vessels in guinea pigs. Since a prostaglandin inhibitor Ro 20-5720 abolished the regulatory effects of PRP, we suggested that the activity of PRP could be mediated by prostaglandins (Zimecki et al., 1978; Wieczorek et al., 1979). Further studies demonstrated that PRP was able to induce maturation of thymocytes and to change the phenotype of mature T cells from the spleen. That property of the polypeptide was revealed in the model of semiallogeneic graft-versus-host reaction (GvH) in mice where cortisone-sensitive, immature thymocytes acquired the ability to induce GvH reaction following incubation with PRP. In addition, these thymocytes converted into cortisone-resistant, mature thymocytes upon incubation with PRP (Zimecki et al., 1982a). In another study, Zimecki et al. (1982b) found that PRP alters the ability of thymocytes and splenocytes to form rosettes with autologous erythrocytes. That property was similar to those of thymosine and a calf thymus extract (TFX). In the next series of articles (Zimecki et al., 1984a, b; Lisowski et al., 1988), PRP was shown to differentially act on two major thymocyte subpopulations: glass nonadherent, peanut agglutinin-positive (PNA^+) and glass-adherent, peanut agglutinin-negative (PNA^-) cells. In PNA^+ cells, PRP generated cells expressing helper activity in the humoral immune response to SRBC, and in the PNA^- subpopulation, it generated cells of suppressor activity. Of interest, PNA^+ cells became PNA^- and PNA^- cells acquired PNA^+ phenotype following incubation with PRP. Lisowski et al. (1988) concluded that bidirectional effects of PRP on PNA binding ability, sensitivity to hydrocortisone, and helper-suppressor function make that polypeptide unique among known immunomodulators. In the later stage of our investigations on the interaction of PRP with thymocytes, we (Janusz et al., 1986) found that murine thymocytes bear a specific receptor for PRP. In that study, PRP, which was covalently linked to cellulose discs or Affigel 702 or adsorbed on polystyrene latex beads, showed activity similar to PRP activity in solution. PRP contact with the cell surface was sufficient to induce maturation of thymocytes. Also, PRP adsorbed on polystyrene latex beads formed rosettes with thymocytes, which were inhibited by the addition of soluble PRP. These results suggested the presence of a specific receptor for PRP on thymocytes. More recently, Sokal et al. (1998) showed that PRP stimulation of β-galactosidase in PNA^{high}, immature thymocytes may be involved in the transformation of PNA^{low} cells. Lastly, we demonstrated that PRP interacts with the minor subpopulation of immature thymocytes bearing the phenotype CD4$^-$, CD8$^-$, CD3$^-$, theta^{low} (Wieczorek et al., 1989). Incubation of these T cell precursors with PRP led to the acquisition of a mature T helper cell phenotype, i.e., CD4$^+$, CD3$^+$, TCR α/β$^+$. These cells exhibited, in addition, a helper function in the model of humoral immune response to SRBC *in vitro*. Interestingly, PRP may also control the humoral immune response to a T cell-independent antigen polyvinylpyrrolidone (PVP) (Zimecki et al., 1983). In this case PRP-induced inhibition of the anti-PVP response was shown to be mediated by intrinsic suppressor T cells. In addition, PRP increased the generation of a specific suppressor T cell precursor, induced by a low-molecular form of PVP. The involvement of suppressor T cells, generated by PRP, was also implicated in the case of experimental autoimmune response to erythrocytes (Hraba et al., 1986) and in suppressive effects of PRP on the development of hemolytic anemia in New Zealand Black mice (Zimecki et al., 1991).

Apart from effects of PRP on the maturation and function of T cells in the generation of antigen-specific immune response, PRP was found to express mitogenic activity toward T and B cells (Zimecki et al., 1987). It appeared that PRP was not mitogenic for thymocytes, however, at doses between 0.1–50 μ/mL, it augmented concanavalin A-induced proliferation of thymocytes in a similar fashion as interleukin 1. At doses higher than 10 μ/mL, the polypeptide induced the proliferation of lymph node cells and splenocytes and T cells from lymph nodes. However, it did not cause a significant proliferation of B cells. Interestingly, PRP's action on cell maturation is not confined only to lymphocytes. In a recent report Kubis et al. (2005) found that PRP may affect the early stages of maturation/differentiation of a premonocytic HL-60 cell line.

The above-described results suggest that PRP delivered in colostrum and absorbed by the gut-associated lymphoid tissue may accelerate the maturation of the immune system cells of the offspring. It seems, however, that the immunotropic properties of PRP are predominantly directed to T cell lineage.

PRP's immunoregulatory activities could be also demonstrated by certain sequences of the polypeptide as well as by synthetic analogs of active PRP fragments. When PRP was subjected to chymotrypsin digestion and separated by gel filtration, three fractions were obtained (Staroscik et al., 1983). Although all three peptides exhibited immunological activities, the shortest one, a nonapeptide (Val-Glu-Ser-Tyr-Val-Pro-Leu-Phe-Pro), retained the activity in all assays performed. Subsequent studies (Kubik et al., 1984) using a series of chemically synthesized peptides revealed that the minimal sequence still demonstrating the immunoregulatory activity is the following: Val-Pro-Leu-Phe-Pro, which represents a C-terminal fragment of the nonapeptide. An attempt was also undertaken to rigidify PRP hexapeptide (Tyr-Val-Pro-Leu-Phe-Pro) by the azo-bridge between Tyr^1 and Phe^5 residues (Szewczuk et al., 1988). The peptide showed significantly better immunoregulatory activity compared with the linear hexapeptide, suggesting that the biologically active conformation of the PRP hexapeptide requires the close proximity of both aromatic rings (Tyr^1 and Phe^5). In order to design biodegradation-resistant analogs of PRP, three analogs with D-amino acid substituents at positions 1 and 5 of the PRP-hexapeptide were synthesized (Kubik et al., 1988). One of the analogs (Tyr-Val-Pro-Leu-D-Phe-Pro) was found to have immunoregulatory activity in the humoral immune response *in vivo* and *in vitro*. To determine the role of consecutive amino acid residues in the shortest, active fragment of PRP, a series of analogs substituted by L-alanine in successive positions of the peptide chain was synthesized (Szewczuk et al., 1991) and tested in the models of the humoral immune response to SRBC and PVP and autologous rosette formation. The results showed that the analog containing alanine instead of proline in position 5 of PRP-pentapeptide was active in all tests. The authors concluded that the side chain of the Pro^5 residue does not have direct functions in mediating biological activity, and this residue may serve as a spacer. The loss of the activity in the analogs where Leu^3 and Phe^4 were replaced by Ala indicates the importance of these residues in the biological activity of PRP-pentapeptide.

Studies Associated with Perspectives for Clinical Application of Colostrinin in the Treatment of Neurodegenerative Disorders

It soon became evident that PRP may be of potential therapeutic value in the treatment of patients with neurodegenerative disorders such as Alzheimer's disease (Inglot et al., 1996). Inglot first proposed the term "Colostrinin" to indicate the source of isolation of the active proline-rich peptides (colostrum). We will use this name instead of PRP. In the first report on the biological activity of Colostrinin (CLN) in the human model (Inglot et al., 1996), the authors showed that CLN was a modest inducer of IFN-γ and TNF-α production in whole blood cell cultures. On the other hand, mitogen-stimulated cytokine production was inhibited by CLN, particularly in Alzheimer patients. In addition, oral administration of CLN-containing tablets led to a transient hyporeactivity of peripheral blood leukocytes in terms of cytokine production.

Within a few years, it appeared clear that the CLN preparation was more heterogenous than previously thought (Kruzel et al., 2001). The authors demonstrated that 32 peptides can be obtained from CLN subjected to HPLC and indicated significant homology of the peptides to three protein precursors:

annexin, β-casein, and a hypothetical β-casein homologue. Several selected peptides were tested for their ability to induce cytokine production in the cultures of human leukocytes and were shown to be active. More recently, a new, simple, two-step extraction/purification method that consists of methanol extraction and ammonium sulfate precipitation was described (Kruzel et al., 2004). When compared with the original material, CLN isolated by this method showed (1) a similar pattern of peptides in SDS PAGE, (2) identical amino acid analysis, (3) a similar pattern of HPLC profiles, and (4) its ability to induce IFN-γ and TNF-α. In addition, the production of high-quality CLN could be accomplished in less than 48 hours.

Further studies in the animal models were designed to support the assumption that CLN may enhance the cognitive processes. In a study by Popik et al. (1999), CLN was administered intraperitoneally into young (3-month-old) and old (13-month-old) Wistar rats. CLN facilitated the acquisition of spatial learning of aged but not young rats and improved incidental learning in aged rats. The authors suggested that CLN might have beneficial effects on cognitive functioning, particularly in old subjects. In another study (Popik et al., 2001), Colostrinin-derived nonapeptide did not change the searching pattern in the Morris water maze test; however, it delayed the extinction of spatial memory. The enhancement of long-term memory by CLN was also checked in one-day-old chickens in the model of the single, one-trial learning paradigm—avoidance of a bitter-tasting substance (methylanthranilate, MeA) (Stewart & Banks, 2006). Birds presented with a bead coated with 100% MeA avoided pecking it 24 hours later, but birds trained with beads coated with 10% MeA pecked the bead 24 hours later, thus demonstrating the lack of long-term memory for the task. However, when CLN was injected intracranially, into a region important for memory formation, prior to training with 10% MeA, the chickens exhibited strong memory retention at 24 hours, similar to those trained on 100% MeA.

Further studies were aimed at the demonstration of various, protective effects of CLN, relevant to its action on the brain tissue, in the *in vitro* models employing cells lines. It was shown (Bacsi et al., 2005) that medullary pheochromocytoma PC12 cells cease to proliferate and extend neurites in a similar way as upon exposure to nerve growth factor. The arrest of CLN-treated PC12 cells in the G1 phase of the cell cycle was associated with an increase in the phosphorylation of p53 at serine^{15} and expression of p21^{WAF1}. The authors concluded that "CLN induces delicate cassettes of signaling pathways common to cell proliferation and differentiation, and mediates activities that are similar to those of hormones and neutrophins, leading to neurite outgrowth."

The alterations in the metabolism of the amyloid beta precursor protein and the formation of A beta plaques are the main cause of neuronal death in Alzheimer patients (Ling et al., 2003). These plaques also promote oxidative stress-induced injury in the brain of these patients. Schuster et al. (2005) used optical and electron microscopy to demonstrate that CLN prevented the aggregation of beta-amyloid peptide A beta (1–40) *in vitro*. These observations were

Table 1 Biological Properties of PRP

Biological Effects of PRP	Reference Number
Promotion of T cell maturation	Lisowski et al., 1988; Sokal et al., 1998; Wieczorek et al., 1989; Zimecki et al., 1982a; Zimecki et al., 1983; Zimecki et al., 1984a; Zimecki et al., 1987.
Helper activity in T cell-dependent immune response	Lisowski et al., 1988; Wieczorek et al., 1979; Zimecki et al., 1984a.
Induction of suppressor cells in T cell-independent immune response	Zimecki et al., 1991
Mitogenic activity	Zimecki et al., 1984b
Suppression of the autoimmune response	Hraba et al., 1986, Zimecki et al., 1991
Regulation of cytokine production *in vitro* and *in vivo* in humans	Inglot et al., 1996
Antimutagenic action	Basci et al., 2005
Antioxidant properties	Boldogh et al., 2003
Enhancement of cognitive processes in animals	Popik et al., 1999, Popik et al., 2001; Stewart & Banks 2006
Prevention of pathological processes in the central nervous system	Basci et al., 2006; Bourhim et al., 2006; Schuster et al., 2005
Clinical trials	Bilikiewicz & Gaus 2004; Leszek et al., 1999; Leszek et al., 2002

compared to the effect of CLN on the neurotoxic activity of beta-amyloid peptides in the culture of SHSY-5Y neuroblastoma cells. The authors showed that the reduction of fibrils of beta-amyloid peptides by CLN was concomitant with the reduction of the cytotoxic effects of beta-amyloid on SHSY-5Y neuroblastoma cells. CLN's ability to reduce fibril formation was utilized to develop a quick method of determining CLN's biological activity (Bourhim et al., 2006). The antioxidant property of CLN has also been investigated (Boldogh et al., 2003). The authors demonstrated that CLN lowered intracellular concentrations of reactive oxygen species (ROS), as evidenced by a decrease of 2',7'-dichlorodihydro-fluorescein-mediated fluorescence, reduced the abundance of 4-hydroxynonenal (4HNE)-protein adducts, inhibited 4HNE-mediated glutathione metabolism, and inhibited 4HNE-induced activation of c-Jun N terminal kinase. Collectively, these results suggest that CLN downregulates 4HNE-induced lipid peroxidation and its product-induced signaling, which may lead to pathological changes. Lastly, CLN was shown to significantly lower the mutation frequency that developed spontaneously or was induced by ROS, chemical, or physical agents (Bacsi et al., 2006). CLN itself had no mutagenic property. The authors conclude that CLN may be used in human therapies for the prevention of diseases associated with sequence alterations in genomic or mitochondrial DNA.

Colostrinin in Clinical Trials

In the first clinical trial (Leszek et al., 1999), 46 Alzheimer patients were divided into three groups and randomly assigned to receive orally (1) Colostrinin (100 µg per tablet), (2) commercially available bioorganic selenium (100 µg of selenium per tablet), or (3) placebo tablets, every second day. Each patient received 10 three-week cycles of treatment, with each cycle separated by a two-week hiatus. Outcomes were assessed by psychiatrists blinded to the treatment assignment. Eight of the 15 AD patients treated with CLN improved, and in seven others the disease had stabilized. In contrast, none of the 31 patients receiving selenium or placebo improved. A subsequent long-term (16–28-month-long) study (Leszek et al., 2002) enrolled 33 patients with mild to moderate severe AD in the trial. The functional abilities of the patients were evaluated using the Mini-Mental State Examination (MMSE) scale. The results showed that CLN induced slight but statistically significant improvement or stabilization of the health status. The adverse reactions observed, if any, were remarkably mild, including anxiety, logorrhea, and insomnia and lasted briefly (3 to 4 days). These trials were then followed by a multicenter study involving six psychiatric centers in Poland (Bilikiewicz & Gaus, 2004) and showed a statistically significant benefit of treatment of AD patients with CLN.

Conclusions

Laboratory and clinical studies revealed that colostrum may serve as the source of substances exhibiting sometimes unexpected, therapeutic properties. Colostrinin, a mixture of peptides, is an example of such an active fraction. Originally postulated to promote maturation of the immune system of newborns, CLN was subsequently found effective in enhancing pro-cognitive functions in animal models and Alzheimer patients. The therapeutic properties of CLN were supported by the laboratory observations revealing antioxidant and neuroprotective actions. The therapeutic efficacy of oral administration of CLN opens CLN for wide applications in diminishing the progress of neurodegenerative disorders.

References

Bacsi, A., Stanton, J. G., Hughes, T. K., Kruzel, M., & Boldogh, I. (2005). Colostrinin-driven neurite outgrowth requires p53 activation in PC12 cells. *Cellular and Molecular Neurobiology, 25*, 1123–1139.

Bacsi, A., Aguilera-Aguirre, L., German, P., Kruzel, M., & Boldogh, I. (2006). Colostrinin decreases spontaneous and induced mutation frequencies at the Hprt locus in Chinese hamster V79 cells. *Journal of Experimental Therapeutics and Oncology, 5*, 249–259.

Bilikiewicz, A., & Gaus, W. (2004). Colostrinin (a naturally occurring, proline-rich, polypeptide mixture) in the treatment of Alzheimer's disease. *Journal of Alzheimer's Disease, 6*, 17–26.

Boldogh, I., Liebenthal, D., Hughes, K., Juelich, T. L., Georgiades, J. A., Kruzel, M. L., & Stanton, G. J. (2003). Modulation of 4HNE-mediated signaling by proline-rich peptides from ovine colostrum. *Journal of Molecular Neuroscience, 20*, 125–133.

Bourhim, M., Kruzel, M., Srikrishnan, T., & Nicotera, T. (2006). Linear quantitation of Aβ aggregation using Thioflavin T: Reduction of fibril formation by Colostrinin. *Journal of Neuroscience Methods*, in press.

Hraba, T., Wieczorek, Z., Janusz, M., Lisowski, J., & Zimecki, M. (1986). Effect of proline-rich polypeptide on experimental autoimmune response to erythrocytes. *Archivum Immunologiae et Therapiae Experimentalis, 34*, 437–443.

Inglot, A. D., Janusz, M., & Lisowski, J. (1996). Colostrinine: A proline-rich polypeptide from ovine colostrum is a modest cytokine inducer in human leukocytes. *Archivum Immunologiae et Therapiae Experimentalis, 44*, 215–224.

Janusz, M., Staroścık, K., Zimecki, M., Wieczorek, Z., & Lisowski, J. (1978). Physicochemical properties of a proline-rich polypeptide (PRP) from ovine colostrum. *Archivum Immunologiae et Therapiae Experimentalis, 26*, 17–21.

Janusz, M., Staroścık, K., Zimecki, M., Wieczorek, Z., & Lisowski, J. (1981). Chemical and physical characterization of a proline-rich polypeptide from sheep colostrum. *Biochemical Journal, 199*, 9–15.

Janusz, M., Staroścık, K., Zimecki, M., Wieczorek, Z., & Lisowski, J. (1986). A proline-rich polypeptide (PRP) with immunoregulatory properties isolated from ovine colostrums. Murine thymocytes have on their surface a receptor specific for PRP. *Archivum Immunologiae et Therapiae Experimentalis, 34*, 427–436.

Kruzel, M., Janusz, M., Lisowski, J., Fischleigh, R. V., & Georgiades, J. A. (2001). Towards an understanding of biological role of Colostrinin peptides. *Journal of Molecular Neuroscience, 17*, 379–389.

Kruzel, M. L., Polanowski, A., Wilusz, T., Sokoowska, A., Pacewicz, M., Bednarz, R., & Georgiades, J. A. (2004). The alcohol-induced conformational changes in casein micelles: A new challenge for the purification of Colostrinin. *The Protein Journal, 23*, 127–133.

Kubik, W., Kliś, A., Szewczuk, Z., & Siemion, I. Z. (1984). Proline-rich polypeptide (PRP)—A new peptide immunoregulator and its partial sequences. *Peptides, 31*, 457–460.

Kubik, A., Szewczuk, Z., Siemion, I. Z., Wieczorek, Z., Spiegel, K., Zimecki, M., Janusz, M., & Lisowski, J. (1988). Configurational requirements of aromatic amino acid residues for the activity of PRP-hexapeptide. *Collection Czechoslovak Chemical Communications, 53*, 2583–2590.

Kubis, A., Marcinkowska, E., Janusz, M., & Lisowski, J. (2005) Studies on the mechanism of action of a proline-rich polypeptide complex (PRP): Effect on the stage of cell differentiation. *Peptides, 26*, 2188–2192.

Leszek, J., Inglot, A. D., Janusz, M., Lisowski, J., Krukowska, K., & Georgiades, J. A. (1999). Colostrinin: A proline-rich polypeptide (PRP) complex isolated from ovine colostrum for treatment of Alzheimer's disease. A double-blind, placebo-controlled study. *Archivum Immunologiae et Therapiae Experimentalis, 47*, 377–385.

Leszek, J., Inglot, A. D., Janusz, M., Byczkiewicz, F., Kiejna, A., Georgiades, J. A., & Lisowski, J. (2002). Colostrinin proline-rich polypeptide complex from ovine colostrum—A long-term study of its efficacy in Alzheimer's disease. *Medical Science Monitor, 8*, P193–P196.

Ling, Y., Morgan, K., & Kalsheker, N. (2003). Amyloid precursor protein (APP) and the biology of proteolytic processing: Relevance to Alzheimer's disease. *International Biochemistry and Cell Biology, 35*, 1505–1535.

Lisowski, J., Wieczorek, Z., Janusz, M., & Zimecki, M. (1988). Proline-rich polypeptide (PRP) from ovine colostrum. Bi-directional modulation of binding of peanut agglutinin. Resistance to hydrocortisone, and helper activity in murine thymocytes. *Archivum Immunologiae et Therapiae Experimentalis, 36,* 381–393.

Popik, P., Bobula, B., Janusz, M., Lisowski, J., & Vetulani, J. (1999). Colostrinin, a polypeptide isolated from early milk, facilitates learning and memory in rats. *Pharmacology Biochemistry and Behavior, 64,* 183–189.

Popik, P., Galoch, Z., Janusz, M., Lisowski, J., & Vetulani, J. (2001). Cognitive effects of colostral-Val nonapeptide in aged rats. *Behavioural Brain Research, 118,* 201–208.

Schuster, D., Rajendran, A., Wen Hui, S., Nicotera, T., Srikrishnan, T., & Kruzel, M. (2005). Protective effect of Colostrinin on neuroblastoma cell survival is due to reduced aggregation of β-amyloid. *Neuropeptides, 39,* 419–426.

Sokal, I., Janusz, M., Miecznikowska, H., Kupryszewski, G., & Lisowski, J. (1998) Effect of colostrinin, an immunomodulatory proline-rich polypeptide from ovine colostrum, on sialidase and β-galactosidase activities in murine thymocytes. *Archivum Immunologiae et Therapiae Experimentalis, 46,* 193–198.

Starościk, K., Janusz, M., Zimecki, M., Wieczorek, Z., & Lisowski, J. (1983). Immunologically active nonapeptide fragment of a proline-rich polypeptide from ovine colostrum: Amino acid sequence and immunoregulatory properties. *Molecular Immunology, 20,* 1277–1282.

Stewart, M. G., & Banks, D. (2006.) Enhancement of long-term memory retention by Colostrinin in one-day-old chicks trained on a weak passive avoidance learning paradigm. *Neurobiology of Learning and Memory, 86,* 66–71.

Szewczuk, Z., Kubik, A., Siemion, I. Z., Wieczorek, Z., Spiegel, K., Zimecki, M., & Lisowski, J. (1988). Conformational modification of the PRP-hexapeptide by a direct covalent attachment of aromatic side chain groups. *International Journal of Peptide Protein Research, 32,* 98–103.

Szewczuk, Z., Kubik, A., & Gocka, G. (1991). New analogs of the immunoregulatory PRP-pentapeptide. *Peptides, 12,* 487–492.

Wieczorek, Z., Zimecki, M., Janusz, M., Starościk, K., & Lisowski, J. (1979). Proline-rich polypeptide from ovine colostrums: Its effect on skin permeability and on the immune response. *Immunology, 36,* 875–881.

Wieczorek, Z., Zimecki, M., Spiegel, K., Lisowski, J., & Janusz, M. (1989). Differentiation of T cells from immature precursors: Identification of a target cell for a proline-rich polypeptide (PRP). *Archivum Immunologiae et Therapiae Experimentalis, 37,* 313–322.

Zimecki, M., Janusz, M., Starościk, K., Wieczorek, Z., & Lisowski, J. (1978). Immunological activity of a proline-rich polypeptide from ovine colostrum. *Archivum Immunologiae et Therapiae Experimentalis, 26,* 23–29.

Zimecki, M., Janusz, M., Starościk, K., Lisowski, J., & Wieczorek, Z. (1982a). Effect of a proline-rich polypeptide on donor cells in graft-versus-host reaction. *Immunology, 47,* 141–147.

Zimecki, M., Starościk, K., Janusz, M., Lisowski, J., & Wieczorek, Z. (1982b). Effect of PRP on autologous rosette formation in mice. *Archivum Immunologiae et Therapiae Experimentalis, 31,* 7–13.

Zimecki, M., Starościk, K., Lisowski, J., & Wieczorek, Z. (1983). The inhibitory activity of a proline-rich polypeptide (PRP) on the immune response to polyvinylpyrrolidone (PVP). *Archivum Immunologiae et Therapiae Experimentalis, 31,* 895–903.

Zimecki, M., Lisowski, J., Hraba, T., Wieczorek, Z., Janusz, M., & Starościk, K. (1984a). The effect of a proline-rich polypeptide (PRP) on the humoral immune response. I. Distinct effect of PRP on the T cell properties of mouse glass-nonadherent (NAT) and glass-adherent (GAT) thymocytes in thymectomized mice. *Archivum Immunologiae et Therapiae Experimentalis, 32,* 191–195.

Zimecki, M., Lisowski, J., Hraba, T., Wieczorek, Z., Janusz, M., & Starościk, K. (1984b). The effect of a proline-rich polypeptide (PRP) on the humoral immune response. II. PRP induces differentiation of helper cells from glass-nonadherent thymocytes (NAT) and suppressor cells from glass-adherent thymocytes (GAT). *Archivum Immunologiae et al., Therapiae Experimentalis, 32*, 197–201.

Zimecki, M., Pierce, C. W., Janusz, M., Wieczorek, Z., & Lisowski, J. (1987). Proliferative response of T lymphocytes to a proline-rich polypeptide (PRP): PRP mimics mitogenic activity of IL-1. *Archivum Immunologiae et Therapiae Experimentalis, 35*, 339–349.

Zimecki, M., Hraba, T., Janusz, M., Lisowski, J., & Wieczorek, Z. (1991). Effect of a proline-rich polypeptide (PRP) on the development of hemolytic anemia and survival of New Zealand Black mice. *Archivum Immunologiae et Therapiae Experimentalis, 39*, 461–467.

III
Milk Peptides

Milk Peptides and Immune Response in the Neonate

Ioannis Politis and Roubini Chronopoulou

Abstract Bioactive peptides encrypted within the native milk proteins can be released by enzymatic proteolysis, food processing, or gastrointestinal digestion. These peptides possess a wide range of properties, including immunomodulatory properties. The first months of life represent a critical period for the maturation of the immune system because a tolerance for nutrient molecules should be developed while that for pathogen-derived antigens is avoided. Evidence has accumulated to suggest that milk peptides may regulate gastrointestinal immunity, guiding the local immune system until it develops its full functionality. Our data using the weaning piglet as the model suggest that several milk peptides can downregulate various immune properties at a time (one to two weeks after weaning) that coincides with immaturity of the immune system. The protein kinase A system and/or the exchange protein directly activated by cyclic AMP (Epac-1) are implicated in the mechanism through which milk peptides can affect immune function in the early postweaning period. Despite the fact that the research in this field is in its infancy, the evidence available suggests that milk protein peptides may promote development of neonatal immune competence.

Milk contains a variety of components that provide immunological protection and facilitate the development of neonatal immune competence. Two main categories of milk compounds are thought to be associated with immunological activity. The first category includes cytokines, which neonates do not produce efficiently. Cytokines present in milk are thought to be protected against intestinal proteolysis and could alleviate immunological deficits, aiding immune system maturation (Kelleher & Lonnerdal, 2001; Bryan et al., 2006). The second category of milk compounds includes milk protein peptides. Milk peptides may affect mucosal immunity possibly by guiding local immunity until it develops its full functionality (Baldi et al., 2005). This chapter focuses on

---

I. Politis
Department of Animal Science, Agricultural University of Athens, 75 Iera Odos, 11855 Athens, Greece
e-mail: i politis@aua.gr

the effects of milk peptides on immune function and attempts to provide an overview of the knowledge available in this field.

The Origin of Milk Peptides

The major proteins present in bovine milk include the family of caseins, β-lactoglobulin, α-lactalbumin, immunoglobulins, lactoferrin, and various minor whey proteins such as transferrin and serum albumin. Milk protein peptides are inactive within the native milk protein. They become active once they are released from the parent protein by enzymatic proteolysis. Milk proteins are susceptible to proteolytic breakdown during gastric processing and later upon exposure of the milk proteins with indigenous or intestinal bacteria-derived enzymes in the gut.

Milk Peptides: Induction or Suppression

A great number of studies demonstrating the ability of milk peptides to enhance or suppress immune function were performed in the last 25 years. Some excellent reviews of these studies are available (Clare & Swaisgood, 2000; Gill et al., 2000; Baldi et al., 2005). Therefore, we will emphasize very recent studies and a selected number of earlier studies to illustrate the point that milk peptides can suppress immune function in certain instances (tolerance to "harmless" antigens) and induce immune function toward pathogen-derived antigens.

Milk Peptides: Induction of Immune Function

A number of recent studies have examined the effects of fermented milks on immune function. These effects can be attributed to the live bacteria present in the fermented milk but also to the presence of metabolites such as peptides or exopolysaccharides produced during fermentation. In certain cases, the effect can be attributed to the generation of specific peptides; in other cases, the effect is multifactorial.

The main mechanism of protection against pathogen-derived antigens at the mucosal level is mediated through IgA(+) cells and secretory IgA, which is capable of neutralizing and, thus, preventing the entry of potentially harmful antigens in the host. Vinderola et al. (2006) investigated the immunomodulatory activity of products derived from milk fermentation by kefir microflora in a murine model. Feeding mice with products of milk fermentation resulted in an increase in interleukin-6 (IL-6) secretion, which is necessary for terminal differentiation of B cells to IgA(+)-secreting cells in the gut. The increase in the IgA(+) cells in the gut was accompanied by an increase in the number of IL-4(+), IL-6(+) cells capable of producing pro-inflammatory cytokines but also IL-10(+) cells, which produce IL-10, which can act as an immunosuppressant in the small intestine. Consumption of products generated by milk fermentation supported the maintenance of intestinal homeostasis by enhancing IgA production at both the small and large intestine levels. In an earlier study, Vinderola et al. (2005) found that consumption of kefir-containing viable bacteria resulted in a Th-1 response in mice that was controlled by Th-2 cytokines. The consumption of pasteurized kefir, which did not contain viable bacteria, would induce both a Th-1 and a Th-2 response. Thus, the presence of viable bacteria and/or metabolites could modulate the immune response in the gut.

de Moreno de LeBlanc et al. (2006) investigated the immunomodulatory activity of kefir in a murine hormone-dependent breast cancer model. They reported that consumption of kefir increased IL-10 in serum and decreased IL-6(+) cells in the mammary gland. IL-6 has been implicated in estrogen synthesis. Furthermore, feeding mice kefir reduced the growth of tumors. These data, taken collectively, indicate that compounds generated by milk fermentation modulate immune function and possess antitumor properties. The same group reported similar results in an earlier publication using milk fermented with the microorganism *Lactobacillus helveticus* (de Moreno de LeBlanc et al., 2005).

Rachid et al. (2006) investigated the beneficial effects of consumption of milk fermented with *Lb. helveticus* on a murine model. They reported that consumption of fermented milk delayed the development of tumors in the mammary gland. The effect was mediated by increased apoptosis and decreased production of the pro-inflammatory cytokine IL-6.

Olivares et al. (2006) investigated the beneficial effects from the consumption of fermented milk with various *Lactobacillus* strains on the immune function. They found that consumption of fermented milk resulted in an increase in the number of phagocytic cells as well as their phagocytic activity. Furthermore, they observed an increase in the proportion of natural killer cells and in the IgA concentrations indicative of enhanced immunity.

LeBlanc et al. (2002) investigated the effect of peptides released during fermentation of milk with the microorganism *Lb. helveticus* on humoral immunity. Three fractions of peptides were generated by size-exclusion HPLC and were then fed to mice. All three peptides were capable of increasing the number of IgA(+) cells in the intestine and reduced the growth of fibrosacroma. Thus, bioactive peptides released in milk following fermentation possess immunoenhancing and antitumor properties.

Matar et al. (2001) investigated the effect of peptides released during fermentation of milk with *Lb. helveticus* on humoral immunity. They reported that feeding mice fermented milk increased the number of IgA(+) cells in the small intestine and the bronchial tissues. The increase in the cells both in the intestine and in the bronchial tissues indicated activation of the IgA cycle.

The notion that peptides were indeed responsible for this effect was strengthened by the finding that a protease-deficient derivative of *Lb. helveticus* was ineffective.

Milk Peptides: Suppression of Immune Function

Prioult et al. (2004) investigated whether the microorganism *Lactobacillus paracasei* was capable of suppressing immune function by generating peptides from the hydrolysis of β-lactoglobulin (β-lg). They reported that peptidases generated by *Lb. paracasei* were capable of further hydrolyzing peptides generated initially by hydrolysis of β-lg by trypsin-chymotrypsin. Furthermore, these peptides suppressed lymphocyte proliferation and increased IL-10 production, which acts as a major immunosuppressant. They concluded that *Lb. paracasei* was capable of inducing oral tolerance to β-lg by producing peptidases that can hydrolyze β-lg.

Pecquet et al. (2000) researched the effect of peptides generated by tryptic hydrolysis of β-lg on various immune functions in mice. They reported that mice fed β-lg hydrolysates or fractions of the hydrolysate developed a tolerance to β-lg. Specific serum and intestinal IgE levels were reduced. Furthermore, delayed-type hypersensitivity and proliferative responses were inhibited.

Pessi et al. (2001) studied whether the microorganism *Lb. rhamnosus* GG was capable of suppressing immune function by generating peptides from the hydrolysis of casein. They reported that digests of casein by peptidases produced by *Lb. rhamnosus* inhibited protein kinase C translocation and downregulated IL-2 expression. Taken together, these results indicate suppression of T cell activation by casein digests.

Milk Peptides and Immune Function: The Evidence from the Weaning Piglet Model

A number of experiments have been performed in our laboratory at the Agricultural University of Athens looking at the effect of milk peptides on immune function using the weaning piglet as a model. All piglets are born with an immature immune system; for this reason, the pig's disease resistance is very limited for the first three to four weeks after birth. The immune system is further compromised when piglets are subjected to the social, environmental, and nutritional stresses at weaning. The antigenic composition of the intestinal contents at weaning changes dramatically as a result of the changing diet and the occurrence of various strains and bacteria species. It is apparent that changes in human lifestyle and in the husbandry of animals have resulted in weaning occurring earlier and becoming much more abrupt than previously in evolution, thus increasing the number of antigens that neonates must simultaneously evaluate (Bailey et al., 2005).

Several studies reported reduced mononuclear cell proliferation, phagocyte activation, and depressed potential for the production of IL-2 at weaning and during the early postweaning period (Bailey et al., 1992; Wattrang et al., 1998; Fragou et al., 2004). We have used the weaning piglet as a model because we can obtain cells from the immature period (first two weeks after weaning) and the mature period (one month later). This allows for a comparison of the effectiveness of milk peptides with cells obtained during these two distinct time periods.

Peptide Fractionation: Testing of Immunomodulating Ability in Vitro

Bovine milk samples were obtained and subjected to *in vitro* digestion as described by Kapsokefalou et al. (2005). This *in vitro* model simulates the gastrointestinal digestion by subjecting milk samples to incubation for 4.5 hours at 37 °C, at different pH values, in the presence of peptic enzymes. The fraction containing the low-molecular-weight soluble compounds was collected and subjected to liquid chromatography through Sephadex G-50 or Sephadex G-25 columns. Typical elution profiles are presented in Fig. 1. Successful peptide separation was obtained with the Sephadex G-25 column, as shown by the appearance of three distinct peaks (A, B, C in Figure 1b). The three fractions corresponding to peaks A, B, and C were further separated using size-exclusion HPLC with the method described by LeBlanc et al. (2002). Each of the three peaks generated three new peaks: peak I (high-MW peptides), peak II (medium-MW peptides), and peak III (low-MW peptides) (Fig. 2).

The immunomodulating activities of the nine fractions (peaks A, B, C × peaks I, II, III) were tested *in vitro* on phagocytes obtained from piglets at two distinct periods: one to two weeks after weaning (immature period) and five to six weeks after weaning (mature period). Two assay systems were utilized: membrane-bound urokinase plasminogen activator (u-PA) and SA production in activated blood monocyte-macrophages and neutrophils. Chronopoulou et al. (2006) described both methodologies. The u-PA present on the cell membrane of phagocytes is an important enzyme for neutrophil diapedesis (Fragou et al., 2004). The u-PA system and SA production were selected as the outcome measures because both systems are altered as the immune system of the weaning piglet moves from immaturity to maturity.

The effect of the peptidic fractions on membrane-bound u-PA activity and SA production of monocyte-macrophages and neutrophils isolated from piglets one to two weeks after weaning are presented in Tables 1 and 2. Of the nine peptidic fractions tested, only the low-MW fraction (fraction III) obtained from further separation of the fraction corresponding to peak C of the liquid chromatography through the Sephadex G-25 column was effective. More specifically, this fraction

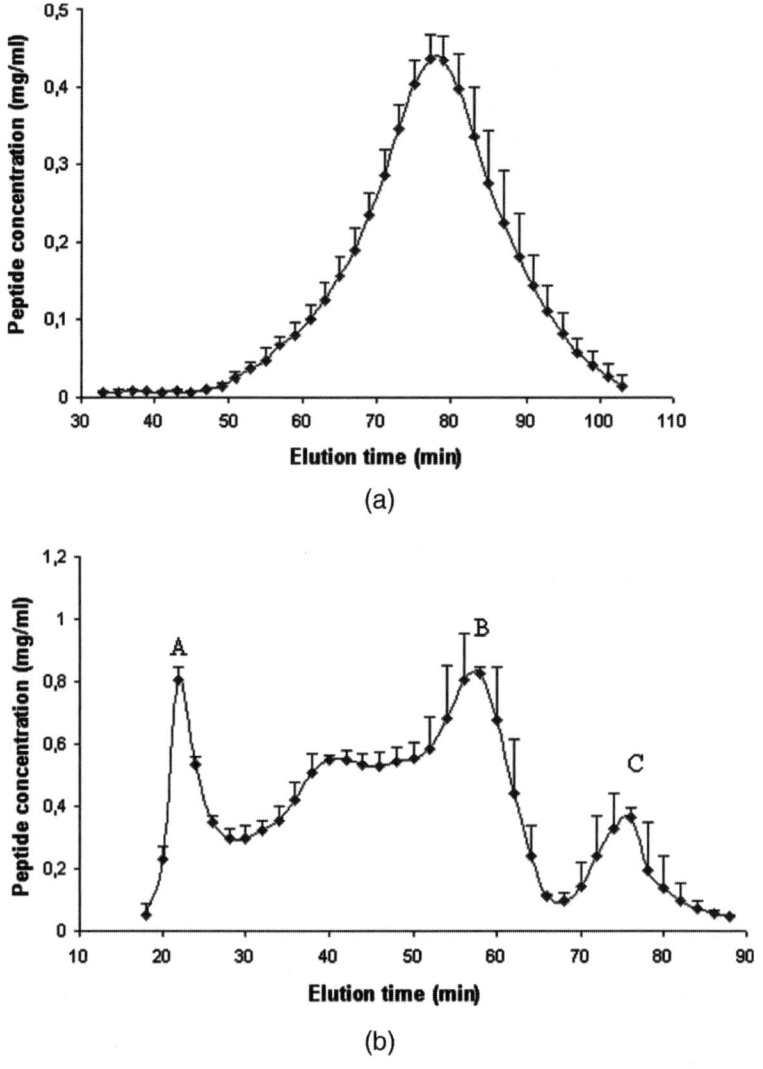

Fig. 1 Peptide fractionation profile obtained with liquid chromatography using (a) Sephadex G-50 and (b) Sephadex G-25. Fractionation was achieved following digestion of milk samples using an *in vitro* model simulating the gastrointestinal digestion process in the presence of peptic enzymes

decreased both membrane-bound u-PA activity and SA production of monocyte-macrophages and neutrophils isolated one to two weeks after weaning. The effectiveness of the peptide was dose-related. None of the peptides affected membrane-bound u-PA activity and SA production of monocyte-macrophages and neutrophils isolated five to six weeks after weaning (data not shown).

Fig. 2 Peptide fractionation profile obtained with size-exclusion HPLC of fractions corresponding to peaks A, B, and C obtained with liquid chromatography using Sephadex G-25

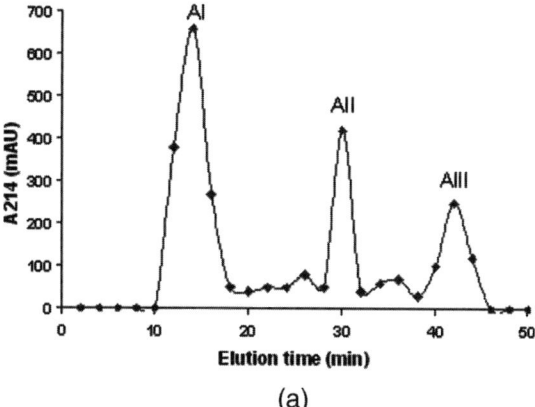

(a)

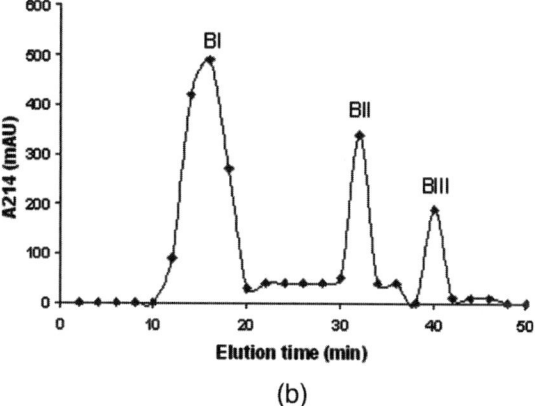

(b)

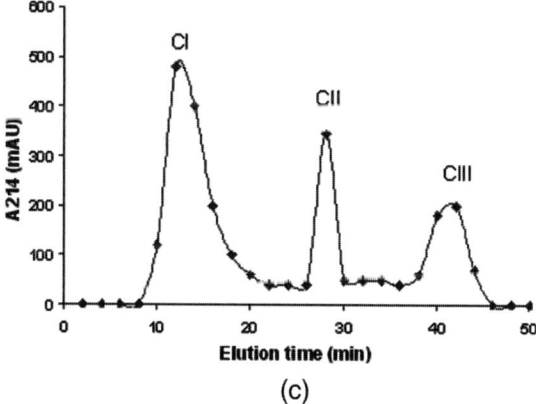

(c)

Table 1 Effect of the Peptidic Fractions Obtained with Size-Exclusion HPLC* on Membrane-Bound u-PA Activity of Porcine Monocyte-Macrophages and Neutrophils Isolated from Weaned Piglets Obtained During the First Two Weeks After Weaning

Treatment	Membrane-Bound u-PA Activity ($\Delta A/h$)			
	Macrophages		Neutrophils	
	Mean	SD	Mean	SD
Control	0.158[a]	0.055	0.254[a]	0.111
Fraction AI (1 mg/mL)	0.166[a]	0.048	0.260[a]	0.116
Fraction AI (10 mg/mL)	0.180[a]	0.053	0.266[a]	0.100
Fraction AII (1 mg/mL)	0.158[a]	0.027	0.272[a]	0.105
FractionA II (10 mg/mL)	0.166[a]	0.057	0.246[a]	0.090
Fraction AIII (1 mg/mL)	0.154[a]	0.055	0.250[a]	0.113
Fraction AIII (10 mg/mL)	0.170[a]	0.071	0.246[a]	0.102
Control	0.146[a]	0.050	0.258[a]	0.098
Fraction BI (1 mg/mL)	0.140[a]	0.038	0.236[a]	0.085
Fraction BI (10 mg/mL)	0.152[a]	0.061	0.270[a]	0.112
Fraction BII (1 mg/mL)	0.142[a]	0.039	0.246[a]	0.082
Fraction BII (10 mg/mL)	0.178[a]	0.052	0.248[a]	0.113
Fraction BIII (1 mg/mL)	0.144[a]	0.044	0.260[a]	0.085
Fraction BIII (10 mg/mL)	0.144[a]	0.046	0.248[a]	0.095
Control	0.144[a]	0.055	0.254[a]	0.112
Fraction I (1 mg/mL)	0.150[a]	0.054	0.256[a]	0.111
Fraction I (10 mg/mL)	0.158[a]	0.058	0.266[a]	0.100
Fraction II (1 mg/mL)	0.148[a]	0.060	0.272[a]	0.110
Fraction II (10 mg/mL)	0.150[a]	0.067	0.256[a]	0.116
Fraction III (1 mg/mL)	0.092[a,b]	0.035	0.188[a,b]	0.074
Fraction III (10 mg/mL)	0.062[b]	0.025	0.138[b]	0.060

[a, b] Mean values within a column with unlike superscript letters are significantly different ($P < 0.01$) according to LSD multiple-range test.
* The three peptidic fractions I, II, and III were obtained with size-exclusion HPLC of the fractions corresponding to peaks A, B, and C obtained with liquid chromatography using Sephadex G-25.

It is apparent that the low-MW peptidic fraction is effective, causing a downregulation of both parameters only in the very early postweaning period at a time that coincides with immaturity of the immune system; it is not effective one month later, when the immune system has presumably gained its full functionality.

Milk Peptides: The Immunomodulating Activity in Vivo

The effect of milk peptides on the function of phagocytes was tested *in vivo* in the early postweaning period in piglets. For this purpose, 27 piglets were

Table 2 Effect of the Peptidic Fractions Obtained with Size-Exclusion HPLC* on Superoxide Anion (SA) Production by Porcine Monocyte-Macrophages and Neutrophils Isolated from Weaned Piglets Obtained During the First Two Weeks After Weaning

Treatment	SA Production (nmol/10^6 Cells)			
	Macrophages		Neutrophils	
	Mean	SD	Mean	SD
Control	1.520^a	0.420	2.420^a	0.683
Fraction AI (1 mg/mL)	1.620^a	0.356	2.340^a	0.665
Fraction AI (10 mg/mL)	1.560^a	0.304	2.680^a	0.947
Fraction AII (1 mg/mL)	1.780^a	0.715	2.360^a	0.450
Fraction AII (10 mg/mL)	1.620^a	0.268	2.300^a	0.703
Fraction AIII (1 mg/mL)	1.660^a	0.602	2.500^a	0.681
Fraction AIII (10 mg/mL)	1.520^a	0.319	2.580^a	0.622
Control	1.640^a	0.550	2.700^a	0.748
Fraction BI (1 mg/mL)	1.780^a	0.746	2.720^a	0.589
Fraction BI (10 mg/mL)	1.660^a	0.439	2.500^a	0.930
Fraction BII (1 mg/mL)	1.620^a	0.238	2.260^a	0.709
Fraction BII (10 mg/mL)	1.720^a	0.614	3.020^a	0.906
Fraction BIII (1 mg/mL)	1.680^a	0.661	2.580^a	0.620
Fraction BIII (10 mg/mL)	1.660^a	0.439	2.880^a	0.939
Control	1.560^a	0.577	$2.540^{a,b}$	0.750
Fraction CI (1 mg/mL)	1.560^a	0.658	$2.560^{a,b}$	0.541
Fraction CI (10 mg/mL)	1.800^a	0.557	2.960^a	0.950
Fraction CII (1 mg/mL)	1.560^a	0.503	$2.560^{a,b}$	0.756
Fraction CII (10 mg/mL)	1.600^a	0.787	$2.660^{a,b}$	0.802
Fraction CIII (1 mg/mL)	$0.960^{a,b}$	0.385	$1.700^{a,b}$	0.777
Fraction CIII (10 mg/mL)	0.536^b	0.308	1.320^b	0.567

*The three peptidic fractions I, II, and III were obtained with size-exclusion HPLC of the fractions corresponding to peaks A, B, and C obtained with liquid chromatography using Sephadex G-25.
a, b Mean values within a column with unlike superscript letters are significantly different ($P < 0.01$) according to LSD multiple-range test.

allowed a three-day adaptation period after weaning and were then assigned to one of three experimental groups: control (no peptide supplementation), low level of peptide supplementation (300 mg/day), and high level of peptide supplementation (600 mg/day). Supplementation lasted for three weeks. The peptide fed to piglets was the peptidic fraction corresponding to the unique peak obtained with liquid chromatography using the Sephadex G-50 column (Fig. 1a). This crude peptidic fraction was highly effective (data not shown) and provided sufficient peptide concentration for the *in vivo* experiments. Piglets were fed a standard growing diet (Fragou et al., 2004). Blood samples were collected at days 7, 14, and 21 of the experimental period. Blood monocyte-macrophages and neutrophils were isolated, and membrane-bound u-PA activity as well as SA production were measured in resting and activated cells (Fragou et al., 2004; Chronopoulou et al., 2006).

Peptide supplementation had no effect on the u-PA system (Table 3) and superoxide anion production (Table 4) of activated macrophages and neutrophils at all sampling points. In contrast, peptide supplementation increased membrane-bound u-PA activity (Table 5) and SA production (Table 6) of resting macrophages and neutrophils in nearly all sampling points. However, the extent of the increase is very limited, and it is highly unlikely that it has any real biological meaning.

There is an apparent contradiction between the *in vitro* and *in vivo* evidence presented here. A downregulation of the immune parameters was the key observation in the *in vitro* experiments for the early postweaning period while the lack of an effect was the key observation in the *in vivo* experiments during the same time period. There are two logical explanations for this discrepancy. First, we were unable to achieve the necessary concentration of peptides *in vivo*. Second, we considered it possible that peptides were degraded or unable to reach the general circulation. In contrast to this notion, the peptides were apparently effective toward resting cells despite the fact that the effect was considered of limited biological value. Our future studies will focus on local effects of the peptides and mainly on those related to gastrointestinal immunity.

Immunomodulating Effects of Specific β-Casein Peptides

A number of experiments were performed at the Agricultural University of Athens to look at the effects of specific, chemically synthesized peptides on immune function at the postweaning period in piglets. The two peptides synthesized were the tripeptide leucine-leucine-tyrosine (LLY) corresponding to residues 191–193 of bovine β-casein and the hexapeptide proline-glycine-proline-isoleucine-proline-asparagine (PGPIPN) corresponding to residues 63–68 of the bovine β-casein. Two specific areas of β-casein (residues 60–70 and 191–202) are considered critical sites because many biologically active peptides are generated (Clare & Swaiswood, 2000). The immunomodulating activities of the two peptides were tested *in vitro* against phagocytes obtained from piglets at two distinct periods: one to two weeks after weaning (immature period) and five to six weeks after weaning (mature period). Two assay systems were utilized: membrane-bound u-PA and SA production in activated blood monocyte-macrophages and neutrophils. All details are as described by Chronopoulou et al. (2006). Data indicated that both peptides suppressed the u-PA system and SA production by activated macrophages isolated from piglets one to two weeks after weaning. Only the tripeptide LLY suppressed the u-PA system and SA production by activated neutrophils during the same time period (Chronopoulou et al., 2006). None of the

Table 3 Effect of Feeding Weaned Piglets with Milk Peptides* on Membrane-Bound u-PA Activity of Porcine Monocyte-Macrophages and Neutrophils Activated by Phorbol Myristate Acetate (PMA, 80 μM)

Treatment	u-PA Activity (ΔA/h)											
	Days After Weaning											
	7				14				21			
	Monocytes-Macrophages		Neutrophils		Monocytes-Macrophages		Neutrophils		Monocytes-Macrophages		Neutrophils	
	Mean	SD	Mean	SD	Mean	SD	Mean	SD	Mean	SD	Mean	SD
Peptide Concentration (mg/day)												
0	0.151	0.057	0.270	0.069	0.181	0.068	0.302	0.077	0.299	0.233	0.355	0.084
300	0.150	0.030	0.272	0.080	0.183	0.058	0.304	0.102	0.210	0.057	0.357	0.116
600	0.149	0.060	0.266	0.080	0.175	0.067	0.297	0.084	0.209	0.085	0.357	0.104

*Weaned piglets were fed 300 or 600 mg/day for three weeks. Supplementation started immediately after weaning. The peptidic fraction fed to weaned piglets corresponds to fractions constituting the unique peak obtained with liquid chromatography using Sephadex G-50 (see Fig. 1a).

Table 4 Effect of Feeding Weaned Piglets with Milk Peptides* on Membrane-Bound u-PA Activity of Resting Porcine Monocyte-Macrophages and Neutrophils

	u-PA Activity ($\Delta A/h$)											
	Days After Weaning											
	7				14				21			
	Monocytes-Macrophages		Neutrophils		Monocytes-Macrophages		Neutrophils		Monocytes-Macrophages		Neutrophils	
Treatment	Mean	SD	Mean	SD	Mean	SD	Mean	SD	Mean	SD	Mean	SD
Peptide Concentration (mg/day)												
0	0.011^a	0.003	0.023	0.005	0.013^a	0.003	0.022^a	0.005	0.012^a	0.005	0.021	0.005
300	0.010^a	0.005	0.021	0.010	0.014^a	0.008	0.022^a	0.010	0.012^a	0.006	0.040	0.022
600	0.020^b	0.005	0.031	0.005	0.025^b	0.007	0.037^b	0.018	0.026^b	0.009	0.039	0.013

* Weaned piglets were fed 300 or 600 mg/day for three weeks. Supplementation started immediately after weaning. The peptidic fraction fed to weaned piglets corresponds to fractions constituting the unique peak obtained with liquid chromatography using Sephadex G-50 (see Fig.1a).
a, b Mean values within a column with unlike superscript letters are significantly different ($P < 0.01$) according to LSD multiple-range test.

Table 5 Effect of Feeding Weaned Piglets with Milk Peptides* on Superoxide Anion (SA) Production by Porcine Monocyte-Macrophages and Neutrophils Activated by Phorbol Myristate Acetate (PMA, 80 μM)

	SA Production (nmol/10^6 Cells)											
	Days After Weaning											
	7				14				21			
	Monocytes-Macrophages		Neutrophils		Monocytes-Macrophages		Neutrophils		Monocytes-Macrophages		Neutrophils	
Treatment	Mean	SD	Mean	SD	Mean	SD	Mean	SD	Mean	SD	Mean	SD
Peptide Concentration (mg/day)												
0	1.336	0.552	2.133	0.479	1.378	0.465	2.215	0.341	1.644	0.515	2.506	0.387
300	1.315	0.538	2.051	0.911	1.463	0.271	2.155	0.805	1.593	0.228	2.446	0.763
600	1.334	0.482	2.051	0.656	1.386	0.446	2.180	0.563	1.496	0.376	2.480	0.619

*Weaned piglets were fed 300 or 600 mg/day for three weeks. Supplementation started immediately after weaning. The peptidic fraction fed to weaned piglets corresponds to fractions constituting the unique peak obtained with liquid chromatography using Sephadex G-50 (see Fig. 1a).

Table 6 Effect of Feeding Weaned Piglets with Milk Peptides* on Superoxide Anion (SA) Production by Resting Porcine Monocyte-Macrophages and Neutrophils

Treatment	SA Production (nmol/10^6 Cells)											
	Days After Weaning											
	7				14				21			
	Monocytes-Macrophages		Neutrophils		Monocytes-Macrophages		Neutrophils		Monocytes-Macrophages		Neutrophils	
Peptide Concentration (mg/day)	Mean	SD	Mean	SD	Mean	SD	Mean	SD	Mean	SD	Mean	SD
0	0^a	0	0^a	0	0^a	0	0^a	0	0^a	0	0^a	0
300	0.001^a	0.002	0.001^a	0.002	0^a	0	0^a	0	0.002^a	0.003	0.005^a	0.009
600	0.017^b	0.017	0.028^b	0.020	0.014^b	0.013	0.018^b	0.016	0.016^b	0.011	0.020^b	0.015

*Weaned piglets were fed 300 or 600 mg/day for three weeks. Supplementation started immediately after weaning. The peptidic fraction fed to weaned piglets corresponds to fractions constituting the unique peak obtained with liquid chromatography using Sephadex G-50 (see Fig. 1a).
a, b Mean values within a column with unlike superscript letters are significantly different ($P < 0.01$) according to LSD multiple-range test.

peptides tested was effective against phagocytes isolated from the same piglets one month later. It is clear that the two chemically synthesized peptides behave in a manner analogous to peptidic fractions obtained through Sephadex G-25 and G-50 liquid chromatography followed by size-exclusion HPLC. It is also apparent that the peptides' effectiveness depends on the cell type, the time since weaning, and/or the state of differentiation of these cells. The key observation, however, remains the same: Peptides are effective against cells obtained at a time that coincides with the immune system's immaturity.

Milk Peptides and Immune System: Mechanism of Action

The cAMP signaling pathway is part of an important mechanism that has been implicated in regulating immune function. Several reports are available suggesting that cAMP has mainly inhibitory effects on various functions of alveolar macrophages. cAMP has been implicated in the inhibition of phagocytosis (Aronoff et al., 2004), the production of reactive oxygen species (Dent et al., 1994), and the production of various inflammatory mediators (Rowe et al., 1997). cAMP acts as a second messenger capable of activating PKA. However, a number of PKA-independent targets for cAMP have been described such as cyclic nucleotide gated channels and the guanine exchange proteins directly activated by cAMP (Epac-1, Epac-2).

A number of experiments were performed at the Agricultural University of Athens to investigate whether milk protein peptides (LLY and PGPIPN) suppress the u-PA system and SA production by porcine phagocytes through PKA and/or Epac-1. Using cAMP analogs that are highly specific activators of the PKA or Epac-1, we found that activation of PKA, but not Epac-1, was responsible for the downregulation of the u-PA system, whereas activation of the PKA and/or Epac-1 was responsible for the downregulation of SA production in both porcine macrophages and neutrophils during the early postweaning period (Chronopoulou et al., 2006).

Conclusions

Milk protein represents the exclusive protein supply for the newborn during the first few months of life. It remains the main protein supply throughout the transition of the immune system from immaturity to maturity. Using the weaning piglet model, we have demonstrated that milk protein peptides are capable of exercising mostly inhibitory effects in the early postweaning period. The most reasonable interpretation of these findings is that milk peptides may guide the immune system until it develops its full functionality. The actions of the peptides are mediated through activation of PKA and/or Epac-1.

Acknowledgment This work was supported by the European Social Fund (75%) and national funds (Ministry of Development—General Secretariat of Research and Technology, 25%).

References

Aronoff, D. M., Canetti, C., & Peters-Golden, M. (2004). Prostaglandin E2 inhibits alveolar macrophage phagocytosis through an E-prostanoid 2 receptor-mediated increase in intracellular cyclic AMP. *Journal of Immunology, 173,* 559–563.

Bailey, M., Clarke, C. J., Wilson, A. D., Williams, N. A., & Stokes, C. R. (1992). Depressed potential for interleukin-2 production following early weaning of piglets. *Veterinary Immunology and Immunopathology, 34,* 197–207.

Bailey, M., Haverson, K., Inman, C., Harris, C., Jones, P., Corfield, G., Miller, B., & Stokes, C. (2005). The development of the mucosal immune system pre- and post-weaning: Balancing regulatory and effector function. *Proceedings of Nutritional Society, 64,* 451–457.

Baldi, A., Ioannis, P., Chiara, P., Eleonora, F., Roubini, C., & Vittorio, D. (2005). Biological effects of milk proteins and their peptides with emphasis on those related to the gastrointestinal ecosystem. *Journal of Dairy Research, 72,* 66–72.

Bryan, D. L., Forsyth, K. D., Gibson, R. A., & Hawkes, J. S. (2006). Interleukin-2 in human milk: A potential modulator of lymphocyte development in the breastfed infant. *Cytokine, 33,* 289–293.

Chronopoulou, R., Xylouri, E., Fegeros, K., & Politis, I. (2006). The effect of two bovine β-casein peptides on various functional properties of porcine macrophages and neutrophils: Differential roles of protein kinase A and exchange protein directly activated by cyclic AMP-1. *British Journal of Nutrition, 96,* 553–561.

Clare, D. A., & Swaisgood, H. E. (2000). Bioactive milk peptides: A prospectus. *Journal of Dairy Science, 83,* 1187–1195.

de Moreno de LeBlanc, A., Matar, C., Theriault, C., & Perdigon, G. (2005). Effects of milk fermented by *Lactobacillus helveticus* R389 on immune cells associated to mammary glands in normal and a breast cancer model. *Immunobiology, 210,* 349–358.

de Moreno de LeBlanc, A., Matar, C., Farnworth, E., & Perdigon, G. (2006). Study of cytokines involved in the prevention of a murine experimental breast cancer by kefir. *Cytokine, 34,* 1–8.

Dent, G., Giembycz, M. A., Rabe, K. F., Wolf, B., Barnes, P. J., & Magnussen, H. (1994). Theophylline suppresses human alveolar macrophage respiratory burst through phosphodiesterase inhibition. *American Journal of Respiratory Cell and Molecular Biology, 10,* 565–572.

Fragou, S., Fegeros, K., Xylouri, E., Baldi, A., & Politis, I. (2004). Effect of vitamin E supplementation on various functional properties of macrophages and neutrophils obtained from weaned piglets. *Journal of Veterinary Medicine. A, Physiology, Pathology, Clinical Medicine, 51,* 1–6.

Gill, H. S., Doull, F., Rutherfurd, K. J., & Cross, M. L. (2000). Immunoregulatory peptides in bovine milk. *British Journal of Nutrition, 84,* S111–S117.

Kapsokefalou, M., Alexandropoulou, I., Komaitis, M., & Politis, I. (2005). *In vitro* evaluation of iron solubility and dialyzability of various iron fortificants and of iron-fortified milk products targeted for infants and toddlers. *International Journal of Food Sciences and Nutrition, 56,* 293–302.

Kelleher, S. L., & Lonnerdal, B. (2001). Immunological activities associated with milk. *Advances in Nutritional Research, 10,* 39–65.

LeBlanc, J. G., Matar, C., Valdez, J. C., LeBlanc, J., & Perdigon, G. (2002). Immunomodulating effects of peptidic fractions issued from milk fermented with *Lactobacillus helveticus*. *Journal of Dairy Science, 85,* 2733–2742.

Matar, C., Valdez, J. C., Medina, M., Rachid, M., & Perdigon, G. (2001). Immunomodulating effects of milks fermented by *Lactobacillus helveticus* and its non-proteolytic variant. *Journal of Dairy Research, 68,* 601–609.

Olivares, M., Diaz-Ropero, M. P., Gomez, N., Lara-Villoslada, F., Sierra, S., Maldonado, J. A., Martin, R., Rodriguez, J. M., & Xaus, J. (2006). The consumption of two new probiotic strains, *Lactobacillus gasseri* CECT 5714 and *Lactobacillus coryniformis* CECT 5711, boosts the immune system of healthy humans. *International Microbiology, 9,* 47–52.

Pecquet, S., Bovetto, L., Maynard, F., & Fritsche, R. (2001). Peptides obtained by tryptic hydrolysis of bovine β-lactoglobulin induce specific oral tolerance in mice. *Journal of Allergy and Clinical Immunology, 105,* 514–521.

Pessi, T., Isolauri, E., Sutas, Y., Kankaanranta, H., Moilanen, E., & Hurme, M. (2001). Suppression of T-cell activation by *Lactobacillus rhamnosus* GG-degraded bovine casein. *International Immunopharmacology, 1,* 211–218.

Prioult, G., Pecquet, S., & Fliss, I. (2004). Stimulation of interleukin-10 production by acidic β-lactoglobulin-derived peptides hydrolyzed with *Lactobacillus paracasei* NCC2461 peptidases. *Clinical and Diagnostic Laboratory Immunology, 11,* 266–271.

Rachid, M., Matar, C., Duarte, J., & Perdigon, G. (2006). Effect of milk fermented with a *Lactobacillus helveticus* R389(+) proteolytic strain on the immune system and on the growth of 4T1 breast cancer cells in mice. *FEMS Immunology and Medical Microbiology, 47,* 242–253.

Rowe, J., Finlay-Jones, J. J., Nicholas, T. E., Bowden, J., Morton, S., & Hart, P. H. (1997). Inability of histamine to regulate TNF-α production by human alveolar macrophages. *American Journal of Respiratory Cell and Molecular Biology, 17,* 218–223.

Vinderola, C. G., Duarte, J., Thangavel, D., Perdigon, G., Farnworth, E., & Matar, C. (2005). Immunomodulating capacity of kefir. *Journal of Dairy Research, 72,* 195–202.

Vinderola, C. G., Perdigon, G., Duarte, J., Farnworth, E., & Matar, C. (2006). Effects of the oral administration of the products derived from milk fermentation by kefir microflora on immune stimulation. *Journal of Dairy Research, 73,* 472–479.

Wattrang, E., Wallgren, P., Lindberg, A., & Fossum, C. (1998). Signs of infections and reduced immune functions at weaning of conventionally reared and specific pathogen free pigs. *Zentralblatt für Veterinärmedizin. Reihe B, 45,* 7–17.

Protective Effect of Milk Peptides: Antibacterial and Antitumor Properties

Iván López-Expósito and Isidra Recio

Abstract There is no doubt that milk proteins provide excellent nutrition for the suckling. However, apart from that, milk proteins can also exert numerous physiological activities benefiting the suckling in a variety of ways. These activities include enhancement of immune function, defense against pathogenic bacteria, viruses, and yeasts, and development of the gut and its functions. Besides the naturally occurring, biologically active proteins present in milk, a variety of bioactive peptides are encrypted within the sequence of milk proteins that are released upon suitable hydrolysis of the precursor protein. A large range of bioactivities has been reported for milk protein components, with some showing more than one kind of biological activity (Korhonen & Pihlanto, 2006). This chapter reviews the most important antimicrobial and antitumor peptides derived from milk proteins, especially those that may have a physiological significance to the suckling neonate. Antimicrobial peptides present in milk that are not derived from milk proteins are also considered. Special attention is given to the generation of these peptides by the action of different proteolytic enzymes and the origin of these enzymes since, if present in the digestive tract, it is likely that the peptides might play a role in the host defense system. Finally, the most relevant *in vivo* studies carried out with this kind of bioactive peptides are discussed.

Antimicrobial Peptides

Milk Protein–Derived Peptides

The antibacterial properties of milk have been known for a long time. In fact, the incidence of diseases like diarrhea or respiratory infections is significantly lower in breastfed infants than in formula-fed infants; a variety of protective

---

I. Recio
Instituto de Fermentaciones Industriales (CSIC), Juan de la Cierva 3, 28006 Madrid, Spain
e-mail: recio@ifi.csic.es

factors in human milk are thought to be responsible for this effect. During the first few days postpartum, the specific activity of immunoglobulins is the dominant factor for immunity (van Hooijdonk et al., 2000). Nonspecific antimicrobial proteins are also important for the host defense system and probably act together with the specific antibodies. The most important of these are lysozyme, lactoperoxidase, and lactoferrin (Pakkanen & Aalto, 1997). In addition to these naturally occurring antimicrobial proteins present in milk, a variety of antibacterial peptides are encrypted within the sequence of milk proteins that are released upon suitable hydrolysis of the precursor protein that could also act as components of innate immunity.

Several peptides with antimicrobial activity have been found within the sequences of whey proteins and caseins (for recent reviews, see Floris et al., 2003; López-Expósito & Recio, 2006); nevertheless, only those peptides having a possible physiological meaning are treated in this chapter.

Whey Protein–Derived Antimicrobial Peptides

There is no doubt that peptides derived from lactoferrin are the antibacterial peptides from milk proteins that have attracted the most attention during the last decade (for a recent review, see Wakabayashi et al., 2003, 2006). In 1992, Bellamy et al. found an antimicrobial domain in the N-terminal region of the human and bovine lactoferrin molecule. These antibacterial domains corresponded to bovine lactoferrin f(17-41) and human lactoferrin f(1-47) and were named bovine and human lactoferricin, respectively. These active peptides were released by pepsin digestion and revealed higher antimicrobial effectiveness than their precursor protein. This observation gave rise to a new mechanism for the antibacterial action of lactoferrin, independent of its iron binding properties (Yamauchi et al., 1993). Lactoferricin has revealed a broad spectrum of activity against Gram-positive and -negative bacteria (Bellamy et al., 1992), fungi (Muñoz & Marcos, 2006), and parasites (León-Sicarios et al., 2006). Furthermore, lactoferricin has been shown to have antiviral (Pietrantoni et al., 2006), antitumor (Iigo et al., 1999), and anti-inflammatory properties (Levay & Viljoen, 1995).

When hydrolyzing bovine lactoferrin with pepsin, in addition to lactoferricin and longer peptides containing lactoferricin, other cationic peptides corresponding to a region of the N-lobe and in spatial proximity to lactoferricin are released. These peptides correspond to lactoferrin f(277-288), lactoferrin f(267-285), and lactoferrin f(267-288) (Recio & Visser, 1999b). A peptide corresponding to this latter domain of lactoferrin, f(268-284), and designated as lactoferrampin has been chemically synthesized and has demonstrated candidacidal activity and antibacterial activity against *Bacillus subtilis*, *Escherichia coli*, and *Pseudomonas aeruginosa* (van der Kraan et al., 2004). Apart from porcine pepsin, other proteolytic enzymes, such as chymosin, can produce analogous fragments to lactoferricin (Hoek et al., 1997).

Lactoferrin from species other than bovines and humans has also been used as precursors of antibacterial peptides. Fragments obtained by the chemical synthesis of residues 17–41 of murine and caprine lactoferrin have demonstrated antibacterial activity, although to a lesser extent than bovine lactoferricin (Vorland et al., 1998). Hydrolysis of caprine lactoferrin with pepsin resulted in antibacterial hydrolysates and a homologous peptide to lactoferricin, corresponding to fragment 14-42, was identified. Caprine lactoferricin showed lower antibacterial activity than bovine lactoferricin against *Escherichia coli* but comparable activity against *Micrococcus flavus*. On the contrary, the region corresponding to lactoferricin within the sequence of ovine lactoferrin was hydrolyzed by the action of pepsin; hence, the activity observed in the ovine lactoferrin hydrolysate could be caused by other lactoferrin fragments (Recio & Visser, 2000). Recently, a 20-residue fragment corresponding to porcine lactoferricin was synthesized [porcine lactoferricin f(17-41)]. The porcine lactoferricin displayed antimicrobial activity against *Escherichia coli, Staphylococcus aureus,* and *Candida albicans*. Porcine lactoferricin was four times more effective than human lactoferricin, but slightly less effective than bovine lactoferricin (Chen et al., 2006b).

The structure of lactoferricin and other lactoferrin-derived peptides, such as lactoferrampin, shares structural features with other well-known antimicrobial peptides, i.e., the presence of positively charged residues and a hydrophobic region that contains tryptophan, a residue involved in membrane insertion (Schiffer et al., 1992). The generally accepted antibacterial mechanism proposed for most cationic peptides of different origins is through disruption of the cytoplasmic membrane (Zasloff, 2002), although several antibacterial peptides have also been shown to have additional intracellular targets. The mode of action of lactoferricin and analogs has been exhaustively studied; a detailed description of lactoferricin's antibacterial activity and other biological activities related to the protection of the host can be found elsewhere (Wakabayashi et al., 2003). The antibacterial activity of lactoferricin starts by electrostatic interaction with the negatively charged membranes of bacteria (Bellamy et al., 1993). In this initial binding, lipopolysaccharide and teichoic acid have been identified as binding sites in Gram-negative and -positive bacteria, respectively (Vorland et al., 1999). However, lactoferricin does not lyse susceptible bacteria but is able to translocate across the cytoplasmic membrane of both Gram-positive and -negative bacteria (Haukland et al., 2001). Ulvatne et al. (2004) demonstrated that once the peptide reaches the cytoplasm, the bacterial protein synthesis is inhibited, although the exact mechanism by which this biosynthesis of macromolecules is inhibited is not known.

Many studies aimed at the research of the structure-activity relationships of lactoferricin have been undertaken during the last decade (for reviews, see Strøm et al., 2002; Vogel et al., 2002). Microcalorimetry and fluorescence spectroscopy data regarding the interaction of the peptide with model membranes show that binding to net negatively charged bacterial and cancer cell membranes is preferred over neutral eukaryotic membranes. In addition, it has been suggested that while the antimicrobial, antifungal, antitumor, and antiviral properties of lactoferricin can be related to the tryptophan-/arginine-rich proportion of the peptide, the anti-inflammatory and immunomodulating properties are more related to a positively charged region of the molecule (Vogel et al., 2002).

Lactoferricin might have a role from a physiological point of view. Orally administered lactoferrin is partially degraded to fragments that contain the lactoferricin sequence (Kuwata et al., 1998a). Furthermore, Kuwata et al. (1998b) demonstrated that these fragments would survive transit through the gastrointestinal tract, although they could not have been detected in portal blood (Wakabayashi et al., 2004). Interestingly, it was shown recently that bovine lactoferrin and its derived peptide, lactoferricin, acted synergistically against *E. coli* and *St. epidermidis*. This synergistic effect could increase hosts' defenses against invading microorganisms (López-Expósito et al., 2007b).

Together with lactoferrin, lysozyme is one of the most extensively studied antibacterial milk proteins. Because lysozyme is present in large amounts in chicken-egg white (1–3 g/L), most studies are performed using lysozyme of this source. A review of the most important experiments carried out with lysozyme can be found elsewhere (Ibrahim, 2003; Pellegrini, 2003; Masschalck & Michiels, 2003). Pellegrini and co-workers (1997) synthesized the helix-loop-helix domain that contains the peptide 98-112, an antibacterial peptide obtained after hydrolyzing lysozyme. The helix-loop-helix domain corresponding to human lysozyme was the fragment 87-115, which exhibited microbicidal activity against Gram-positive and -negative bacteria and the fungus *Candida albicans* (Ibrahim et al., 2001).

Tryptic or chymotryptic digestion of bovine α-lactalbumin and β-lactoglobulin yielded several polypeptide fragments with a moderate antibacterial activity against Gram-positive bacteria (Pellegrini et al., 1999, 2001). Later on, the antimicrobial activity of ovine whey proteins and of their peptic hydrolysates was measured against different pathogenic microbial strains. The peptic hydrolysates inhibited the growth of *Escherichia coli* HB101, *E. coli* Cip812, *Bacillus subtilis* Cip5265, and *Staphylococcus aureus*, but no peptide identification was carried out (El-Zahar et al., 2004). Recently, the *in vitro* digestion of caprine whey proteins was investigated by a two-step degradation assay, using human gastric juice at pH 2.5 and human duodenal juice at pH 7.5. The protein degradation and antibacterial activity obtained were compared with those obtained after treatment with commercial enzymes, by using pepsin and a mixture of trypsin and chymotrypsin. The two methods resulted in different caprine protein and peptide profiles. Active growing cells of *E. coli* were inhibited by the digestion products from caprine whey obtained after treatment with human gastric juice and human duodenal juice. Cells of *Bacillus cereus* were inhibited only by whey proteins obtained after reaction with human gastric juice, while the products after further degradation with human duodenal juice demonstrated no significant effect (Almaas et al., 2006).

Casein-Derived Antimicrobial Peptides

Caseins have traditionally been considered proteins with nutritional and calcium-modulation functions. However, in the last two decades, a number of bioactive peptides encrypted within the primary structure of caseins have been described (Hernández-Ledesma et al., 2006). Among these, several peptides with antibacterial activity were found within the amino acid sequence of this group of milk proteins by employing different enzymatic strategies. Fragments, origin, and other biological activities of the most relevant antibacterial peptides are summarized in Table 1.

Isracidin was the first peptide with antimicrobial properties identified in the sequence of bovine α_{s1}-casein (Hill et al., 1974). It was obtained by chymosin digestion of bovine casein and corresponded to the N-terminal fragment α_{s1}-casein f(1-23). Isracidin was found to inhibit the growth of lactobacilli *in vitro* and of a variety of Gram-positive bacteria, but only at high concentrations. One characteristic of isracidin is that, *in vivo*, much smaller quantities are required to exert a protective effect prior to bacteria challenge than to inhibit bacterial growth *in vitro* (Lahov & Regelson, 1996).

Recently, McCann et al. (2006) isolated and identified a novel fragment from bovine α_{s1}-casein. This cationic peptide corresponded to residues 99–109 of bovine α_{s1}-casein. This peptide was obtained by hydrolysis with pepsin of bovine sodium caseinate. Fragment 99-109 of bovine α_{s1}-casein has shown a broad spectrum of activity against Gram-positive and -negative microorganisms. According to the amino acid sequences of sheep, goat, and water buffalo α_{s1}-casein, peptides corresponding to residues 99–109 have 10 of the total 11 amino acids that are identical to residues 99–109 of bovine α_{s1}-casein. These findings suggest that the peptides corresponding to residues 99–109 of sheep, goat, and water buffalo α_{s1}-casein might also have antibacterial properties. Because this peptide was derived from digestion of bovine casein with pepsin, it might be released in the stomach and contribute to protection against microbial infection in the gastrointestinal tract.

In the same year, two potent α_{s1}-casein antibacterial peptides were obtained by fermentation of a bovine sodium caseinate with *Lactobacillus acidophilus* DPC6026 (Hayes et al., 2006). The peptides denoted as caseicins A and B corresponded to f(21-29) and f(30-38) of bovine α_{s1}-casein. Both peptides revealed a potent activity against *Enterobacter sakazakii*, a strain that can be present in milk-based infant formulas (Nazarowec-White & Farber, 1997) and is responsible for a distinct syndrome of meningitis in neonates. Therefore, these peptides could be an important way of protection against this microorganism by producing a casein-based milk ingredient through fermentation, although *in vivo* studies are necessary.

Another peptide that has received particular attention is that corresponding to the f(183-207) of bovine α_{s2}-casein. This fragment was identified together with the f(164-179) of bovine α_{s2}-casein in a peptic hydrolysate of the same protein (Recio & Visser, 1999a). Both fragments showed an important

Table 1 Fragments, Origin, and Other Biological Activities of the Most Relevant Antibacterial Peptides[a,b]

Fragment[a,b]	Antibacterial Activity	Isolation	Others Activities	References
α_{s1}-casein f(1-23)	Gram-positive bacteria, fungi and yeast; in vitro and in vivo studies	Bovine casein digested with chymosin	Immunomodulatory	Lahov and Regelson (1996)
α_{s1}-casein f(99-109)	Several Gram-positive and -negative bacteria	Bovine sodium caseinate digested with pepsin	N.R.	McCann et al. (2006)
α_{s1}-casein f(21-29) α_{s1}-casein f(30-38)	Several Gram-positive and -negative bacteria	Bovine sodium caseinate fermented with Lactobacillus acidophilus DPC6026	N.R.	Hayes et al. (2006)
α_{s2}-casein f(164-169) α_{s2}-casein f(183-207)	Several Gram-positive and -negative bacteria	Bovine α_{s2}-casein digested with pepsin	Growth promoter	Recio and Visser (1999b) Smith and Wilkinson (1997)
°α_{s2}-casein f(165-170) °α_{s2}-casein f(165-181) °α_{s2}-casein f(184-208) °α_{s2}-casein f(203-208)	Several Gram-positive and -negative bacteria	Ovine α_{s2}-casein digested with pepsin	Antihypertensive Antioxidant	López-Expósito et al. (2007b) Recio et al. (2005)
[h]κ-casein f(43-97)	Several Gram-positive and -negative bacteria, yeasts	Human milk digested with pepsin	N.R.	Liepke et al. (2001)
κ-casein f(106-169)	S. mutans P. gingivalis E. coli	Bovine casein digested with chymosin	Bifidogenic Immunomodulatory	Malkoski et al. (2001) Prouxl et al. (1992) Brody (2000)
κ-casein f(18-24) κ-casein f(30-32) κ-casein f(139-146)	Several Gram-positive and -negative bacteria	Bovine κ-casein digested with pepsin	N.R.	López-Expósito et al. (2007c)
[h]β-casein f(184-210)	Several Gram-positive and -negative bacteria	Human β-casein digested with a proteinase of Lactobacillus helveticus PR4	N.R.	Minervini et al. (2003)

Table 1 (continued)

Fragment[a,b]	Antibacterial Activity	Isolation	Others Activities	References
LF f(17-41/42)	Several Gram-positive and -negative bacteria, viruses, fung., parasites	Bovine LF digested with pepsin or chymosin	Antitumor Antiinflamatory	Bellamy et al. (1992) Hoek et al. (1997) Iigo et al. (1999) Levay and Viljoen (1995)
[H]LF f(1-47)	Several Gram-positive and Gram-negative bacteria	Human LF digested with pepsin	N.R.	Bellamy et al. (1992)
LF f(1-48) LF f(1-47) LF f(277-288) LF f(267-285) LF f(267-288)	*Micrococcus flavus* Simple Para	Bovine LF digested with pepsin	N.R.	Recio and Visser (1999)
[c]LF f(14-42)	*Micrococcus flavus* *Escherichia coli*	Caprine LF digested with pepsin	N.R.	Recio and Visser (2000)
α-Lac f(1-5) α-Lac f(17-31)S-S(109-114) α-Lac f(61-68)S-S(75-80)	Several Gram-positive bacteria	Bovine α-Lac digested with chymotrypsin	N.R.	Pellegrini et al. (1999)
β-Lg f(15-20) β-Lg f(25-40) β-Lg f(78-83) β-Lg f(92-100)	Several Gram-positive bacteria	Bovine β-Lg digested with trypsin	N.R.	Pellegrini et al. (2001)

[a]Peptides obtained by chemical synthesis are not included.
[b]Unless otherwise indicated, all the proteins are from bovine origin.
Superscripts refer to the origin of the precursor protein being o = ovine, h = human, and c = caprine origin.
LF = lactoferrin; β-Lg = β-lactoglobulin; α-Lac = α-lactalbumin.
N.R. = nonreported.

antibacterial activity against Gram-positive and -negative bacteria, with MIC values ranging from 25 to 100 μM in the case of f(164-179), and from 8–16 μM in f(183-207). Recently, López-Expósito et al. (2007b) demonstrated the synergistic effect between the f(183-207) and lactoferrin against *Escherichia coli*, *Staphylococcus epidermidis,* and *Listeria monocytogenes*. If the α_{s2}-casein peptide could be generated upon enzymatic hydrolysis in the suckling gastrointestinal tract, both compounds could coexist in the gastrointestinal tract of a breastfed infant. Therefore, this synergism might have physiological meaning and could play a role in the host defenses against invading microorganisms. The first approaches related to the mechanism of action of the α_{s2}-casein peptide f(183-207) have recently been studied (López-Expósito et al., 2007a). Results showed that initial binding sites of the peptide were lipoteichoic acid in Gram-positive bacteria and lipopolysaccharide in Gram-negative. The peptide is able to permeabilize the outer and inner membranes. Moreover, the α_{s2}-casein peptide f(183-207) generated pores in the outer membrane of Gram-negative bacteria and in the cell wall of Gram-positive. In the Gram-negative bacteria, the f(183-207) originated the cytoplasm condensation, and in the Gram-positive bacteria, the cytoplasmic content leaked to the extracellular medium. In addition to the antimicrobial activity, it has been shown that fragment f(183-207) derived from the C-terminal end of the α_{s2}-casein could act as cell growth promoter (Smith et al., 1997).

The search for antibacterial activity from α_{s2}-casein has been extended to milk from other species. Four antibacterial peptides could be identified from a pepsin hydrolysate of ovine α_{s2}-casein (López-Expósito et al., 2006a). The peptides corresponded to α_{s2}-casein f(165-170), f(165-181), f(184-208), and f(203-208). Fragments f(165-181) and f(184-208) were homologous to those previously identified in the bovine protein. However, in contrast to bovine fragments, where f(183-207) exhibited higher antibacterial activity than f(164-179), in this study ovine α_{s2}-casein f(165-181) showed the higher antibacterial activity against all bacteria tested than ovine α_{s2}-casein f(184-208). Peptides from ovine α_{s2}-casein showed less potent antibacterial activity than those of bovine origin against Gram-negative bacteria. For the Gram-positive bacteria, all the peptides assayed revealed a strong activity with log-cycle reduction values from 1.1 to 6.0. A peptide identified in this casein digest, ovine α_{s2}-casein f(203-208), is a good example of a multifunctional peptide because it exhibited not only antimicrobial activity but also potent antihypertensive and antioxidant activity (Recio et al., 2005; López-Expósito et al., 2007c).

The κ-casein molecule is also a precursor of some antimicrobial peptides. Liepke et al. (2001) identified an antimicrobial peptide corresponding to the nonglycosylated portion 63-117 of human κ-casein. This peptide was obtained after acidification of human milk and incubation with pepsin for 2 hours at 37 °C. The spectrum of chemically synthesized f(63-117) includes growth inhibition of several Gram-positive and -negative bacteria and yeasts. These results had physiological relevance because they strongly supported the hypothesis that antimicrobially active peptides are released from human milk during infant digestion and, through this method, may play an important role in the host defense system of the newborn. Kappacin is another example of an antimicrobial peptide derived from κ-casein (Malkoski et al., 2001). Kappacin corresponds to the nonglycosylated, phosphorylated form of caseinmacropeptide (CMP). In order to characterize the active region of kappacin, the peptide was subjected to hydrolysis with endoproteinase Glu-C, given that the peptide Ser (P)$^{149}\kappa$-casein-A(138-158) was the active form with antimicrobial activity against *Streptococcus mutans, Escherichia coli,* and *Porphyromonas gingivalis*. It is important to emphasize that the active form is the phosphorylated and nonglycosylated form, since it has been demonstrated that the nonphosphorylated and glycosylated form does not reveal any activity against *Streptococcus mutans*. Molecular modeling and secondary-structure predictions of kappacin revealed that the peptide contained residues that could form an amphipatic helical structure. At pH 6.5, kappacin increased the permeability of liposomes, indicating that kappacin has a membranolytic mode of action; it has been proposed that the peptide could aggregate to form an anionic pore. In addition, phosphorylation, which is essential for activity, can produce a change in the conformation of the peptide through electrostatic repulsion or by divalent metal ion binding; in this form it could adopt a specific conformation to interact with the bacterial cell membrane (Dashper et al., 2005). CMP has been detected in the stomach, duodenum, and jejunum of humans after milk ingestion (Ledoux et al., 1999). The release of kappacin in the stomach could therefore be a mechanism to limit gastrointestinal tract infection in the developing neonate. Besides its antimicrobial activity, CMP has other biological activities (Brody, 2000), including the ability to bind cholera toxin (Kawasaki et al., 1992) and *E. coli* enterotoxins (Isoda et al., 1999). Also, CMP inhibits the bacterial adhesion to salivary pellicle (Schupbach et al., 1996) and the human influenza virus adhesion (Kawasaki et al., 1993). CMP can inhibit *Helicobacter pylori* infection (Hirmo et al., 1998) and the gastric secretions stimulated by cholecystokinin (Beucher et al., 1994). Furthermore, CMP promotes bifidobacterial growth (Proulx et al., 1992) and modulates immune system responses (Brody, 2000). Other antibacterial peptides derived from κ-casein that could have a physiological importance because they are obtained with a gastric enzyme are those reported by López-Expósito et al. (2006b). Six peptides with antibacterial activity were identified in a peptic digest of κ-casein. Of the peptides identified, the most active corresponded to κ-casein f(18-24), f(139-146), and f(30-32).

Although β-casein is one of the most abundant proteins in human milk (35% of the whole casein) (Fox, 2003), few studies search for antibacterial peptides from this protein. An antimicrobial sequence derived from human β-casein was obtained after hydrolysis of human milk with a purified proteinase of *Lactobacillus helveticus* PR4 (Minervini et al., 2003). The peptide corresponded to human β-casein f(184-210) and showed a large inhibition spectrum against Gram-positive and -negative bacteria, including species of potential clinical interest. In addition, these authors demonstrated that, once generated, the peptide was resistant to further degradation by trypsin and chymotrypsin.

Defensins and Cathelicidins

Besides the enzymatically produced peptide sequences, in milk other peptides with properties implicated in the defense of the newborn are directly expressed in an active form by the mammary gland cells. Host defense peptides are recognized components of the immune system and are conserved across plants, animals, and insects. Initially it was believed that their sole role in innate immunity was to kill invading microorganisms, but evidence now suggests that these peptides play diverse and complex roles in the immune response. To date, only two categories of these antimicrobial peptides have been identified in human milk: cathelicidins and defensins. The role of these host defense peptides has been extensively studied and reviewed (Zasloff, 2002; Epand & Vogel, 1999; Bowdish et al., 2005), but this section considers only those peptides found in human milk. Cathelicidins and defensins present in human milk, together with their concentration and principal biological activities, are summarized in Table 2.

Cathelicidins

Cathelicidins are synthesized as large precursor peptides containing an N-terminal signal peptide sequence, i.e., a conserved cathelin-like domain (pro-peptide), followed by a variable C-terminal antimicrobial domain. The heterogeneity in the C-terminal domain that encodes the mature peptide can

Table 2 Cathelicidins and Defensins Present in Human Milk, Concentration, and Biological Activities

Peptide	Concentration Range	Biological Activity	References
Human cathelicidin (LL37)	0–160.6 µg/mL	Antimicrobial Antitumoral Immunomodulant Reepitheliazation	Murakami et al. (2005) Okamura et al. (2004) Bals et al. (1999) Zametti et al. (2004)
Human β-defensin 1 (hBD1)	0–23 µg/mL	Antimicrobial Immunomodulant	Armogida et al. (2004) Jia et al. (2001) Yang et al. (1999)
Human β-defensin 2 (hBD2)	8.5–56 µg/mL	Antimicrobial	Armogida et al. (2004) Lehrer and Ganz (2002)
Human β-defensin 5 (hBD5)	0–11.8 µg/mL	Antimicrobial	Armogida et al. (2004) Porter et al. (1997)
Human β-defensin 6 (hBD6)			
Human neutrophil-derived -peptide (hNP1-3)	5–43.5 µg/mL	Antimicrobial	Armogida et al. (2004) Ganz and Weiss (1997)

range between 12 and 80 amino acids, thus yielding peptides with different bactericidal potential (Dommett et al., 2005).

In human milk, only one cathelicidin, LL37, has been identified. This peptide was derived by proteolysis from the C-terminal end of the human CAP18 protein (hCAP18) (Gudmundsson et al., 1996). This precursor protein, hCAP18, is thought to be inactive. After processing, the N-terminal cathelin protein also has antimicrobial and protease-inhibitor activity (Zaiou et al., 2003). Neither the N-terminal cathelin protein nor the precursor hCAP18 has been found in human milk; however, the peptide LL37 is relatively abundant in human milk, reaching concentrations of 32 μM (Murakami et al., 2005). At this concentration, synthetic LL37 peptide has demonstrated antimicrobial activity against a wide range of microbes (Turner et al., 1998). Although factors such as pH, sodium chloride concentration, and binding to other macromolecules in the soluble solution are critical determinants of the ability of antimicrobial peptides to interact with the microbial target membrane (Goldman et al., 1997), it has been confirmed that LL37 has direct bactericidal activity on bacteria in human milk solution (Murakami et al., 2005). The *in vivo* activity could be further augmented by the synergistic presence of other antimicrobial compounds in human milk, such as sIgA, lactoferrin, lysozyme, fatty acids, or glycans (Newburg, 2005; Isaacs, 2005). In addition to its antimicrobial activity, the peptide LL37 binds and neutralizes lipopolysacaride and protects against endotoxic shock in a murine model of septicemia (Bals et al., 1999). Furthermore, it is chemotactic for neutrophils, monocytes, mast cells, and T cells, induces degranulation of mast cells, alters transcriptional responses in macrophages, stimulates wound vascularization and the reepithelialization of healing skin (Zanetti, 2004), and has antitumor activity (Okumura et al., 2004).

Defensins

Defensins are a group of naturally occurring small peptides containing 29–45 amino acid residues that display antibiotic and nonspecific cytotoxic properties. Their antimicrobial spectra include both Gram-positive and -negative bacteria, mycobacteria, fungi, and some enveloped viruses (Lehrer et al., 1993). Two structurally distinct defensin peptide families have been identified in humans: α-defensins, found in phagocytic cells and Paneth cells of the small intestine, and β-defensins, expressed in epithelial tissues. To date, several defensin peptides have been identified in human milk or mammary gland tissues in significant concentrations (in the range of μg/mL): human β-defensin-1 (hBD-1), human β-defensin-2 (hBD-2), human neutrophil–derived-α-peptide (hNP1-3), human α-defensin-5 (hD-5), and human α-defensin-6 (hD-6) (Armogida et al., 2004).

hBD-1 was first found in mammary gland epithelia in both lactating and nonlactating women (Tunzi et al., 2000). Shortly thereafter, peptide hBD-1 was detected in human breast milk samples in concentrations ranging from 1 to 10 µg/mL (Jia et al., 2001). The concentrations reported for hBD-1 in human milk exceed those reported at other mucosal surfaces. It has been shown that urinary concentrations of hBD-1 were greater in pregnant women than in other subjects, suggesting that the hormone milieu of pregnancy may regulate hBD-1 expression (Valore et al., 1998). The production and secretion of hBD-1 by mammary gland epithelia may be increased during lactation and offer a new example of the protective effects of breastfeeding. Furthermore, hBD-1 could also have a protective function for the mother. Due to the broad spectrum of activity showed against bacteria, fungi, and enveloped viruses, this peptide may exert microbicidal effects that influence colonization or infection of the nasopharynx or upper gastrointestinal tract. hBD-1 might also act together with other microbicidal proteins present in breast milk. In addition to the direct microbicidal effect, hBD-1 might have an impact on neonatal immunity through immunomodulatory effects. hBD-1 was shown to be chemotactic for immature dendritic cells and memory T cells (Yang et al., 1999). hBD-1 in breast milk might promote the priming of adaptive immune responses in the newborn's nasopharynx or gastrointestinal tract by recruiting dendritic cells and T cells to these mucosal surfaces.

hBD-2 mRNA was first found in breast tissue by RNA dot-blot and *in situ* hybridization (Bals et al., 1998). The peptide hBD-2 has been detected in human milk in concentrations ranging from 8.5 to 56 µg/mL (Armogida et al., 2004), although other studies have failed to detect the presence of hBD-2 transcripts (Tunzi et al., 2000; Jia et al., 2001; Murakami et al., 2005). This apparent conflict may be due to the instability of the peptide hBD-2. In addition, it is possible that the mammary expression of antimicrobial peptides occurs at both a constitutive and inducible level. hBD-2 has potent activity against Gram-negative bacteria and *Candida* (Lehrer & Ganz, 2002), and its expression can be augmented during infection and inflammation.

hNP1-3 has been detected in human milk at concentrations ranging from 5 to 43.5 µg/mL (Armogida et al., 2004). This α-defensin has been shown to possess antimicrobial activity against *Candida albicans* [lethal dose 50 (LD_{50}) = 2.2 µg/mL] and *Escherichia coli* and *Streptococcus faecalis* with an LD_{50} value $\geq$ 10 µg/mL (Ganz & Weiss, 1997; Harder et al., 1997).

Concentrations of HD-5 and HD-6 in human milk are between 0–11.8 µg/mL (Armogida et al., 2004). HD-5 has demonstrated a broad antimicrobial spectrum of activity against Gram-positive and -negative bacteria and yeast (Porter et al., 1997). The α-defensins HD-5 and HD-6, although present in lower amounts in milk than HNP-1 and HBD-2, may still be of critical importance in the defense of the neonatal gastrointestinal tract. Inadequate levels of HD-5 and HD-6 may contribute to the increased risk of necrotizing enterocolitis, a serious gastrointestinal disorder of unknown etiology in premature infants

(Salzman et al., 1998). It is important to highlight that HD-5 and HD-6 concentrations are significantly higher in colostrum than in mature milk.

The precise mechanism for the antimicrobial activity of defensins has not yet been clearly defined. It has been proposed that defensins permeabilize membranes through the formation of multimeric pores. In order to form pores within the cell wall, antimicrobial peptides must fulfill structural criteria that allow (1) binding to bacterial membrane, (2) aggregation within the membrane, and (3) channel formation. It has been suggested that the cationic characteristic and amphipathic nature of these antimicrobial peptides allow binding and direct interaction with the lipid bilayer of the cell membrane, leading to leakage of the internal aqueous contents of the cell (Chen et al., 2006a).

Antitumor Peptides

Based on various cytochemical studies, there is an increasing evidence for the possible involvement of milk-derived peptides as specific signals that can trigger the viability of cancer cells by inducing apoptosis. Milk protein–derived inducers of apoptosis may be of significance in the control of malignant cell proliferation. A vast majority of tumor promoters are potent inhibitors of apoptosis, and therefore apoptosis-inducing peptides can be classified as probable human anticarcinogens. Effects on both cell viability and immune cell function may be a mechanism through which bioactive peptides exert protective effects in cancer development (Meisel, 2005).

In 1999, Roy et al. reported that bovine skimmed milk digested with purified protease B from *Saccharomyces cerevisae* inhibited proliferation of human leukemia cells (HL-60) by an apoptotic mechanism. In addition, purified peptides corresponding to bioactive sequences of casein were identified. Apoptosis of HL-60 cells was induced by the opioid peptide β-casomorphin-7 [β-casein f(60-66)] and the phosphopeptide β-casein f(1-25)4P (Hata et al., 1998). Furthermore, peptides corresponding to bovine α_{s1}-casein f(1-3), f(101-103), and f(104-105) are reported to induce necrosis of several kinds of animal lymphocytes including leukemic T and B cell lines in serum-free medium (Otani & Suzuki, 2003). With respect to κ-casein, a pentapeptide called κ-casecidin [corresponding to bovine κ-casein f(17-21)], apart from its antimicrobial activity, also displayed cytotoxic activity toward some mammalian cells, including human leukemic cells lines, probably due to apoptosis (Matin & Otani, 2002). Together with the apoptotic mechanism, it has been suggested that casein-derived peptides may exert their antitumor activity by a mechanism partly involving opioid receptors. Indeed, Kampa et al. (1997) described that several casomorphin peptides derived from both α- and β-casein could decrease the proliferation of prostatic cancer cell lines through this mechanism.

Peptides derived from the N-terminal end of lactoferrin have also been studied in order to identify sequences with antitumor activity. The fragment corresponding to residues 17–38 of bovine lactoferrin induced apoptosis in HL-60 cells (Roy et al., 2002). In the same year, Eliassen et al. (2002) found that bovine lactoferricin [bovine lactoferrin f(17-41)] exerted cytotoxic activity against fibrosarcoma (Meth A), melanoma, and colon carcinoma cells lines *in vitro*, significantly reducing the size of solid Meth A tumors. Apart from that, lactoferricin displayed antitumor activity against breast cancer cells (Furlong et al., 2006) and cytotoxic activity *in vitro* and *in vivo* against neuroblastoma cells (Eliassen et al., 2006). Recently, Mader et al. (2006) showed that bovine lactoferricin has the capacity of inhibit angiogenesis, a process necessary for tumor growth because of the tumor's need for oxygen and nutrient supply, as well as waste removal. Interestingly, the structural parameters that describe the antitumor effects of lactoferricin are very similar to those that described the antibacterial activity (Gifford et al., 2005). In addition to a certain net positive charge and hydrophobicity, the ability to adopt an amphipatic conformation is critical for antitumor activity (Yang et al., 2004). Besides, it is known that the cytotoxic activity of lactoferricin against tumor cells is located within the amino acid sequence FKCRRWQWRM (Mader et al., 2005).

In Vivo Studies of Milk Antibacterial Peptides

Several *in vivo* studies dealing with host defense peptides in mammals (defensins and cathelicidins) have been performed (for recent reviews, see Bowdish et al., 2005; Brodgen et al., 2004). Animal models indicate that host defense peptides are crucial for the prevention and clearance of infection. In mammals, conditions at many *in vivo* sites are such that several of these peptides probably have little if any direct microbicidal activity, but instead may have multiple immunomodulatory effects. Studies of these additional effects are in their early stages and have largely been performed *in vitro*. Innovative *in vivo* modeling approaches will be required to dissect the constitutive components of the host response that can be assigned to these peptides as well as the significance of each component. Thus, whether host defense peptides have meaningful microbicidal or immunoregulatory activities *in vivo* must be examined by considering two fundamental issues: (1) the environment in which these activities are assessed and (2) the concentrations at which such peptides are found *in vivo*.

On the contrary, only a few studies about the *in vivo* activity of antimicrobial milk protein–derived peptides have been performed using animal models or in clinical trials in humans. Few studies using animal models have reported a protective effect of orally administered lactoferricin against infection by methicillin-resistant *Staphylococcus aureus* (Nakasone et al., 1994) and against infection caused by the parasite *Toxoplasma gondii* (Isamida et al., 1998). Most of these studies are performed to demonstrate antibacterial effects of the entire protein lactoferrin after oral administration. However, because orally administered lactoferrin is partially degraded to fragments that contain the lactoferricin sequence (Kuwata et al., 1998a, b), some of the effects demonstrated for lactoferrin can probably be attributed to lactoferrin fragments or to the combined action of lactoferrin and its derived peptides. Lactoferrin has been shown to suppress the intestinal overgrowth and bacterial translocation of enterobacteria in mouse (Teraguchi et al., 1994, 1995). Orally administered lactoferrin also has a protective effect against infection caused by methicillin-resistant *Staphylococcus aureus* and *Candida albicans* (Bhimani et al., 1999). Recently, Lee et al. (2005) demonstrated that human lactoferrin decreases the hepatic colonization of *Listeria monocytogenes*, hepatic necrosis, and expression of inflammatory cytokines in mice infected orally with *Listeria monocytogenes*.

In humans, a study with low-birth-weight infants fed with a lactoferrin-enriched infant formula concluded that lactoferrin contributes to the formation of bifidobacteria-rich flora (Kawaguchi et al., 1989). Di Mario et al. (2003) showed that lactoferrin is effective in the eradication of *Helicobacter pylori* when used as a supplement to an antibiotic treatment. Other *in vivo* studies on lactoferrin to investigate its antiviral and immunomodulatory effects, and other host-protective activities such as cancer prevention, as well as clinical applications of lactoferrin have been reviewed (Tomita et al., 2002; Marshall, 2004).

There are some studies about the *in vivo* activity of isracidin. In mice, it exerts a protective effect against *Listeria monocytogenes, Streptococcus pyogenes,* and *Staphylococcus aureus*. Protection of rabbits, guinea pigs, and sheep against *Staphylococcus aureus* has also been achieved. In cows with mastitis, isracidin obtained a success rate of over 80% in the treatment of chronic streptococcal infection. Furthermore, it has been demonstrated that isracidin possesses immunomodulant properties. Isracidin had a significant effect on the production of IgG, IgM, and antibody-forming cells and also increased the cell-mediated immunity when injected to mice (Lahov & Regelson, 1996).

Another interesting finding was made with a tryptic casein hydrolysate for treatment and prophylaxis of newborn calf colibacillosis (Biziulevicius et al., 2003). The casein hydrolysate showed high therapeutic and prophylactic efficacies comparable to Fermosob, a veterinary antimicrobial preparation widely used to treat colibacillosis (Biziulevicius & Zukaite, 1999). The hydrolysate revealed 93.0% therapeutic and 93.5% prophylactic efficacy in addition to not only an antimicrobial effect, but also immnunostimulatory activity. However, in this study the peptides presumed to be responsible for these activities were not identified.

Recently, a product obtained from bovine colostrum rich in immunoglobulins, growth factors, antibacterial peptides, and nutrients reduced the number of evacuations of stools per day in patients with HIV-associated diarrhea. Also, an increase in hemoglobin and albumin was achieved, the patients' fatigue was alleviated, and their body weight increased. Moreover, there was a rise in the $CD4^+$ count (Florén et al., 2006).

Concluding Remarks

Cohort studies have provided evidence of the health benefits of breastfeeding; such benefits are mainly related to the protection against infection. The presence of antimicrobial molecules in milk could play a relevant role in this protective effect for the newborn, and they can also be of importance for protection of the mammary gland against infection and development of mastitis and protection of milk from microbial proliferation after secretion. In spite of other milk components, milk contains a wide array of proteins that provide biological activities ranging from antimicrobial effects to immunostimulatory functions. In addition, peptides formed from human milk proteins during digestion can inhibit the growth of pathogenic bacteria and viruses and, therefore, protect against infection. Other peptides are secreted into milk in an active form, like cathelicidins and defensins. The ability of these peptides to directly confer protection against bacterial colonization of epithelial surfaces in the gut, lung, and skin has been shown. It must be emphasized that milk proteins also provide adequate amounts of essential amino acids to the growing infant. It is therefore suggested that mammals possess a highly adapted digestive system, which would allow the survival of some proteins and peptides in the upper gastrointestinal tract, allowing amino acid utilization from these proteins and peptides further down in the gut. This knowledge could be transferred to other fields. There is increasing interest from the industry in the application of functional food proteins and peptides, especially in the application of antimicrobial proteins and peptides that might contribute to human and animal well-being. It is now possible to produce proteins and peptides with biological activity on a large scale at a low cost, by recombinant procedures, or, in the case of peptides, by enzymatic hydrolysis or bacterial fermentation. These proteins and peptides could be used without much purification and without safety concerns in animal and human diets and foods.

Acknowledgment Projects AGL-2005-03381 from the Ministerio de Educación y Ciencia and PIF 2005-70F0111 from Consejo Superior de Investigaciones Científicas are acknowledged for financial support.

References

Almaas, H., Holm, H., Langrud, T., Flengsrud, R., & Vegarud, G.E. (2006). *In vitro* studies of the digestion of caprine whey proteins by human gastric and duodenal juice and the effects on selected microorganisms. *British Journal of Nutrition, 96*, 562–569.

Armogida, S. A., Yannaras, N. M., Melton, A. L., & Srivastava, M. (2004). Identification and quantification of innate immune system mediators in human breast milk. *Allergy and Asthma Proceedings, 25*, 297–304.

Bals, R., Wang X., Wu, Z., Freeman, T., Bafna, V., Zasloff, M., & Wilson, J. M. (1998). Human β-defensin 2 is a salt-sensitive peptide antibiotic expressed in human lung. *Journal of Clinical Investigation, 102,* 874–880.

Bals, R., Weiner, D. J., Moscioni, A. D., Meegalla, R. L., & Wilson, J. M. (1999). Augmentation of innate host defense by expression of a cathelicidin antimicrobial peptide. *Infection and Immunity, 67,* 6084–6089.

Bellamy, W., Takase, M., Yamauchi, K., Wakabayashi, H., Kawase, K., & Tomita, M. (1992). Identification of the bactericidal domain of lactoferrin. *Biochimica et Biophysica Acta, 1121,* 130–136.

Bellamy, W., Wakabayashi, H., Takase, M., Kawase, K., Shimamura, S., & Tomita, M. (1993). Killing of *Candida albicans* by lactoferricin-B, a potent antimicrobial peptide derived from the N-terminal region of bovine lactoferrin. *Medical Microbiology and Immunology, 182,* 97–105.

Beucher, S., Levenez, F., Yvon, M., & Corring, T. (1994). Effect of caseinomacropeptide (CMP) on choleocystokinin (CCK) release by intestinal cells in rat. *Journal of Nutritional Biochemistry, 5,* 578–584.

Bhimani, R. S., Vendrov, Y., & Furmanski, P. (1999). Influence of lactoferrin feeding and injection against systemic staphylococcal infections in mice. *Journal of Applied Microbiology, 86,* 135–144.

Biziulevicius, G. A., & Zukaite, V. (1999). Lysosubtilin modification, Fermosob, designed for polymeric carrier-mediated intestinal delivery of lytic enzymes: Pilot-scale preparation and evaluation of this veterinary medicinal product. *International Journal of Pharmacology, 189,* 43–55.

Biziulevicius, G. A., Zukaite, V., Normatiene, T., Biziuleviciene, G., & Arestov, I. (2003). Non-specific immunity-enhancing effects of tryptic casein hydrolysate versus Fermosob for treatment/prophylaxis of newborn calf colibacillosis. *FEMS Immunology and Medical Microbiology, 39,* 155–161.

Bowdish, D. M. E., Davidson, D. J., & Hancock, R. E. W. (2005). A re-evaluation of the role of host defence peptides in mammalian immunity. *Current Protein and Peptide Science, 6,* 35–51.

Brody, E. P. (2000). Biological activities of bovine glycomacropeptide. *British Journal of Nutrition, 84,* S39–S46.

Brogden, K. A., Ackermann, M., Zabner, J., & Welsh, M. J. (2004). Antimicrobial peptides suppress microbial infection and sepsis in animal models. In R. E. W. Hancock & D. Devine (Eds.), *Mammalian Host Defense Peptides* (pp. 189–229). New York: Cambridge University Press.

Chen, H., Xu, Z., Peng, L., Fang, X., Yin, X., Xu, N., & Cen, P. (2006a). Recent advances in the research and development of human defensins. *Peptides, 27,* 931–940.

Chen, H. L., Yeng, C. C., Lu, C. Y., Yu, C. H., & Chen, C. M. (2006b). Synthetic porcine lactoferricin with a 20-residue peptide exhibits antimicrobial activity against *Escherichia coli, Staphylococcus aureus* and *Candida albicans. Journal of Agricultural and Food Chemistry, 54,* 3277–3282.

Dashper, S. G., O'Brien-Simpson, N. M., Cross, K. J., Paolini, R. A., Hoffman, B., Catmull, D. V., Malkoski, M., & Reynolds, E. C. (2005). Divalent metal cations increase the activity of the antimicrobial peptide kappacin. *Antimicrobial Agents and Chemotherapy, 49,* 2322–2328.

Di Mario, F., Aragona, G., Dal Bo, N., Cavestro, G. M., Cavallaro, L., Iori, V., Comparato, G., Leandro, G., Pilotto, A., & Franze, A. (2003). Use of bovine lactoferrin for *Helicobacter* eradication. *Digestive and Liver Disease, 35,* 706–710.

Dommett, R., Zilbauer, M., George, J. T., & Bajaj-Elliot, M. (2005). Innate immune defence in the human gastrointestinal tract. *Molecular Immunology, 42,* 903–912.

Eliassen, L. T., Berge, G., Sveinbjornsson, B., Svendsen, J. S., Vorland, L. H., & Rekdal, Ø. (2002). Evidence for a direct antitumor mechanism of action of bovine lactoferricin. *Anticancer Research, 22,* 2703–2710.

Eliassen, L. T., Berge, G., Leknessund, A., Wikman, M., Lindin, I., Løkke, C., Pontham, F., Johnsen, J. I., Sveinbjørnsson, B., Kogner, P., Flægstad, T., & Rekdal, Ø. (2006). The antimicrobial peptide, Lactoferricin B, is cytotoxic to neuroblastoma cells *in vitro* and inhibits xenograft *in vivo*. *International Journal of Cancer, 119*, 493–500.

El-Zahar, K., Sitohy, M., Choiset, Y., Métro, F., Haertlé, T., & Chobert, J. M. (2004). Antimicrobial activity of ovine whey protein and their peptic hydrolysates. *Milchwissenschaft, 59*, 653–656.

Epand, R. M., & Vogel, H. J. (1999). Diversity of antimicrobial peptides and their mechanisms of action. *Biochimica et Biophysica Acta, 1462*, 11–28.

Florén, C. H., Chinenye, S., Elfstrand, L., Hagman, C., & Ihse, I. (2006). Coloplus, a new product based on bovine colostrums, alleviates HIV-associated diarrhoea. *Scandinavian Journal of Gastroenterology, 41*, 682–686.

Floris, R., Recio, I., Berkhout, B., & Visser, S. (2003). Antibacterial and antiviral effects of milk proteins and derivatives thereof. *Current Pharmaceutical Design, 9*, 1257–1275.

Fox, P. F. (2003). Milk proteins: General and historical aspects. In P. F. Fox & P. L. H. McSweeney (Eds.), *Advanced Dairy Chemistry 1. Proteins* (pp. 1–49). New York: Kluwer Academic/Plenum Publishers.

Furlong, S. J., Mader, J. S., & Hoskin, D. W. (2006). Lactoferricin-induced apoptosis in estrogen-nonresponsive MDA-MB-435 breast cell cancer cells is enhanced by C6 ceramide or tamoxifen. *Oncology Reports, 15*, 1385–1390.

Ganz, T., & Weiss, J. (1997). Antimicrobial peptides of phagocytes and epithelia. *Seminars of Hematology, 34*, 343–354.

Gifford, J. L., Hunter, H. N., & Vogel, H. J. (2005). Lactoferricin: A lactoferrin-derived peptide with antimicrobial, antiviral, antitumor and immunological properties. *Cell and Molecular Life Science, 62*, 2588–2598.

Goldman, M. J., Anderson, G. M., Stolzenberg, E. D., Kari, U. P., Zasloff, M., & Wilson, J. M. (1997). Human β-defensin-1 is a salt-sensitive antibiotic in lung that is inactivated in cystic fibrosis. *Cell, 88*, 553–560.

Gudmundsson, G. H., Agerberth, B., Odeberg, J., Bergman, T., Olsson, B., & Salcedo, R. (1996). The human gene FALL39 and processing of the cathelin precursor to the antibacterial peptide LL-37 in granulocytes. *European Journal of Biochemistry, 238*, 325–332.

Harder, J., Bartels, J., Christophers, E., & Schroder, J. M. (1997). A peptide antibiotic from human skin. *Nature, 387*, 861.

Hata, I., Higashiyama, S., & Otani, H. (1998). Identification of a phosphopeptide in bovine α_{s1}-casein digests as a factor influencing proliferation and immunoglobulin production in lymphocyte cultures. *Journal of Dairy Research, 65*, 569–578.

Haukland, H. H., Ulvatne, H., Sandvik, K., & Vorland, L. H. (2001). The antimicrobial peptides lactoferricin B and magainin-2 cross over the bacterial cytoplasmic membrane and reside in the cytoplasm. *FEBS Letters, 508*, 389–393.

Hayes, M., Ross, R. P., Fitzgerald, G. F., Hill, C., & Stanton, C. (2006). Casein-derived antimicrobial peptides generated by *Lactobacillus acidophilus* DPC6026. *Applied and Environmental Microbiology, 72*, 2260–2264.

Hernández-Ledesma, B., López-Expósito, I., Ramos, M., & Recio, I. (2006). Bioactive peptides from milk proteins. In R. Pizzano (Ed.), *Immunochemistry in Dairy Research* (pp. 37-60). Kerala, India: Trivandrum.

Hill, R. D., Lahov, E., & Givol, D. (1974). A rennin-sensitive bond in alpha and beta casein. *Journal of Dairy Research, 41*, 147–153.

Hirmo, S. Kelm, S., Iwersen, M., Hotta, K., Goso, Y., Ishihara, K., Suguri, T., Morita, M., Wadström, T., & Schauer, R. (1998). Inhibition of *Helicobacter pylori* sialic acid-specific haemagglutination by human gastrointestinal mucins and milk glycoproteins. *FEMS Immunology and Medical Microbiology, 20*, 275–281.

Hoek, K., Milne, J. M., Grieve, P. A., Dionoysius, D. A., & Smith, R. (1997). Antibacterial activity of bovine lactoferrin-derived peptides. *Antimicrobial Agents and Chemotherapy*, *41*, 54–59.

Ibrahim, H. R. (2003). Hen egg white lysozyme and ovotransferrin: Mystery, structural role and antimicrobial function. *Proceedings of the 10th European Symposium on the Quality of Eggs and Egg Products*. Saint-Brieuc, France, September, pp. 1113–1128.

Ibrahim, H. R., Thomas, U., & Pellegrini, A. (2001). A helix-loop-helix peptide at the upper lip of the active site cleft of lysozyme confers potent antimicrobial activity with membrane permeabilization action. *Journal of Biological Chemistry*, *276*, 43767–43774.

Iigo, M., Kuhara, T., Ushida, Y., Sekine, K., Moore, M. A., & Tsuda, H. (1999). Inhibitory effects of bovine lactoferrin on colon carcinoma 26 lung metastasis in mice. *Clinical & Experimental Metastasis*, *17*, 35–40.

Isaacs, C. E. (2005). Human milk inactivates pathogens individually, additively and synergistically. *Journal of Nutrition*, *135*, 1286–1288.

Isamida, T., Tanaka, T., Omata, Y., Yamauchi, K., Shimazaki, K., & Saito, A. (1998). Protective effect of lactoferricin against *Toxoplasma gondii* infection in mice. *Journal of Veterinary Medical Science*, *60*, 241–244.

Isoda, H., Kawasaki, Y., Tanimoto, M., Dosako, S., & Idota, T. (1999). Use of compounds containing or binding sialic acid to neutralize bacterial toxins. European patent application no. 385112.

Jia, H. P., Starner, T., Ackerman, M., Kirby, P., Tack, B. F., & McCray, P. B. (2001). Abundant human β-defensin-1 expression in milk and mammary gland epithelium. *Journal of Pediatrics*, *138*, 109–112.

Kampa, M., Bakogeorgou, E., Hatzoglou, A., Damianaki, A., Martin, P. M., & Castanas, E. (1997). Opioid alkaloids and casomorphin peptides decrease the proliferation of prostatic cells lines (LNCaP, PC3 and DU145) through a partial interaction with opioid receptors. *European Journal of Pharmacology*, *335*, 255–265.

Kawaguchi, S., Hayashi, T., Masano, H., Okuyama, K., Suzuki, T., & Kawase, K. (1989). Effect of lactoferrin-enriched infant formula on low birth weight infants [in Japanese]. *Shuusnakiigaku*, *19*, 125–130.

Kawasaki, Y., Isoda, H., Tanimoto, M., Dosako, S., Idota, T., & Ahiko, K. (1992). Inhibition by lactoferrin and κ-casein glycomacropeptide of binding of cholera toxin to its receptor. *Biotechnology and Biochemistry*, *56*, 195–198.

Kawasaki, Y., Isoda, K., Shinmoto, H., Tanimoto, M., Dosako, S., Idota, T., & Nakajima, I. (1993). Inhibition by κ-casein glycomacropeptide and lactoferrin of influenza virus hemaglutination. *Bioscience, Biotechnology and Biochemistry*, *57*, 1214–1215.

Korhonen, H., & Pihlanto, A. (2006). Bioactive peptides: Production and functionality. *International Dairy Journal*, *16*, 945–960.

Kuwata, H., Yip, T. T., Tomita, M., & Hutchens, T. W. (1998a) Direct evidence of the generation in human stomach of an antimicrobial peptide domain (lactoferricin) from ingested lactoferrin. *Biochimica et Biophysica Acta*, *1429*, 129–141.

Kuwata, H., Yip, T. T., Yamauchi, K., Teraguchi, S., Hayasawa, H., Tomita, M., & Hutchens, T. W. (1998b). The survival of ingested lactoferrin in the gastrointestinal tract of adult mice. *Biochemistry Journal*, *334*, 321–323.

Lahov, E., & Regelson W. (1996). Antibacterial and immunostimulating casein-derived substances from milk: Casecidin, isracidin peptides. *Federal Chemistry Toxicology*, *34*, 131–145.

Ledoux, N., Mahé, S., Dubarry, M., Bourras, M., Benamouzig, R., & Tomé, D. (1999). Intraluminal immunoreactive caseinomacropeptide after milk protein ingestion in humans. *Nahrung*, *43*, 196–200.

Lee, H. Y., Park, J. H., Seok, S. H., Baek, M. W., Kim, D. J., Lee, B. H., Kang, P. D., Kim, Y. S., & Park, J. H. (2005). Potencial antimicrobial effects of human lactoferrin against oral infection with *Listeria monocytogenes* in mice. *Journal of Medical Microbiology*, 54, 1049–1054.

Lehrer, R. I., & Ganz, T. (2002). Defensins of vertebrate animals. *Current Opinion in Immunology*, 14, 96–102.

Lehrer, R. I., Lichtenstein, A. K., & Ganz, T. (1993). Defensins: Antimicrobial and cytotoxic peptides of mammalian cells. *Annual Reviews in Immunology*, 11, 105–128.

León-Sicarios, N., Reyes-López, M., Ordaz-Pichardo, C., & de la Garza, M. (2006). Microbicidal action of lactoferrin and lactoferricin and their synergistic effect with metronizadole in *Entoamoeba histolytica*. *Biochemistry and Cell Biology*, 84, 327–336.

Levay, P. F., & Viljoen, M. (1995). Lactoferrin, a general review. *Haematologica*, 80, 252–267.

Liepke, C., Zucht, H. D., Forssman, W. G., & Ständker, L. (2001). Purification of novel peptide antibiotics from human milk. *Journal of Chromatography B*, 752, 369–377.

López-Expósito, I., & Recio, I. (2006). Antibacterial activity of peptides and folding variants from milk proteins. *International Dairy Journal*, 16, 1294–1305.

López-Expósito, I., Gómez-Ruiz, J. A., Amigo, L., & Recio, I. (2006a). Identification of antibacterial peptides from ovine α_{s2}-casein. *International Dairy Journal*, 16, 1072–1080.

López-Expósito, I., Minervini, F., Amigo, L., & Recio, I. (2006b). Identification of antibacterial peptides from bovine κ-casein. *Journal of Food Protection*, 69, 2992–2997.

López-Expósito, I. (2007a). Novel peptides with antibacterial activity derived from food proteins. Study of the mode of action and synergistic effect. Dissertation Tesis. Faculty of Science. Universidad Autónoma de Madrid.

López-Expósito, I., Pellegrini, A., Amigo, L., & Recio, I. (2007b). Synergistic effect between different milk-derived peptides and proteins. *Journal of Dairy Science* (submitted).

López-Expósito, I., Quirós, A., Amigo, L., & Recio, I. (2007c). Casein hydrolysates as source of antimicrobial, antioxidant and antihypertensive peptides. *Le Lait* (in press).

Mader, J. S., Salsman, J., Conrad, D. M., & Hoskin, D. W. (2005). Bovine lactoferricin selectively induces apoptosis in human leukemia and carcinoma cells lines. *Molecular Cancer Therapy*, 4, 612–624.

Mader, J. S., Smyth, D., Marshall, J., & Hoskin, D. W. (2006). Bovine lactoferricin inhibits basic fibroblast growth factor- and vascular endothelial growth factor$_{165}$–induced angiogenesis by competing for heparin-like binding sites on endothelial cells. *American Journal of Pathology*, 169, 1753–1766.

Malkoski, M., Dashper, S. G., O'Brien-Simpson, N. M., Talbo, G. H., Macris, M., Cross, K. J., & Reynolds, E. C. (2001). Kappacin, a novel antimicrobial peptide from bovine milk. *Antimicrobial Agents and Chemotherapy*, 45, 2309–2315.

Marshall, K. (2004). Therapeutic applications of whey protein. *Alternative Medicine Review*, 9, 136–156.

Masschalck, B., & Michiels, C. W. (2003). Antimicrobial properties of lysozyme in relation to foodborne vegetative bacteria. *Critical Reviews in Microbiology*, 29, 191–214.

Matin, A., & Otani, H. (2002). Cytotoxic and antibacterial activities of chemically synthesized κ-casecidin and its partial peptide fragments. *Journal of Dairy Research*, 69, 329–334.

McCann, K. B., Shiell, B. J., Michalski, W. P., Lee, A., Wan, J., Roginski, H., & Coventry, M. J. (2006). Isolation and characterisation of a novel antibacterial peptide from bovine α_{s1}-casein. *International Dairy Journal*, 16, 316–323.

Meisel, H. (2005). Biochemical properties of peptides encrypted in bovine milk proteins. *Current Medicinal Chemistry*, 12, 1905–1919.

Minervini, F., Algaron, F., Rizzello, C. G., Fox, P. F., Monnet, V., & Gobetti, M. (2003). Angiotensin I-converting-enzyme-inhibitory and antibacterial peptides from *Lactobacillus helveticus* PR4 proteinase-hydrolyzed caseins of milk from six species. *Applied and Environmental Microbiology*, 69, 5297–5305.

Muñoz, A., & Marcos, J. F. (2006). Activity and mode of action against fungal phytopathogens of bovine lactoferrin-derived peptides. *Journal of Applied Microbiology*, 101, 1199–1207.

Murakami, M., Dorschner, R. A., Stern, L. J., Lin, K. H., & Gallo, R. L. (2005). Expression and secretion of cathelicidin antimicrobial peptides in murine mammary glands and human milk. *Pediatric Research, 57*, 10–15.

Nakasone, Y., Adjei, A., Yoshise, M., Yamauchi, K., Takase, M., Yamauchi, K., Shimamura, S., & Yamamoto, S. (1994). Effect of dietary lactoferricin on the recovery of mice infected with methicillin-resistant *Staphylococcus aureus*. *Abstract Annual Meeting of the Japanese Society of Nutritional Food Science* [in Japanese], p. 50.

Nazarowec-White, M., & Farber, J. M. (1997). Thermal resistance of *Enterobacter sakazakii* in reconstituted dried infant formula. *Letters in Applied Microbiology, 24*, 9–13.

Newburg, D. S. (2005). Innate immunity and human milk. *Journal of Nutrition, 135*, 1308–1312.

Okumura, K., Itoh, A., Isogai, E. Hirose, K., Hosokawa, Y., Abiko, Y., Shibata, T., Hirata, M., & Isogai, H. (2004). C-terminal domain of human CAP18 antimicrobial peptide induces apoptosis in oral squamous cell carcinoma SAS-H1 cells. *Cancer Letters, 212*, 185–194.

Otani, H., & Suzuki, H. (2003). Isolation and characterization of cytotoxic small peptides, α-casecidins, from bovine α_{s1}-casein digested with bovine trypsin. *Animal Science Journal, 74*, 427–435.

Pakkanen, R., & Aalto, J. (1997). Growth factors and antimicrobial factors of bovine colostrums. *International Dairy Journal, 7*, 285–297.

Pellegrini, A. (2003). Antimicrobial peptides from food proteins. *Current Pharmaceutical Design, 9*, 1225–1238.

Pellegrini, A., Thomas, U., Bramaz, N., Klauser, S., Humziker, P., & von Fellenberg, R. (1997). Identification and isolation of a bactericidal domain in chicken egg white lysozyme. *Journal of Applied Microbiology, 82*, 372–378.

Pellegrini, A., Thomas, U, Bramaz, N., Hunziker, P., & Von Fellenberg, R. (1999). Isolation and identification of three bactericidal domains in the bovine α–lactalbumin molecule. *Biochimica et Biophysica Acta, 1426*, 439–448.

Pellegrini, A., Dettling, C., Thomas, U., & Hunziker, P. (2001). Isolation and characterization of four bactericidal domains in the bovine β-lactoglobulin. *Biochimica et Biophysica Acta, 1526*, 131–140.

Piertrantoni, A., Ammendolia, M. G., Tinari, A., Siciliano, R., Valenti, P., & Superti, F. (2006). Bovine lactoferrin peptidic fragments envolved in inhibition of Echovirus 6 *in vitro* infection. *Antiviral Research, 69*, 98–106.

Porter, E. M., Dam, E. V., Valore, E. V., & Ganz, T. (1997). Broad-spectrum antimicrobial activity of human intestinal defensin 5. *Infection and Immunology, 65*, 2396–2401.

Prouxl, M., Gauthier, S. F., & Roy, D. (1992). Effect of casein hydrolysates on the growth of bifidobacteria. *Le Lait, 72*, 393–404.

Recio, I., & Visser, S. (1999a). Identification of two distinct antibacterial domains within the sequence of bovine α_{s2}-casein. *Biochimica et Biophysica Acta, 1428*, 314–326.

Recio, I., & Visser, S. (1999b). Two ion-exchange chromatographic methods for the isolation of antibacterial peptides from lactoferrin. *In situ* enzymatic hydrolysis on an ion-exchange membrane. *Journal of Chromatography A, 831*, 191–201.

Recio, I., & Visser, S. (2000). Antibacterial and binding characteristics of bovine, ovine and caprine lactoferrins: A comparative study. *International Dairy Journal, 10*, 597–605.

Recio, I., Quirós, A., Hernández-Ledesma, B., Gómez-Ruiz, J. A., Miguel, M., Amigo, L., López-Expósito, I., Ramos, M., & Aleixandre, A. (2005). Bioactive peptides identified in enzyme hydrolysates from milk caseins and procedure for their obtention. Spanish patent application ES200501373.

Roy, M. K., Watanabe, Y., & Tamai, Y. (1999). Induction of apoptosis in HL-60 cells by skimmed milk digested with a proteolytic enzyme from the yeast *Saccharomyces cerevisiae*. *Journal of Bioscience Bioengineering, 88*, 426–432.

Roy, M. K., Kuwabara, Y., Hara, K., Watanabe, Y., & Tamai, Y. (2002). Peptides from the N-terminal end of bovine lactoferrin induce apoptosis in human leukemic (HL-60) cells. *Journal of Dairy Science, 85*, 2065–2074.

Salzman, N. H., Polin, R. A., Harris, M. C., Ruchelli, E., Hebra, A., Zirin-Butler, S., Jawad, A., Porter, E. M., & Bevins, C. L. (1998). Enteric defensin expression in necrotizing enterocolitis. *Pediatric Research, 44*, 20–26.

Schiffer, M., Chang, C. H., & Stevens, F. J. (1992). The functions of tryptophan residues in membrane proteins. *Protein Engineering, 5*, 213–214.

Schupbach, P., Neeser, J. R., Golliard, M., Rouvet, M., & Guggenheim, B. (1996). Incorporation of caseinoglycomacropeptide and caseinophosphopeptide into the salivary pellicle inhibits adherence of mutans streptococci. *Journal of Dental Research, 75*, 1779–1788.

Smith, J. A., Wilkinson, M. C., & Liu, Q. M. (1997). Casein fragments having growth promoting activity. International patent WO 97/16460.

Strøm, M. H., Haug, B. E., Rekdal, O., Skar, M. L., Stensen, W., & Svendsen, J. S. (2002). Important structural features of 15 residue lactoferricin derivatives and methods for improvement of antimicrobial activity. *Biochemistry and Cell Biology, 80*, 65–74.

Teraguchi, S., Ozawa, K., Yasuda, S., Shin, K., Fukuwatari, Y., & Shimamura, S. (1994). The bacteriostatic effects of orally administered bovine lactoferrin on intestinal *Enterobacteriaceae* of SPF mice fed bovine milk. *Bioscience, Biotechnology and Biochemistry, 58*, 482–487.

Teraguchi, S., Shin, K., Ogata, T., Kingaku, M., Kaino, A., Miyauchi, H., Fukuwatari, Y., & Shimamura, S. (1995). Orally administered bovine lactoferrin inhibits bacterial translocation in mice fed bovine milk. *Applied and Environmental Microbiology, 61*, 4131–4134.

Tomita, M., Wakabayashi, H., Yamauchi, K., Teraguchi, S., & Hayasawa, H. (2002). Bovine lactoferrin and lactoferricin derived from milk: Production and applications. *Biochemistry and Cell Biology, 80*, 109–112.

Tunzi, C. R., Harper, P. A., Bar-Oz, B., Valore, E. V., Semple, J. L., Watson-MacDonell, J., Ganz, T., & Ito, S. (2000). β-Defensin expression in human mammary gland epithelia. *Pediatric Research, 48*, 30–35.

Turner, J., Cho, Y. Dinh, N. N., Waring, A. J., & Lehrer, R. I. (1998). Activities of LL37, a cathelin-associated antimicrobial peptide of human neutrophils. *Antimicrobial Agents and Chemotherapy, 42*, 2206–2214.

Ulvatne, H., Samuelsen, Ø., Haukland, H. H., Krämer, M., & Vorland, L. H. (2004). Lactoferricin B inhibits bacterial macromolecular synthesis in *Escherichia coli* and *Bacillus subtilis*. *FEMS Microbiology Letters, 237*, 377–384.

Valore, E. V., Park, C. H., Quayle, A. J., Wiles, K. R., McCray, P. B., & Ganz, T. (1998). Human β-defensin-1, an antimicrobial peptide of urogenital tissues. *Journal of Clinical Investigation, 101*, 1633–1642.

van der Kraan, M. I. A., Groenink, J., Nazmi, K., Veerman, E. C. I., Bolscher, J. G. M., & Nieuw Amerongen, A. V. (2004). Lactoferrampin: A novel antimicrobial peptide in the N1-domain of bovine lactoferrin. *Peptides, 25*, 177–183.

van Hooijdonk, A. C. M., Kussendrager, K. D., & Steijns, J. M. (2000). *In vivo* antimicrobial and antiviral activity of components in bovine milk and colostrums involved in non-specific defence. *British Journal of Nutrition, 84*, 127–134.

Vogel, H. J., Schibli, D. J., Weiguo, J., Lohmeier-Vogel, E. M., Epand, R. F., & Epand, R. M. (2002). Towards a structure-function analysis of bovine lactoferricin and related tryptophan and arginine containing peptides. *Biochemistry and Cell Biology, 80*, 49–63.

Vorland, L. H., Ulvatne, H., Andersen, J., Haukland, H. H., Rekdal, Ø., Svendsen, J. S., & Gutteberg, T. J. (1998). Lactoferricin of bovine origin is more active than lactoferricins of human, murine and caprine origin. *Scandinavian Journal of Infectious Diseases, 30*, 513–517.

Vorland, L. H., Ulvatne, H., Rekdal, Ø., & Svendsen, J. S. (1999). Initial binding sites of antimicrobial peptides in *Staphylococcus aureus* and *Escherichia coli*. *Scandinavian Journal of Infectious Diseases, 31*, 467–473.

Wakabayashi, H., Takase, M., & Tomita, M. (2003). Lactoferricin derived from milk protein lactoferrin. *Current Pharmaceutical Design, 9*, 1277–1287.

Wakabayashi, H., Kuwata, H., Yamauchi, K., Teraguchi, S., & Yoshitaka, T. (2004). No detectable transfer of dietary lactoferrin or its multifunctional fragments to portal blood in healthy adults rats. *Bioscience, Biotechnology and Biochemistry, 68*, 853–860.

Wakabayashi, H., Yamauchi, K., & Takase, M. (2006). Lactoferrin: Research, technology and applications. *International Dairy Journal, 16*, 1241–1251.

Yamauchi, K., Tomita, M., Giehl, T. J., & Ellison, R. T., III (1993). Antibacterial activity of lactoferrin and a pepsin-derived lactoferrin peptide fragment. *Infection and Immunity, 61*, 719–728.

Yang, D., Chertov, O., Bykovskaia, S. N., Chen, Q., Buffo, M. J., Shogan, J., Anderson, M., Schroder, J. M., Wang, J. M., Howard, O. M. Z., & Oppenheim, J. J. (1999). Beta-defensins: Linking innate and adaptative immunity through dendritic and T-cell CCR6. *Science, 286*, 525–528.

Yang, N., Strøm, M. B., Mekonnen, S. M., Svendsen, J. S., & Rekdal, Ø. (2004). The effects of shortening lactoferrin derived peptides against tumour cells, bacteria and normal human cells. *Journal of Peptide Science, 10*, 37–46.

Zaiou, M., Nizet, V., & Gallo, R. L. (2003). Antimicrobial and protease inhibitory functions of the human cathelicidin (hCAP18/LL37) prosequence. *Journal of Investigation in Dermatology, 120*, 810–816.

Zanetti, M. (2004). Cathelicidins, multifunctional peptides of the innate immunity. *Journal of Leukocyte Biology, 75*, 39–48.

Zasloff, M. (2002). Antimicrobial peptides of multicellular organisms. *Nature, 415*, 389–395.

Antihypertensive Peptides Derived from Bovine Casein and Whey Proteins

Tadao Saito

Abstract Peptides play an important primary role as a supply of essential amino acids and a source of nitrogen. Recent studies have reported on another role of peptides: having specific amino acid sequences that can express some biological functions *in vivo*. For an exhaustive study and supply of biologically active peptides, a large-scale screening of protein sources is necessary. Various physiologically functional peptides, such as opioid, immunostimulating, mineral carrier, ACE inhibitory, antihypertensive, and antimicrobial peptides, have been derived from milk protein: both caseins and whey proteins (Meisel, 1998; Korhonen & Pihlanto-Leppälä, 2001).

Milk is known to be a rich source for the supply of bioactive peptides compared to other protein sources such as animal and fish meat, wheat, and soybean proteins. Among the bioactive peptides, ACE inhibitory peptides and antihypertensive peptides have been extensively researched worldwide, because hypertension is a major risk factor in cardiovascular disease, such as heart disease (FitzGerald & Meisel, 2000; Kitts & Weiler, 2003).

We discuss the isolation, utilization, and application of bioactive peptides, especially ACE inhibitory peptides and antihypertensive peptides including our recent human studies on their use as a functional food material.

Angiotensin-Converting Enzyme and Its Inhibitors

Angiotensin I-converting enzyme (ACE, kininase II, EC 3.4.15.1) is a carboxy dipeptidyl metallopeptidase. ACE is predominantly expressed as a membrane-bound form in vascular endothelial cells, in epithelial or neuroepithelial cells, and in the brain, and it also exists as a soluble form in blood and numerous body fluids (Skidgel & Erdos, 1993). ACE cleaves the C-terminal dipeptide from

---

T. Saito
Graduate School of Agricultural Science, Tohoku University, Tsutsumidori-Amamiyamachi, Aoba-ku, 981-8555, Sendai, Japan, Tel: 81227178711, Fax: 81227178715
e-mail: tsaito@bios.tohoku.ac.jp

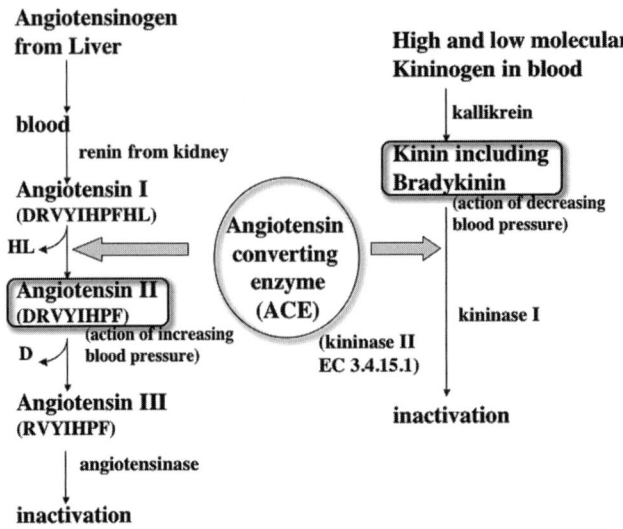

Fig. 1 The regulation system of human blood pressure by the rennin-angiotensin system and the Kallikrein-Kinin system

angiotensin I and bradykinin. The renin-angiotensin system (RAS) includes the key enzyme ACE, which plays an important role in blood pressure regulation.

Renin cleaves the inactive decapeptide angiotensin I from the prohormone angiotensinogen, a noninhibiting member of the serpin superfamily of serine protease inhibitors. Angiotensin II is produced from the cleavage of angiotensin I by the action of ACE. Angiotensin II is a potent vasoconstrictor, acting directly on vascular smooth muscle cells. Angiotensin II works through the mediation of AT1 receptor. ACE regulates the balance between the renin-angiotensin system (increasing blood pressure) and the kallikrein-kinin system (decreasing blood pressure) (Fig. 1). ACE inhibition lowers blood pressure and is a key clinical target for blood pressure control. The first competitive inhibitors to ACE were reported as naturally occurring peptides isolated from snake venom (Ferreira et al., 1970; Ondetti et al., 1971).

ACE inhibitors (ACEi) as drugs (pharmacological substances) are well established in the therapy of hypertension and heart failure and have been shown to exert organ-protective effects (Parmley, 1998). In order to prevent cardiovascular disease, drugs can be used to decrease high blood pressure to within the normal range. There are five categories of antihypertensive drugs: (1) ACE inhibitors, (2) calcium channel antagonists, (3) β- and α-blockers, (4) natriuretic agents, and (5) endothelin receptor antagonists. At the present time, many ACE inhibitors are commercially available in various countries as monotherapeutic drugs. However, ACE inhibitory drugs are known to produce several side effects such as cough and fetal abnormalities, thus provoking the global research and search for natural and safe ACE inhibitors.

Bioactive Peptides Derived from Milk

Characteristics of Milk Proteins (Casein and Whey Protein)

Milk contains two major protein groups: caseins and whey proteins. Caseins, which occupy about 80% of the total protein in bovine milk, exist mainly in macromolecular complexes as casein micelles consisting of more than 1,000 casein submicelles. Caseins are known to be precursors of a number of different bioactive peptides. Casein is a group of phosphoproteins and consists of about 30 different components including genetic variants. Casein consists mainly of α_{s1}-, α_{s2}-, β-, and κ-casein (Swaisgood, 2003).

The whey proteins, which account for about 20% of the total milk proteins in bovine milk, represent an excellent source of both functional and nutritious proteins. The main whey protein constituents, β-lactoblobulin and α-lactalbumin, account for 70–80% of the total whey proteins in bovine milk. Other minor components include bovine serum albumin (BSA), immunoglobulins (Igs) (mainly the G type), lactoferrin (LF), lactoperoxidase (LP), proteose-peptones (PP), and many enzymes (Fox, 2003).

Milk proteins have been identified as an important source of several bioactive peptides. An important and interesting point is that these peptides are in an inactive state within the milk protein molecule in the natural form and can be released during enzymatic digestion *in vitro* and *in vivo*.

ACE Inhibitory Peptides from Milk Proteins

Milk protein-derived peptides have been found to have a variety of specific activities, such as antihypertensive, antimicrobial, immunomodulatory, opioid, and mineral-binding traits. Some peptides show multifunctional activities, i.e., specific peptide sequences may exert several different biological activities. The bioactivity in antihypertension is mainly due to the inhibition of ACE, which plays two important roles in the regulation of blood pressure, as already mentioned. One is the conversion of the inactive decapeptide angiotensin I (DRVYIHPFHL) to the potent vasoconstrictor and salt-retaining octapeptide angiotensin II (DRVYIHPF). The other role is the inactivation of the vasodilator and natriuretic nonapeptide bradykinin. ACE is very important for controlling the balance between blood pressure-increasing and -decreasing systems and the maintenance of homeostasis in the blood system (see Fig. 1). Therefore, we can expect a decrease in blood pressure if some peptides can inactivate ACE located in the body after intake from the intestine.

In recent years, many review articles have been published about milk-borne bioactive peptides (FitzGerald et al., 2004). Milk proteins are known to be a good source of bioactive peptides such as ACE inhibitory peptides. The ACE inhibitory peptides have been produced by the enzymatic hydrolysis of milk proteins or by fermentation with lactic acid bacteria. Many milk peptides have been reported to inhibit ACE *in vitro*. Table 1 shows the ACE inhibitor peptides derived from bovine milk proteins by enzymatic treatment or fermentation by lactic acid bacteria (LAB). Large peptides with more than 13 amino acid residues were omitted from this table.

There is no apparent consensus on the peptide sequence for the expression of ACE inhibitory activity because no common sequence was observed in those bioactive peptides. From research to date, the three residues in the C-terminal region of peptide seem to bind to the active center of ACE, and it seems that high ACE inhibitory activity is observed if the hydrophobic amino acids including aromatic amino acids such as Trp, Tyr, and Phe or the imino acid Pro are located in this position. Moreover, the positive charge from Arg and/or Lys residues may increase the inhibitory activity. Especially in milk proteins, low-molecular-weight peptides containing Pro residues are considered to show very strong ACE inhibitory activity.

Antihypertensive Peptides from Milk Proteins

Several milk peptides have been reported to inhibit ACE clearly *in vitro*. Recently, much research has involved animal experiments with spontaneously hypertensive rats (SHR) to evaluate the effects of antihypertensive peptides. Some *in vitro* ACE inhibitory peptides have been confirmed to show little or no antihypertensive activity with *in vivo* experiments using SHR. Therefore, it is recommended to introduce animal experiments with SHR in the research process at the final stage in order to distinguish ACE inhibitory from antihypertensive peptides. We should categorize peptides as antihypertensive when they decrease blood pressure after SHR experiments. Table 2 shows antihypertensive peptides derived from milk proteins.

From our knowledge of nutritional science, the large peptides such as those larger than tetrapeptides seem to prevent their intake into intestinal epithelial cells during the digestion process because of the lack of specific receptors or transporters. Therefore, antihypertensive peptides with high molecular weight such as more than tridecapeptides (13 amino acid residues) were omitted from the table. Although some active region must exist in these peptides, the determination of the active epitope position is not clear in every large peptide.

Enzymatic Preparation of ACE Inhibitory and Antihypertensive Peptides from Cheese Whey Proteins and Their Application

In the mass manufacturing process, bioactive peptides can be produced from milk proteins by three methods: (1) enzymatic hydrolysis with protease, (2) fermentation of milk by microorganisms with high protease activity, or

Table 1 Angiotensin I-converting Enzyme Inhibitors (ACEI) Derived from Milk Proteins

Protein Source (Origin)	Peptide (a.a. sequence)	Preparation	$IC_{50}(\mu M)$	Notes	References
Casein					
α_{s1}-casein	LW	synthesis	50		1
	YL	proteinase	122		4
	RY	synthesis	10.5		6
	PLW	synthesis	36		1
	VAP	synthesis	2	α_{s1}-casokinin	1
	FVAP	synthesis	10		1
	FFVAP	peptidase	6	CEI_5	1
	AYFYPE	trypsin	16		1
	LAYFYP	LAB fermentation	65		4
	TTMPLW	trypsin	12		6
	YKVPQL	LAB proteinase	22		1
	DAYPSGAW	LAB fermentation	98		4
	RPKHPIKHQ	Gouda cheese manufacture	13.4		6
	FFVAPFPEVFGK	trypsin	18	CEI_{12}	6
α_{s2}-casein	RY	synthesis	10.5		6
	TVY	trypsin	15		6
	IPY	synthesis	206		6
	VRYL	synthesis	24.1		6
	FALPQY	trypsin	4.3	α_{s2}-casokinin	3
	FPQYLQY	trypsin	14		6
	NMAINPSK	trypsin	60		6
β-casein	FP	proteinase K	315		1
	VYP	proteinase K	288		1
	IPA	proteinase K	141		1
	IPP	LAB fermentation	5	β-casokinin	1
	VPP	LAB fermentation	9		1
	VYPFPG	proteinase K	221		1

Table 1 (continued)

Protein Source (Origin)	Peptide (a.a. sequence)	Preparation	IC$_{50}$(μM)	Notes	References
	YQQPVL	fermentation	280		4
	AVPYPQR	trypsin	15		1
	YPFPGPI	proteinase	500	β-casomorphin 7	4
	RDMPIQAF	proteinase of L.helveticus	209		4
	YQQPVLGPVR	proteinase	300	β-casokinin 10	4
	TPVVVPPFLQP	proteinase K	749		1
κ-casein	YP	fermentation	720		6
	VTSTAV	proteinase	52	κ-casokinin	3
	YIPIQYVLSR	trypsin	100	casokisin C	4
Whey proteins					
α-lactalbumin	LF	synthesis	349		6
	YGL	pepsin, trypsin	409		2
	YGLF	pepsin, trypsin, chymotrypsin	733	α-lactorphin	2
	LAHKAL	fermentation	621		2
	WLAHK	trypsin	77		2
	VGINYWLAHK	trypsin	327	lactokinin	2,3
β-lactoglobulin	VFK	trypsin	1029		5
	IPA	proteinase K	141		5
	LAMA	trypsin	1062		5
	YLLF	Pepsin, Trypsin, Chymotrypsin	172	β-lactorphin	2
	ALPMH	Pepsin, Trypsin, Chymotrypsin	521		5

Table 1 (continued)

Protein Source (Origin)	Peptide (a.a. sequence)	Preparation	IC$_{50}$(μM)	Notes	References
	CMENSA	trypsin	788		5
	G L D I Q K	fermentation	580		2
	VAGTWY	fermentation	1682		2
	ALPMHIR	trypsin	43	lactokinin	2,3
	VLDTDYK	Pepsin, Trypsin, Chymotrypsin	946		5
	LDAQSAPLR	trypsin	635		5
serum albumin	FP	proteinase K	315		6
	ALKAWSVAR	trypsin	3	albutensin A	3
others	FL	alkaline proteinase	16		4
	VY	alkaline proteinase	18		4
	IL	alkaline proteinase	21		4
β$_2$-microglobulin	GKP	proteinase K	352		6

references 1(Mizuno and Yamamoto, 2004)
2(Pihlanto-Leppälä, 2001)
3(FitzGerald, Murray and Walsh, 2004)
4(Saito, Nakamura and Itoh, 2000b)
5(Pihlanto-Leppälä et al., 2000)
6(Murray and FitzGerald, 2007)

Table 2 Antihypertensive Peptides Derived from Milk Proteins

Protein Source (Origin)	Peptide Sequence	Preparation	IC$_{50}$(μM)	Dose(mg/kg)	SBP(mmHg)	Notes	References
Casein							
α$_{s1}$-casein	YP	fermentation	720	1	−27.4		1
	TTMPLW	trypsin	16	100	−13.6		1
	YKVPQL	LAB proteinase	22	1	−12.5		1
	RPKHPIKHQ	cheese fermentation	13.4	6.1-7.5	−9.3		1
	FFVAPFPEVFGK	trypsin	77	100	−13		1
α$_{s2}$-casein	TKVIP		400		−9		2
	AMPKPW		580		−5		2
	MKPWIQPK		300		−3		2
β-casein	YP	fermentation	720	1	−27.4		1
	FP	Proteinase K	315	8	−27	β-lactocin A	1
	VPP	LAB fermentation	9	1.6	−32.1		2
	VYP	Proteinase K	288	8	−21	p-peptocin B	2
	IPP	LAB fermentation	5	1	−28.3		2
	LQSW	proteinase	500		−2		2
	VYPFPG	Proteinase K	221	8	−22		1
	KVLPVP	digestive enzyme	5	1	−32.2		1
	KVLPVPQ	proteinase	>1000	1	−31.5		2
	AVPYPQR	trypsin	15	100	−10	CEI$_{β7}$	1
	YPFPGPIPN	cheese fermentation	14.8	6.1-7.5	−7		3
	TPVVVPPFLQP	Proteinase K	749	8	−8	p-peptocin C	2

Table 2 (continued)

Protein Source (Origin)	Peptide Sequence	Preparation	IC$_{50}$(μM)	Dose(mg/kg)	SBP(mmHg)	Notes	References
κ-casein	FFVAPFPEVFGK	trypsin	77	100	−13	CEI$_{12}$	1
	YP	fermentation	720	1	−27.4		1
	IPP	LAB fermentation	5	1	−15.1		1
	IASGQP	pepsin	>1000	6.7-7.1	−22.5		
Whey proteins							
α-lactalbumin	YGLF	pepsin	733	10	−17	α-lactorphin	2
β-lactoglobulin	IPA	Proteinase K	141	8	−31		1
β-microglobulin	GKP	Proteinase K	352	8	−26	β-microcin A	1
serum albumin	FP	Proteinase K	315	8	−27	β-lactocin A	1

references 1 (Yamamoto, Ejiri and Mizuno, 2003)
2 (FitzGerald, Murray and Walsh, 2004)
3 (Huth, DiRienzo and Miller, 2006)

(3) through the action of enzymes derived from proteolytic microorganisms. In case 1, we can use a variety of commercially available food-grade proteases such as exopeptidases and endopeptidases to hydrolyze milk proteins. Many ACE inhibitory peptides have been isolated and identified in enzymatic hydrolysates of bovine casein. The isolation of ACE inhibitory peptides from whey protein is usually limited by the rigid structure of native β-lactoglobulin because it is resistant to digestive enzymes such as pepsin and pancreatin.

In our laboratory, we have recently reported on several ACE inhibitory and antihypertensive peptides prepared by enzymatic digestion of cheese whey proteins as byproducts from the manufacture of cheese (Abubakar et al., 1998; Saito, 2004). Whey protein isolated from cheese whey powder was digested for 24 hours with seven proteases (pepsin, trypsin, chymotrypsin, proteinase K, actinase E, thermolysin, and papain). Seven whey protein digests were submitted for evaluation of ACE inhibitory activity and systolic blood pressure (SBP) in SHR. Table 3 shows the results of ACE inhibitory activity (*in vitro*) and antihypertensive activity (*in vivo*). The strong ACE inhibitory activity of more than 95% was derived after digestion with thermolysin (98.6%) and proteinase K (95.7%). In the SHR experiment, 6 hours after gastric intubation, significant antihypertensive activity was observed after digestion with trypsin (−51 mm Hg), proteinase K (−55 mm Hg), and actinase E (−55 mm Hg). The decreasing effect on SBP was generally maintained 6 to 12 hours after gastric intubation, and the SBP generally returned to the initial value (220 ± 3.0 mm Hg) after 24 hours. Based on these results, proteinase K was selected as the most potent protease that induced the most effective component in both *in vitro* (96.7%) and *in vivo* (−55 mm Hg) evaluations. The sample was fractionated by hydrophobic chromatography using a LiChroprep RP-18 resin, and fraction 5 was selected as that with the highest antihypertensive activity (−46 mm Hg). The fractionation of fraction 5 was performed by HPLC in both RP and

Table 3 The Derivation of the Inhibitory Activity of ACE and Antihypertensive Activity from Whey Protein After Digestion with One of Seven Proteaseses (Abubakar et al., 1998)

Sample[1]	ACE Inhibitory Activity (%)	Decreased SBP (mm Hg)[2]	
		X	SE
whey protein (control)	0	− 38	1.7
Pepsin	83.7	− 47	2.6
Trypsin	56.7	− 51*	3.5
Chymotrypsin	76	− 40	4.4
Proteinase K	95.7	− 55**	2.6
Actinase E	55.7	− 55**	4.4
Thermolysin	98.6	− 42	3.5
Papain	86.5	− 47	3.6

[1] Dose was 8 mg of whey protein hydrolysate/kg per sample.
[2] The mean systolic blood pressure (SBP) for the entire group was about 220 mmHg before administration. The number showed the mean value (n = 3) of the decreasing SBP in SHR.
* Different from control ($P < 0.05$) ** Different from control ($P < 0.01$).

Fig. 2 An elution profile of the fraction 5 by reversed-phase HPLC. Column: Superiorex ODS (4.6 × 150 mm;, Shiseido), elution A(10% CH3CN Containing 0.05% TFA) and B(60% CH3CN containing 0.05% TFA), mobile phase: elution A 100% to elution B 100% within 30 min., flow rate 0.5 ml/min At 40 °C and detection at 220 nm

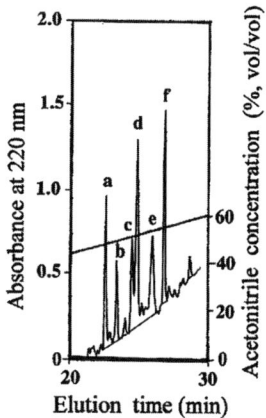

gel-filtration modes. A representative elution profile of fraction 5 by RP-HPLC is shown in Fig. 2. After fractionation by ion-exchange chromatography of the six peaks a–f, six major components (a, b1, b2, c2, d, f) were further purified and fractionated by GPC-HPLC. Finally, six new antihypertensive peptides [VYPFPG (β-casein: f59–64), GKP (β-microglobulin: f18–20), IPA (β-lactoglobulin: f78–80), FP (β-casein: f62–63, f157–158, f205–206, serum albumin: f221–222), VYP (β-casein: f59–61), and TPVVVPPFLQP (β-casein: f80–90)] were identified. It was observed that peptides with high ACE inhibitory activity were relatively short sequences with a Pro residue at the C-terminus.

Between 2002 and 2003, a new type of functional yogurt named "Fitdown" was jointly developed and was on the market under the collaboration of our laboratory and Japan Milk Community Co., Ltd. (Fig. 3). In this yogurt, the

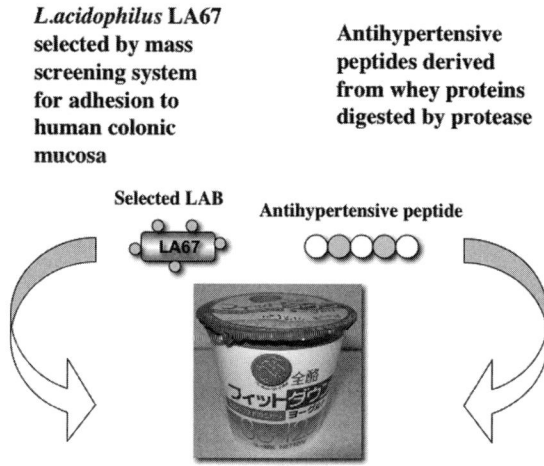

Fig. 3 A new functional yogurt "Fitdown" which expect to deccrease hypertension risk

mixture of antihypertensive peptides derived from cheese whey protein by actinase E digestion was added and fermented with probiotic LAB: *L. acidophilus* LA67, which shows binding affinity to rat and human intestinal mucin (Matsumura et al., 1999). The daily intake of this yogurt resulted in a decrease in blood pressure in hypertensive consumers.

ACE Inhibitory Peptides from Fermented Milk Product

Discovery of Lactotripeptide: IPP and VPP

Fermentation with LAB involves the proteolytic processing of proteins to release peptides for use as a nitrogen source. LAB are suitable microbes for milk fermentation because they have a proteolytic system that decomposes casein, along with lactose hydrolyzing enzymes. A variety of oligopeptides released from casein by an extracellular proteinase of LAB have been reported. Recently, an antihypertensive effect related to ACE inhibitory peptides was found in sour milk produced by *L. helveticus*. Two kinds of bioactive peptides, IPP and VPP, with ACE inhibitory activity were isolated and identified from the sour milk, which had been fermented until pH 3.3 (Nakamura et al., 1995a). The two tripeptides named "lactotripeptide" were confirmed as having antihypertensive activity using SHR (Nakamura et al., 1995b). The ACE inhibitory activities of the two peptides VPP and IPP were very high compared to other reported peptides, and the concentrations of peptides producing 50% inhibition of ACE (IC_{50} value) were 9 and 5 µM, respectively. The amino acid sequences of VPP and IPP were found in the primary structure of bovine β-casein (84–86) (74–76) and κ-casein (108–110), respectively. They were produced during fermentation, but were not found in the hydrolysate of casein after digestion with an extracellular proteinase of *L. helveticus* (Mizuno & Yamamoto, 2004). They may have been processed from the casein molecule by an extracellular proteinase, followed by peptidase action during fermentation. The importance of the extracellular proteinase in the first decomposition of casein and the endopeptidase in the carboxyl terminal processing has been suggested (Yamamoto et al., 1994). Recently, the *L. helveticus* LBK-16H-fermented milk containing IPP and VPP, when consumed in normal daily ingestion, had a blood pressure-lowering effect in hypertensive patients in Japan (Hata et al., 1996) and Finland (Seppo et al., 2003).

Mode of Action of Lactotripeptide

In order to achieve the antihypertensive function *in vivo*, antihypertensive peptides such as IPP and VPP must be absorbed from the intestine in an active form without decomposition. It is well known that small peptides, such as di- and tripeptides, are easily adsorbed in the intestine. In SHR fed with fermented milk containing these two peptides, ACE activities in the aorta, heart, liver, testes, kidney, lung, and brain were measured. Among various organs, ACE activity in the aorta was significantly lower in the animals fed with sour milk than in the control group (Nakamura et al., 1996). Interestingly, the major antihypertensive peptides in the sour milk, VPP and IPP, were detected in a heat-treated solubilized fraction from the abdominal aorta of rats fed with the sour milk, but not in rats fed with unfermented milk (Masuda et al., 1996). The results suggested that a pair of lactotripeptides were absorbed directly, without being decomposed by digestive enzymes, and transported to the abdominal aorta, where they inhibited ACE, producing an antihypertensive effect in SHR. The elimination half-life of an antihypertensive peptide from the blood and organs may affect its antihypertensive activity, and this may be different for each peptide or drug. In the case of the most popular ACE inhibitory drug captopril, the half-life in blood was estimated to be about 1.5 hours and about 3.5 hours for Val-Tyr (Matsui et al., 2002). The antihypertensive effect of a drug such as captopril was observed a few hours after administration, but the effects of lactotripeptides continued for more than 10 hours. The reasons for the difference in half-life between drug and bioactive peptides are still unclear. In research using Caco-2 cells, the transport of VPP across the Caco-2 cell monolayer via paracellular diffusion and quick enzymatic hydrolysis of the transported intracellular VPP in the cell has been suggested (Satake et al., 2002). More information about the elimination half-life and pharmacokinetics of peptides would help in the future understanding of their precise mode of action.

Enzymatic Mass Production of Antihypertensive Peptides Including IPP and VPP from Casein

Recently, an enzyme suitable for release of VPP and IPP from casein was successfully selected from commercially available food-grade proteases (Mizuno et al., 2004). During screening for a suitable enzyme, ACE inhibitory activities were measured after hydrolysis of casein by nine different proteases. Among these enzymes, a hydrolysate with a protease from *Aspergillus oryzae* (Sumizyme FP) showed the highest ACE inhibitory activity per unit weight (μg) of hydrolysate (Table 4). The casein hydrolysate prepared with Sumizyme FP also showed the highest activity in blood pressure-lowering activity among the nine hydrolysates tested. To further understand the *in vitro* and *in vivo* activities of the *A. oryzae* hydrolysate among the other casein hydrolysates, the number of N-terminal ends of all samples was calculated by the OPA method (Mizuno et al., 2004). The average peptide length of the *A. oryzae* hydrolysate was 1.4, which was the shortest of all tested samples; most of the other peptides had more than 4.0 amino acids (Trypsin = 6, Sumizyme CP = 4, Protease S = 4.5,

Table 4 Comparison of ACE Inhibitory Activities and Antihypertensive Effects of Various Enzymatic Casein Hydrolysates (Mizuno et al., 2004)

Protease	Origin	ACE Inhibitory Activity (%/µg)	Decrease in SBP (mm Hg)[1] X	SD
Sumizyme FP	Aspergillus oryzae	4.1	−25.0***	4.3
Trypsin	Porcine pancreas	0.8	− 4.3	5.3
Sumizyme CP	Bacillus subtilis	1.2	not tested	
Protease S	B. stearothermophilus	1.4	− 3.7	6.4
Papain	Carica papaya	1	2.8	5.2
Thermoase	B. stearothermophilus	1.1	not tested	
Neurase F3G	Rhizopus nivenus	0.9	2.9	5.4
Sumizyme RP	Rhizopus delemar	0.7	− 5.1	7
Bromeraine	Pineapple cannery	1.1	− 4.6	6.9
control	no enzyme	not tested	− 2.5	7.5

[1] Mean of changes in SBP at 5 hr after an oral administration to SHR (n = 5).
*** significant difference from the control ($p < 0.001$).
ACE inhibitory activity was measured by the method of Nakamura et al. (1995a).

Papain = 8, Neurase F3G = 5, Sumizyme RP = 4, Thermoase = 4.5, Bromeraine = 5). The result suggests that peptide length may be an important factor connected with ACE inhibitory and antihypertensive activities.

To further characterize the *A. oryzae* hydrolysate, the amino acid sequence of the whole peptide mixture was investigated. Various amino acids were detected at the first C-terminal position, but Pro residues were frequently present at the second and third positions of the peptides in the mixture. The result suggests that the *A. oryzae* peptide mixture with potent antihypertensive activity in SHR mainly contains short peptides of X-Pro and X-Pro-Pro sequences. It has been suggested that strong *in vitro* ACE inhibitory and antihypertensive activity in SHR of the *A. oryzae* hydrolysate may be due to Pro-rich peptides.

ACE Inhibitory and Antihypertensive Peptides from Several Cheeses

Ripened cheese is an important dairy product and contains numerous peptides that originate mainly from casein by proteolysis during the ripening period and contribute to the flavor, taste, and texture of the cheese. Although there have been a few reports on ACE inhibitory peptides in several cheeses, there is little information on antihypertensive substances that exist naturally in cheeses.

We studied seven kinds of ripened cheeses (8- and 24-month-aged gouda, emmental, blue, camembert, edam, and havarti) (Saito et al., 2000). Water-soluble peptides were prepared by hydrophobic chromatography with Wakogel

Table 5 The ACE Inhibitory Activity and Antihypertensive Activity of Free Peptides Prepared from Seven Different Cheeses by Hydrophobic Chromatography (Saito et al., 2000)

Samples	ACE Inhibitory Activity (%)	Decreased SBP (mm Hg)[1]	
		X	SE
Gouda (8 mo)	75.5	− 24.7**	0.3
Gouda (24 mo)	78.2	− 17.2	4.5
Emmental	48.8	− 13	4.1
Blue	49.9	− 20.3*	3.8
Camembert	69.1	− 7.1	3.1
Edam	56.2	− 20.7*	3.3
Havarti	72.7	− 20.0**	2

[1]SBP(mm Hg): After 6h of oral administration of the free peptide sample (2 mg/2ml of distilled water, dosage of 6.1-7.5 mg/kg of body weight) to SHR (n = 3) by gastric intubation, SBP (systolic blood pressure) was measured.
* Significantly different from the control ($P < 0.05$)
** Significantly different from the control ($P < 0.01$)

LP-40C18. The ACE inhibitory and antihypertensive effects of the peptides prepared from six different cheeses were determined (Table 5). The highest inhibitory effect against ACE was detected in gouda aged for 24 months. In the SHR experiments, the decrease in SBP (mm Hg) was statistically significant for four cheeses (gouda aged for 8 months, blue, edam, havarti). The strongest antihypertensive effect was observed in the peptide sample from gouda cheese aged for 8 months (-24.7 ± 0.3 mm Hg, $P < 0.01$). The peptide sample was fractionated by hydrophobic chromatography, and all fractions were tested for ACE inhibitory and antihypertensive activities. The strongest effects in both assays were found in the 45% CH_3OH fraction (ACE inhibitory activity, 58.4%; -29.3 ± 0.9 mm Hg, $P < 0.01$). Figure 4 shows a typical elution profile of the peptide fraction G8–45 (45% CH_3OH fraction from gouda aged for 8 months) by RP-HPLC. Although about 40 peaks were observed in the chromatogram, only the seven major peaks were selected, which were termed A to G. After GPC-HPLC, four peptides (A, B, F, and G) were isolated and were subjected to structural analysis. The N-terminal amino acids of four peptides (A, B, F, and G) were Arg, Arg, Tyr, and Met, respectively, and molecular weights were determined by fast atom bombardment-mass spectrometry (FAB-MS) to be 1141, 1537, 1002, and 1352 (m/z, $M^+ + H$), respectively. From combination analysis with N-terminal amino acid, molecular weight, amino acid composition, and protein sequence data, the primary structures and origin of four peptides in casein components were clarified.

Table 6 summarizes the primary structures, origins, molecular weight, and IC_{50} values for the ACE inhibitory activity and antihypertensive effects of four peptides isolated from gouda aged for 8 months. Peptide A has previously been isolated from cheddar cheese, and peptide B was the major peptide produced during the ripening process. Various water-soluble peptides that mainly originated from the N-terminal portions of α_{s1}-casein or the internal portions of β-casein have been reported in cheddar and feta. Peptide G was a novel peptide

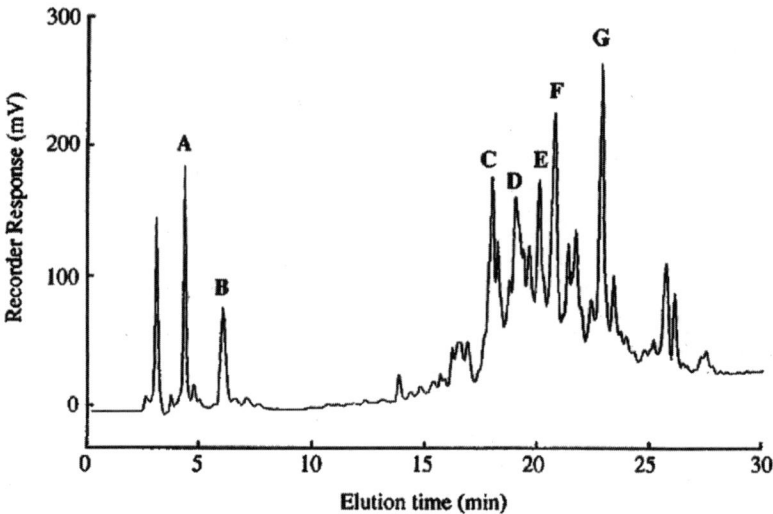

Fig. 4 A typical elution profile of the peptide fraction G8-45 obtained by reversed-phage HPLC Chromatographic conditoins are same as Fig. 2.

that was isolated from general cheeses. All four peptides contained more than two Pro residues located at an internal position.

Chemically synthesized peptides A and F showed potent ACE inhibitory activity (Table 6). Cheung et al. (1979) reported on peptide binding to ACE and showed the importance of hydrophobic (aromatic or branched-chain aliphatic) amino acid residues at each of the three C-terminal positions and the potent inhibitory effect of peptide F containing Ile and Pro at the C-terminus.

Peptides A and F did not have strong antihypertensivity in SHR, and p-peptosin C (undecapeptide, β-casein: f80–90) in our previous research was also weak (−8.0 ± 2.6 mm Hg) in spite of their low IC_{50} values (Abubakar et al., 1998). These peptides are rather large molecules that may require further digestion by intestinal proteases or peptidases before absorption. The antihypertensive activity might be generated after additional digestion in SHR. A deletion study, using peptide derivatives from the N- or C-terminus of the two peptides, would also be very useful to understand the ACE inhibitory mechanism of oligopeptides *in vivo*.

ACE Inhibitory and Antihypertensive Peptides from Commercial Whey Product

Murakami et al. (2004) determined the ACE inhibitory and antihypertensive activities in samples of 12 kinds of protein hydrolysates that are commercially available in Japan. Table 7 lists their origins, average molecular weights, protein contents, and degrees of solubility. A sample of WE80M derived from whey

Table 6 The Primary Structure, Origin, Molecular Mass, Inhibitory Activity of ACE and Antihypertensive Activity of Four Peptides Isolated from Gouda (8 month) Cheese (Saito et al., 2000)

Sample	Sequence	Origin	Molecular Mass (Da)	IC_{50} (μM)[1]	SBP (mm Hg)[2] X	SE
A	RPKHPIKHQ	α_{s1}-casein B-8P (f1-9)	1140	13.4	−9.3	4.8
B	RPKHPIKHQGLPQ	α_{s1}-casein B-8P (f1-13)	1536	not determined		
F	YPFPGPIPD	β-casein A^2-5P (f60-68)	1001	14.8	−7	3.8
G	MPFPKYYPVQPF	β-casein A^2-5P (f109-119)	1351	not determined		

[1] IC_{50}(μM): The 50% inhivitory concentratin (IC_{50}) value is the peptide concentration (μM) that inhibits the activity of ACE(angiotensin I-converting enzyme) by 50%.

[2] SBP(mm Hg): After 6h of oral administration of the free peptide sample (2 mg/2ml of distilled water, dosage of 6.1-7.5 mg/kg of body weight) to SHR (n = 3) by gastric intubation, SBP (systolic blood preasure) was measured.

Table 7 Characteristics of Twelve Commercial Products of Protein Hydrolysate

Peptides[1]	Origin	Average Molecular Weight (Da)	Protein Content(%)	Solubility in Water (g/l)
WE80BG	Whey protein	570	81.5	700
WE80M	Whey protein	3,000	79.4	350
WE90FS	Whey protein	8,500	90.0	–
LE80GF	Whey protein	4,600	77.2	150
CE90STL	Casein	400	86.7	300
CE90GMM	Casein	640	89.9	250
CE90F	Casein	18,500	91.2	200
EE90FX	Ovalbumin	2,000	85.9	150
WGE80GPA	Wheat gluten	660	77.6	200
WGE80GPN	Wheat gluten	670	79.0	400
WGE80GPU	Wheat gluten	6,700	77.5	200
SE50BT	Soybean protein	320	53.2	300

[1] All peptides are commercially available from DMV JAPAN (Osaka) in Japanese market.

protein showed the highest level of ACE inhibitory activity (78.2%). Samples of four peptides derived from milk proteins (WE80BG, WE80M, CE90STL, CE90GMM) showed strong antihypertensive effects (>-18.0 mm Hg) with medium levels of ACE inhibitory activities (>51.0%). A sample of WE80BG was selected and subjected to hydrophobic chromatography with Wakogel LP40C18 resin. The hydrophobic peptides were further fractionated by gel filtration into five fractions according to molecular weight. Fraction 4 showed

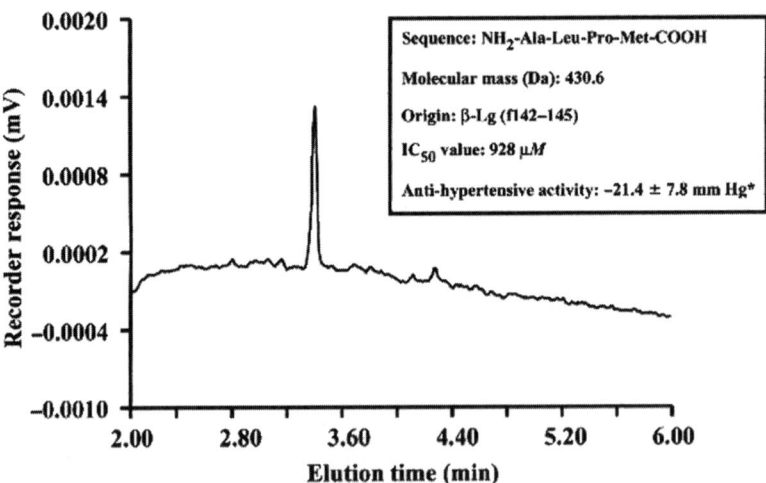

Fig. 5 Elution profile of peptide d by capillary electrophoresis isolated by RP and gel permeation HPLC
The data was recorded by a BioFocus 3000 (Bio-Rad Laboratories) equipped with a coated column (25 μm × 17 cm) in 0.1 M phosphate buffer (pH 2.5) at 10.00 kV for 15 min.

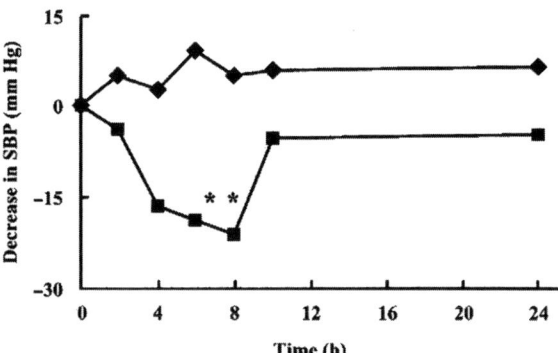

Fig. 6 Changes in systolic blood pressure (SBP) were measured in Spontaneously hypertensive rats (SHR, n = 5) at 2,4,6,8,10 and 24 hr after administration The data was 1mg of ALPM peptide/1 mL of distilled water. Control = ◆, ALPM = ■, *Significant difference form the control ($P<0.05$).

the strongest antihypertensive activity (-18.8 ± 6.8 mm Hg) with a medium level of ACE inhibitory activity (42.5%). Fraction 4 was further purified by RP-HPLC and divided into five fractions (a–e). Peptide d, which was eluted as a single peak, showed the highest level of ACE inhibitory activity (55.2%) and was subjected to the structural analysis. The chemical structure of peptide d was identified as shown in Fig. 5, from amino acid composition analysis, N-terminal analysis by Edman degradation, and molecular mass analysis by FAB-MS.

The tetrapeptide was thought to be originated from β-lactoblobulin (f142–145). The chemically synthesized peptide d (ALPM) showed an IC_{50} value of 928 µM (Murakami et al., 2004). Changes in SBP were measured in SHR at 2, 4, 6, 8, 10, and 24 hours after administration of the peptide (Figure 6). At 6 and 8 hours after administration, SBP was decreased to -19.0 ± 8.6 and -21.4 ± 7.8 mm Hg, respectively, with a significant difference ($P < 0.05$) compared with that of the control.

Mullally et al. (1997) reported that the ACE inhibitory heptapeptide "lactokinin" (ALPMHIR: β-Lg f 142–148) derived by tryptic hydrolysis of whey proteins showed a low IC_{50} value of 42.6 µM. Pihlanto-Leppälä et al. (2000) reported that a pentapeptide (ALPMH: β-Lg f 142–146) had an IC_{50} value of 521 µM and that a tripeptide (HIR: β-Lg f 146–148) had an IC_{50} value of 953 µM. Although ALPM (928 µM) and HIR (953 µM) showed medium levels of ACE inhibitory activity, ALPMHIR showed strong ACE inhibitory activity (42.6 µM).

The importance of hydrophobic amino acid residues in peptides for ACE inhibitory activity has been discussed. Cheung et al. (1980) showed that hydrophobic amino acid residues such as aromatic (Trp, Tyr, and Phe) or imino (Pro) amino acids at each of the three C-terminal positions contribute to the expression of ACE inhibitory activity in peptides. Ondetti and Cushman (1982) reported that the C-terminal tripeptide residues could interact with three subsites of ACE. Meisel (1998) suggested that a positive charge such as that of the guanidine group of Arg is important for ACE inhibition. For the first purification step of antihypertensive peptides from WE80BG, we selected hydrophobic peptides. ALPM is composed of hydrophobic amino acids, and HIR has Arg and the hydrophobic amino acid Ile. The hydrophobicity of both ALPM and HIR is considered important for the expression of ACE inhibitory activity.

FOSHU System in Japan and FOSHU Products with Antihypertensive Effect: "Amiel-S" and "Peptio" as FOSHU

Japan is credited with creating the term "functional foods" in the late 1980s. Functional foods are now a distinct category within the Japanese food supply. Japan is the only nation that has legally defined functional foods, and the Japanese functional food market is now one of the most advanced in the world. In 1991, Japan's Ministry of Health, Labor and Welfare instituted an approval of "Foods for Specified Health Use," or the "FOSHU" system, for functional foods located between normal food and medical supplies. The new system was intended to help promote the manufacture of foods designed to remedy serious health problems. This system was unique to Japan. A special label is used to show that particular products have FOSHU approval [Fig. 7 (center)].

As of June 2007, 678 FOSHU products were permitted for sale in Japan (http://www.mhlw.go.jp/topics/bukyoku/iyaku/syoku-anzen/hokenkinou/hyouz

"Amiel S" from
Calpis Co., Ltd.

"Peptio" from
Kanebo Co., Ltd.

Fig. 7 The label mark of FOSHU and representative two FOSHU products which expect to decrease the blood pressure

iseido-1.html). Several FOSHU products containing bioactive peptides derived from milk proteins exert antihypertensive effects following daily intake by persons with high blood pressure. In Japan, normal blood pressure is defined as pressure below 130 (systolic) over 85 (diastolic pressure) mm Hg. Under this definition of hypertension, 25% of the Japanese population is thought to be hypertensive.

Among many functional peptides derived from milk proteins, those possessing hypotensive activity are thought to be very useful as functional food materials for patients with high blood pressure (Meisel, 2005). Figure 7 shows two well-known antihypertensive products with FOSHU approval currently on the market in Japan. Daily intake of these products is recommended for people with hypertension. "Amile-S" is a pasteurized fermented acid milk produced by Calpis Co., Ltd. and contains the antihypertensive peptides lactotripeptide IPP and VPP (FOSHU-approved in 1999). The Casein DP "Peptio" drink is a soft drink produced by Kanebo Co., Ltd. and contains the casein dodecapeptide (DP) (FFVAPFPQVFGK) prepared from bovine casein (FOSHU-approved in 2000).

Conclusions

The main ACE inhibitory and antihypertensive peptides currently available are generated from milk proteins (casein and whey protein). Milk fermented by *L. helveticus* and enzymatic hydrolysis of casein and whey proteins by proteinase K, trypsin, and actinase E have demonstrated antihypertensive effects following SHR and human studies. Functional peptides derived from milk proteins having beneficial effects on hypertensive subjects without any side effects should have the potential to reduce the risk of cardiovascular disease.

References

Abubakar, A., Saito, T., Kitazawa, H., Kawai, Y., & Itoh, T. (1998). Structural analysis of new antihypertensive peptides derived from cheese whey protein by proteinase K digestion. *Journal of Dairy Science, 81*, 3131–3138.

Cheung, H., Wang, F., Ondetti, M. A., Sabo, E. F., & Cushman, D. W. (1980). Binding of peptide substrates and inhibitors of angiotensin-converting enzyme. Importance of the COOH-terminal dipeptide sequence. *Journal of Biological Chemistry, 255*, 401–407.

Ferreira, S. H., Bartet, D. C., & Greene, L. J. (1970). Isolation of bradykinin potentiating peptides from *Bothrops jararaca* venom. *Biochemistry, 9*, 2583–2593.

FitzGerald, R. J., & Meisel, H. (2000). Milk protein-derived peptide inhibitors of angiotensin-I-converting enzyme. *British Journal of Nutrition, 84*, S33–S37.

FitzGerald, R. J., Murray, B. A., & Walsh, D. J. (2004). Hypotensive peptides from milk proteins. *Journal of Nutrition, 134*, 980S–988S.

Fox, P. F. (2003). Milk proteins: General and historical aspects. In P. F. Fox and P. L. H. MsSeeney (Eds.), *Advanced Dairy Chemistry, Vol. 1, Proteins* (pp. 1–48). New York: Kluwer Academic/Plenum Press.

Geerlings, A., Villar, I. C., Zarco, F. H., Sanchez, M., Vera, R., Gomez, A. Z., Boza, J., & Duarte, J. (2006). Identification and characterization of novel angiotensin-converting enzyme inhibitors obtained from goat milk. *Journal of Dairy Science, 89*, 3326–3335.

Hata, Y., Yamamoto, M., Ohni, M., Nakajima, K., & Nakamura, Y. (1996). A placebo-controlled study of the effect of sour milk on blood pressure in hypertensive subjects. *American Journal of Clinical Nutrition, 64*, 767–771.

Huth, P. J., DiRienzo, D. B., & Miller, G. D. (2006). Major scientific advances with dairy foods in nutrition and health. *Journal of Dairy Science, 89*, 1207–1221.

Kitts, D. D., & Weiler, K. (2003). Bioactive proteins and peptides from food sources. Applications of bioprocesses used in isolation and recovery. *Drugs and Pharmaceuticals, Current Pharmaceutical Design, 9(16)*, 1309–1323.

Korhonen, H., & Pihlanto-Leppälä, A. (2001). Milk protein-derived bioactive peptides. Novel opportunities for health promotion. *Bulletin of the IDF, 363*, 17–26.

Masuda, O., Nakamura, Y., & Takano, T. (1996). Antihypertensive peptides are present in aorta after oral administration of sour milk containing these peptides to spontaneously hypertensive rats. *Journal of Nutrition, 126*, 3063–3068.

Matsui, T., Tamaya, K., Seki, E., Osajima, K., Matsumoto, K., & Kawasaki, T. (2002). Val-Tyr as a natural antihypertensive dipeptide can be absorbed into the human circulatory blood system. *Clinical and Experimental Pharmacology and Physiology, 29*, 204–208.

Matsumura, A., Saito, T., Arakuni, M., Kitazawa, H., Kawai, Y., & Itoh, T. (1999). New binding assay and preparative trial of cell-surface lectin from *Lactobacillus acidophilus* group lactic acid bacteria. *Journal of Dairy Science, 82*, 2525–2529.

Meisel, H. (1998). Overview on milk protein-derived peptides. *International Dairy Journal, 8*, 363–373.

Meisel, H. (2005). Biochemical properties of peptides encrypted in bovine milk proteins. *Current Medicinal Chemistry, 12*, 1905–1919.

Mizuno, S., & Yamamoto, N. (2004). Antihypertensive peptides from food proteins. *Current Topics in Biotechnology, 1*, 43–54.

Mizuno, S., Nishimura, S., Matsuura, K., Gotou, T., & Yamamoto, Y. (2004). Release of short and proline-rich antihypertensive peptides from casein hydrolysate with an *Aspergillus oryzae* protease. *Journal of Dairy Science, 87*, 3183–3188.

Mullally, M. M., Meisel, H., & FitzGerald, R. J. (1997). Identification of a novel angiotensin-I-converting enzyme inhibitory peptide corresponding to a tryptic fragment of bovine β-lactoblobulin. *FEBS Letters, 402*, 99–101.

Murakami, M., Tonouchi, H., Takahashi, R., Kitazawa, H., Kawai, Y., Negishi, H., & Saito, T. (2004). Structural analysis of a new anti-hypertensive peptide (β-Lactosin B) isolated from a commercial whey product. *Journal of Dairy Science, 87*, 1967–1974.

Nakamura, Y., Yamamoto, N., Sakai, K., Okubo, A., Yamazaki, S., & Takano, T. (1995a). Purification and characterization of angiotensin I-converting enzyme inhibitors from sour milk. *Journal of Dairy Science, 78*, 777–783.

Nakamura, Y., Yamamoto, N., Sakai, K., & Takano, T. (1995b). Antihypertensive effect of sour milk and peptides isolated from it that are inhibitors to angiotensin I-converting enzyme *Journal of Dairy Science, 78*, 1253–1257.

Nakamura, Y., Masuda, O., & Takano, T. (1996). Decrease of tissue angiotensin I-converting enzyme activity upon feeding sour milk in spontaneously hypertensive rats. *Bioscience, Biotechnology, and Biochemistry, 60*, 488–489.

Ondetti, M. A., & Cushman, D. W. (1982). Enzymes of the renin-angiotensin system and their inhibitors. *Annual Review of Biochemistry, 51*, 283–308.

Ondetti, M. A., Williams, N. J., Sabo, E. F., Pluscec, J., Weaver, E. R., & Kocy, O. (1971). Angiotensin-converting enzyme inhibitors from the venom of *Bothrops jararaca*. Isolation, elucidation of structure, and synthesis. *Biochemistry, 10,* 4033–4039.

Parmley, W. W. (1998). Evolution of angiotensin-converting enzyme inhibition in hypertension, heart failure, and vascular protection. *American Journal of Medicine, 105,* 27S–31S.

Pihlanto-Leppälä, A., Koskinen, P., Piilola, K., Tupasela, T., & Korhonen, H. (2000). Angiotensin I-converting enzyme inhibitory properties of whey protein digest: Concentration and characterization of active peptides. *Journal of Dairy Research, 67,* 53–64.

Pin, J.J., & Keenan, J.M. (2006). Effects of whey peptides on cardiovascular disease risk factors. *Journal of Clinical Hypertension, 8,* 775–782.

Saito, T. (2004). Selection of useful probiotics lactic acid bacteria from the *Lactobacillus acidophilus* group and their applications to functional foods. *Animal Science Journal, 75,* 1–13.

Saito, T., Nakamura, T., Kitazawa, H., Kawai, Y., & Itoh, T. (2000). Isolation and structural analysis of antihypertensive peptides that exist naturally in gouda cheese, *Journal of Dairy Science, 83,* 1434–1440.

Satake, M., Enjoh, M., Nakamura, Y., Takano, T., Kawamura, Y., Arai, S., & Shimizu, M. (2002). Transepithelial transport of the bioactive tripeptide, Val-Pro-Pro, in human intestinal Caco-2 cell monolayers. *Bioscience, Biotechnology, and Biochemistry, 66,* 378–384.

Seppo, L., Jauhiainen, T., Poussa, T., & Korpela, R. (2003). A fermented milk high in bioactive peptides has a blood pressure-lowering effect in hypertensive subjects. *American Journal of Clinical Nutrition, 77,* 326–330.

Skidgel, R. A., & Erdos, E. (1993). Biochemistry of angiotensin I-converting enzyme. In J. I. S. Robertson and M. G. Nicholls (Eds.), *The Renin-Angiotensin System* (pp. 10.1–10.10). New York: Raven Press.

Swaisgood, H. E. (2003). Chemistry of the caseins. In P. F. Fox and P. L. H. McSeeney (Eds.), *Advanced Dairy Chemistry, Vol. 1, Proteins, Part A* (pp. 139–201). New York: Kluwer Academic/Plenum Press.

Yamamoto, N., Akino, A., & Takano, T. (1994). Antihypertensive effect of the peptides dereived from casein by an extracellular proteinase from *Lactobacillus helveticus* CP790. *Journal of Dairy Science, 77,* 917–922.

IV
Induced Biologically Active Components from the Milk of Livestock Animals

Targeted Antibodies in Dairy-Based Products

Lennart Hammarström and Carina Krüger Weiner

Introduction

Bovine antibodies consist of IgM (Mousavi et al., 1998), IgD (Zhao et al., 2002), three IgG subclasses: IgG1 (Butler et al., 1972a, 1972b), IgG2 (Kacskovics & Butler, 1996), IgG3 (Rabbani et al., 1997; Zhao et al., 2003), IgA (Brown et al., 1997), and IgE (Mousavi et al., 1997; Zhao et al., 2006). Colostrum is an extremely rich source of antibodies, but all immunoglobulins decrease within a few days to a total immunoglobulin concentration of 0.7–1.0 mg/mL, with IgG1 representing the major Ig class in milk throughout the lactation period.

Bovine IgG1 antibodies are transported from the plasma of the cow to the colostrum and milk via an active transport mechanism, mainly during the last three weeks before parturition (Butler, 1983). As the placental IgG transport in cows is markedly less efficient than that in humans, passive immunization through colostrum and milk postpartum is extremely important for the calves (Butler, 1983). The high concentration of antibodies in colostrum also makes it suitable as a source for antibodies for oral therapy in humans, since the cow can be immunized with antigens from specific pathogens. Hence, colostrum from immunized animals may have more than a 100-fold increase in antibody titers compared to colostrum from nonimmunized animals (Jansson et al., 1994). Also, as the cow produces about 1–1.5 kg of immunoglobulins in the first few days after calving, it is attractive for large-scale antibody production. Because of the rapid drop in immunoglobulin concentration immediately after partum (Butler, 1983), colostrum is preferable to mature milk.

An increasing number of controlled clinical studies, using colostrum from immunized cows, have shown both prophylactic and therapeutic effects against

---

L. Hammarström
Division of Clinical Immunology, Department of Laboratory Medicine, Karolinska University Hospital Huddinge, SE-141 86, Stockholm, Sweden, Ph: +46 8 524 835 86
e-mail: lennart.hammarström@ki.se

oral and gastrointestinal pathogens in humans (reviewed in Weiner et al., 1999; Korhonen et al., 2000). In order to be therapeutically active against intestinal pathogens, oral hyperimmune bovine immunoglobulin concentrate (BIC) must survive its passage through the oral and gastrointestinal tracts. As with human antibodies, bovine antibodies are partially digested during the passage through the human intestinal tract (Bogstedt et al., 1997; Hilpert et al., 1987; Leresche et al., 1972; McClead et al., 1988; Roos et al., 1995), but up to 20% of ingested bovine immunoglobulins may be found intact in stool from healthy babies (Zinkernagel et al., 1972). Kelly et al. (1997) investigated the activity of anti-C. difficile toxin antibodies (BIC) after passage through the human intestinal tract and found that oral administration of BIC in enteric capsules was the most successful delivery vehicle, with high fecal bovine IgG levels and a retained neutralizing activity against toxins A and B (Warny et al., 1999). The resistance of bovine IgG1 to low pH and luminal proteolysis (Tzipori et al., 1986) makes it functionally similar to human IgA. Bovine immunoglobulins have been used successfully for both prophylaxis and therapy in different animal systems (Table 1) and in human studies (Tables 2 and 3), suggesting a potential role in clinical practice. We give a more detailed analysis of their use in humans.

Treatment of Oral Candidiasis

Different Candida species constitute the majority of isolates found in deep fungal infections in humans; Candida albicans *(C. albicans)* alone accounts for approximately 75% of these cases. *Candida* infections are usually of endogenous origin and are most often derived from the GI tract. Organ and bone marrow-transplanted patients are frequently colonized with *Candida* and are thus highly susceptible to invasive fungal infections. Mortality is high, as up to 70% of bone marrow-transplanted patients with candidemia succumb. In a study performed in bone marrow-transplanted patients, bovine antibodies from cows immunized with whole *Candida* organisms and purified mannan (Tollemar et al., 1999) were administered orally three times per day before and after transplantation. The immunoglobulin product had an estimated purity of approximately 50% IgG, and 59 patients received 3.3 g powder/dose. The results suggested a treatment-related reduction in *Candida* colonization in a majority (7/10) of patients, and one patient became culture negative. No side effects were noted.

Antibody Therapy Against Streptococcus mutans

Streptococcus mutans (S. mutans) is considered to be the main bacteriological agent of human dental caries. The major route for early acquisition is vertical transmission from mother to child (Aaltonen et al., 1990). Bovine antibodies

Table 1 Transfer of Immunity by Oral Administration of Bovine Antibodies in Animal Species

Animal	Pathogen	Preparation	Efficacy	References
Chicken	E. acervulina	Colostrum HBC	Inhibited parasite development and reduced severity of parasite-related gut lesions	Fayer and Jenkins, 1992
Chicken	C. jejuni	Colostrum	Prophylactic and therapeutic effect with lower bacterial counts	Tsubokura et al., 1997
Cow	C. parvum	Colostrum	No protection	Harp et al., 1989
Cow	C. parvum	Colostrum	Less days with shed oocysts and diarrhea	Fayer et al., 1989
Cow	E. coli	Colostrum	No protection	Acres et al., 1979
Cow	E. coli	Colostrum	Prevented diarrheal disease	Snodgrass et al., 1982
Cow	Rotavirus	Colostrum	No protection	Zaane et al., 1986
Cow	Rotavirus	IgG	No therapeutic effect, some protection, some preventive effect soon after calving	Osame et al., 1991
Guinea pig	C. wrairi	HBC	No effect	Hoskins et al., 1991
Hamster	C. difficile	Colostrum	Prophylactic protection against disease	Lyerly et al. (1991)
Mouse	C. parvum	Colostrum	Partial protection	Fayer et al. (1989)
Mouse	C. parvum	IgG (HBC)	Reduction in parasite number	Fayer et al. (1990)
Mouse	C. parvum	Colostrum	Neutralized sporozoites and partially protected against oral challenge with C. parvum oocytes	Perryman et al. (1990)
Mouse	—	Milk	Protected mice from infection and disease	Yolken et al. (1985)
Mouse	C. parvum	HBC	Partial (50%) protection against cryptosporidiosis in immunosuppressed mice	Jenkins et al. (1999)
Mouse	C. parvum	Cclostrum	High level of protection against infection	Martin-Gomez et al. (2005)
Mouse	E. coli	Colostrum	Prevented indigenous infection after pharmacological impairment of intestinal microflora	Nomoto et al. (1992)
Mouse	E. coli	Colostrum	Resulted in rapid decrease in the bacteria numbers and inhibited bacterial attachment	Funatogawa et al. (2002)

Table 1 (continued)

Animal	Pathogen	Preparation	Efficacy	References
Mouse	H. felis	Colostrum	Decreased colonization	Marnila et al. (2003)
Mouse	H. felis	Colostrum	Reduced colonization of H. felis in gastric antrum	Marnila et al. (1996)
Mouse	K. pneumoniae	Lactoglobulin	Prevented infection of K. pneumoniae	Soboleva et al. (1991)
Mouse	P. aeruginosa	Colostrum	Prevented infection	Stephan et al. (1990)
Mouse	S. pullorum	Colostrum	Delayed death	Campbell and Petersen (1959)
Mouse	S. typhimurium	Colostrum	Prevented infection	Stephan et al. (1991)
Pig	E. coli	Colostrum	Protected against fatal diarrhea in suckling pigs	Isaacson et al. (1980)
Pig	E. coli	IgG	Passive immune protection	Cordle et al. (1991)
Pig	Rotavirus	Colostrum	Protected piglets from the clinical effects of a porcine rotavirus	Bridger and Brown (1981)
Pig	Rotavirus	Colostrum	Protected agammaglobulinemic piglets from infection	Lecce et al. (1991)
Pig	Rotavirus	IgG	Viral shedding and diarrhea were reduced or eliminated in a dose-dependent manner	Schaller et al. (1992)
Pig	—	Colostrum	Cow colostrum or diets containing porcine gamma globulin protected infected piglets	Lecce et al. (1976)
Rabbit	V. cholera	IgG	Reduced diarrhea	Boesman-Finkelstein et al. (1989)
Rabbit	V. cholera	IgG	Decreased mortality and intestinal fluid response	McClead and Gregory (1984)
Rat	C. difficile	Colostrum	Decreased enterotoxic symptoms	Kelly et al. (1996)
Rat	S. mutans	IgG	Lower caries development	Michalek et al. (1987)
Rat	S. mutans	IgG	Less caries development	Mitoma et al. (2001)
Snake	C. parvum	HBC	Cleared infection in 3/12 snakes. Decreased number of oocysts and stool	Graczyk et al. (1998)

Table 2 Human Oral Prophylactic Studies Using Bovine Antibodies

Reference	Type of Study	Number of Patients Volunteers or "Active"/Placebo	Type of Infection	Source of Antibodies	Dose	Ig Contents or Titer	Outcome
Brunser et al. (1992)	Double-blind cohort	Children; 117/115	EPEC	MIC	6–7 months of an infant formula supplemented with antibody concentrate	500 mg/100 mg formula	No effect
Tacket et al. (1988)	Challenge; double-blind, controlled	Adults; 10/10	ETEC	BIC	3.6 g tid for 7 days	38% of powder	0/10 of "active" sick vs. 9/10 given placebo
Tacket et al. (1992)	Randomized, double-blind	Adults; 11/11/7	Shigella flexeri	HIC	10g tid for 7 days	Titer 1;2,560	None of the HIC group became ill vs. 45% in control
Freedman et al. (1998)	Challenge, randomized, double-blind, controlled	Adults; 15/10	ETEC	BIC	3 doses/day for 4–7 days	Titer 20,000 (LT, LPS)-640,000 (CFA)	1/15 of "active" developed diarrhea vs. 7/10 in placebo group
Brunser et al. (1992)*	Double-blind cohort	Children; 117/115	rotavirus	MIC	6–7 months of an infant formula supplemented with antibody concentrate	500 mg/100 mg formula	No effect
Ebina et al. (1985)	Randomized, open	6/7	rotavirus	HBC	20 mL for 3 days	600 mg/day	Mitigated disease
Davidson et al. (1989)	Randomized, controlled	55/65	rotavirus	HBC	50 mL for 1 day	1500 mg	Total protection
Turner and Kelsey (1993)	Randomized, controlled	Infants; 31/33	rotavirus	HBC	>360-mL formula Per day for max. 6 months	200 ug IgG/mL formula	Fewer days with diarrhea

Table 2 (continued)

Reference	Type of Study	Number of Patients Volunteers or "Active"/Placebo	Type of Infection	Source of Antibodies	Dose	Ig Contents or Titer	Outcome
Davidson et al. (1994)	Randomized, double-blind	Children; 50/102	rotavirus	HBC	5 g for 4–10 days	40% (2g/day)	Total protection
Brinkworth and Buckley (2003)	Double-blind, controlled	Adult males; 93 and 81	URTI	CBC	60 g/day for 8 weeks	–	Fewer symptoms of URTI
Tawfeek et al. (2003)	Double-blind randomized	125 infants	EPEC	HBC	0.5 g/kg body weight	–	Lower incidence of diarrhea
Filler et al. (1991)	Randomized	Healthy volunteers; 9/10	*S. mutans*	Whey fraction	Rinse twice daily 2 g × 2	–	Reduced S. mutans level in plaque
Loimaranta et al. (1999)	Opened	Volunteers	*S. mutans/ S. sobrinus*	Whey protein	Rinse three times daily for three days	37% of IgG	Decreased the relative numbers of mutans streptococci
Shimazaki et al. (2001)	Randomized	Volunteers	*S. mutans*	Immune milk	Rinse twice per day for 14 days 10 mL i2	–	Inhibited recolonization of S. mutans in plaque and saliva

*This is one study where the enrolled "active" group received a formula supplemented with antibodies against both EPEC and rotavirus. EPEC = enteropathogenic E. coli; ETEC = enterotoxigenic E. coli; HIC = hyperimmune colostrums; BIC = bovine immunoglobulin concentrate; CFA = anticolonization factor antigen; HBC = hyperimmune bovine colostrums; LT = antiheat-labile toxin; LPS = antilipopolysaccharide; MIC = milk immunoglobulin concentrate; tid = three times a day; URTI = upper respiratory tract infection; CBC = concentrated colostrum protein.

Table 3 Human Therapeutic Studies Using Bovine Antibodies

Reference	Type of Study	Number or Type of Patients "Active"/Placebo	Type of Infection	Source of Antibodies	Dose	Ig Contents or Titer	Outcome
Rump et al. (1992)	Open, uncontrolled	29	HIV with unspecified diarrhea	LIG	10 g daily for 10 days	—	Normalized stool frequency in 21 of 29; cryptosporidiosis disappeared in five patients
Tollemar et al. (1999)	Open	59	C. albicans	HBC	10 g per day in 3 doses for 32 days	4.2 g per day	Reduced concentration of fungi in saliva
Ungar et al. (1990)	Case report	1	C. parvum	HBC	20 mL/h for 7 days	Titer 1:200 000	Resolved diarrhea and oocyst excretion
Saxon and Weinstein (1987)	Case series	3	C. parvum	NIC	1–5 L/day for 5–7 days	Titer 1:10 to 1:40	No effect
Plettenberg et al. (1993)	Open, uncontrolled	25 (7 confirmed infection, 18 absence of pathogen)	C. parvum	NIC	10 g daily for 10 days		Complete remission in three and partial in two (of seven patients)
Nord et al., 1990	Pilot study	3/2	C. parvum	HBC/NIC	20 mL/h for 10 days	30 mg/mL	Reduced diarrhea and oocyst excretion in one; reduced diarrhea volume in controls
Greenberg and Cello (1996)	open	20	C. parvum	BIC	Powder 10 g (n = 12) or 10-g capsules (n = 8) qid for 21 days	Purity 20% of powder/capsules (i.e., 8 g/day)	Reduced stool weight in 12, no effect in 8
Tzipori et al. (1987)	Case series	3 years (CHG), 40 years (HIV), 4 years (ALL)	C. parvum	HBC	200–500 mL/day for 10–21 days	IFAT: 12,000–100,000	All recovered
Tzipori et al. (1986)	Case report	3 years old with CHG	C. parvum	HBC	Infusion of 200 mL/24 h for 21 days + 50 mL/day for 4 days	No information	Diarrhea resolved, no oocysts

Table 3 (continued)

Reference	Type of Study	Number or Type of Patients "Active"/Placebo	Type of Infection	Source of Antibodies	Dose	Ig Contents or Titer	Outcome
van Dissel et al. (2005)	Open	16	C. difficile	IP	5 g/100 mL mineral water	No information	Prevented relapse of disease
Lodinový et al. (1987)	Open, controlled	56 and 29 (premature and term infants)	E. coli	HBC	3-5 mL 6 times daily for 5 days	No information	Reduced need for antibiotics, dehydration, and pathogen in stool
Mietens et al. (1979)	Open	53 infants	E. coli	MIC	1 g/kg and day for 10 days	Purity 40% of powder	Negative stool culture in 84%
Huppertz et al. (1999)	Double-blind	13 and 14	EHEC	NIC	7 g tid for 14 days	52% mainly IgG	Lower frequency of loose stools
Casswall et al. (2000)	Randomized, placebo-controlled	32/31	EPEC/ETEC	BIC	20 g daily for 4 days	45% for EPEC, 40% for ETEC	No therapeutic effect
Ando and Nakamura (1991)	Open	9 adults	H. pylori	Colostrum	1.7 g bid i28 days	Purity 65% of powder	Infection eradicated in all
Ando and Nakamura (1991)	Open	11 adults	H. pylori	Colostrum	1.5g bid i28 days	Purity 65% of powder	Infection eradicated in all
Tarpila et al. (1994)	Open	9 adults	H. pylori	Colostrum	20 g/day i28 days	Purity 25% of powder	Infection eradicated in one
Casswall et al. (2002)	Open	8 adults	H. pylori	HBC	4 g bid i28 days	Purity 55% of powder	Infection eradicated in two
Casswall et al. (1999)	Open	10 adults	H. pylori	HBC	4 g bid i14 days + omeprazole	Purity 55% of powder	No eradication of infection
Casswall et al. (1998)	Double-blind	Children; 10 and 9	H. pylori	HBC	1 g/day i30 days	Purity 56% of powder	No eradication of infection
Oona et al. (1994)	Open	13 children	H. pylori	Colostrum	Colostrum daily for 21 days	Not given	No eradication, but lower Hp density

Table 3 (continued)

Reference	Type of Study	Number or Type of Patients "Active"/Placebo	Type of Infection	Source of Antibodies	Dose	Ig Contents or Titer	Outcome
Opekun et al. (1999)	Open	3 + 6 + 6 adults	H. pylori	BIC	14 doses of 3.7–7.8 g of either antiurease, adhesion or whole cell 12 days	1:3200–1:6400	No eradication in any group as measured with 13C-UBT
Ebina et al. (1985)	Open	6–7	Rotavirus	HBC	20 mL for 3 days	30 mg/mL	Mitigated disease
Mitra et al. (1995)	Double-blind	35 and 33	Rotavirus	HBC	300 mL for 3 days	9 g/day	Mitigated disease
Ylitalo et al. (1998)	Double-blind	42 and 83 (nonimmunized colostrum and milk)	Rotavirus	HBC	100 mL qid for 4 days	IFAT 1:597 (HBC)–1:128 (colostrum)	Trends of mitigated disease
Sarker et al. (1998)	Double-blind	40 and 40 (children)	Rotavirus	HBC	2.5 qid for 4 days	3.6 g/day	Mitigated disease
Hilpert et al. (1987)	Open, controlled	30 and 43	Rotavirus	MIC	2 g/kg for 5 days	100 mg/mL of titer 1:6000	Reduced viral shedding
Hilpert et al. (1987)	Open, controlled	45 and 66	Rotavirus	MIC	2 g/kg for 5 days	100 mg/mL of titer 1:330–1:1100	No effect
McClead et al. (1988)	Double-blind	23 and 42 (adults)	V. cholerae	IgG	2 g i2-8 doses	44–76 mg antiCT IgG	No effect

ALL = acute lymphoblastic anemia; BIC = bovine milk immunoglobulin concentrate; bid = twice daily; CHG = congenital hypogammaglobulinemia; C. parvum = Cryptosporidium parvum; CT = cholera toxin; HBC = hyperimmune bovine colostrums; IP = immune product; HIV = human immunodeficiency virus; IFAT = immunofluorescence antibody titer; LIG = Lactobin; MIC = milk immunoglobulin concentrate; NIC = nonimmunized colostrums; qid = four times daily; rx = therapy.

have been successfully used in animal studies and have proven effective in several *in vitro* studies (reviewed in Koga et al., 2002). Three different studies in humans used bovine antibodies against *S. mutans* (Table 2).

In the first study, a mouth rinse of bovine immune milk containing antibodies to *S. mutans* resulted in an initial reduction in the numbers of recoverable bacteria in a group of nine healthy individuals. In addition, after culture, the Streptococci recovered from the dental plaques from subjects who used the immune bovine milk rinse formed smaller colonies than those from pretreatment plaques and from all plaques of subjects who used the control rinse (Filler et al., 1991). In the second study, a short-term clinical trial was performed using immune colostrum containing antibodies against *S. mutans* and *S. sobrinus* as a mouth rinse for three days. The treatment resulted in a higher resting pH in the dental plaque and a lower proportion of caries-associated Streptococci (Loimaranta et al., 1999). The third study examined the effect of bovine milk produced after immunization with PacA-GB, a fusion protein of the saliva-binding alanine-rich region (PacA) of Pac and the glucan binding (GB) domain of Glucosyltransferase I, GTF-I, both important factors for *S. mutans* colonization in humans. Eight adult subjects were included in the study and four rinsed their mouths with immune milk, which significantly inhibited the recolonization of *S. mutans* in saliva and plaque (Shimazaki et al., 2001). These results suggest that milk produced from immunized cows may be useful for controlling the number of *S. mutans* in the human oral cavity.

Bovine Antibody Therapy in Clostridium difficile Infection

Clostridium difficile (C. difficile) is the causative agent of antibiotic-associated diarrhea and pseudo-membranous colitis. It is also a common cause of nosocomial infection (Cloud et al., 2007). Pathogenic strains produce two exotoxins: Toxin A, first described in 1980 (Abrams et al., 1980), is responsible for intestinal inflammation, mucosal damage, and fluid secretion, whereas toxin B is cytotoxic (Lyerly et al., 1982). The efficacy of both active and passive immunization against *C. difficile* has been well established in animal models (Ward et al., 1999; Lyerly et al., 1991) (Table 1). Previously, bovine antibodies against *C. difficile* were shown to resist digestion in the human upper gastro-intestinal tract (Kelly et al., 1997), and specific anti-C. difficile toxin A binding and neutralizing activity was retained (Warny et al., 1999).

A pilot study evaluated the feasibility of using an immune whey protein concentrate (40%; immune WPC-40) to prevent relapse of *C. difficile* diarrhea (van Dissel et al., 2005). Immune WPC-40 was produced from milk after immunization of Holstein-Frisian cows with *C. difficile*-inactivated toxins and killed whole cell *C. difficile*. To obtain preliminary data in humans, 16 patients (10 male; median age: 57 years) with toxin- and culture-confirmed *C. difficile* diarrhea were enrolled in an uncontrolled cohort study. After completion of standard antibiotic treatment, the patients received immune WPC-40 for two weeks; it was well tolerated and no treatment-related adverse effects were observed. In all but one case, *C. difficile* toxins disappeared from the feces upon completion of treatment. During a follow-up period of a median of 333 days (range: 35 days to 1 year), none of the patients suffered another episode of *C. difficile* diarrhea. These preliminary data suggest that immune bovine antibodies may prevent relapse of *C. difficile* diarrhea.

Bovine Antibody Treatment of Cryptosporidium parvum-Induced Diarrhea

The obligate parasite *Cryptosporidium parvum (C. parvum)* is a protozoan that infects the GI tract and, rarely, the biliary tract of humans. This organism can cause debilitating diarrhea associated with dehydration and malnutrition and is often found in immunocompromised patients. The potential efficacy of immune colostrum in treating and/or preventing infection has been demonstrated in several animal models (Riggs, 1997; Jenkins, 2004) (Table 1).

Seventy-two *C. parvum*-infected, immunocompromised patients were treated with either nonspecific or *C. parvum*-specific bovine immunoglobulin (Stephan et al., 1990; Saxon et al., 1987; Plettenberg et al., 1993; Tzipori et al., 1986, 1987; Ungar et al., 1990; Nord et al.; 1990; Greenberg & Cello, 1996) (Table 2). Antibodies were administered by an oral, nasogastric, or nasoduodenal route, and the treatment lasted 1 to 21 days. Most patients experienced a reduction or elimination of clinical symptoms, but nonspecific colostrum, from nonimmunized animals, failed to provide any clinical improvement (Saxon et al., 1987). However, 25 HIV-infected patients with chronic refractory diarrhea and cryptosporidiosis ($n=7$), or absence of pathogenic microorganism ($n=18$), were treated with a daily oral dose of 10 g of colostrum from nonimmunized animals over a period of 10 days (Plettenberg et al., 1993). Complete remission was achieved in three, and partial remission in two, of the seven patients with cryptosporidiosis. Among the 18 patients with diarrhea and negative stool cultures, complete remission of diarrhea was obtained in 7 and partial remission in 4 (Plettenberg et al., 1993). Ungar and co-workers (1990) described an HIV patient with severe diarrhea who received hyperimmune bovine colostrum to cryptosporidium by duodenal infusion. During infusion, the patient's fecal output decreased and, 48 hours after treatment, stools were formed and oocysts to *Cryptosporidium* were absent. The patient remained asymptomatic for three months. A previous study using a specific bovine immunoglobulin preparation for treatment of *C. parvum* diarrhea associated with AIDS in eight patients (Greenberg & Cello, 1996) saw a decrease in both stool frequency and stool output both at the end of treatment and one month after completing treatment.

Clinical Effect of Bovine Antibodies Against Rotavirus Infection

Rotavirus is the most common infective agent causing diarrhea in infants and young children, with a high mortality rate in the developing world, causing 700,000 deaths annually (Santos et al., 2005).

A number of studies in humans use bovine antibodies both as prophylaxis (Table 2) and as therapy (Table 3). Several of these studies have shown a total protection or mitigation of disease when used therapeutically in diarrhea induced by rotavirus (Hilpert et al., 1987; Davidson et al., 1989, 1994; Mitra et al., 1995; Turner & Kelsey, 1993; Sarker et al., 1993). Ebina and co-workers (1985) showed a prophylactic effect using hyperimmune colostrum administered to children in an orphanage. However, in a prophylactic study of Chilean children, no effect was seen when antibodies from hyperimmunized cows were introduced to the children's formula (Brunser et al., 1992). This was probably caused by dilution of the relevant antibodies in the milk formula by immunoglobulins from nonimmunized cows. Mitigated disease was seen in a double-blind study where 300 mL of immune bovine colostrum were given daily for three days to infants with rotavirus diarrhea (Mitra et al., 1995). In contrast, no effect was seen in a Japanese study where children received 20-50 mL of colostrum (250–4,200 mg of IgG/100 mL) for three days after the onset of diarrhea (Ebina et al., 1985). The same preparation had shown a prophylactic effect in children in an orphanage (Ebina et al., 1985), suggesting that the antibody titer may have been too low for a therapeutic effect. In a therapeutic study from Finland (Ylitalo et al., 1998), a trend of mitigated disease could be seen using 100 mL of hyperimmunized colostrum four times daily for four days in 134 children. However, only the duration of diarrhea and number of stools were measured. Furthermore, the duration of diarrhea before the start of therapy (mean: 74.4 hours) may have been too long to observe a therapeutic effect, as rotavirus infection is a self-limiting disease. Hence, the antibodies may have had a neglible effect at this late stage. Furthermore, only antibodies toward one serotype of rotavirus were used. Thus, cross-reactivity to other rotavirus strains, which would have been desirable, may have been limited. In a double-blind, placebo-controlled study performed by Sarker and co-workers (1998), 80 children with rotavirus-induced diarrhea received immunuglobulins purified from immunized bovine colostrum (IIBC), containing high titers of antibodies against four serotypes of rotavirus. These children had a significantly lowered stool output and required a smaller amount of oral rehydration solution than the placebo group. Finally, diarrhea induced by rotavirus in Western countries may be less severe compared to the situation in developing countries where malnutrition prevails. Thus, differences in outcome may be easier to observe than in studies performed in Western countries.

Bovine Antibodies in Infections by Escherichia coli (E. coli)

Diarrhea may result from infection with one or more strains of *E. coli*. Enterotoxigenic *E. coli* (ETEC) strains are the most common cause of diarrhea in travelers to less developed countries (Freedman et al., 1998). The effectiveness of bovine antibodies in protecting volunteers against oral challenge has been demonstrated in two separate, double-blind studies (Table 2). In the first study (Tacket et al., 1988), a multiple-antigen *E. coli*-specific milk immunoglobulin concentrate was completely effective in preventing diarrhea in 10 volunteers orally challenged with one of the ETEC strains used for immunization, whereas 9 of 10 controls developed diarrhea. In the second trial (Freedman et al., 1998), the efficacy of bovine antibodies against purified ETEC colonization factor antigens (CFA) was evaluated. The milk-derived antibodies protected 14 of 15 subjects from clinical diarrhea, whereas 7 of the 10 volunteers receiving a placebo preparation developed symptoms.

Enteropathogenic *E. coli* (EPEC) are important causes of acute infectious diarrhea in young children throughout the world, and a major contributor to infant mortality in developing countries. Supplemental bovine antibody preparations in infant feeding formulas have been used to treat active EPEC disease in infants in one of the earliest controlled studies using bovine antibodies (Mietens et al., 1979). Sixty hospitalized infants (10 days to 18 months of age) with ongoing EPEC-induced diarrhea were treated for 10 days with 1 g/kg bodyweight/day of a specific milk immunoglobulin concentrate (MIC). Among 51 patients infected with E. coli strains also present in the vaccine, which was used for the production of the bovine antibody preparation, 84% became stool culture-negative following treatment, whereas only one of the nine infants infected with strains not included in the vaccine became negative. In the study by Lodinová-Zàdinikovà and co-workers (1987), partly purified colostral antibodies against enteropathogenic E. coli were given six times daily for five days to premature and full-term infants with diarrhea. The results showed that 78% of the premature children and 83% of the infants recovered after oral administration of antibodies. The therapeutic efficacy of an oral bovine immunoglobulin milk concentrate (BIC) from cows hyperimmunized with ETEC and EPEC strains was also evaluated in a randomized, placebo-controlled study in children with *E. coli*-induced diarrhea in Bangladesh. Eighty-six children between 4–24 months of age received orally administered BIC (20 g) containing anti-ETEC/EPEC antibodies (Casswall et al., 2000). In contrast to the previous studies, no significant therapeutic benefit was observed. A prophylactic, double-blind, placebo-controlled field study in 232 infants receiving bovine antibodies was also not successful in reducing the incidence of diarrhea (Brunser et al., 1992). An inadequate amount of relevant antibodies, due to dilution with immunoglobulins from milk from nonimmunized cows, may have accounted for the negative results in the latter study.

Enterohemorrhagic *E. coli* (EHEC) are recognized as key pathogens in the development of extraintestinal sequelae due to their production of Shiga-like toxins (Lissner et al., 1996). As the prevalence of EHEC in cattle is high, colostral antibodies are thought to contribute to the protection of newborn calves. Nonimmunized colostrum containing antibodies against Shiga toxins, intimin, and EHEC-hemolysin was thus used as treatment for EHEC-induced diarrhea in children. Twenty-seven children were treated for 14 days; a reduction in stool frequency from three to one per day was observed during the study period. However, although bovine colostrum was well tolerated, no effect of therapy on the carriage of the pathogens or on complications of the infection could be demonstrated (Huppertz et al., 1999).

Oral Therapy with Bovine Antibodies Against Helicobacter pylori

Helicobacter pylori (H. pylori) infection is present in the stomach of more than 50% of the human population worldwide (Mitchell, 1999), causing chronic gastritis, peptic ulcer disease, and gastric adenocarcinoma (Howden, 1996). In 1994, the World Health Organization declared *H. pylori* as a class I carcinogen. Bactericidal activity against *H. pylori* has been shown in bovine colostrum, indicating its potential for passive immunization (Casswall et al., 2002; Korhonen et al., 1995). In clinical trials on adults with chronic gastritis and *H. pylori* infection (Tarpila et al., 1995), a daily treatment with an immune colostral preparation for 28 days attenuated symptoms. The colonization load decreased in most patients, but eradication was achieved in only one case out of nine (Tarpila et al., 1995). In a small pilot study in Sweden, eight adult patients with *H. pylori*-induced gastritis were treated with 4 g of immune bovine immunoglobulin concentrate (BIC) daily for 28 days. Eradication of infection was noted in two of the eight patients (Casswall et al., 2002). In *H. pylori*-infected children receiving 12 g of immune preparation daily for 21 days, Oona et al. (1997) observed a reduced degree of inflammation in the gastric antrum. *H. pylori* was not eradicated, however, in any of the 20 children treated in the study of Casswall et al. (1998), and Opekun et al. (1999) reported that treatment with *H. pylori* immune milk neither eradicated nor decreased an established *H. pylori* colonization in infants or adults. In Casswall et al. (1998), antibodies against *H. pylori* were derived from hyperimmunized cows and administered orally to 24 infants in rural Bangladesh for 30 days. None of the children treated with bovine antibodies cleared their *H. pylori* infection. However, transient infection is common among infants in high endemic areas, as is reinfection after clearance.

Discussion and Future Aspects

Bovine colostrum has been used therapeutically in humans for decades. However, as the active component resides in the IgG fraction, it is being gradually replaced by purified immunoglobulin preparations. Specific bovine antibodies bind to virulence factors on target pathogens, but the interactions between whey preparations and human lymphocytes and granulocytes are yet not well known. Bovine colostral whey proteins from cows immunized with *S. mutans/ S. sobrinus* were found to block adherence and to promote aggregation of cariogenic bacteria (Loimaranta et al., 1998), and a colostral whey protein preparation from hyperimmunized cows was found to activate human leucocytes by opsonizing the targeted bacteria (Loimaranta et al., 1999). These results show that bovine colostral whey proteins are able to support an interaction between human phagocytes and pathogenic microorganisms and that this property is related to specific antibodies in the whey preparations (Loimaranta et al., 1999). Bovine antibodies may thus be able to prevent cariogenic bacteria from colonizing the oral cavity and to influence the activation of human phagocytes against pathogenic microbes.

Xu et al. (2006) demonstrated that specific IgG against 17 strains of pathogenic diarrheagenic bacteria had a strong inhibitory activity on *in vitro* growth and colonization by agglutinating the bacteria and destroying the cell walls. In these studies, normal IgG, purified from nonimmunized bovine colostrum, was incapable of eliciting the same consequences as specific IgG. Specific IgG has also been shown to prevent enteroinvasive *E. coli-/Salmonella typhi*-induced diarrhea by enhancing splenic NK cell activity, elevating IL-2 levels, and inhibiting excessive release of TNF-α in mice (Xu et al., 2006).

Rokka and co-workers (2001) evaluated the effect of a commercial bovine colostral whey preparation on the complement-mediated immune responses of calves. Groups fed colostrum had two to three times higher bacteriolytic activity than the control group of both the classic and alternative complement pathways. This effect is obviously not caused solely by the antibodies ingested but also involves other unknown colostral factors, possibly complement factors or lectins, which will increase the complement activity. Thus, the antibody-independent complement activity of serum can be increased substantially by feeding colostral whey concentrate to calves during their first days of life.

Increased secretory IgA levels were demonstrated in a small study performed in athletes who received a supplement of nonimmunized bovine colostrum for 12 weeks. These results correlated to protection against upper respiratory tract infection (URTI) (Crooks et al., 2006) and point to a need for future investigations into the mechanism of action of colostral products.

Nonsteroidal anti-inflammatory drugs (NSAIDs) are effective analgesics but cause gastrointestinal injury. Bovine colostrum from nonimmunized cows has been shown to reduce this effect in rats and mice (Playford et al., 1999). Furthermore, Playford and co-workers examined whether spray-dried, defatted colostrum could reduce the rise in gut permeability (a noninvasive marker of intestinal injury) caused by NSAIDs in volunteers and patients taking NSAIDs for clinical reasons. Indomethacin caused a threefold increase in gut permeability (Playford et al., 2001), whereas no significant increase in permeability was seen when colostrum was co-administered. These studies provide preliminary evidence that bovine colostrum, which is currently available as an over-the-counter preparation in some countries, may provide a novel approach to the prevention of NSAID-induced gastrointestinal damage. Lissner et al. (1997) evaluated another product from nonimmunized colostrum, Lactobin®, for its activity against *Yersinia enterocoliticia*. A strong reactivity was shown in vitro, but only small amounts of bovine immunoglobulins, without antibody reactivity, were detected in stool, probably because of absorption and degradation. This is in contrast to hyperimmune immunoglobulins, which, when administered in enteric capsules, kept their neutralizing activity after passage through the human intestinal tract (Kelly et al., 1997).

Specific antibodies against a variety of pathogens are present in nonimmunized cows, but the amount of antibodies is clearly not sufficient for a therapeutic effect in humans. Through immunization, a 100-fold or greater increase in the amount of relevant antibodies can be achieved. High antibody titers in the blood and colostrum have been achieved using a combination of intramuscular and intramammary inoculations (Schaller et al., 1992). In most cases, the antibody response has depended on the nature of the adjuvant used. In experimental studies, Freund's complete or incomplete adjuvant has been found to induce the strongest humoral immune response (Schaller et al., 1992; Korhonen et al., 1994), but its use is limited by concerns about possible side effects. This has led to the use of "safer" aluminum hydroxide-based adjuvants for the immunization of farm animals with a lower titer of specific antibodies as a consequence. Thus, we need new strategies for optimizing the immunization procedure for improving the concentrations of specific antibodies in colostrum to be able to use bovine immunoglobulins for passive immunization against human diseases cost-efficiently.

Large-scale production of human IgG in transgenic cows is an attractive strategy. Kuroiwa and co-workers (2002) introduced a human artificial chromosome (HAC) vector into bovine primary fetal fibroblasts and, after further selection, produced transchromosomic (Tc) calves. Human immunoglobulins were detected in the blood of these animals, and vaccination increased the levels of relevant antibodies. This work represents an important step toward a system for the production of therapeutic human polyclonal antibodies in cows. Pharmacokinetic studies showed that human IgG had $\sim$33 days' serum half-life in both normal and transchromosomic calves, which is more than twice that of its bovine counterpart (Kacskovics et al., 2006). This finding probably reflects the high affinity between human IgG and the bovine neonatal Fc receptor (bFcRn) (Kacskovics et al., 2006). Thus, the bFcRn (Kacskovics et al., 2000) plays a potential role in regulating not only the transport of IgG from maternal plasma to colostrum (Mayer et al., 2005) and intestinal uptake of

IgG but also IgG homeostasis in serum. A future strategy to increase the amount of specific antibodies after immunization could be to improve the expression or affinity of the bFcRn, as there is clearly a need for upgrading the production efficacy and yield of antibodies for therapeutic applications in humans. To our knowledge, ingestion of milk-derived IgG has not been related to any side effects in children or adults.

Summary

Bovine colostrum contains immunoglobulins that provide the newborn calf with protection against microbial infections until its own immune system matures. The concentration of antibodies in colostrum against pathogens can be raised by immunizing cows with pathogens or their antigens. Bovine colostrum-based immune milk products have proven efficacy in prophylaxis and treatment against various infectious diseases in humans such as diarrheal diseases caused by various pathogens like *E. coli* and rotavirus. Still, future attempts are needed to increase the yield and the concentration of antibodies in order to achieve full protection.

References

Aaltonen, A. S., Tenovuo, J., Lehtonen, O. P., & Saksala, R. (1990). Maternal caries incidence and salivary close-contacts with children affect antibody levels to *Streptococcus mutans* in children. *Oral Microbiology and Immunology, 5,* 12–18.

Abrams, G. D., Allo, M., Rifkin, G. D., Fekety, R., & Silva, J., Jr. (1980). Mucosal damage mediated by clostridial toxin in experimental clindamycin-associated colitis. *Gut, 21,* 493–499.

Acres, S. D., Isaacson, R. E., Babiuk, L. A., & Kapitany, R. A. (1979). Immunization of calves against enterotoxigenic *Colibacillosis* by vaccinating dams with purified K99 antigen and whole cell bacterins. *Infectious Immunology, 25,* 121–126.

Ando, K., & Nakamura, T. (1991). A method for producing a new medicine for both treating and preventing peptic ulcer diseases and gastritis and thus formulated medicines. European patent application no. 91310049.1.

Bogstedt, A. K., Hammarström, L., & Robertson, A. K. (1997). Survival of immunoglobulins from different species through the gastrointestinal tract in healthy adult volunteers: Implications for human therapy. *Antimicrobial Agents of Chemotherapy, 41,* 2320.

Brinkworth, G. D., & Buckley, J. D. (2003). Concentrated bovine colostrum protein supplementation reduces the incidence of self-reported symptoms of upper respiratory tract infection in adult males. *European Journal of Nutrition, 42,* 228–232.

Brock, J. H., Ortega, F., & Pineiro, A. (1975). Bactericidal and haemolytic activity of complement in bovine colostrum and serum: Effect of proteolytic enzymes and ethylene glycol tetraacetic acid (EGTA). *Annals of Immunology, 126C,* 439–451.

Brown, W. R., Rabbani, H., Butler, J. E., & Hammarström, L. (1997). Characterization of the bovine Cα gene. *Immunology, 91,* 1–6.

Brunser, O., Espinoza, J., Figueroa, G., Araya, M., Spencer, E., Hilpert, H., Link-Amster, H., & Brüssow, H. (1992). Field trial of an infant formula containing anti-rotavirus and

anti-*Escherichia coli* milk antibodies from hyperimmunized cows. *Journal of Pediatric Gastroenterology and Nutrition, 15,* 63–72.
Butler, J. E. (1983). Bovine immunoglobulins: An augmented review. *Veterinary Immunology and Immunopathology, 4,* 43–152.
Butler, J. E., Kiddy, C. A., Pierce, C. S., & Rock, C. A. (1972a). Quantitative changes associated with calving in the levels of bovine immunoglobulins in selected body fluids. I. Changes in the levels of IgA, IgG1 and total protein. *Canadian Journal of Comparative Medicine, 36,* 234–242.
Butler, J. E., Maxwell, C. F., Pierce, C. S., Hylton, M. B., Asofsky, R., & Kiddy, C. A. (1972b). Studies on the relative synthesis and distribution of IgA and IgG1 in various tissues and body fluids of the cow. *Journal of Immunology, 109,* 38–46.
Butler, J. E., Seawright, G. L., McGivern, P. L., & Gilsdorf, M. (1986). Preliminary evidence for a diagnostic immunoglobulin G1 antibody response among culture-positive cows vaccinated with *Brucella abortus* strain 19 and challenge exposed with strain 2308. *American Journal of Veterinary Research, 47,* 1258–1264.
Casswall, T. H., Sarker, S. A., Albert, M. J., Fuchs, G. J., Bergström, M., Björk, L., & Hammarström, L. (1998). Treatment of *Helicobacter pylori* infection in infants in rural Bangladesh with oral immunoglobulins from hyperimmune bovine colostrum. *Alimentary Pharmacology and Therapy, 12,* 563–568.
Casswall, T. H., Sarker, S. A., Faruque, S. M., Weintraub, A., Albert, M. J., Fuchs, N., Dahlström, A. K., Link, H., Brüssow, H., & Hammarström, L. (2000). Treatment of enterotoxigenic and enteropathogenic *Escherichia coli* induced diarrhoea in children with bovine immunoglobulin milk concentrate from hyperimmunized cows: A double-blind, placebo-controlled, clinical trial. *Scandinavian Journal of Gastroenterology, 7,* 711–718.
Casswall, T. H., Nilsson, H. O., Björck, L., Sjöstedt, S., Xu, L., Nord, C. K., Boren, T., Wadström, T., & Hammarström, L. (2002). Bovine anti-*Helicobacter pylori* antibodies for oral immunotherapy. *Scandinavian Journal of Gastroenterology, 37,* 1380–1385.
Cloud, J., & Kelly, C. P. (2007). Update on *Clostridium difficile* associated disease. *Current Opinion in Gastroenterology, 23,* 4–9.
Crooks, C. V., Wall, C. R., Cross, M. L., & Rutherford-Markwick, K. J. (2006). The effects of bovine colostrum supplementation on salivary IgA in distance runners. *International Journal of Sport Nutrition and Exercise Metabolism, 16,* 47–64.
Davidson, G. P., Daniels, E., Nunan, H., Moore, A. G., Whyte, P. B. D., Franklin, K., McCloud, P. I., & Moore, D. J. (1989). Passive immunisation of children with bovine colostrum containing antibodies to human rotavirus. *Lancet, 2,* 709–712.
Davidson, G. P., Tam, J., & Kirubacaran, C. (1994). Passive protection against symptomatic hospital acquired rotavirus infection in India and Hong Kong. *Journal of Pediatric Gastroenterology and Nutrition, 19,* 351.
Ebina, T., Sato, A., Umezu, K., Ishida, N., Ohyama, S., Oizumi, A., Aikawa, K., Katagiri, S., Katsushima, N., Imai, A., Kitaoka, S., Suzuki, H., & Konno, T. (1985). Prevention of rotavirus infection by oral administration of cow colostrum containing antihumanrotavirus antibody. *Medical Microbiology and Immunology, 174,* 177–185.
Fayer, R., & Jenkins, M. C. (1992). Colostrum from cows immunized with *Eimeria acervulina* antigens reduces parasite development *in vivo* and *in vitro*. *Poultry Science, 71,* 1637–1645.
Fayer, R., Andrews, C., Ungar, B. L., & Blagburn, B. (1989). Efficacy of hyperimmune bovine colostrum for prophylaxis of cryptosporidiosis in neonatal calves. *Journal of Parasitology, 75,* 393–397.
Fayer, R., Guidry, A., & Blagburn, B. L. (1990). Immunotherapeutic efficacy of bovine colostral immunoglobulins from a hyperimmunized cow against cryptosporidiosis in neonatal mice. *Infectious Immunology, 58,* 2962–2965.
Filler, S. J., Gregory, R. L., Michalek, S. M., Katz, J., & McGhee, J. R. (1991). Effect of immune bovine milk on *Streptococcus mutans* in human dental plaque. *Archives in Oral Biology, 36,* 41–47.

Freedman, D. J., Tacket, C. O., Delehanty, A., Maneval, D. R., Nataro, J., & Crabb, J. H. (1998). Milk immunoglobulin with specific activity against purified colonization factor antigens can protect against oral challenge with enterotoxigenic *Escherichia coli*. *Journal of Infectious Diseases, 177*, 662–667.

Funatogawa, K., Ide, T., Kirikae, F., Saruta, K., Nakano, M., & Kirikae, T. (2002). Use of immunoglobulin enriched bovine colostrum against oral challenge with enterohaemorrhagic *Escherichia coli* O157:H7 in mice. *Medical Microbiology and Immunology, 46*, 761–766.

Graczyc, T. K., Cranfield M. R., Helmer, P., Fayer, R., & Bostwick, E. F. (1998). Therapeutic efficacy of hyperimmune bovine colostrum treatment against clinical and subclinical *Cryptosporidium serpentis* infections in captive snakes. *Veterinary Parasitology, 31*, 123–132.

Greenberg, P. D., & Cello, J. P. (1996). Treatment of severe diarrhea caused by *Cryptosporidium parvum* with oral bovine immunoglobulin concentrate in patients with AIDS. *Journal of AIDS, 13*, 348–354.

Harp, J. A., Woodmansee, D. B., & Moon, H. W. (1989). Effects of colostral antibody on susceptibility of calves to *Cryptosporidium parvum* infection. *American Journal of Veterinary Research, 50*, 2117–2119.

Hilpert, H., Brüssow, H., Mietens, C., Sidoti, J., Lerner, L., & Werchau, H. (1987). Use of bovine milk concentrate containing antibody to rotavirus to treat rotavirus gastroenteritis in infants. *Journal of Infectious Diseases, 156*, 158–166.

Hoskins, D., Chrisp, C. E., Suckow, M. A., & Fayer, R. (1991). Effect of hyperimmune bovine colostrum raised against *Cryptosporidium parvum* on infection of guinea pigs by *Cryptosporidium wrairi*. *Journal of Protozoology, 38*, 185S–186S.

Howden, C. W. (1996). Clinical expressions of *Helicobacter pylori* infection. *American Journal of Medicine, 100*, 27S–32S.

Huppertz, H. I., Rutkowski, S., Busch, D. H., Eisebit, R., Lissner, R., & Karch, H. (1999). Bovine colostrum ameliorates diarrhea in infection with diarrheagenic *Escherichia coli*, Shiga toxin-producing *E. coli*, and *E. coli* expressing intimin and hemolysin. *Journal of Pediatric Gastroenterology and Nutrition, 29*, 452–456.

Isaacson, R. E., Dean, E. A., Morgan, R. L., & Moon, H. W. (1980). Immunization of suckling pigs against enterotoxigenic *Escherichia coli*-induced diarrheal disease by vaccinating dams with purified K99 or 987P pili: Antibody production in response to vaccination. *Infectious Immunology, 19*, 824–826.

Jansson, A., Nava, S., Brüssow, H., Mahanalabis, D., & Hammarström, L. (1994). Titers of specific antibodies in immunized and non-immunized colostrum; implications for their use in the treatment of patients with gastrointestinal infections. In *Indigenous Antimicrobial Agents of Milk—Recent Developments* (pp. 221–228). Brussels: International Diary Federation Press.

Jenkins, M. (2004). Present and future control of cryptosporidiosis in humans and animals. *Expert Review of Vaccines, 3*, 669–671.

Jenkins, M. C., O'Brien, C., Trout, J., Guidry, A., & Fayer, R. (1999). Hyperimmune bovine colostrum specific for recombinant *Cryptosporidium parvum* antigen confers partial protection against cryptosporidiosis in immunosuppressed mice. *Vaccine, 17*, 2453–2460.

Kacskovics, I., & Butler, J. E. (1996). The heterogeneity of bovine IgG2. VIII. The complete cDNA sequence of bovine IgG2a (A2) and an IgG1. *Molecular Immunology, 33*, 189–195.

Kacskovics, I., Wu, Z., Simister, N. E., & Hammarström, L. (2000). Cloning and characterization of the bovine MHC class I-like Fc receptor. *Journal of Immunology, 164*, 1889–1897.

Kacskovics, I., Kis, Z., Mayer, B., West, A. P., Jr., Tiangco, N. E., Tilahun, M., Cervenak, L., Bjorkman, P. J., Goldsby, R. A., Szenci, O., & Hammarström, L (2006). FcRn mediates elongated serum half-life of human IgG in cattle. *International Immunology, 18*, 525–536.

Kelly, C. P., Chetham, S., Keates, S., Bostwick, E. F., Roush, A. M., Castagliuolo, I., LaMont, J. T., & Pothoulakis, C. (1997). Survival of anti-*Clostridium difficile* bovine immunoglobulin concentrate in the human gastrointestinal tract. *Antimicrobial Agents of Chemotherapy, 41*, 236–241.

Kelly, P., & Farthing, M. J. (1998). Bacterial infections of the gut (excluding enteric fever). *Current Opinion in Infectious Diseases, 11*, 577–582.

Koga, T., Oho, T., Shimazaki, Y., & Nakano, Y. (2002). Immunization against dental caries. *Vaccine, 20*, 2027–2044.

Korhonen, H., Syväoja, E.-L., Ahola-Luttila, H., Sivelä, S., Kopola, S, Husu, J., & Kosunen, T. U. (1994). *Helicobacter pylori*-specific antibodies and bactericidal activity in serum, colostrum and milk of immunised and non immunised cows. *Indigenous Antimicrobial Agents of Milk—Recent Developments. IDF Special Issue 9404, 4*, 151–163.

Korhonen, H., Syvaoja, E. L., Ahola-Luttila, H., Sivela, S., Kopola, S., & Husu, J. (1995). Bactericidal effect of bovine normal and immune serum, colostrums and milk against *Helicobacter pylori*. *Journal of Applied Bacteriology, 78*, 655–662.

Korhonen, H., Marnila, P., & Gill, H. S. (2000). Bovine antibodies for health. *British Journal of Nutrition, 84*, S135–S146.

Kuroiwa, Y., Kasinathan, P., Choi, Y. J., Naeem, R., Tomizuka, K., Sullivan, E. J., Knott, J. G., Duteau, A., Goldsby, R. A., Osborne, B. A., Ishida, I., & Robl, J. M. (2002). Cloned transchromosomic calves producing human immunoglobulin. *Nature Biotechnology, 20*, 889–894.

Leresche, E., Paunier, L., & Andrejevic, G. (1972). Stabilité dans le tractus digestif du nourrisson et tolerance clinique d'immunoglobulines bovines à teneur anticorps eleveé contre les *E. Coli* pathogenes, administrées par voie orale. *Comp Immun Microbiol Infect Dis, 1*, 205–220.

Lissner, R., Schmidt, H., & Karch, H. (1996). A standard immunoglobulin preparation produced from bovine colostra shows antibody reactivity and neutralization activity against Shiga-like toxins and EHEC-hemolysin of *Escherichia coli* O157:H7. *Infection, 24*, 378–383.

Lissner, R., Thürmann, P. A., Merz, G., & Karch, H. (1997). Antibody reactivity and fecal recovery of bovine immunoglobulins following oral administration of a colostrum concentrate from cows (Lactobin) to healthy volunteers. *International Journal of Clinical Pharmacological Therapy, 36*, 239–245.

Lodinová-Zàdnikovà, R., Korych, B., & Bartakova, Z. (1987). Treatment of gastrointestinal infections in infants by oral administration of colostral antibodies. *Die Nahrung, 31*, 465–467.

Loimaranta, V., Tenovuo, J., Virtanen, S., Marnila, P., Syvaoja, E. L., Tupasela, T., & Korhonen, H. (1997). Generation of bovine immune colostrum against *Streptococcus mutans* and *Streptococcus sobrinus* and its effect on glucose uptake and extracallular polysaccharide formation by mutans streptococci. *Vaccine, 15*, 1261–1268.

Loimaranta, V., Carlen, A., Olsson, J., Tenovuo, J., Syvaoja, E. L., & Korhonen, H. (1998). Concentrated bovine colostral whey proteins from *Streptococcus mutans/Strep. sobrinus* immunized cows inhibit the adherence of *Strep. mutans* and promote the aggregation of mutans streptococci. *Journal of Dairy Research, 65*, 599–607.

Loimaranta, V., Nuutila, J., Marnila, P., Tenovuo, J., Korhonen, H., & Lilius, E. M. (1999). Colostral proteins from cows immunised with *Streptococcus mutans/S. sobrinus* support the phagocytosis and killing of mutans streptococci by human leukocytes. *Journal of Medical Microbiology, 48*, 917–926.

Lyerly, D. M., Lockwood, D. E., Richardson, S. H., & Wilkins, T. D. (1982). Biological activities of toxins A and B of *Clostridium difficile*. *Infectious Immunology, 35*, 1147–1150.

Lyerly, D. M., Bostwick, E. F., Binion, S. B., & Wilkins, T. D. (1991). Passive immunization of hamsters against disease caused by *Clostridium difficile* by use of bovine immunoglobulin G concentrate. *Infectious Immunology, 59*, 2215–2218.

Martin-Gomez, S., Alvarez-Sanchez, M. A., & Rojo-Vazquez, F. A. (2005). Oral administration of hyperimmune anti-*Cryptosporidium parvum* ovine colostral whey confers a high level of protection against cryptosporidiosis in newborn NMRI mice. *Journal of Parasitology, 91,* 674–678.

Mayer, B., Doleschall, M., Bender, B., Bartyik, J., Bosze, Z., Frenyo, L. V., & Kacskovics, I. (2005). Expression of the neonatal Fc receptor (FcRn) in the bovine mammary gland. *Journal of Dairy Research, 72,* 107–112.

McClead, R. E., & Gregory, S. A. (1984). Resistance of bovine colostral anti-cholera toxin antibody to *in vitro* and *in vivo* proteolysis. *Infectious Immunology, 44,* 474–478.

McClead, R. E., Butler, T., & Rabbani, G. H. (1988). Orally administered bovine colostral anti-cholera toxin antibodies: Results of two clinical trials. *American Journal of Medicine, 85,* 811–816.

Mietens, C., Keinhorst, H., Hilpert, H., Gerber, H., Amster, H., & Pahud, J. J. (1979). Treatment of infantile *E. coli* gastroenteritis with specific bovine anti-*E. coli* milk immunoglobulins. *European Journal of Pediatrics, 132,* 239–252.

Mitchell, H. M. (1999). The epidemiology of *Helicobacter pylori*. *Current Topics in Microbiology and Immunology, 24,* 11–30.

Mitra, A. K., Mahalanabis, D., Ashraf, H., Unicomb, L., Eeckels R., & Tzipori, S. (1995). Hyperimmune cow colostrum reduces diarrhoea due to rotavirus: A double-blind, controlled clinical trial. *Acta Paediatrics, 84,* 996–1001.

Mousavi, M., Rabbani, H., & Hammarström, L. (1997). Characterization of the bovine ε gene. *Immunology, 92,* 369–373.

Mousavi, M., Rabbani, H., Pilström, L., & Hammarström, L. (1998). Characterization of the gene for the membrane and secretory form of the IgM heavy-chain constant region gene ($C\mu$) of the cow (*Bos taurus*). *Immunology, 93,* 581–588.

Nomoto, K., Matsuoka, K., Hayakawa, K., Ohwaki, M., Kan, T., & Yoshikai, Y. (1992). Antibacterial effect of bovine milk antibody against *Escherichia coli* in a mouse indigenous infection model. *Medical Microbiology and Immunology, 181,* 87–98.

Nord, J., Ma, P., DiJohn, D., Tzipori, S., & Tacket, C. O. (1990). Treatment with bovine hyperimmune colostrum of cryptosporidial diarrhea in AIDS patients. *AIDS, 4,* 581–584.

Oona, M., Rägo, T., Maaroos, H., Mickelsaar, M., Loivukene, K., Salminen, S., & Korhonen, K. (1997). *Helicobacter pylori* children with abdominal complaints: Has immune bovine colostrum some influence on gastritis? *Alpe Adr Microbiol Journal, 6,* 49–57.

Opekun, A. R., El-Zaimaity, H. M., Osato, M. S., Gilger, M. A., Malaty, H. M., Terry, M., Headon, D. R., & Graham, D. Y. (1999). Novel therapies for *Helicobacter pylori* infection. *Alimentary Pharmacology and Therapeutics, 13,* 35–41.

Osame, S., Ichijo, S., Ohta, C., Watanabe, T., Benkele, W., & Goto, H. (1991). Efficacy of colostral immunoglobulins for therapeutic and preventive treatments of calf diarrhea. *Journal of Veterinary Medical Science, 53,* 87–91.

Perryman, L. E., Riggs, M. W., Mason, P. H., & Fayer, R. (1990). Kinetics of *Cryptosporidium parvum* sporozoite neutralization by monoclonal antibodies, immune bovine serum, and immune bovine colostrum. *Infectious Immunology, 58,* 257–259.

Playford, R. J., Floyd, D. N., Macdonald, C. E., Calnan, D. P., Adenekan, R. O., Johnson, W., Goodlad, R. A., & Marchbank., T. (1999). Bovine colostrum is a health food supplement which prevents NSAID induced gut damage. *Gut, 44,* 653–658.

Playford, R. J., MacDonald, C. E., Calnan, D. P., Floyd, D. N., Podas, T., Johnson, W., Wicks, A. C., Bashir, O., & Marchbank, T. (2001). Co-administration of the health food supplement, bovine colostrum, reduces the acute non-steroidal anti-inflammatory drug-induced increase in intestinal permeability. *Clinical Science, 100,* 627–633.

Plettenberg, A., Stoehr, A., Stellbrink, H. J., Albrecht, H., & Meigel, W. (1993). A preparation from bovine colostrum in the treatment of HIV positive patients with chronic diarrhea. *Clinical Investigations, 71,* 42–45.

Rabbani, H., Brown, W. R., Butler, J. E., & Hammarström, L. (1997). Polymorphism of the IGHG3 gene in cattle. *Immunogenetics, 46,* 326–331.

Riggs, M. W. (1997). Immunology: Host response and development of passive immunotherapy and vaccines In R. Fayer (Ed.), *Cryptosporidium and Cryptosporidiosis* (pp. 129–162). Boca Raton, FL: CRC Press.

Rokka, S., Korhonen, B. H., Nousiainen, J., & Marnila, P. (2001). Colostral whey concentrate supplement increases complement activity in the sera of neonatal calves. *Journal of Dairy Research, 68,* 357–367.

Roos, N., Mahé, S., Benamouzig, R., Sick, H., Rautureau, J., & Tomé, D. (1995). ^{15}N-labeled immunoglobulins from bovine colostrum are partially resistant to digestion in human intestine. *Journal of Nutrition, 125,* 1238–1244.

Rump, J. A., Arndt, R., Arnold, A., Bendick, C., Dichtelmüller, H., Franke, M., Helm, B. B., Jäger, H., Kampmann, B., Kolb, P., Kreuz, W., Lissner, R., Meigel, W., Ostendorf, P., Peter, H. H., Plettenberg, A., Schedel, I., Stellbrink, H. W., & Stephan, W. (1992). Treatment of diarrhoea in human immunodeficiency virus-infected patients with immunoglobulins from bovine colostrum. *Clinical Investigations, 70,* 588–594.

Sarker, S. A., Casswall, T. H., Mahalanabis, D., Alam, N. H., Albert, M. J., Brüssow, H., Fuchs, G. J., & Hammarström, L. (1998). Successful treatment of rotavirus diarrhea in children with immunoglobulin from immunized bovine colostrum. *Pediatric Infectious Disease Journal, 17,* 1149–1154.

Saxon, A., & Weinstein, W. (1987). Oral administration of bovine colostrum anti-*Cryptosporidia* antibody fails to alter the course of human *Cryptosporidios*. *Journal of Parasitology, 73,* 413–415.

Schaller, J. P., Saif, L. J., Cordle, C. T., Candler, E., Jr., Winship, T. R., & Smith, K. L. (1992). Prevention of human rotavirus-induced diarrhea in gnotobiotic piglets using bovine antibody. *Journal of Infectious Diseases, 165,* 623–630.

Shimazaki, Y., Mitoma, M., Oho, T., Nakano, Y., Yamashita, Y., Okano, K., Nakano, Y., Fukuyama, M., Fujihara, N., Nada, Y., & Koga, T. (2001). Passive immunization with milk produced from an immunized cow prevents oral recolonization by *Streptococcus mutans*. *Clinical and Diagnostic Lab Immunology, 8,* 1136–1139.

Snodgrass, D. R., Nagy, L. K., Sherwood, D., & Campbell, I. (1982). Passive immunity in calf diarrhea: Vaccination with K99 antigen of enterotoxigenic *Escherichia coli* and rotavirus. *Infectious Immunology, 37,* 586–591.

Stephan, W., Dichtelmüller, H., & Lissner, R. (1990). Antibodies from colostrum in oral immunotherapy. *Journal of Clinical and Chemical Biochemistry, 28,* 19–23.

Tacket, C. O., Losonsky, G., Link, H., Hoang, Y., Guesry, P., Hilpert, H., & Levine, M. M. (1988). Protection by milk immunoglobulin concentrate against oral challenge with enterotoxigenic *Escherichia coli*. *New England Journal of Medicine, 318,* 1240–1243.

Tacket, C. O., Binion, S., Bostwick, E., Losonsky, G., Roy, M. J., & Edelman, R. (1992). Efficacy of bovine milk immunoglobulin concentrate in preventing illness after *Shigella flexneri* challenge. *American Journal of Tropical Medicine and Hygiene, 47,* 276–283.

Tarpila, S., Korhonen, H., & Salminen, S. (1995). Immune colostrum in the treatment of *Helicobacter pylori* gastritis. In *Abstract Book of the 24th International Dairy Congress. Melbourne, Australia* (pp. 18–22)..

Tawfeek, H. I., Najim, N. H., & Al-Mashikhi, S. (2003). Efficacy of an infant formula containing anti-*Escherichia coli* colostral antibodies from hyperimmunized cows in preventing diarrhea in infants and children: A field trial. *International Journal of Infectious Diseases, 7,* 120–128.

Tollemar, J., Gross, N., Dolgiras, N., Jarstrand, C., Ringdén, O., & Hammarström, L. (1999). Passive immunity against *Candida* in bone-marrow transplanted patients: Reduction of salivary yeast counts by oral administration of bovine anti-*Candida* antibodies. *Bone Marrow Transplant, 23,* 283–290.

Tsubokura, K., Berndtson, A., Bogstedt, A., Kaijser, B., Kim, M., Ozeki, M., & Hammarstrom, L. (1997). Oral administration of antibodies as prophylaxis and therapy in *Campylobacter jejuni*-infected chickens. *Clinical Experiments in Immunology, 108*, 451–455.

Turner, R. B., & Kelsey, D. K. (1993). Passive immunization for prevention of rotavirus illness in healthy infants. *Pediatric Infectious Disease Journal, 12*, 718–722.

Tzipori, S., Robertson, D., & Chapman, C. (1986). Remission of diarrhoea due to cryptosporidiosis in an immunodeficient child treated with hyperimmune bovine colostrum. *British Medical Journal, 293*, 1276–1277.

Tzipori, S., Robertson, D., Cooper, D., & White, L. (1987). Chronic cryptosporidial diarrhoea and hyperimmune cow colostrum. *Lancet, 2*, 344–345.

Ungar, B. L., Ward, D. J., Fayer, R., & Quinn, C. A. (1990). Cessation of *Cryptosporidium*-associated diarrhea in an acquired immunodeficiency syndrome patient after treatment with hyperimmune bovine colostrum. *Gastroenterology, 98*, 486–489.

van Dissel, J. T., de Groot, N., Hensgens, C. M., Numan, S., Kuijper, E. J., Veldkamp, P., & van't Wout, J. (2005). Bovine antibody-enriched whey to aid in the prevention of a relapse of *Clostridium difficile*-associated diarrhoea: Preclinical and preliminary clinical data. *Journal of Medical Microbiology, 54*, 197–205.

van Zaane, D., Ijzerman, J., & De Leeuw, P. W. (1986). Intestinal antibody response after vaccination and infection with rotavirus of calves fed colostrum with or without rotavirus antibody. *Veterinary Immunology and Immunopathology, 11*, 45–63.

Ward, S. J., Douce, G., Dougan, G., & Wren, B. W. (1999). Local and systemic neutralizing antibody responses, induced by intranasal immunization with the nontoxic binding domain of toxin A from *Clostridium difficile*. *Infectious Immunology, 67*, 5124–5132.

Warny, M., Fatimi, A., Bostwick, E. F., Lebel, F., LaMont, J. T., Pothoulakis, C., & Kelly C. P. (1999). Bovine immunoglobulin concentrate-*Clostridium difficile* retains *C. difficile* toxin neutralising activity after passage through the human stomach and small intestine. *Gut, 44*, 212–217.

Weiner, C., Pan, Q., Hurtig, M., Borén, T., Bostwick, E., & Hammarström, L. (1999). Passive immunity against human pathogens using bovine antibodies. *Clinical Experiments in Immunology, 116*, 193–205.

Xu, L. B., Chen, L., Gao, W., & Du, K. H. (2005). Bovine immune colostrum against 17 strains of diarrhea bacteria and *in vitro* and *in vivo* effects of its specific IgG. *Vaccine, 24*, 2131–2140.

Ylitalo, S., Uhari, M., Rasi, S., Pudas, J., & Leppaluoto, J. (1998). Rotaviral antibodies in the treatment of rotaviral gastroenteritis. *Acta Paediatrics, 87*, 264–267.

Yolken, R., Losonsky, G. A., Vonderfecht, S., Leister, F., & Wee, S.-B. (1988). Antibody to human rotavirus in cow's milk. *New England Journal of Medicine, 312*, 605–610.

Zhao, Y., Kacskovics, I., Pan, Q., Liberles, D. A., Geli, J., Davis, S. K., Rabbani, H., & Hammarström, L. (2002). Artiodactyl IgD: The missing link. *Journal of Immunology, 169*, 4408–4416.

Zhao, Y., Kacskovics, I., Rabbani, H., & Hammarström, L. (2003). Physical mapping of the bovine immunoglobulin heavy chain constant region gene locus. *Journal of Biological Chemistry, 278*, 35024–35032.

Zhao, Y., Jackson, S. M., & Aitken, R. (2006). The bovine antibody repertoire. *Developmental and Comparative Immunology, 30*, 175–186.

Zinkernagel, R. M., Hilpert, H., & Gerber, H. (1972). The digestion of colostral bovine immunoglobulins in infants. *Experientia, 28*, 741.

Manipulation of Milk Fat Composition Through Transgenesis

A. L. Van Eenennaam and J. F. Medrano

Introduction

The lipids that comprise bovine milk fat contain a large number of different fatty acids (FA) and give rise to one of the most complex naturally occurring fats, but this observation must be set against the much-repeated generalization that milk fat is unhealthy since it is rich in saturated FA. The typical FA profile of bovine milk fat is 70% saturated FA, 25% monounsaturated fatty acids (MUFA), and 5% polyunsaturated fatty acids (PUFA). This high level of saturated FA in ruminant milk can be attributed in part to the process of rumen biohydrogenation that rapidly hydrogenates dietary unsaturated FA (Grummer, 1991). Although human dietary consumption of fats containing high levels of medium-chain (12:0–16:0) saturated FA has been linked to increased serum cholesterol and coronary heart disease, the FA present in milk are not uniform in structure or biological effect, and some may even have beneficial health effects (Bauman et al., 2006).

It has been suggested that the ideal nutritional milk fat for human consumption would contain 8% saturated FA, 82% MUFA, and 10% PUFA (O'Donnell, 1989). One option that has been pursued to increase the unsaturated FA content of milk fat has been to apply different physical or chemical treatments to the fats fed to ruminants to "protect" dietary unsaturated FA from biohydrogenation in the rumen (Lock & Bauman, 2004). An alternative approach would be to genetically engineer animals to produce their own desaturase enzymes to allow for the endogenous desaturation of FA in the mammary gland. Although there has been significant progress in the technologies to generate transgenic livestock and much has been written about the potential for transgenesis to modify milk fat composition to improve its nutritional quality, this area of research has received less attention in

---

A.L. Van Eenennaam
University of California, Davis, Department of Animal Science, One Shields Ave., Davis, CA 95616-8521
e-mail: alvaneenennaam@ucdavis.edu

laboratories than it has in review articles (Houdebine, 2005; Melo et al., 2007; Pintado & Gutierrez-Adan, 1999; Wall et al., 1997). Very few laboratories worldwide are involved in the application of genetic engineering for animal agricultural applications; consequently, only a few publications report the successful modification of milk fat composition through transgenesis. This is perhaps surprising given that modification of milk fat composition would seem to be a valuable phenotype that is well suited to a gain-of-function transgenic approach and the fact that traditional animal breeders are actively pursuing other QTL-based approaches based on naturally occurring variation in milk fat composition to achieve this goal (Morris et al., 2007).

Fatty Acid Composition

As discussed in the chapter by Bernard, Lenux and Chilliard, animals have an endogenous stearoyl-CoA desaturase (SCD) enzymatic activity that introduces a cis double bond into a broad spectrum of fatty acyl-CoA substrates. The bovine SCD gene carries out extensive desaturation of palmitate (16:0) and stearic acid (18:0) in the mammary gland and acts on vaccenic acid (18:1 *trans*-11) to form *cis*-9, *trans*-11conjugated linoleic acid (CLA). This CLA, also known as rumenic acid, represents 75 to 90% of the total CLA in milk fat and is principally derived from endogenous desaturation of vaccenic acid by SCD in the mammary gland (Bauman et al., 2006). Not unexpectedly, this gene was one of the first targeted for transgenic overexpression in the ruminant mammary gland as a part of a project aimed at providing a permanent and heritable means to improve the healthfulness of milk (Reh et al., 2004). In this study, the rat SCD was placed under the control of the bovine β-lactoglobulin promoter and expressed in the mammary gland of transgenic goats. Expression of this transgene changed the composition of the milk to a less saturated and more monounsaturated FA profile in some of the founders and increased the levels of *cis*-9, *trans*-11 CLA in one of the lines. However, the distinct milk FA phenotype was not maintained throughout the entire lactation and had diminished by day 30, which the authors attributed to unstable expression of the transgene.

Further desaturation of oleic acid (OA, 18:1 *n*-9) cannot occur in mammals because vertebrates lack the $\Delta 12$ and $\Delta 15$ FA desaturases responsible for producing linoleic acid (LA, 18:2 *n*-6) and α-linolenic acid (ALA, 18:3 *n*-3), respectively; hence, these two PUFA are essential components of vertebrate diets (Wallis et al., 2002). Genes encoding the $\Delta 12$ and $\Delta 15$ FA desaturases have been identified in various plant systems and in simple organisms including the free-living nematode *Caenorhabditis elegans*, which synthesizes a wide range of PUFA and possesses the only known example of a $\Delta 15$ FA desaturase enzyme and one of the few known examples of a $\Delta 12$ FA desaturase enzyme in the animal kingdom (Fig. 1; Peyou-Ndi et al., 2000; Spychalla et al., 1997; Watts & Browse, 2002). Among the three kingdoms, three types of FA desaturases are found: acyl-ACP, acyl-lipid, and acyl-CoA desaturases (Pereira et al., 2003). Acyl-ACP desaturases are found in plant plastids in a soluble form, and desaturate FA linked to an acyl carrier protein (ACP). The acyl-lipid desaturases introduce unsaturated double bonds into membrane lipid-bound FA and are usually found in plants, fungi, and cyanobacteria. The FA desaturases found in most animals (e.g., SCD) desaturate FA that are esterified to Coenzyme A (CoA).

To circumvent the requirement to obtain n-6 and n-3 FA from the diet, researchers have been working to enable their endogenous production by developing transgenic animals expressing enzymes with $\Delta 12$ and $\Delta 15$ FA desaturase activities. Transient expression of a fungal acyl-lipid $\Delta 12$ desaturase in mouse L cells resulted in a significant increase in the amount of LA at the expense of OA. There was also an increase in the endogenous n-6 PUFA (20:3 *n*-6 and 20:4 *n*-6). Significantly, the increase in n-6 FA was seen only in the cellular phospholipids and not in the cellular triglyceride pool (Kelder et al., 2001).

The only published report of a transgenic animal expressing a $\Delta 12$ FA desaturase is that of pigs expressing an acyl-lipid desaturase from spinach under the control of the mouse *aP2* promoter (Saeki et al., 2004). Although the milk FA composition of these pigs was not analyzed, this paper is significant because it did demonstrate that it is possible to obtain the functional expression of a plant FA desaturase gene in mammals. Two founder pigs, one male and one female, were found to have white adipose tissue-specific transgene expression. Levels of LA in adipose tissues from these transgenic pigs were 1.2-fold higher than those from wild-type animals. This limited augmentation of LA was disappointing, and the authors suggested that it may have been related to substrate availability and the fact that the spinach gene is known to act only on acyl lipids, whereas all known mammalian FA desaturases recruit acyl-CoA substrates. The major pathway for synthesis of triglycerides involves the transfer of an acyl moiety from an acyl-CoA glycerol backbone via the Kennedy pathway (Kennedy, 1961). It may be that the LA accumulated in the phospholipid fraction *in vivo* and was therefore not available as an acyl-CoA substrate for triglyceride biosynthesis.

There is some support for this concept in that expression of a $\Delta 5$ acyl-CoA desaturase gene in transgenic soybean seeds resulted in a large increase of its novel desaturated FA product (C20:1 $\Delta 5$) in the triglyceride, but not the membrane, fraction of the seed (Cahoon et al., 2000). It was postulated that one of the reasons for the success of this experiment was that the novel unsaturated FA did not spend any time in the membrane, as it was made from a CoA substrate and hence was available to be incorporated into the triglyceride pool via the Kennedy pathway (Voelker & Kinney, 2001). Given this finding, it would be of interest to examine the effects of expressing a $\Delta 12$ acyl-CoA desaturase gene in transgenic animals. At present, no gene encoding a $\Delta 12$ desaturase that uses an acyl-CoA substrate has been cloned, although intriguingly, there is a report of an enzyme with such activity in the house cricket, *Acheta domesticus* (Borgeson et al., 1990; Cripps et al., 1990). From the perspective of lactation, an interesting side note with regard to the spinach $\Delta 12$ FA desaturase transgenic founder female pig was that she was unable to support her piglets due to agalactia.

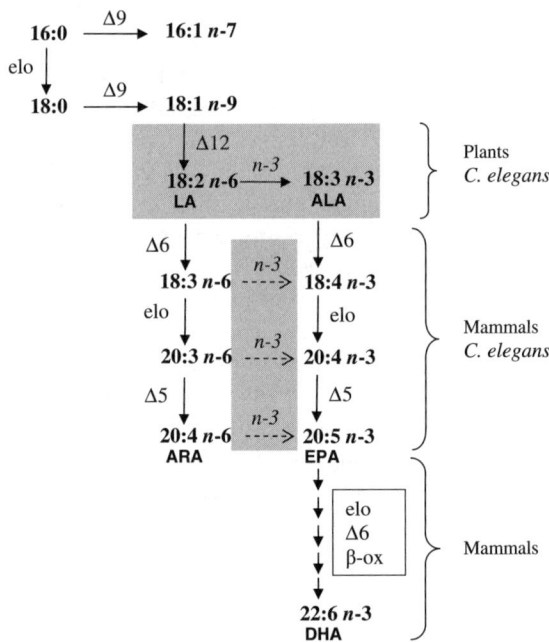

Fig. 1 Pathways for the synthesis of polyunsaturated fatty acids (PUFA) in *C. elegans*, plants, and mammals. The steps are denoted by arrows with the enzymatic activity, desaturase ($\Delta\_$), or elongase (elo), shown next to or above the arrow. Genes encoding $\Delta 12$ and *n*-3 FA desaturases are responsible for producing linoleic acid (LA, 18:2 *n*-6) and a-linolenic acid (ALA, 18:3 *n*-3), respectively, in plants and *C. elegans*. LA and ALA are essential nutritional components of mammals and can only come from diet, since mammals lack the enzyme to synthesize them. The horizontal dotted arrows on the shaded rectangle denote a unique activity of the *C. elegans* n-3 desaturase in the synthesis of PUFA. The synthesis of DHA in mammals requires several steps involving elongation, $\Delta 6$ desaturase, and β-oxidation (β-ox). Important mammalian PUFA are arachidonic acid (AA, 20:4 *n*-6), eicosapentaenoic acid (EPA, 20:5 *n*-3), and docosahexaenoic acid (DHA, 22:6 *n*-3). [Figured modified from Wallis et al. (2002) and Watts and Browse (2002).]

Since the discovery of the *C. elegans* $\Delta 15$ FA desaturase enzyme (Spychalla et al., 1997), a small number of research laboratories have worked on the transgenic expression of this gene in other animal species. The enzyme encoded by this gene is capable of introducing a double bond at the *n*-3 position of both C18 and C20 PUFA and is therefore more correctly termed an *n*-3 FA desaturase (Fig. 1). The high efficiency with which this *C. elegans* n-3 FA enzyme desaturates LA in membrane lipids suggests that it may act on acyl-lipid substrates (Spychalla et al., 1997) rather than the acyl-CoA substrates typically targeted by animal FA desaturases. Adenovirus-mediated introduction of this *n*-3 FA desaturase gene into mammalian cells was able to quickly and effectively elevate the cellular *n*-3 PUFA content and decrease the ratio of *n*-6/*n*-3 PUFA, providing proof that this gene can function in mammalian cells (Kang et al., 2001).

Subsequent constitutive expression of a codon-optimized version of this gene in transgenic mice resulted in a dramatic increase in n-3 FA and a reduction in n-6 FA in many tissues including milk (Kang et al., 2004). No data on the actual FA composition or the n-6/n-3 ratio of phospholipids versus triglycerides were provided in this brief report. However, the finding that the n-6/n-3 ratio of milk fat was significantly decreased, combined with the fact that FA in milk are present principally as triglycerides which comprise greater than 99% of mouse milk lipids, suggests that the FA compositions of both the triglyceride and phospholipid fractions were impacted despite the fact that the desaturase is thought to act only on acyl-lipid substrates. FA are known to be apportioned to membrane phospholipids, triglycerides, or oxidation according to the metabolic demands of the tissue (German et al., 1997). Considered together, these data suggest that over the course of their lifetime these transgenic mice were able to remove or edit the endogenously produced n-3 FA from membrane phospholipids, such that they then became available as acyl-CoA substrates for the milk triglyceride biosynthetic pathway. This same *C. elegans* n-3 FA desaturase gene construct was also used to produce transgenic pigs. Transgenic animals clearly showed a decreased n-6/n-3 ratio when compared to age-matched samples from control pigs. The ratio was 8.5 in the controls versus 1.6 in the transgenic animals in tail samples, and in muscle samples the ratio was 2.8 as compared to 13.5 in control pigs (Lai et al., 2006).

Increasing the n-3 PUFA content of ruminant milk will be more complex than the expression of a single n-3 or $\Delta15$ FA desaturase gene. As a result of rumen biohydrogenation, very little dietary LA reaches the bovine mammary gland. It will therefore be necessary to have both $\Delta12$ and n-3 FA desaturase activities in ruminants to achieve high levels of n-3 PUFA in ruminant products. The Van Eenennaam laboratory has been working toward the eventual goal of genetically engineering ruminants to produce high n-3 PUFA milk by enabling *de novo* $\Delta12$ and n-3 FA desaturation in the mammary gland. Modifying bovine milk fat composition to increase the n-3 PUFA content would help to improve the nutritional composition of an important component of the Western diet. To establish initial proof of concept for this idea, the cDNA coding sequences of *C. elegans* $\Delta12$ and n-3 FA desaturases were each placed under the control of separate constitutive eukaryotic promoters and simultaneously introduced into HC11 mouse mammary epithelial cells by adenoviral transduction. Phospholipids from transduced cells showed a significant decrease in the ratios of both MUFA/PUFA and n-6/n-3 FA relative to control cultures. The FA composition of triglycerides derived from transduced cells was similarly, but less dramatically, affected (Morimoto et al., 2005).

Following the work in cultured cells, lines of transgenic mice expressing either the *C. elegans* $\Delta12$ or the n-3 FA desaturase under the control of the mammary gland-specific goat β-casein promoter (Roberts et al., 1992) were produced with the intent of eventually crossing the two transgenic lines to produce a double transgenic line able to endogenously produce n-3 PUFA from OA. Milk phospholipids from n-3 fatty desaturase transgenic mice were found to contain significantly decreased levels of LA and arachidonic acid (AA, 20:4 n-6) and increased levels of ALA and eicosapentaenoic acid (EPA, 20:5 n-3), reflecting the known activities of the *C. elegans* n-3 FA desaturase (Kao et al., 2006b). Levels of n-3 PUFA were also significantly increased in the milk triglyceride fraction, although the changes were less dramatic. These results again suggest a limited progression of newly synthesized n-3 FA from the phospholipid to the milk triglyceride fraction. Interestingly, despite the fact that the absolute levels of docosahexaenoic acid (DHA, 22:6 n-3) in milk were not significantly impacted by transgenic expression of the n-3 fatty acid desaturase, the FA composition of brains from pups nursing on these transgenic dams revealed greatly elevated levels of DHA relative to pups nursing on nontransgenic control dams (Kao et al., 2006a).

At the present time, efforts have not been successful in obtaining a $\Delta 12$ FA desaturase transgenic mouse line with a high n-6 milk FA phenotype. Although several lines have been generated expressing the transgene, the FA composition of the milk derived from these lines was not found to differ from controls (A. L. Van Eenennaam, unpublished data). Additionally, an agalactic phenotype has been observed in several of these lines. Lactating females appear to have difficulty producing enough milk to nurse their pups, and the consistency of the milk itself is highly viscous. The similarity of this phenotype to the spinach $\Delta 12$ FA desaturase transgenic founder female pig is intriguing, although a larger sample size will be needed to determine if this phenotype is in some way related to the $\Delta 12$ FA desaturase activity. Certainly such an eventuality could be envisioned, given the importance of phospholipid membrane composition on cellular metabolism and the association of PUFA status with many chronic disease states, prostaglandin synthesis, and fertility. Although agalactia is an interesting phenotype, efforts continue to pursue the development of additional transgenic lines of $\Delta 12$ FA desaturase transgenic mice with the objective of finding a line that produces milk with elevated levels of n-6 PUFA.

Triglyceride Structure

The triglyceride structure also has important human health implications. The positional distribution of FA on the glycerol backbone has a significant effect on the digestibility and subsequent metabolism of FA in monogastric animals. Long-chain FA, especially 16:0 and 18:0, are poorly absorbed when placed at the sn-1 and sn-3 positions (Ramirez et al., 2001). This is due to the fact that in the digestive process pancreatic lipase cleaves the sn-1 and sn-3 FA off the triglycerides, thereby determining whether FA are taken up as 2-monoacylglycerol (the FA in the sn-2 position) or less readily as free FA (i.e., FA cleaved from the sn-1 and sn-3 positions). Fat absorption has been linearly related to the amount of palmitic acid (PA, 16:0) in the sn-2 position (Kubow, 1996). Newborn infants given a formula consisting of lard with PA in the *sn*-2 position absorbed 95% of all FA, whereas those fed lard with PA randomized in all three positions of the triglycerides absorbed only 72% of the FA (Carnielli et al., 1996). Dietary saturated FA, especially C12:0–C16:0, have been shown to increase circulating concentrations of serum cholesterol in adults when dietary cholesterol is high. Numerous studies have concluded that dietary intake of saturated FA is a risk factor in coronary heart disease; from a human health perspective, it would be desirable to decrease the consumption of animal products containing *sn*-2 saturated FA.

The triglyceride structure of bovine milk fat is distinctive in that it has a high proportion of short-chain FA in the *sn*-3 position and a predisposition for medium- and long-chain saturated FA (C12:0-C18:0) in the *sn*-2 position. In milk fats of placental mammals other than ruminants, PA is heavily concentrated at the *sn*-2 position, and OA at the *sn*-1 and *sn*-3 positions (Anderson et al., 1985). The finding that saturated FA in the *sn*-2 position of triglycerides predominate in the milk fat of almost all mammals suggests that maybe there is a selective advantage in this characteristic being conserved (Grigor, 1980). Indeed, the presence of PA in the *sn*-2 position in triglycerides in human milk fat is thought to be one of the reasons that fat from human milk is better absorbed than fat contained in infant formulas (Tomarelli et al., 1968). It has been suggested that this type of conservation, even during extreme dietary manipulation, could point to a possible evolutionary selection for milk fat triglycerides synthesized with saturated FA at the *sn*-2 position (German et al., 1997).

The first step of triglyceride biosynthesis is the acylation of glycerol 3-phosphate at the *sn*-1 position by glycerol 3-phosphate acyl transferase (GPAT) to form lysophosphatidic acid. AGPAT (1-acyl *sn*-glycerol-3-phosphate acyltransferase), also known as LPAAT (lysophosphatidic acid acyltransferase), catalyzes the conversion of lysophosphatidic acid to phosphatidic acid. This product is then dephosphorylated to form diacylglycerol, which is acylated in the *sn*-3 position by DGAT (diacylglycerol acyltransferase) to form triglyceride. The stereospecific distribution of acyl groups in the triglycerides can arise either via the specificity of the acyltransferases, or as a result of the distribution of acyl groups in the fatty acyl-CoA pool. The relative importance of these two factors is unclear, and studies carried out on the acyl chain specificities of acyltransferases have generated conflicting results (Dircks & Sul, 1999). The microsomal glycerol-3 phosphate acyltransferase (GPAT) has generally been shown to utilize long-chain saturated and unsaturated fatty acyl-CoAs equally, suggesting that the substrate pool of fatty acyl-CoAs determines the FA composition at the *sn*-1 position (German et al., 1997).

Many AGPAT isoforms have been identified in mammals; there have been conflicting reports regarding their specificities and the impact this has on the *sn*-2 FA composition of the fat in different tissues (Dircks & Sul, 1999). In most naturally occurring fats, there is a preponderance of unsaturated FA at the *sn*-2 position. The notable exception is milk, and it has been shown in MAC-T bovine mammary gland cell lines that the bovine mammary AGPAT has a greater affinity for saturated fatty acyl-CoAs (Morand et al., 1998a). It is not yet known whether regulation during lactation alters the substrate specificity of AGPAT, or whether mammary tissue differentiation leads to a developmentally regulated expression of an AGPAT isoform that is specific to the mammary gland. Although mammary gland–specific gene regulation of biosynthetic enzymes for milk fat production has been reported (Safford et al., 1987), the acyltransferase specificity of MAC-T bovine mammary gland cell lines supports the claim that the AGPAT in the mammary gland is a mammary tissue-specific isoform (Morand et al., 1998a). A study reporting the cloning of the bovine AGPAT from mammary gland tissue did not examine the substrate specificity of the enzyme (Mistry & Medrano, 2002). A mammary AGPAT with substrate specificity for unsaturated FA could have beneficial human health effects for dairy consumers. One study reports how site-directed mutagenesis of an *Escherichia coli* AGPAT increased the *in vitro* substrate specificity for unsaturated FA (Morand et al., 1998b). Overexpression of such a gene in the bovine mammary gland could lead the way for the production of milk triglycerides with an increased proportion of bioavailable monounsaturated *sn*-2 monoglycerides, at the expense of the less healthful saturated *sn*-2 monoglycerides typically found in bovine milk.

In this regard, the AGPAT from the echidna (*Tachyglossus aculeatus*), a very primitive monotreme mammal of New Guinea and Australia, provides an intriguing case study of an approach to optimize milk fat structure for human

Table 1 Positional Distribution of Major FA in Triglycerides from Bovine, Echidna, and Human Milk

FA Position (mol %)	Bovine *sn* Position			Echidna *sn* Position			Human *sn* Position		
	1	2	3	1	2	3	1	2	3
4:0	5.0	2.9	43.3						
6:0	3.0	4.8	10.8						
8:0	0.9	2.3	2.2						
10:0	2.5	6.1	3.6				0.2	0.2	1.8
12:0	3.1	6.0	3.5				1.3	2.1	6.1
14:0	10.5	20.4	7.1	1.7	0.9	0.4	3.2	7.3	7.1
14:1				1.3	0.7	0.2			
15:0				0.8	0.2	0.1			
15:1				0.4	0.1	0.2			
16:0	35.9	**32.8**	10.1	31.5	**9.0**	27.9	16.1	**58.2**	49.7
16:1	2.9	2.1	0.9	7.1	7.0	8.0	3.6	4.7	7.3
17:0				1.5	0.4	1.6			
17:1				0.7	0.8	0.6			
18:0	14.7	6.4	4.0	16.8	2.1	14.3	15.1	3.3	2.0
18:1	20.6	**13.7**	14.9	3.1	**57.6**	39.8	46.1	**12.7**	49.7
18:2	1.2	2.5	<1.0	4.1	18.3	4.9	11.0	7.3	14.7
18:3				1.0	2.9	2.0	0.4	0.6	1.6

health. The positional distribution of triglyceride FA in echidna milk is very different from that of any other mammal. The major saturated FA (C16:0 and C18:0) are almost equally distributed between the sn-1 and sn-3 positions on the glycerol moiety, whereas unsaturated FA (C18:1, C18:2, and C18:3) are preferentially esterified at the sn-2 position (Table 1). As such, echidna milk triglycerides have a FA distribution similar to that found in vegetable oils (Parodi, 1982; Parodi & Griffiths, 1983). An examination of the relative abundance of fatty acids in milk triglycerides and those found specifically as 2-monoglycerides suggests that the echidna is the only known mammal where PA is preferentially excluded from the sn-2 position of milk triglycerides (Grigor, 1980). This example highlights how information derived from comparative studies may help to guide researchers as they strive to develop transgenic approaches to produce innovative dairy products.

Conclusions and Perspectives

This chapter reviewed the studies that have directly focused on using transgenic approaches to modify the composition of milk fat. Important targets for modification include altering the FA composition to increase the proportion of beneficial FA, particularly CLA and n-3 PUFA, and altering the triglyceride structure to reduce the proportion of saturated FA in the sn-2 position. Due to the innate physiological and metabolic attributes of ruminants, such as rumen microbial biohydrogenation and mammals' inability to synthesize LA or ALA, it is difficult to achieve significant progress toward these intractable goals using conventional selection and breeding approaches. Transgenesis offers a powerful approach to overcome these biological obstacles and thereby enable the development of dairy products that have beneficial human health effects for consumers.

An aspect that should be highlighted is the importance of comparative and functional genomics to identify promising gene targets to modify metabolic processes by transgenic approaches. A gain-of-function transgenic approach offers the only way to effect an enzymatic pathway or function that is not present in the organism of interest. The increasing availability of genomic sequences from a multitude of organisms, prokaryotes, and eukaryotes and the bioinformatics tools to identify them will facilitate this effort.

Despite the fact that the manipulation of milk fat composition through transgenesis could result in products with long-awaited consumer benefits, the prognosis for such applications appears somewhat bleak, at least in North America and Europe. The animal biotechnology industry is faced with regulatory indecision, decreased funding support, and a negative public perception. Although transgenesis offers a unique opportunity to introduce novel traits into livestock production systems, no transgenic agricultural animals have been successfully commercialized to date. There is no doubt that transgenesis could result in the optimization of milk fat composition for human health. However, the question remains as to whether the dairy industry and consumers are willing to accept this approach to achieve that goal.

Acknowledgment Investigators acknowledge the support of NIH Grant 1R03HD047193–01 (to A.L.V.) and the University of California Davis Agricultural Experimental Station.

References

Anderson, R. R., Collier, R. J., Guidry, A. J., Heald, C. W., Jenness, R., Larson, B., & Tucker, H. A. (1985). *Lactation*. Ames: Iowa State University Press.

Bauman, D. E., Mather, I. H., Wall, R. J., & Lock, A. L. (2006). Major advances associated with the biosynthesis of milk. *Journal of Dairy Science, 89,* 1235–1243.

Borgeson, C. E., de Renobales, M., & Blomquist, G. J. (1990). Characterization of the delta 12 desaturase in the American cockroach, *Periplaneta americana*: The nature of the substrate. *Biochimica et Biophysica Acta, 1047,* 135–140.

Cahoon, E. B., Marillia, E. F., Stecca, K. L., Hall, S. E., Taylor, D. C., & Kinney, A. J. (2000). Production of fatty acid components of meadowfoam oil in somatic soybean embryos. *Plant Physiology, 124,* 243–251.

Carnielli, V. P., Wattimena, D. J., Luijendijk, I. H., Boerlage, A., Degenhart, H. J., & Sauer, P. J. (1996). The very low birth weight premature infant is capable of synthesizing arachidonic and docosahexaenoic acids from linoleic and linolenic acids. *Pediatric Research, 40,* 169–174.

Cripps, C., Borgeson, C., Blomquist, G. J., & de Renobales, M. (1990). The delta 12-desaturase from the house cricket, *Acheta domesticus* (Orthoptera: Gryllidae): Characterization and form of the substrate. *Archives in Biochemistry and Biophysics, 278,* 46–51.

Dircks, L., & Sul, H. S. (1999). Acyltransferases of *de novo* glycerophospholipid biosynthesis. *Progress in Lipid Research, 38,* 461–479.

German, J. B., Morand, L. Z., Dillard, C. J., & Xu, R. (1997). Milk fat composition: Targets for alteration of function and nutrition. In R. A. S. Welch, D. J. W. Burns, S. R. Davis, A. I. Popay, and C. G. Prosser (Eds.), *Milk Composition, Production and Biotechnology* (pp. 35–72). Cambridge, UK: CAB International.

Grigor, M. R. (1980). Structure of milk triacylglycerols of five marsupials and one monotreme—Evidence for an unusual pattern common to marsupials and eutherians but not found in the echidna, a monotreme. *Comparative Biochemistry and Physiology B—Biochemistry & Molecular Biology, 65,* 427–430.

Grummer, R. R. (1991). Effect of feed on the composition of milk fat. *Journal of Dairy Science, 74,* 3244–3257.

Houdebine, L. M. (2005). Use of transgenic animals to improve human health and animal production. *Reproduction in Domestic Animals, 40,* 269–281.

Jensen, R. G. (1989). *The Lipids of Human Milk*. Boca Raton, FL: CRC Press.

Kang, J. X., Wang, J., Wu, L. & Kang, Z. B. (2004). Transgenic mice: Fat-1 mice convert n-6 to n-3 fatty acids. *Nature, 427,* 504.

Kang, Z. B., Ge, Y. L., Chen, Z. H., Cluette-Brown, J., Laposata, M., Leaf, A., & Kang, J. X. (2001). Adenoviral gene transfer of *Caenorhabditis elegans* n-3 fatty acid desaturase optimizes fatty acid composition in mammalian cells. *Proceedings of the National Academy of Sciences USA, 98,* 4050–4054.

Kao, B. T., DePeters, E. J., & Van Eenennaam, A. L. (2006a). Mice raised on milk transgenically enriched with n-3 PUFA have increased brain docosahexaenoic acid. *Lipids, 41,* 543–549.

Kao, B. T., Lewis, K. A., DePeters, E. J., & Van Eenennaam, A. L. (2006b). Endogenous production and elevated levels of long-chain *n*-3 fatty acids in the milk of transgenic mice. *Journal of Dairy Science, 89,* 3195–3201.

Kelder, B., Mukeji, P., Kirchner, S., Hovanec, G., Leonard, A. E., Chuang, L. T., Kopchick, J. J., & Huang, Y. S. (2001). Expression of fungal desaturase genes in cultured mammalian cells. *Molecular and Cellular Biochemistry, 219,* 7–11.

Kennedy, E. P. (1961). Biosynthesis of complex lipids. *Federation Proceedings, 20,* 934–940.

Kubow, S. (1996). The influence of positional distribution of fatty acids in native, interesterified and structure-specific lipids on lipoprotein metabolism and atherogenesis. *Journal of Nutritional Biochemistry, 7,* 530–541.

Lai, L. X., Kang, J. X., Li, R. F., Wang, J. D., Witt, W. T., Yong, H. Y., Hao, Y. H., Wax, D. M., Murphy, C. N., Rieke, A., Samuel, M., Linville, M. L., Korte, S. W., Evans, R. W., Starzl, T. E., Prather, R. S., & Dai, Y. F. (2006). Generation of cloned transgenic pigs rich in omega-3 fatty acids. *Nature Biotechnology, 24,* 435–436.

Lock, A. L., & Bauman, D. E. (2004). Modifying milk fat composition of dairy cows to enhance fatty acids beneficial to human health. *Lipids, 39,* 1197–1206.

Melo, E. O., Canavessi, A. M. O., Franco, M. M., & Rumpf, R. (2007). Animal transgenesis: State of the art and applications. *Journal of Applied Genetics, 48,* 47–61.

Mistry, D. H., & Medrano, J. F. (2002). Cloning and localization of the bovine and ovine lysophosphatidic acid acyltransferase (LPAAT) genes that codes for an enzyme involved in triglyceride biosynthesis. *Journal of Dairy Science, 85,* 28–35.

Morand, L. Z., Morand, J. N., Matson, R., & German, J. B. (1998a). Effect of insulin and prolactin on acyltransferase activities in MAC-T bovine mammary cells. *Journal of Dairy Science, 81,* 100–106.

Morand, L. Z., Patil, S., Quasney, M., & German, J. B. (1998b). Alteration of the fatty acid substrate specificity of lysophosphatidate acyltransferase by site-directed mutagenesis. *Biochemical and Biophysical Research Communications, 244,* 79–84.

Morimoto, K. C., Van Eenennaam, A. L., DePeters, E. J., & Medrano, J. F. (2005). Hot topic: Endogenous production of *n*-3 and *n*-6 fatty acids in mammalian cells. *Journal of Dairy Science, 88,* 1142–1146.

Morris, C. A., Cullen, N. G., Glass, B. C., Hyndman, D. L., Manley, T. R., Hickey, S. M., McEwan, J. C., Pitchford, W. S., Bottema, C. D., & Lee, M A. (2007). Fatty acid synthase effects on bovine adipose fat and milk fat. *Mammalian Genome, 18,* 64–74.

O'Donnell, J. (1989). Milk fat technologies and markets-A summary of the Wisconsin Milk Marketing Board 1988 Milk Fat Roundtable. *Journal of Dairy Science, 72,* 3109–3115.

Parodi, P. W. (1982). Positional distribution of fatty acids in triglycerides from milk of several species of mammals. *Lipids, 17,* 437–442.

Parodi, P. W., & Griffiths, M. (1983). A comparison of the positional distribution of fatty-acids in milk triglycerides of the extant monotremes platypus (*Ornithorhynchus-Anatinus*) and echidna (*Tachyglossus-Aculeatus*). *Lipids, 18,* 845–847.

Pereira, S. L., Leonard, A. E., & Mukerji, P. (2003). Recent advances in the study of fatty acid desaturases from animals and lower eukaryotes. *Prostaglandins, Leukotrienes and Essential Fatty Acids, 68,* 97–106.

Peyou-Ndi, M. M., Watts, J. L., & Browse, J. (2000). Identification and characterization of an animal Delta(12) fatty acid desaturase gene by heterologous expression in *Saccharomyces cerevisiae*. *Archives of Biochemistry and Biophysics, 376,* 399–408.

Pintado, B., & Gutierrez-Adan, A. (1999). Transgenesis in large domestic species: Future development for milk modification. *Reproduction Nutrition Development, 39,* 535–544.

Ramirez, M., Amate, L., & Gil, A. (2001). Absorption and distribution of dietary fatty acids from different sources. *Early Human Development, 65 (Suppl),* S95-S101.

Reh, W. A., Maga, E. A., Collette, N. M. B., Moyer, A., Conrad-Brink, J. S., Taylor, S. J., DePeters, S. J., Oppenheim, S., Rowe, J. D., BonDurant, R. H., Anderson, G. B., &

Murray, J. D. (2004). Hot topic: Using a stearoyl-CoA desaturase transgene to alter milk fatty acid composition. *Journal of Dairy Science, 87*, 3510–3514.

Roberts, B., Ditullio, P., Vitale, J., Hehir, K., & Gordon, K. (1992). Cloning of the goat beta-casein-encoding gene and expression in transgenic mice. *Gene, 121*, 255–262.

Saeki, K., Matsumoto, K., Kinoshita, M., Suzuki, I., Tasaka, Y., Kano, K., Taguchi, Y., Mikami, K., Hirabayashi, M., Kashiwazaki, N., Hosoi, Y., Murata, N., & Iritani, A. (2004). Functional expression of a $\Delta 12$ fatty acid desaturase gene from spinach in transgenic pigs. *Proceedings of the National Academy of Sciences USA, 101*, 6361–6366.

Safford, R., de Silva, J., Lucas, C., Windust, J. H., Shedden, J., James, C. M., Sidebottom, C. M., Slabas, A. R., Tombs, M. P., & Hughes, S. G. (1987). Molecular cloning and sequence analysis of complementary DNA encoding rat mammary gland medium-chain S-acyl fatty acid synthetase thio ester hydrolase. *Biochemistry, 26*, 1358–1364.

Spychalla, J. P., Kinney, A. J., & Browse, J. (1997). Identification of an animal omega-3 fatty acid desaturase by heterologous expression in Arabidopsis. *Proceedings of the National Academy of Sciences USA, 94*, 1142–1147.

Tomarelli, R. M., Meyer, B. J., Weaber, J. R., & Bernhart, F. W. (1968). Effect of positional distribution on absorption of fatty acids of human milk and infant formulas. *Journal of Nutrition, 95*, 583.

Voelker, T., & Kinney, A. J. (2001). Variations in the biosynthesis of seed-storage lipids. *Annual Reviews of Plant Physiology and Plant Molecular Biology, 52*, 335–361.

Wall, R. J., Kerr, D. E., & Bondioli, K. R. (1997). Transgenic dairy cattle: Genetic engineering on a large scale. *Journal of Dairy Science, 80*, 2213–2224.

Wallis, J. G., Watts, J. L., & Browse, J. (2002). Polyunsaturated fatty acid synthesis: What will they think of next? *Trends in Biochemical Science, 27*, 467–473.

Watts, J. L., & Browse, J. (2002). Genetic dissection of polyunsaturated fatty acid synthesis in *Caenorhabditis elegans*. *Proceedings of the National Academy of Sciences USA, 99*, 5854–5859.

Producing Recombinant Human Milk Proteins in the Milk of Livestock Species

Zsuzsanna Bösze, Mária Baranyi, and C. Bruce A. Whitelaw

Abstract Recombinant human proteins produced by the mammary glands of genetically modified transgenic livestock mammals represent a special aspect of milk bioactive components. For therapeutic applications, the often complex posttranslational modifications of human proteins should be recapitulated in the recombinant products. Compared to alternative production methods, mammary gland production is a viable option, underlined by a number of transgenic livestock animal models producing abundant biologically active foreign proteins in their milk. Recombinant proteins isolated from milk have reached different phases of clinical trials, with the first marketing approval for human therapeutic applications from the EMEA achieved in 2006.

Introduction

Recombinant DNA technology has revolutionized the production of therapeutic proteins. Even before the sequence of the Human genome became known, genes of a great number of human proteins have already been identified and cloned, including clotting factors VII (hfVII), VIII (hfVIII), and IX (hfIX), growth hormone (hGH), protein C (hPC), insulin (hI), insulin-like growth factor-1 (hIGF-1), interleukin-2 (hIL 2), antithrombin III (hAT-III), tissue plasminogen activator (htPA), α-1 antiprypsin (hα$_1$AT), lactoferrin (hLF), extracellular superoxide dismutase (hEC-SOD), and erythropoietin (hEPO). Human proteins have been used in medicine for many years, but the supply was limited by the availability of the human tissue from which they were extracted (e.g., sourcing hGH from pituitary glands of human cadavers). The first attempts toward producing therapeutic proteins from cloned genes were made in microorganisms, such as yeast and bactcria. Unfortunately, for many

---

Z. Bösze
Agricultural Biotechnology Center, P.O. Box 411, H-2100, Gödöllő, Hungary
e-mail: bosze@abc.hu

proteins this is not a viable option, because microorganisms are not capable of reproducing the posttranslational modifications necessary for protein activity and stability (Swartz, 2001). Furthermore, lower eukaryotic systems such as yeast, filamentous fungi, and unicellular algae are often limited by their ability to duplicate human patterns of protein production and yield recombinant products that are immunogenic and lack activity (Dyck et al., 2003). Insect cell systems have unique glycosylation patterns and are usually restricted to use at the laboratory scale (Farrell et al., 1998). Although mammalian cell culture systems provide the required complex posttranslational modifications, they are expensive and technically demanding when used at a commercial scale. The use of transgenic animals as "bioreactors" overcomes these problems. The expression of human proteins in the mammary gland of livestock (e.g., rabbits, pigs, sheep, and goats) provides a practically unlimited source of correctly processed, active, and stable proteins for clinical use at lower costs than mammalian cell culture, though they still have many regulatory hurdles to cross.

Producing Transgenic Animals

The Targeted Tissue

The main objective in biopharming is the economical production of valuable complex human therapeutic proteins using transgenic livestock species. Ideally, protein production in an animal should allow collection of the product in significant amounts without killing the animal and be isolated such that practical and cheap purification methods can be applied. Harvesting proteins from body fluids (blood, milk, urine) rather than from solid tissue is desirable because the fluids are renewable and most of the biomedically important proteins are secreted into body fluids. Collecting the protein from the bloodstream is possible by targeting expression to the liver or kidney, or to the blood lymphocytes. The main drawback, however, is that high circulating levels of biologically active proteins may have an adverse effect on the health of the animals. Therefore, the most obvious and viable tissue to target the expression of foreign proteins is the mammary gland. Milk is readily collected in a repeatable manner and is available in large quantities. The presence of large amounts of active foreign proteins in milk usually does not interfere with the health of the lactating animals. In some cases adverse effects have been observed, e.g., with human erythropoietin in transgenic rabbits (Massoud et al., 1996) and with hGH in transgenic mice (Devinoy et al., 1994); however, in these animals high concentrations of the recombinant proteins were detected in the blood during lactation. Since it is very unlikely that lactogenic hormones enhance the ectopic expression of the transgenes, the most probable cause is "leaking" of the recombinant protein from the mammary epithelium into the blood.

Milk contains a relatively small number of major protein components, which are secreted exclusively by the mammary gland and belong to either the caseins (α_{S1}-, α_{S2}-, β-, and κ-CN) or the whey proteins (BLG, α-LA, and WAP). The milk protein genes of several species have been cloned and characterized (Mercier & Vilotte, 1993). These single-copy genes are transcribed at high levels specifically in the mammary gland during pregnancy and lactation. Using promoter sequences from different milk protein genes allows high expression of foreign proteins into the milk of transgenic animals to be achieved.

The Gene Construct

The first step toward creating a transgenic animal is to engineer a DNA construct that will target the expression of the candidate protein specifically to the mammary gland. The recombinant protein expression must be restricted to milk to avoid any deleterious side effects on the animal's health. Therefore, the construct usually consist of a milk protein–specific promoter linked to the coding sequence of the desired protein. The regulatory elements that control the temporal and spatial expression of a gene are usually located within a few kilobases of the 5'-end of the transcribed region of the gene. Heterologous gene expression can be targeted specifically to the lactating mammary gland by using promoters isolated from different milk-specific genes. Additional regulatory elements, e.g., insulators, may be added to the construct to ensure high-level and/or position-independent expression of the transgene (Fig. 1). Genomic sequences coding for the candidate proteins was found to be expressed at higher levels than cDNA sequences (Whitelaw et al., 1991; Brinster et al., 1988), although at least the goat β-casein (Ziomek, 1998) and the mouse whey acidic protein promoter (Velander et al., 1992) seem to be capable of directing high-level expression of some cDNA-based constructs.

Large DNA fragments, e.g., BAC or YAC, ensure integrated copy number–dependent tissue and developmental specific expression of the coded genes, which was the case for human and goat α-lactalbumin (Fujiwara et al., 1997; Stinnakre et al., 1999) and porcine whey acidic protein (Rival-Gervier et al., 2002) genes in transgenic mice. Those BAC or YAC vectors could also be used to express human genes at a high level in milk (Fujiwara et al., 1999; Soulier et al., 2003).

Milk protein promoters have been isolated and well characterized from several species, including mice, rats, guinea pigs, rabbits, goats, sheep, and cattle. These promoters usually work well across species. However, the most commonly used promoters in commercial transgenic pharmaceutical production are murine and rabbit whey acidic protein, bovine α_{S1}-casein, goat β-casein, and ovine β-lactoglobulin. Nevertheless, an ideal mammary gland–specific vector still remains to be designed.

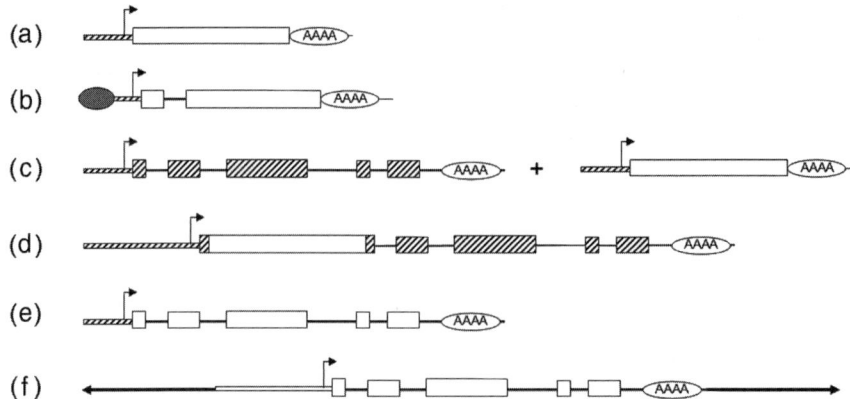

Fig. 1 The most typical types of transgene constructs. (a) cDNA-based gene constructs containing a homologue or heterologue promoter sequence attached to the cDNA sequence of the desired protein. (b) Different sequence elements can be added to the cDNA construct to enhance expression and/or tissue specificity, for, e.g., homologue or heterologue intron sequences, scaffold or matrix attachment elements [SAR/MAR], locus control regions [LCR], insulator elements. (c) Transgene expression can be "rescued" by co-injection with a high-expressing transgene, (d) or the cDNA can also be inserted into the genomic sequence of a highly expressed transgene. (e) Genomic gene constructs containing the whole coding region of the desired protein driven by a homologue or heterologue promoter sequence. (f) Gene constructs containing large genomic sequences with all the endogenous regulatory elements are mainly used with artificial chromosomes [BAC/YAC]

The Choice of Livestock Species

The next step in creating a transgenic animal is the process of introducing the transgene construct into the fertilized eggs of the species of interest. The key consideration for choosing a species for protein production is the quantity of protein product required and the timescale for production (Table 1). The feasibility and the costs of keeping and breeding the animals should also be considered.

The smallest and easiest to keep are rabbits, with about 1 L of milk per lactation, up to 8–10 lactations per year, and a minimum of 6 months to produce a lactating animal. Cattle with the longest timeline are the most costly. Up to 10,000 L of milk per cow can be collected, but a minimum of 2.5 years is necessary to produce a lactating cow (with an additional 2 years if the founder was a bull). Sheep, goats, and pigs are between these two extremes (Table 2). From sheep and goats, several hundred liters of milk can be obtained per lactation and about 1.5 years are needed to reach the first lactation. The generation of pigs can be established in about 15 months, and usually 100–150 L of milk can be collected per lactation (two lactations per year). Milk collection from ruminants is achieved with milking machines.

Table 1 Demand of Some Human Proteins for Clinical Use in the United States and the Estimated Number of Livestock Animals Needed for Production Calculated with an Average Recombinant Protein Expression of 1 g/L of Milk

	Factor VIII(hfVIII)	Factor IX(hFIX)	Protein C(hPC)	Antithrombin III(hAT-III)	Fibrinogen(hFib)	Albumin(hAlb)	Annual Milk Yield (L)
Amount needed (kg/year)	0.3	4	10	21	150	315,000	
Rabbit	60	800	2,000	4,200	30,000	63,000,000	5
Pig	1	14	34	70	500	1,050,000	300
Sheep	1	8	20	42	300	630,000	500
Goat	1	5	13	27	188	393,750	800
Cattle	1	1	2	3	19	39,375	8,000

Table 2 Parameters to Be Considered When Choosing Animal Species for Transgenic Milk Expression

Species	Milk Yield per Lactation(L)	Gestation(Months)	Maturation(Months)	Elapsed Time from Microinjection to First Lactation (Months)
Rabbit	1–1.5	1	4–6	6–8
Pig	100–300	4	7–8	15–16
Sheep	400–600	5	6–8	16–18
Goat	800–1,000	5	6–8	16–18
Cattle	Up to 10,000	9	12–15	30–33

Methods to Create Transgenic Animals

Several methods are available for the generation of transgenic mice; however, they differ in their usefulness when working with livestock species. We describe some of these methods here.

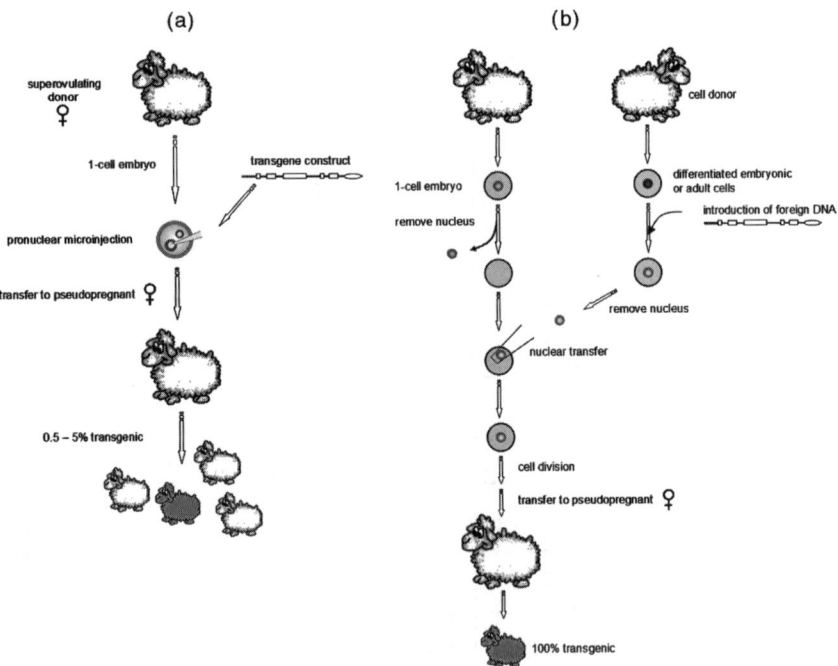

Fig. 2 Methods for producing transgenic livestock animals. (a) Pronuclear microinjection and (b) nuclear transfer. (*See* color plate 4)

Pronuclear microinjection (Fig. 2a) was considered the first successful mammalian transgenic technique (Gordon et al., 1980). It involves the direct injection of a foreign DNA sequence into the pronucleus of a fertilized egg, followed by a surgical implantation into the reproductive tract of a hormonally primed recipient foster mother. Successful gene transfers by this route have been described for all of the major livestock species, including rabbits (Bühler et al., 1990), pigs (Hammer et al., 1985), sheep (Simons et al., 1988), goats (Ebert et al., 1991), and cattle (Krimpenfort et al., 1991). The main assets of this method are as follows: It is well described, relatively simple to carry out for a skilled individual, relatively cost-effective, and DNA sequences of up to several hundred kilobases can be integrated. Further considerations include the low efficiency of generating transgenic founders (usually less than 10% of the animals carry the transgene) and the random integration of the transgene. When a transgene is integrated in a silent region of a chromosome, its product will be poorly or not expressed at all unless otherwise protected by regulatory elements. This phenomenon is called the "position effect." Pronuclear microinjection, however, is very labor-intensive, requiring special skills to carry out the micromanipulations. Nevertheless, despite its limitations, pronuclear microinjection remains the most straightforward and consistently successful means of gene transfer into the mammalian germline.

Retroviral transfection was first used in 1985 to create transgenic animals (Huszar et al., 1985; Jahner et al., 1985; van der Putten et al., 1985). The genetic material of the retroviruses is capable of stable integration into the chromosomes of the infected cells. The replication-defective viral vector construct to be used must contain not only the transgene but also regulating sequences for viral integration and packaging. Also necessary is the use of a "packaging" helper cell line, which allows assembly of the transgene-containing virus. After infection, the few days old embryos are implanted into recipient foster mothers. No micromanipulation is needed. The advantages of this method are its efficiency (nearly 100%) and the one-copy integration (Soriano et al., 1986), but since the viral infection does not occur at the one-cell zygote stage, the resulting animals will always be mosaic (not all cells carry the transgene). Only DNA sequences smaller than 8 kb can be integrated; therefore, in general, only cDNA constructs can be used. There is also the theoretical risk of recombination events leading to the development of new retroviruses. Combined with microinjection, this method has been adapted for cattle (Chan et al., 1998). Because of the above-mentioned limitations, the use of this method is restricted. More recently, the ability of lentivirus vectors to efficiently introduce transgenes has rekindled some interest in viral transgenesis (Clark & Whitelaw, 2003; Whitelaw, 2004).

Direct in vivo transfection of the mammary gland has been proposed as a faster and cheaper alternative to target the expression of a heterologous gene to the secretory mammary epithelial cells. Targeting of transgene carrying replication-defective retroviruses directly to the mammary secretory epithelia cells in lactating goats to produce foreign proteins in the milk has been demonstrated (Archer et al., 1994). Following trials through several ways of transducing the mammary epithelium, recent publications point to a special advantage of the direct instillation of a recombinant adenoviral vector. This method allowed efficient secretion of human growth hormone (Sanchez et al., 2004) and human erythropoietin (Toledo et al., 2006) at levels of up to 2 g/L in the milk of mice and goats. Direct transduction of mammary epithelial cells by means of a recombinant adenovirus could be a suitable alternative to transgenic technology, especially for the production of potentially toxic proteins in milk, at levels high enough for their purification and biological characterization.

Sperm-mediated gene transfer has a history of increased transgenesis efficiency claims. Following controversial results, lactoferrin-producing transgenic rabbits were created through dimethylsulfoxide-treated sperm transfection (Li et al., 2006). The expression of the human lactoferrin (LF) gene was controlled by the goat β-casein gene 5' flanking sequence. Eighty-nine rabbit offspring were produced, with 46 of these being transgenic. Nevertheless, stable transmission of the transgene and expression levels in the consecutive generations has not been reported yet.

Embryonic stem (ES) cells are widely used at present to manipulate the mouse genome. These cells are isolated from the inner cell mass (ICM) of mouse blastocysts. They are undifferentiated cells, which, in the presence of the necessary growth factors, can be cultured without losing their pluripotency. Genetic modification of ES cells can be performed *in vitro,* making targeted transgene integration through homologous recombination possible. When injected back to a host blastocyst or using aggregation with normal diploid embryos (Nagy et al., 1990), their descendants contribute to the tissues of the resulting chimera including the germline. The main drawbacks of the ES technology include that it needs cell culture capabilities, it is very labor-intensive, a good ES cell line is needed, and only the second generation gives germline transgenic animals. Its main assets are that, in theory, construct expression can be tested prior to generating the animal and some advantages do result from site-specific recombination. Unfortunately, despite intensive efforts, no validated ES cells have been described for rabbits, pigs, sheep, goats, or cattle yet.

The major bottleneck in producing transgenic livestock is the low efficiency of generating transgenic founders. A radical improvement would be to carry out the required genetic manipulations not on the zygotes, but in conventionally cultured cells that could then be used to generate animals. Wilmut and co-workers (1997) introduced a new breakthrough in transgenic technology. *Cloning via nuclear transfer* enables viable animals to be created when nuclei from differentiated embryonic or somatic cells kept in *in vitro* culture are transferred into enucleated oocytes. Contradicting scientific dogma, the genetic material of the adult cell is capable of directing the growth and development of the oocyte into a healthy animal (Wilmut et al., 1997). This method is important because of the cloning technique, and it enables the creation of transgenic animals through the genetic manipulation of the donor cells in culture (Fig. 2b). The first transgenic sheep created using this method expressed human factor IX in milk

(Schnieke et al., 1997). Since then transgenic goats and cattle producing recombinant human proteins in their milk have been created by nuclear transfer (Parker et al., 2004). This method also allows gene targeting and the knocking in of transgenes to specified genomic loci, first demonstrated by the production of human α1-antitrypsin in sheep milk (McCreath et al., 2000). More recently, goat fetal fibroblasts have been gene-targeted by inserting the exogenous htPA cDNA into the β-casein locus through homologous recombination (Shen et al., 2006). The targeted insertion of the coding region of recombinant proteins into target loci will ensure more predictable expression levels in the future.

The advantages of cloning via nuclear transfer are that all animals are transgenic, the creation of transgenic animals can be shortened by one generation, the cultured cells can be stored almost indefinitely, and site-specific integration can be accomplished (McCreath et al., 2000). Though problems remain to be resolved, such as low efficiency and high mortality after birth, somatic cloning is already the preferred choice for producing transgenic ruminants. Notably, at the end of 2006, the U.S. FDA announced that food from cloned animals is safe to enter the food chain, although the debate about labeling continues.

Posttranslational Modifications

Many proteins require so-called posttranslational modifications, including signal peptide removal, forming of disulfide bonds, amino acid modifications, proteolytic processing, and subunit assembly. Bacteria, yeast, insect cell systems, or transgenic plants cannot provide all of the necessary modifications, which results in a lack of activity or immunogenicity of products. Mammalian cells and transgenic animals are the choice for recombinant protein production when complex posttranslational modifications are needed for the biologically active protein. Some of the amino acid modifications (e.g., glycosylation, carboxylation) are essential for the biological activity and/or stability of the proteins and are a key point in producing biologically active pharmaceuticals by recombinant organisms. It is by no means certain that the mammary gland is always capable of performing these modifications correctly. Since the recombinant proteins isolated from milk do not always have the same structure as their native counterparts, the possible differences and their effects must be evaluated case by case.

Glycosylation is the process (or result) of the addition of a glycosyl group to asparagine, hydroxylysine, serine, or threonine, resulting in a glycoprotein. The process is one of the principal co-translational and posttranslational modification steps in the synthesis of membrane and secreted proteins, and the majority of proteins synthesized in the rough endoplasmic reticulum undergo glycosylation. It is an enzyme-directed, site-specific process with a very important role in affecting the folding, solubility, stability, biological

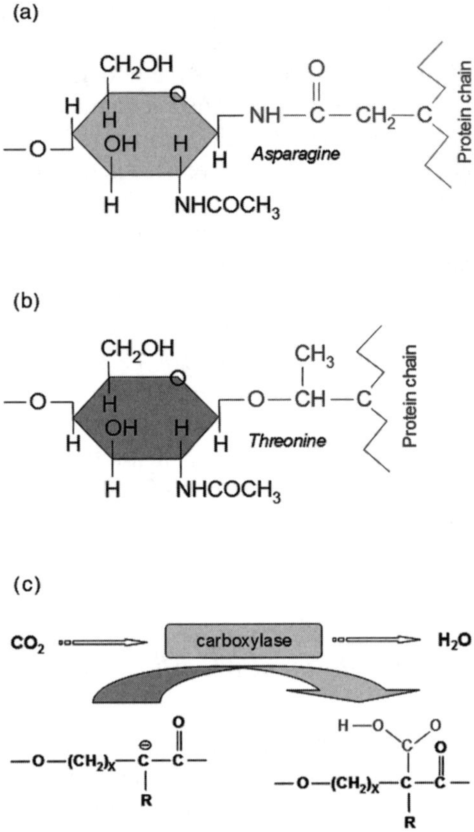

Fig. 3 Representative types of posttranslational modifications. Basic types of glycosylation: (a) N-linked glycosylation to the amide nitrogen of asparagine side chains; (b) O-linked glycosylation to the hydroxy oxygen of serine and threonine side chains. (c) Carboxylation. (*See* color plate 5)

activity, and immunogenicity of proteins. Two types of glycosylation exist: N-linked glycosylation to the amide nitrogen of asparagine side chains and O-linked glycosylation to the hydroxy oxygen of serine and threonine side chains (Figs 3a and b). Glycosylation is undoubtedly one of the most important posttranslational events for therapeutic proteins. It is essential for the stability of many proteins in blood circulation, required for the biological activity of gonadotropins, to some extent for antibodies, and often necessary for correct protein folding, conformation, intracellular transport, or tissue targeting.

The mammary gland naturally secrets N- or O-glycosylated proteins; although the mammary cells are capable of carrying out these modifications, the recombinant proteins isolated from milk are not always glycosylated in the appropriate manner. If this does not adversely affect the activity or stability, and does not cause ill side effects, it is of little consequence for the utilization of the recombinant protein. Detailed characterization of the recombinant human C1 inhibitor produced in the milk of transgenic rabbits showed an overall similarity of N-glycan structures with only the degree of sialylation and core fucosylation being lower (Koles et al., 2004a). The first crystal structure of a recombinant protein produced in milk confirmed that the slightly modified glycosylation pattern of the recombinant human lactoferrin did not alter the protein's structural integrity (Thomassen et al., 2005). Systematic studies on the glycosylation capabilities of the mammary glands of the different species, including the effects of the stage of lactation and individual variations, are still to be performed.

Carboxylation is the introduction of a carboxyl group or carbon dioxide into a compound with formation of a carboxylic acid (Fig. 3c), e.g., the vitamin K–dependent blood clotting factors and regulatory proteins (e.g., hFVII, hFIX, hFX, hPC) require the conversion of glutamic acid (Glu) residues to γ-carboxyglutamic acid (Gla). The γ-carboxylated amino acid residues bind calcium, which is essential for their activity. Species-specific differences were observed in the ability of mammary epithelial cells to carboxylate heterologous recombinant proteins. Usually, carboxylation of a protein present at a low level is not adversely affected, while in some species high-level expression of the protein leads to the reduction in the amount of fully γ-carboxylated, biologically active components. This may result from the saturation of the cellular γ-carboxylase machinery. The species-specific differences may reflect differences in enzyme levels and/or substrate specificity.

Proteolytic processing is also of great importance. The first step in protein maturation is the removal of the signal peptide. Signal peptides are short peptide sequences (usually 13–30 amino acids) at the N-terminal part of proteins that direct the posttranslational transport of the proteins (which are synthesized in the cytosol) to certain organelles for further processing. In case of certain proteins (e.g., vitamin K–dependent plasma proteins) that are first synthesized as inactive preproteins, the removal or cleavage of some other parts of the precursor is essential for the development of the final structure and activity of the mature protein. The significance of the presence of the pro-protein processing enzyme furin has been confirmed experimentally in CHO cells and in transgenic mice. In both cases the expression of furin led to an increased level of mature recombinant human coagulation factor IX (rhFIX) (Lubon & Paleyanda, 1997). Significant species-specific differences were observed regarding the proteolytic processing capacity of the mammary gland.

It has been proven that the mammary gland is able to associate protein subunits in a correct fashion in a number of different cases (e.g., collagen type I, fibrinogen, and superoxide dismutase). Therefore, subunit assembly does not seem to be rate-limiting in recombinant protein production. The most impressive example in this regard is the recombinant fibrinogen produced at a high level in sheep's milk (Garner & Colman, 1998): Fibrinogen comprises six polypeptide chains of dimeric α, β, and γ chains. The recombinant fibrinogen isolated from sheep's milk was functional in clotting assay and was produced at a 1000-fold greater level than that achieved in cell culture.

Purification of Recombinant Proteins

The purification of recombinant proteins from milk for laboratory testing is not particularly problematic, although the purification procedure has to be adapted for each expressed recombinant protein individually. Milk has only a few main protein components, and simple procedures for removal of caseins, the major milk proteins, have been established. However, milk is a complex biological fluid, and eliminating some of its components requires more complex methods; chromatography can produce a high purity of the protein (Wright & Colman, 1997). Furthermore, if the recombinant protein has a high degree of similarity to an abundant milk protein, e.g., human serum albumin (hSA), separation from the equivalent host protein can be difficult.

For commercial products, since the recombinant proteins produced in transgenic livestock will be administered to humans, the therapeutic products must be purified to a very high degree and free of viral and prion proteins. The purification procedures usually involve the combination of several steps and methods like skimming, filtration, precipitation (e.g., selective precipitation by polyethylene glycol, enrichment by barium/citrate precipitation), viral inactivation, and, if necessary, chromatography (usually based on ion exchange, hydrophobic interaction, or immunoaffinity).

Clinical Trials

Before administering new therapeutic products to human patients, the safety and effectiveness must be proven. A series of preclinical and clinical trials must be completed. During preclinical trials, biochemical, toxicity, and pharmacokinetic properties will be tested and detailed information will be collected on the source and means of production. The clinical trials consist of three phases. Phase I is conducted on a small number of healthy volunteers to test if there are any adverse effects. Phase II is carried out on patient and control groups to further evaluate safety and efficiency. In phase III, the product is evaluated in much bigger patient groups and controls to set the proposed use and dosage. Only new products that pass all three trials are licensed to be marketed commercially. The aim of phase IV—following the permission for human therapeutic application—is to evaluate the effect of long-term application on patients. The first transgenic product ATryn® (antithrombin III) approved for human therapeutic use is produced in transgenic goats; marketing approval was granted in 2006 to GTC Biotherapeutics (http://www.gtc-bio.com).

Livestock Species as Bioreactors

Rabbits

Among the transgenic livestock species, rabbits are the smallest and easiest to obtain and maintain. Because of their short generation time and large litter size, they are an attractive alternative to large dairy animals, where the major drawback is the time required to generate the transgenic animals and to deliver a product to the market. The rabbit has well-described laboratory breeds, it is easy to superovulate, and the manipulation of the embryos is quite simple. It is ideal as a model animal but can also be used as a bioreactor if the amount of expressed protein does not need to be more than a few kilograms, as in case of the blood clotting factors (Brem et al., 1998). Other advantages are that it can be kept pathogen-free, it can be easily milked, and its milk composition is well described (Dayal et al., 1982; Baranyi et al., 1995), containing about four times as much protein as cow's milk. Although the efficiency of generating transgenic rabbits by microinjection is lower than that in mice (about 1–2%), it still is at least equivalent to other livestock species (Brem et al., 1998). The ratio of mosaics among the founders is usually quite high, around 30%, resulting in a reduced rate of transgene transmission to the offspring (Castro et al., 1999). This may result from the fact that the embryonic development of rabbits is significantly faster than that of mice, with transgene integration often occurring after the first round of cellular division. Recombinant proteins and peptides that have been produced in the milk of transgenic rabbits are described in Table 3.

Pigs

Although pigs are not conventional dairy animals, they have a distinct advantage over ruminant animal models: Sows have a relatively short generation time of one year and produce two litters and about 20 offspring per year. Transgenic pigs can be created either by microinjection or by somatic cloning. The efficiency of transgenesis is influenced by the fact that pigs differ from many other types of livestock, because unless there are at least four viable fetuses in the womb, pregnancies fail to go to term. In pigs, the mammary gland has no cisternae, with stored milk ejected by an active process and up to 300 L of milk obtained annually from a sow. Table 4 summarizes the recombinant proteins produced in the milk of transgenic pigs. The mammary gland's capacity for performing posttranslational modifications has been compared in transgenic pigs producing rhPC and rhFIX (Van Cott et al., 1999). The γ-carboxylation of the two recombinant proteins was rate-limiting and showed

Table 3 Expression of Recombinant Human Proteins in the Milk of Transgenic Rabbits

Protein	Promoter	Expression Level	Status	Company	Reference
hIL-2	Rabbit β-casein	0.43 μg/mL			Bühler et al. (1990)
htPA	Bovine α_{S1}-casein	50 μg/mL			Riego et al. (1993)
hIGF-1	Bovine α_{S1}-casein	1 mg/mL			Brem et al. (1994)
hGH	Mouse WAP	50 μg/mL			Limonta et al. (1995)
hα_1AT	Goat β-casein	4.0 mg/mL	In development	GTC Biotherapeutics	Genzyme Transgenics (1996)
hPC	Ovine-BLG	0.7 mg/mL		GTC Biotherapeutics	Genzyme Transgenics (1996)
hEPO	Rabbit-WAP	50 μg/mL			Massoud et al. (1996)
hEC-SOD	Murine WAP	2.9 mg/mL			Stromqvist et al. (1997)
hEPO	Bovine BLG	0.5 mg/mL			Korhonen et al. (1997)
sCT*	Ovine β-LG	1–2.1 mg/mL	Phase II finished	PPL	McKee et al. (1998)
hαGLU	Bovine α_{S1}-casein	8 mg/mL			Bijvoet et al. (1999)
hNGF-β	Bovine α_{S1}-casein	250 μg/mL			Coulibaly et al. (1999)
hFVIII	Murine WAP	0.083 IU/mL			Hiripi et al. (2003)
hGH	Rat WAP(6xHisTyr)	Cleavage by trombin to activate			Lipinski et al. (2003)
hTNAP	Rabbit WAP	826 IU/mL			Bodrogi et al. (2006)
hLF	Goat β-casein	153 μg/mL			Li et al. (2006)
hFVII			In development	GTC Biotherapeutics/LFB Biotechnologies	http://www.transgenics.com/products/prod.html
hC1INH	Bovine α_{S1}-casein	12 mg/mL	Phase III	Pharming	Pharming literature online

*sCT salmon calcitonin: The human calcitonin aggregates; therefore, the piscine equivalent was produced.

Table 4 Expression of Recombinant Human Proteins in the Milk of Transgenic Pigs

Protein	Promoter	Expression Level	Company	Reference
hPC	Mouse WAP	1 mg/mL	GTC Biotherapeutics	Velander et al. (1992)
hPC	Ovine BLG	0.75 mg/mL	PPL Therapeutics	PPL literature
hFVIII	Mouse WAP	3 μg/mL		Paleyanda et al. (1997)
hEPO	Mouse WAP	878 IU/mL		Park et al. (2006)

differences between them, resulting in varying degrees of posttranslational modifications.

Sheep and Goats

The length of time to milk production is obviously the major factor in the choice of species; however, the disease status of animals, litter size, and volume of milk should also be taken into consideration. For proteins produced in sheep's milk, a special concern is related to animal health due to the issue of prions responsible for scrapie. Therefore, if sheep is the animal of choice for recombinant protein production, only animals from countries such as New Zealand should be used, because of their scrapie-free status. It was stated a decade ago that high-level (35–45-g/L) production of α1-antitrypsin in the milk of transgenic sheep does not seem to be at the expense of the production of other milk proteins (Colman, 1996).

The prototype of a transgenic goat producing recombinant protein was published in 1991 (Ebert et al., 1991). Ten years later a special dwarf breed of goat (BELE: breed early lactate early) was adapted to produce transgenic goats with nuclear cloning. The early sexual maturity of BELE goats shortens the generation time for producing recombinant proteins (Keefer et al., 2001). The type and expression levels of recombinant proteins produced in the milk of transgenic sheep and goats are described in Tables 5 and 6, respectively. More recently extension of the gene therapy techniques resulted in high-level expression of recombinant proteins in the milk through direct transduction of the mammary epithelium of goats (Sanchez et al., 2004; Toledo et al., 2006). In the future, the direct introduction of a foreign gene into a mammary gland could dramatically reduce the time of production of pharmaceuticals in milk from years to weeks.

Cattle

Transgenic cattle are the most economic choice if the aim is to produce huge amounts of recombinant protein. This species is attractive as a bioreactor

Table 5 Expression of Recombinant Human Proteins in the Milk of Transgenic Sheep

Protein	Promoter	Expression Level	Company	Reference
hα_1AT	Ovine BLG	35 mg/mL		Wright et al. (1991)
hFVII	Ovine BLG	2 mg/mL	PPL Therapeutics	PPL literature
hFVIII	Ovine BLG	6 ng/mL		Niemann et al. (1999)
hFIX	Ovine BLG	25 ng/mL		Simons et al. (1988)
hFIX	Ovine BLG	5 ng/mL		Clark et al. (1989)
hFIX	Ovine BLG	1.0 mg/mL	PPL Therapeutics	Schnieke et al. (1997)
hFIB	Ovine BLG	5.0 mg/mL	PPL Therapeutics	Garner and Colman (1998)
hFIB	Ovine BLG	5 mg/mL		Butler et al. (1997)
hPC	Ovine BLG	0.3 mg/mL	PPL Therapeutics	Garner and Colman (1998)

given the development of many established embryological techniques for cattle. Due to the high value of dairy cattle, "slaughterhouse-derived" oocytes are used for microinjection following *in vitro* oocyte maturation and fertilization. Alternatively, high-quality oocytes can be obtained via ovum pickup. Embryos are individually cultured, and a multiplex PCR analysis can be performed on biopsies to identify the males and the transgenics; since the aim is to produce foreign protein in the milk, preferably only the female transgenic embryos are selected for transfer into synchronized recipient heifers. Pregnancy initiation can be confirmed by ultrasound detection of a fetal heartbeat at ~28 days.

Herman, the world's first transgenic bull, was created in 1991 through microinjection of transgene αs1-casein promoter and the cDNA sequence for human lactoferrin (Krimpenfort et al., 1991). Today transgenic cattle can be generated far more efficiently via nuclear transfer (Table 7). Due to the use of female totipotent cells for genetic manipulation and the subsequent selection of transgenic cells before nuclear transfer, all calves born will be female and transgenic. Since calving-induced lactation will not occur until the animal is ~2 years old, to speed up the selection process, hormonal induction of lactation can be used when the animal is between 2–6 months of age. In the most optimal case, the elapsed time to obtain a lactating female for recombinant protein production is 48–56 months depending on the sex of the founder created by microinjection, which could be reduced to 33 months if nuclear transfer was applied.

Table 6 Expression of Recombinant Human Proteins in the Milk of Transgenic Goats

Protein	Promoter	Expression Level	Status	Company	Reference
htPA	Murine WAP	3 μg/mL 610,000 IU/mg			Ebert et al. (1991), Denman et al. (1991)
htPA	Goat β-casein	3 mg/mL (6 mg/mL?)			Ebert et al. (1994)
hAT-III	Goat β-casein	20 mg/mL	ATryn® EU: approved US: prelaunch	GTC Biotherapeutics/LEO Pharma	Genzyme Transgenics (1996), GTC literature; Edmunds et al. (1998); Baguisi et al. (1999)
hAT-III	Goat β-casein	5.8 mg/mL			Archer et al. (1994)
hGH	Retrovirus	60 ng/mL			Sanchez et al. (2004)
hGH	Adenovirus	0.3 mg/mL			
hα₁AT	Goat β-casein	14 mg/mL		GTC Biotherapeutics	Genzyme Transgenics (1996), GTC literature
hαFP	Goat β-casein		Phase II (2004)	GTC Biotherapeutics/Merrimack Pharmaceuticals	Parker et al. (2004), http://www.transgenics.com/products/novel.html, http://www.clinicaltrials.gov/ct/show/NCT00147329?order=1
hEPO	Adenovirus	2 mg/mL			Toledo et al. (2006)
hLF	Adenovirus	2.6 mg/mL			Han et al. (2007)

Table 7. Expression of Recombinant Human Proteins in the Milk of Transgenic Cattle

Protein	Promoter	Expression Level	Method	Status	Company	Reference
hα-LA		2.4 mg/mL	Microinjection		PPL Therapeutics	Krimpenfort et al. (1991)
hα-LA	Human α-LA	2.4 µg/mL?		Preclinical	Pharming	PPL literature
hCOL				Preclinical	Pharming	
hFIB	Bovine α$_{S1}$-casein	3 mg/mL	Nuclear transfer			
hGH	Bovine α$_{S1}$-casein	5 mg/mL	Nuclear transfer		Bio Sidus SA, Buenos Aires, Argentina	Salamone et al. (2006)
hLF	Bovine α$_{S1}$-casein		Microinjection		PPL Therapeutics	McKee et al. (1998)
hLF	Bovine α$_{S1}$-casein	2.8 mg/mL	Microinjection	Phase I completed	Pharming	van Berkel et al. (2002)
hSA	Bovine α$_{S1}$-casein			In development	GTC Biotherapeutics/TransOva Genetics	GTC literature

Human Recombinant Proteins Produced in Transgenic Livestock

Blood Clotting Factors (hFVII, hFVIII, hFIX)

The blood clotting or coagulation factors are key players in blood coagulation. They are generally serine proteases, with some exceptions; for example, FVIII and FV are glycoproteins and factor XIII is a transglutaminase. Serine proteases act by cleaving other proteins at specific sites. Both plasma-derived and recombinant products are currently used in treating hemophilia. Clotting factors have been used for many years in medicine; initially isolated from human plasma, transgenic animals now provide a promising alternative to these limited recourses. Several blood clotting factors are currently produced in transgenic livestock.

Factor VIIa (also known as proconvertin, serum prothrombin conversion accelerator, or cothromboplastin) is an extrinsic endopeptidase with Gla residues that activates factors IX and X in the blood coagulation cascade. GTC Biotherapeutics, in collaboration with LFB Biotechnologies (http://www.lfb.fr), has developed a transgenically produced recombinant form of human factor VIIa (rhFVIIa) expressed in the milk of transgenic rabbits.

Factor VIII (also known as antihemophilic factor A or antihemophilic globulin) is an intrinsic protein co-factor of factor IX, with which it forms the tenase complex (the tenase complex is formed on a phospholipid surface in the presence of calcium and is responsible for the activation of factor X). FVIII is synthesized predominantly in hepatocytes as a single-chain macromolecule. Congenital X-linked deficiency of hFVIII (hemophilia A) is the most common human bleeding disorder and affects approximately 1 of every 5,000 males. Recombinant hFVIII is currently produced in cell culture system for replacement therapy of hemophilia A patients. The restrictive costs associated with cell culture–produced rFVIII have provided the incentive to develop an alternative production system. At first, transgenic pigs using the regulatory sequences of the mouse whey acidic protein gene and the human FVIII cDNA were created (Paleyanda et al., 1997), with the identity of processed heterodimeric rFVIII confirmed using specific antibodies, by thrombin digestion, and by activity assays. The secretion of 2.7 µg/mL of rFVIII in milk was detected. With an hFVIII cDNA/murine metallothionein I hybrid gene construct containing the ovine β-lactoglobulin promoter, transgenic sheep have been created by microinjection that produce up to 6 ng/mL of hFVIII in their milk (Niemann et al., 1999). The same hFVIII cDNA/murine metallothionein I hybrid gene coupled with the murine whey acidic protein promoter gave low expression levels in transgenic rabbits (Hiripi et al., 2003).

Factor IX (also known as Christmas factor, antihemophilic factor B, or plasma thromboplastin component) is an intrinsic endopeptidase. Through forming the tenase complex with factor VIII, it activates factor X. Its deficiency results in hemophilia B, which can be treated with FIX. Using the ovine-βlactoglibulin promoter, several groups have created transgenic sheep expressing hFIX in their mammary gland (Schnieke et al., 1997; Clark et al., 1989; Simons et al., 1988). Usually, expression levels were in the ng/mL range.

Protein C (hPC)

Human protein C (hPC) is a regulator of hemostasis, a zymogen of a serine protease that is activated by thrombin. hPC has a complex structure, containing nine γ-carboxylated glutamic acid residues that bind calcium at about 1 to 3 mM. Gamma-carboxylation is a vitamin K–dependent posttranslational modification. Transgenic pigs were generated that produced human protein C in their milk at up to 1 g/L. The gene construct consisted of the cDNA for human protein C inserted into the first exon of the mouse whey acidic protein gene (Velander et al., 1992). A monoclonal antibody that binds an epitope in the glutamic acid domain of hPC in the absence of calcium was used to study the conformational behavior of immunopurified rhPC. Immunopurified rhPC from higher-expressing pigs gave a less calcium-dependent response, suggesting that a rate limitation in γ-carboxylation by the mammary gland occurs at expression levels about >500 µg/mL in pigs (Subramanian et al., 1996). These studies provide evidence that γ-carboxylation can occur at high levels in the mammary gland of a pig.

The effects of rhPC expression levels on endogenous immunoglobulin and transferrin content of the milk of different lineages of transgenic pigs were studied. The levels of rhPC in the milk ranged from 40 to 1,200 µg/mL. Transgenic pigs with rhPC expression levels lower than 500 µg/mL had no significant differences in milk protein composition with respect to nontransgenic pigs. A line of transgenic pigs having rhPC expression levels of 960–1,200 µg/mL had two- to threefold higher IgG, IgM, and secretory IgA concentrations compared to other transgenic and nontransgenic pig groups. Since IgG, IgM, secretory IgA, and transferrin are transported into the milk by transcytosis, higher levels of these proteins indicate that transcyctosis in the mammary epithelial cell was likely upregulated in pigs having high rhPC expression levels (Van Cott et al., 2001).

Growth Hormone (hGH)

Human growth hormone (hGH) is not only a valuable recombinant therapeutic protein for hormone deficiency indications, but also is an extensively characterized molecule both from recombinant bacterial systems and as circulating in human blood. Treatment of growth hormone (GH) deficiency via parenteral administration of recombinant hGH has greatly benefited from recombinant DNA technology allowing the production of practically unlimited amounts of the pure hormone. An unwanted side effect of using recombinant human growth hormone, in combination with other products (e.g., androgens, erythropoietin), is for doping in sports. Although its effectiveness in enhancing physical performance is still unproved, the compound is likely used for its potential anabolic effect on muscle growth.

There have been several attempts in the last 15 years to produce recombinant hGH in the milk of transgenic livestock animals as an alternative to production in recombinant *Escherichia coli*. One of the early models was a direct transfer of the *hGH* gene into the mammary gland by using replication-defective retrovirus vector in goats (Archer et al., 1994). Since then transgenic rabbits and transgenic cows producing high levels of rhGH in milk have been generated (Limonta et al., 1995; Salamone et al., 2006; Lipinski et al., 2003). Transgenic rabbits were created through microinjection with a chimeric gene comprising 5' sequences from mouse whey acidic protein gene linked to the *hGH* gene. The foreign protein was detected in the milk and serum of these animals at levels of up to 50 μg/mL and 0.6 ng/mL, respectively. In the milk of cloned transgenic cows, up to 5 g/L of rhGH were detected. The hormone is identical to that currently produced by expression in *Escherichia coli*. In addition, the hematological and somatometric parameters of the cloned transgenic cattle are within the normal range for the breed and display normal fertility, being capable of producing normal offspring. In the future, transgenic cattle could be used as a cost-effective alternative for the production of rhGH.

Serum Albumin (hSA)

Human serum albumin (hSA), a protein currently derived from pooled human plasma, is used therapeutically to maintain osmotic pressure in the blood. Approximately 440 metric tons of plasma-derived albumin are used annually worldwide, with annual sales of approximately US$1.5 billion; 5,400 cows would be needed to produce 100,000 kg of hSA (Rudolph, 1999). The use of the recombinant form, produced in *Saccharomyces cerevisiae*, is limited to excipient applications. Animal bioreactor approaches have been established, with hSA produced in the milk of transgenic cows by GTC Biotherapeutics and Genzyme Biotech (http://www.genzyme.com); GTC Biotherapeutics has formed a joint venture with Fresenius (http://www.fresenius.de) to expand the commercial development opportunities of recombinant hSA. The aim is to manage the development of hSA for both the blood expander market and the use of hSA in the excipient market.

Lactoferrin (hLF)

Human lactoferrin (hLF) is a natural protein that helps to fight and prevent infections and excessive inflammation and strengthens the human defense system. The protein is present in significant amounts in numerous human biological fluids and mucus secretions, including tears and lung secretions, and has been shown to fight bacteria that cause infections of the eye and lungs. In addition, hLF is present in substantial quantities in mother's milk and plays an important role in the defense system of infants as well as that of adults. Lactoferrin promotes the health of the gastrointestinal system by improving the intestinal microbial balance. Since the protein has the ability to bind iron, it is a natural antibacterial, antifungal, and antiviral agent. It is also an antioxidant and has immunomodulatory properties; large groups of people might benefit from orally administered hLF.

Pharming (http://www.pharming.com) has a patent on hLF from the Japanese Patent Office, which covers the production and purification of hLF with Pharming's technology as well as its use in sports and food formulations. In Japan, bovine lactoferrin is currently used as an additive in food products and as a nutritional supplement. Pharming is producing human lactoferrin for use as pharmaceuticals (for infection and inflammatory diseases) and as nutraceuticals, using the bovine α_{S1}-casein promoter to direct expression into the milk of transgenic cows; a method that fits functional food development very well as cow's milk is a common food source worldwide. Comparing the recombinant protein with its native counterpart, a slight difference in the molecular weights was identified due to differences in N-linked glycosylation (van Berkel et al., 2002). Natural hLF contains only complex-type glycans, while in recombinant hLF, oligomannose- and/or hybrid-type glycans were also found. The substitution of some of the galactose with N-acetylgalactosamine has also been observed in other transgenic systems, e.g., hAT-III produced in goat's milk (Edmunds et al., 1998). Importantly, the two most important functional activities of hLF, namely iron binding and release and antibacterial activity, were not influenced by these differences. Pharming has filed a GRAS (Generally Recognized As Safe) notification for its recombinant hLF in the United States and has completed clinical trials phase I.

Human lactoferrin has also been produced in goats. Directly transfecting the lactating mammary glands with a replication-defective adenovirus vector containing the human lactoferrin cDNA resulted in high-level expression of up to 2.6 mg/mL of the recombinant protein (Han et al., 2007).

Alpha-1-Antitrypsin (hα_1AT)

Alpha-1-antitrypsin (hα_1AT), also known as α-1-proteinase inhibitor, is an enzyme produced by the liver and released into the bloodstream. One of the primary roles of hα_1AT is to protect the lungs from neutrophil elastase, an enzyme released by white blood cells. Neutrophil elastase can attack healthy lung tissue if not controlled by hα_1AT.

Proteinase-antiproteinase imbalances are recognized in several diseases, including the two most common lethal hereditary disorders of white populations, hα_1AT deficiency and cystic fibrosis (CF). In hα_1AT deficiency, the type

Z variant of hα$_1$AT forms polymers in the endoplasmic reticulum of hepatocytes, resulting in childhood liver disease. In CF, chronic bacterial lung infections due to impaired mucociliary clearance lead to a vigorous influx of neutrophils in the airways. Serine proteinases released from the neutrophils, particularly elastase, exceed the antiproteinase capacity of endogenous serine proteinase inhibitors in the airways. Strategies to augment the antiproteinase defenses in the airways of patients with severe hα$_1$AT deficiency or CF include the intravenous or aerosol administration of serine proteinase inhibitors. Studies in both patient groups using plasma-derived or transgenic recombinant secretory leukoprotease inhibitors or synthetic elastase inhibitors show promising results concerning drug safety and efficacy.

Wright et al. (1991) reported the generation of five sheep transgenic for a fusion of the ovine β-lactoglobulin gene promoter to the hα$_1$AT genomic sequence as one of the earliest successes in using transgenic farm animals as bioreactors. Analysis of the expression of hα$_1$AT in the milk of three of these females showed that all expressed the human protein at levels greater than 1 g/L. hα$_1$AT purified from the milk of these animals appeared to be fully N-glycosylated and had a biological activity indistinguishable from human plasma-derived material.

Transgenic hα$_1$AT has gone through phase II clinical trials in the cystic fibrosis patient, delivering it in aerosol form to assess its safety and efficacy. Trends toward a reduction in neutrophil elastase activity were observed in patients treated with 500 mg and 250 mg of recombinant hα$_1$AT compared to placebo. Although significant differences between recombinant hα$_1$AT and placebo for neutrophil elastase activity were not observed, some improvements were found for secondary efficacy variables. Results show that nebulized recombinant hα$_1$AT is safe and well tolerated but has a limited effect on neutrophil elastase activity and other markers of inflammation (Martin et al., 2006).

GTC Biotherapeutics also has established founder transgenic animals that express rhα$_1$AT. Using the goat β-casein promoter, they achieved a 4-mg/mL recombinant protein expression in rabbits, while that in goats was 14 mg/mL. GTC believes that rhα$_1$AT may also be developed as an effective treatment for other diseases, potentially including cystic fibrosis, chronic obstructive pulmonary disease, acute respiratory syndrome, and severe asthma.

Extracellular Superoxide Dismutase (hEC-SOD)

EC-SOD is the major SOD isoenzyme in plasma, lymph, and synovial fluids. Studies with SOD molecules have indicated a number of interesting therapeutic actions, including acute pancreatitis, cardiovascular disease, and renal transplantation. hEC-SOD has been produced at up to 3 mg/mL in rabbit's milk (Stromqvist et al., 1997). The milk-derived hEC-SOD was purified and compared to the native and CHO cell-produced proteins. All proteins were glycosylated, tetrameric metalloproteins. Since each homotetramer contains one copper ion per monomer, production of hEC-SOD at a high level in milk means that the mammary gland will need to collect large amounts of copper, presumably from the blood.

Erythropoietin (hEPO)

Erythropoietin (hEPO) is a glycoprotein hormone that is a cytokine for erythrocyte precursors in the bone marrow. Also called hematopoietin or hemopoietin, hEPO is mainly produced by the adult kidney and circulates in blood plasma, a small portion is synthesized by the liver, and possibly by macrophages in the bone marrow, and it is the hormone that regulates red blood cell production. At present, hEPO is available as a therapeutic agent only through production by recombinant DNA technology in mammalian cell culture (Jacobs et al., 1985; Krantz, 1991; Kim et al., 2005). It is used in treating anemia resulting from chronic renal failure or from cancer chemotherapy. It is also effective as a blood doping agent that is believed to be common in endurance sports such as cycling, triathlons, and marathons.

Trangenic rabbit (Korhonen et al., 1997; Massoud et al., 1996), transgenic pig (Park et al., 2006), and nontransgenic goat (Toledo et al., 2006) animal systems have been developed for large-scale production of recombinant hEPO. In rabbits the rabbit WAP and bovine BLG promoters were used to direct expression into the mammary gland. High expression of biologically active hEPO into the mammary gland had adverse affects on the lactating female (Massoud et al., 1996); thus, only an expression level of 50 µg/mL could be achieved with the rabbit whey acidic protein promoter driving a genomic hEPO gene construct. Trying to compensate for unwanted side effects, a bovine β-lactoglobulin promoter driving a hEPO cDNA fusion protein with lower biological activity was expressed at a level of 500 µg/mL (Korhonen et al., 1997). The biological activity of the bovine β-lactoglobulin promoter linked to hEPO cDNA was less than 10–20% of that of the native hEPO due to different glycosylation. Upon digestion with IgA protease, the normal biological activity could be recovered (Korhonen et al., 1997). In spite of decreased biological activity, transgenic females expressing the fusion protein showed elevated hematocrit values (up to 80%) during lactation.

hEPO can be successfully produced in transgenic pigs (Park et al., 2006). hEPO-expressing pigs were created via microinjection and use of a mouse whey acidic protein-driven hEPO genomic construct. Expression level was up to 900 IU/mL (EPO levels in normal humans are between 10 and 30 IU/mL). The transgenic animals were generally healthy, except for a few examples of physiological problems (e.g., low sperm quality with erectile dysfunction in some males, and elevated reticulocyte counts and hematocrit levels in both sexes).

A high level (2 mg/mL) of hEPO expression in the milk of nontransgenic goats has been achieved by direct transduction of the lactating mammary gland without causing any harm to the animals (Toledo et al., 2006). A replication-defective adenovirus vector containing the hEPO cDNA was used. The recombinant hEPO had low *in vivo* hematopoietic activity due to underglycosylation.

Tissue Plasminogen Activator (htPA)

Tissue plasminogen activator (htPA) is a serine protease that converts plasminogen to plasmin and can trigger the degradation of extracellular matrix proteins. The glycosylation variant of htPA designated longer-acting tissue-type plasminogen activator (LAtPA) has been produced in the milk of transgenic goats and rabbits (Ebert et al., 1991, 1994). Recombinant htPA was extensively purified from the milk of a transgenic goat by a combination of acid fractionation, hydrophobic interaction chromatography, and immunoaffinity chromatography. Although the early availability of this product had been predicted, recent indications suggest that predicted recombinant htPA, produced by GTC Biotherapeutics for the treatment of coronary clots, is not close to market.

Tissue-Nonspecific Alkaline Phosphatase (hTNAP)

Alkaline phosphatase is a promising therapeutic agent in the Gram-negative bacterial lipopolysaccharide-mediated acute and chronic diseases. Contrary to other alkaline phosphatase isozymes, purified tissue-nonspecific alkaline phosphatase (hTNAP) is not available in large quantities from tissue sources that would enable us to analyze its efficacy in animal sepsis models. Two transgenic rabbit lines were created by pronuclear microinjection with the whey acidic protein promoter-hTNAP minigene (Bodrogi et al., 2006). Alkaline phosphatase enzymatic activity was two orders of magnitude higher compared to normal human serum levels. As indicated by fractionation of milk samples, the recombinant hTNAP was associated with the membrane of milk fat globules.

The production of cystic fibrosis transmembrane conductance regulator in the milk fat globules of transgenic mice was the first report on the expression of a membrane-bound protein, but its biological activity was not examined (DiTullio et al., 1992). Therefore, the milk of transgenic rabbits could be a source of membrane receptors to define their structure after crystallization. This approach may be essential to define synthetic molecules acting on the receptors.

Acid α-Glucosidase (hαGLU)

The clinical spectrum of glycogen storage disease type II/Pompe disease comprises infants, children, and adults. All patients characteristically have acid α-glycosidase deficiency and suffer from progressive skeletal muscle weakness. Affected infants die of cardiorespiratory failure within the first two years of life. Cell culture and transgenic animal technology were explored to produce recombinant human acid α-glucosidase (hαGLU) on a large scale (Van Hove et al., 1996). Transgenic rabbits expressing the human acid α-glucosidase gene under the bovine $\alpha s1$-casein promoter were constructed, resulting in a selected transgenic line producing up to 8 g/L of recombinant protein in milk. The therapeutic efficacy of the product purified from rabbit milk has been demonstrated in clinical trials. The enzyme is transported to lysosomes and lowers the glycogen concentration in the tissues. Phase II clinical trials in patients with classical infantile Pompe disease revealed an overall improvement in cardiac function, skeletal muscle function, and histological appearance of skeletal muscle (Klinge et al., 2005). Long-term intravenous treatment with recombinant hαGLU from milk encourages enzyme replacement therapy for several forms of Pompe disease and underlines that safe and effective medicine can be produced in the milk of mammals (Van den Hout et al., 2004).

Fibrinogen (hFIB)

Fibrinogen (hFIB) is a soluble plasma glycoprotein synthesized by the liver and a key component in blood clotting. In its natural form, fibrinogen is useful in forming bridges between platelets, by binding to their GpIIb/IIIa surface membrane proteins, though the major use of hFIB is as a precursor to fibrin. Processes in the coagulation cascade activate the zymogen prothrombin, producing the serine protease thrombin, which is responsible for converting hFIB into fibrin. Fibrin is then cross-linked by factor XIII to form a clot that serves as an *in vivo* hemostatic plug that prevents further blood loss. hFIB is a hexamer containing two sets of three different chains (α, β, and γ), linked to each other by disulfide bonds. The N-terminal sections of these three chains are evolutionarily related and contain the cysteines that participate in the cross-linking of the chains. Because of this complexity to the protein, expression in bacterial or yeast expression systems has not been feasible. Expression in mammalian cell culture systems has been demonstrated, but this approach is likely to be too expensive for the production of the large amounts of hFIB needed. Since the mammary gland appears to be able to secrete fully assembled recombinant hFIB, the only way to obtain sufficiently large amounts of human fibrinogen safely and cost-effectively is the transgenic production in the milk of large animals. Transgenic livestock—sheep and cattle—have been created for this purpose.

Transgenic sheep were produced (Garner & Colman, 1998; Butler et al., 1997) using the ovine β-lactoglobulin promoter to direct transgene expression to the mammary gland with hFIB expression levels up to 5 mg/mL, while Pharming has used the bovine α_{S1}-casein promoter and nuclear transfer technology to create transgenic cattle producing hFIB at a concentration of 3 mg/mL.

α-Fetoprotein (hαFP)

Alpha-fetoprotein (hαFP) is a serum glycoprotein expressed at high concentrations in the fetal liver, but its concentration drops dramatically after birth. Potential indications for the use of recombinant hαFP include autoimmune diseases such as rheumatoid arthritis, multiple sclerosis, myasthenia gravis (a chronic autoimmune neuromuscular disease), and psoriasis. Since hαFP is produced normally during pregnancy, it is not commercially available from fractionation of the human blood supply. Using the goat β-casein promoter to direct transgene expression, GTC Biotherapeutics has developed transgenic goats that express hαFP in their milk. Since neither glycosylation of hαFP nor any bound ligands are necessary for activity (Semeniuk et al., 1995), to avoid unwanted glycosylation patterns, the single N-linked glycosylation site of the protein was removed (by mutagenesis) from the transgene construct. Characterization of the hαFP in the milk of transgenic goats shows that the structure was indeed not affected by removal of the glycosylation site. Furthermore, the cell binding and pharmacokinetic properties of the recombinant protein were identical to the native protein (Parker et al., 2004). Through cooperation between GTC Biotherapeutics and Merrimack Pharmaceuticals (http://www.merrimackpharma.com), recombinant hαFP purified from goat's milk entered phase II clinical trials in 2004.

Interleukin-2 (hIL-2)

Interleukin-2 (hIL-2), formerly referred to as T cell growth factor, is an immunoregulatory lymphokine that is produced by lectin- or antigen-activated T cells. It is produced not only by mature T lymphocytes on stimulation but also constitutively by certain T cell lymphoma cell lines. It is useful in the study of the molecular nature of T cell differentiation and, like interferons, augments natural killer cell activity. hIL-2 can act as a growth hormone for both B and T lymphocytes. Since hIL-2 and interleukin-2 receptor act as required for the proliferation of T cells, defects in either the ligand or the receptor would be expected to cause severe combined immunodeficiency.

At present, a recombinant form of hIL-2 is manufactured by the Chiron Corporation with the brand name Proleukin (http://www.proleukin.com). It is produced by recombinant DNA technology using genetically engineered *Escherichia coli* containing a modified hIL-2 gene. Transgenic technology has also been used to produce hIL-2. Microinjection and a gene construct containing the rabbit β-casein promoter and the hIL-2 genomic sequence were used to create transgenic rabbits expressing hIL-2 in their milk (Bühler et al., 1990). The recombinant protein, produced at a concentration of up to 430 ng/mL, was stable and biologically active. But to be able to compete with the present method to produce hIL-2, the transgene constructs need to be significantly improved to direct protein production at considerably higher levels.

Insulin-Like Growth Factor-1 (hIGF-1)

The insulin-like growth factors (hIGFs) are polypeptides with high sequence similarity to insulin. They are part of a complex system that cells use to communicate with their physiological environment. This complex system (often referred to as the IGF "axis" or the growth hormone/IGF-1 axis) consists of two cell-surface receptors (IGF-1R and IGF-2R), two ligands (IGF-1 or somatomedin C and IGF-2 or somatomedin A), a family of six high-affinity IGF binding proteins (IGFBP 1-6), as well as associated IGFBP degrading enzymes. hIGF-1 is mainly secreted by the liver as a result of stimulation by hGH. It is important for both the regulation of normal physiology as well as a number of pathological states, including cancer.

Commercially available hIGF-1 has been manufactured recombinantly on a large scale using both yeast and *Escherichia coli*. Transgenic rabbits were also created to produce hIGF-1 (Brem et al., 1994). Brem and co-workers used microinjection to integrate a bovine α_{S1}-casein promoter driving the hIGF-1 cDNA construct into the genome of rabbits. The amount of recombinant protein in the milk of the transgenic rabbits was up to 1 mg/mL. Since the recombinant protein was associated with the casein micelles, purification included extraction with urea and dithioerythritol, gel filtration, and chromatographic enrichment. The recombinant protein was correctly processed and biologically active (Wolf et al., 1997). The local production of hIGF-1 in mammary tissue was found to be associated with increased secretion of IGFBP-2, which may prevent major biological effects by high levels of hIGF-1 on the mammary gland (Zinovieva et al., 1998).

Several companies have evaluated hIGF-1 in clinical trials for several indications, including growth failure, type 1 diabetes, type 2 diabetes, amyotrophic lateral sclerosis (Lou Gehrig's disease), severe burn injury, and myotonic muscular dystrophy. In August 2005, the FDA approved Tercica's (http://www.tercica.com) hIGF-1 drug, Increlex, as replacement therapy for severe primary IGF-1 deficiency. In December 2005, the FDA also approved IPLEX, Insmed's (http://www.insmed.com) IGF-1/IGFBP-3 complex. In the human body, 97 to 99% of hIGF-1 is always bound to one of six hIGF binding proteins, with

IGFBP-3 the most abundant binding protein, accounting for approximately 80% of all hIGF binding. Delivering the drug in a complex achieved the same efficacy as far as growth rates but with fewer side effects and less severe hypoglycemia. The drug is injected once a day versus Tercica's twice-a-day version.

Antithrombin III (hAT-III)

Antithrombin (hAT-III) is a plasma protein with anticoagulant and anti-inflammatory properties. It regulates thrombin, a blood protein that plays a key role in controlling clot formation. Patients with hereditary hAT-III deficiency can have either Type I or Type II deficiency. Type I is a quantitative deficiency characterized by low levels of hAT-III. Type II is a qualitative deficiency characterized by the presence of hAT-III variants that do not function properly. Individuals with Hereditary Antithrombin Deficiency are at risk for blood clots, organ damage, or even death. An acquired form of the disease is also known. It causes disseminated intravascular coagulation, a widespread formation of clots within blood vessels, which is most severe when it occurs in association with sepsis.

GTC Biotherapeutic's lead product, ATryn®, is a recombinant form of hAT-III. hAT-III is produced in the milk of transgenic goats. A goat β-casein promoter-driven hAT-III cDNA transgene was microinjected to create the transgenic animals, resulting in an expression of the transgene as high as 20 mg/mL. The specific activity of the recombinant hAT-III was found to be identical to human plasma-derived AT-III; however, its affinity for heparin was fourfold higher than plasma hAT-III. The recombinant protein was structurally identical to phAT-III except for differences in glycosylation. Oligomannose structures and some GalNAc for galactose substitutions were observed, along with a higher degree of fucosylation and lower degree of sialylation. It was concluded that the increase in affinity of the recombinant protein resulted from the presence of oligomannose-type structures on the Asn155 glycosylation site and differences in sialylation (Edmunds et al., 1998).

Transgenic goats to produce hAT-III have also been created via fetal somatic cell nuclear transfer (Baguisi et al., 1999). Somatic cell lines were generated from 35-day- to 40-day-old fetuses resulting from the mating of hAT-III–expressing transgenic goats (goat β-casein promoter-driven hAT-III cDNA). Analysis of the milk of the transgenic cloned animals showed high-level production of hAT-III (up to 5.8 mg/mL with an activity of 20.5 IU/mL), which was similar to the parental transgenic line.

In 2006, ATryn® was approved for human therapeutic application in the EU. It has completed phase III in the United States and is at the state of prelaunch indicated for Hereditary Antithrombin Deficiency; for other indications (disseminated intravascular coagulation in sepsis), it is in phase II of clinical trials.

C1 Inhibitor (hC1INH)

C1-inhibitor is a serine protease inhibitor (serpin) protein, the main function of which is inhibition of the complement system. It circulates in blood at levels around 0.25–0.45 g/L. Human C1 inhibitor (hC1INH) is used for the treatment of hereditary angioedema (HAE). In the Western world, approximately 1 in 30,000 persons, or some 22,000 people, suffers from HAE, a life-threatening genetic disorder. The shortage of hC1INH results in recurrent attacks of edema, causing painful swelling in the body's soft tissues. The disease seriously affects the quality of life of patients and can even be lethal if attacks in the throat area lead to asphyxiation.

Pharming has developed a method for the easy, quick, and clean production of hC1INH in large quantities, highly suitable for pharmaceutical applications and treatment of HAE. Its recombinant hC1INH is purified from the milk of transgenic rabbits. The DNA construct used contains the bovine α_{S1}-casein promoter sequence functionally linked to the gene encoding hC1INH and directs the expression at levels of 12 mg/mL. The glycosylation pattern of the recombinant protein is essentially similar to the native protein, only the degree of sialylation and core fucosylation was lower (Koles et al., 2004a, b). Pharming is nearing the end of the development program, practically all safety tests in laboratory animals have been finalized, and the product is now in phase III of clinical testing in humans.

Nerve Growth Factor Beta (hNGF-β)

Nerve growth factor beta (hNGF-β), the founder member of the protein family termed neurotrophins, is a protein secreted by a neuron's target. hNGF-β is critical for the survival and maintenance of primary sensory neurons, sympathetic neurons, and cholinergic neurons of the basal forebrain. When hNGF-β is released from the target cells, it binds to and activates its high-affinity receptor (TrkA) and is internalized into the responsive neuron. The hNGF-β/TrkA complex is subsequently trafficked back to the cell body. This movement of hNGF-β from axon tip to soma is thought to be involved in the long-distance signaling of neurons. Secreted pro-hNGF-β has been demonstrated in a variety of neuronal and nonneuronal cell populations. It has been proposed that secreted pro hNGF-β can elicit neuron death in a variety of neurodegenerative conditions, including Alzheimer's disease, following the observation of an increase of pro-hNGF-β in the nucleus basalis of postmortem Alzheimer's brains. Recombinant hNGF-β may be used to treat neuronal dysfunction of the central and peripheral nervous system as well as HIV-related peripheral neuropathy (clinical trials are in progress).

In the past 15 years, mass production of hNGF-β has been carried out by mammalian cell culture systems (now commercially available), and although great progress could be made to increase the yield, transgenic technology offers a more viable solution. Recombinant hNGF-β has been produced in transgenic rabbits (Coulibaly et al., 2002). Using a construct containing the pre-pro-hNGF-β cDNA under the control of bovine α_{S1}-casein promoter, an expression of up to 250 µg/mL of the recombinant protein was achieved. hNGF-β could be purified from the milk by a two-step chromatographic procedure. Biological activity of the purified protein and also of crude defatted milk from transgenic animals demonstrated full biological activity when compared to commercial recombinant hNGF-β.

Collagen Type I (hCOL)

Recombinant human collagen type I (hCOL) is being developed by Pharming and its partner Cohesion Technologies (http://www.cohesiontech.com) for the biomaterials market. hCOL accounts for 85% of the total collagen in humans and is found in almost all collagen-based products on the market. Collagen is the main protein in connective tissue in animals and the most abundant protein in mammals (about 25% of the total protein content). Collagen is partly responsible for skin strength and elasticity, it strengthens blood vessels, and during aging its degradation leads to wrinkles. In crystalline form it is also present in the cornea and lens of the eye. Collagen is a commonly used biomaterial in the medical and pharmaceutical industries based on its structural role and compatibility within the human body. Applications include hemostats, vascular sealants, tissue sealants, implant coatings (orthopedic and vascular), artificial skin, bone graft substitutes, "injectables" for incontinence treatment, dental implants, and (antibiotic) wound dressings. Many additional applications are currently under development, such as engineering of cartilage, bone, skin, artificial tendons, blood vessels, nerve regeneration, and several drug delivery applications.

Pharming has successfully produced recombinant hCOL at high expression levels in transgenic cattle. The recombinant hCOL is indicated as an intermediate for medical devices and aesthetic products. The purified protein is now in the preclinical phase of trials.

Conclusions

Less than 25 years ago the first transgenic livestock animals with altered milk composition were born. Since then a number of valuable animal models have been created and characterized, with improved transgene constructs. The methods of creating transgenic livestock species have also been developed and have become more efficient, with fewer side effects. Some of the recombinant proteins are at advanced stages of clinical trials, and ATryn®, the flagship of pharmaceutical proteins produced in the milk of transgenic animals, was approved in 2006 by the EMEA for human therapeutic application. Notwithstanding issues of public acceptance and authorization hurdles that are specific to pharmaceutical proteins produced by genetically modified mammals, the scientific developments achieved so far make animal bioreactor production systems a viable option. This is especially so when considering large-scale production of human proteins that require complex posttranslational modifications.

References

Archer, J. S., Kennan, W. S., Gould, M. N., & Bremel, R. D. (1994). Human growth hormone (hGH) secretion in milk of goats after direct transfer of the hGH gene into the mammary gland by using replication-defective retrovirus vectors. *Proceedings of the National Academy of Sciences USA, 91,* 6840–6844.

Baguisi, A., Behboodi, E., Melican, D. T., Pollock, J. S., Destrempes, M. M., Cammuso, C., Williams, J. L., Nims, S. D., Porter, C. A., Midura, P., Palacios, M. J., Ayres, S. L., Denniston, R. S., Hayes, M. L., Ziomek, C. A., Meade, H. M., Godke, R. A., Gavin, W. G., Overstrom, E. W., & Echelard, Y. (1999). Production of goats by somatic cell nuclear transfer. *Nature Biotechnology, 17,* 456–461.

Baranyi, M., Brignon, G., Anglade, P., & Ribadeau-Dumas, B. (1995). New data on the proteins of rabbit (*Oryctolagus cuniculus*) milk. *Comparative Biochemistry and Physiology B: Biochemistry and Molecular Biology, 111,* 407–415.

Bijvoet, A. G., Van Hirtum, H., Kroos, M. A., Van de Kamp, E. H., Schoneveld, O., Visser, P., Brakenhoff, J. P., Weggeman, M., van Corven, E. J., Van der Ploeg, A. T., & Reuser, A .J. (1999). Human acid alpha-glucosidase from rabbit milk has therapeutic effect in mice with glycogen storage disease type II. *Human Molecular Genetics, 8,* 2145–2153.

Bodrogi, L., Brands, R., Raaben, W., Seinen, W., Baranyi, M., Fiechter, D., & Bösze, Z. (2006). High level expression of tissue-nonspecific alkaline phosphatase in the milk of transgenic rabbits. *Transgenic Research, 15,* 627–636.

Brem, G., Hartl, P., Besenfelder, U., Wolf, E., Zinovieva, N., & Pfaller, R. (1994). Expression of synthetic cDNA sequences encoding human insulin-like growth factor-1 (IGF-1) in the mammary gland of transgenic rabbits. *Gene, 149,* 351–355.

Brem, G., Besenfelder, U., Castro, F. O., & Muller, M. (1998). In F. O. Castro and J. Janne (Eds.) Mammary Gland Transgenisis: *Therapeutic Protein Production* (pp. 107–142), Berlin: Springer.

Brinster, R. L., Allen, J. M., Behringer, R. R., Gelinas, R. E., & Palmiter, R. D. (1988). Introns increase transcriptional efficiency in transgenic mice. *Proceedings of the National Academy of Sciences USA, 85,* 836–840.

Bühler, T. A., Bruyere, T., Went, D. F., Stranzinger, G., & Bürki, K. (1990). Rabbit beta-casein promoter directs secretion of human interleukin-2 into the milk of transgenic rabbits. *Biotechnology (NY), 8,* 140–143.

Butler, S. P., van Cott, K., Subrumanian, A., Gwazduaskas, F. C., & Velander, W. H. (1997). Current progress in the production of recombinant human fibrinogen in the milk of transgenic animals. *Thrombosis and Haemostasis, 78,* 537–542.

Castro, F. O., Limonta, J., Rodriguez, A., Aguirre, A., de la Fuente, J., Aguilar, A., Ramos B., & Hayes, O. (1999). Transgenic rabbits for the production of biologically-active recombinant proteins in the milk. *Genetic Analysis, 15,* 179–187.

Chan, A. W., Homan, E. J., Ballou, L. U., Burns, J. C., & Bremel, R. D. (1998). Transgenic cattle produced by reverse-transcribed gene transfer in oocytes. *Proceedings of the National Academy of Sciences USA, 95*, 14028–14033.

Clark, A. J., Bessos, H., Bishop, J. O., Brown, P., Harris, S., Lathe, R., McClenaghan, M., Prowse, C., Simons, J. P., Whitelaw, C. B. A., & Wilmut, I. (1989). Expression of human anti-hemophilic factor IX in the milk of transgenic sheep. *Bio/Technology, 7*, 487–492 (abstract).

Clark, J., & Whitelaw, B. (2003). A future for transgenic livestock. *Nature Review Genetics, 4*, 825–833.

Colman, A. (1996). Production of proteins in the milk of transgenic livestock: Problems, solutions, and successes. *American Journal of Clinical Nutrition, 63*, 639S–645S.

Coulibaly, S., Besenfelder, U., Fleischmann, M., Zinovieva, N., Grossmann, A., Wozny, M., Bartke, I., Togel, M., Muller, M., & Brem, G. (1999). Human nerve growth factor beta (hNGF-β): Mammary gland specific expression and production in transgenic rabbits. *FEBS Letters, 444*, 111–116.

Coulibaly, S., Besenfelder, U., Miller, I., Zinovieva, N., Lassnig, C., Kotler, T., Jameson, J.L., Gemeiner, M., Muller, M., & Brem, G. (2002). Expression and characterization of functional recombinant bovine follicle-stimulating hormone (boFSHα/β) produced in the milk of transgenic rabbits. *Molecular Reproduction and Development, 63*, 300–308.

Dayal, R., Hurlimann, J., Suard, Y. M., & Kraehenbuhl, J. P. (1982). Chemical and immunochemical characterization of caseins and the major whey proteins of rabbit milk. *Biochemistry Journal, 201*, 71–79.

Denman, J., Hayes, M., O'Day, C., Edmunds, T., Bartlett, C., Hirani, S., Ebert, K. M., Gordon, K., & McPherson, J. M. (1991). Transgenic expression of a variant of human tissue-type plasminogen activator in goat milk: Purification and characterization of the recombinant enzyme. *Biotechnology (NY), 9*, 839–843.

Devinoy, E., Thepot, D., Stinnakre, M. G., Fontaine, M. L., Grabowski, H., Puissant, C., Pavirani, A., & Houdebine, L. M. (1994). High level production of human growth hormone in the milk of transgenic mice: The upstream region of the rabbit whey acidic protein (WAP) gene targets transgene expression to the mammary gland. *Transgenic Research, 3*, 79–89.

DiTullio, P., Cheng, S. H., Marshall, J., Gregory, R .J., Ebert, K. M., Meade, H. M., & Smith, A. E. (1992). Production of cystic fibrosis transmembrane conductance regulator in the milk of transgenic mice. *Biotechnology (NY), 10*, 74–77.

Dyck, M. K., Lacroix, D., Pothier, F., & Sirard, M. A. (2003). Making recombinant proteins in animals—Different systems, different applications. *Trends in Biotechnology, 21*, 394–399.

Ebert, K. M., Selgrath, J. P., DiTullio, P., Denman, J., Smith, T. E., Memon, M. A., Schindler, J. E., Monastersky, G. M., Vitale, J. A., & Gordon, K. (1991). Transgenic production of a variant of human tissue-type plasminogen activator in goat milk: Generation of transgenic goats and analysis of expression. *Biotechnology (NY), 9*, 835–838.

Ebert, K. M., DiTullio, P., Barry, C. A., Schindler, J. E., Ayres, S. L., Smith, T. E., Pellerin, L. J., Meade, H. M., Denman, J., & Roberts, B. (1994). Induction of human tissue plasminogen activator in the mammary gland of transgenic goats. *Biotechnology (NY), 12*, 699–702.

Edmunds, T., Van Patten, S. M., Pollock, J., Hanson, E., Bernasconi, R., Higgins, E., Manavalan, P., Ziomek, C., Meade, H., McPherson, J. M., & Cole, E. S. (1998). Transgenically produced human antithrombin: Structural and functional comparison to human plasma-derived antithrombin. *Blood. 91*, 4561–4571.

Farrell, P. J., Lu, M., Prevost, J., Brown, C., Behie, L., & Iatrou, K. (1998). High-level expression of secreted glycoproteins in transformed lepidopteran insect cells using a novel expression vector. *Biotechnology and Bioengineering, 60*, 656–663.

Fujiwara, Y., Miwa, M., Takahashi, R., Hirabayashi, M., Suzuki, T., & Ueda, M. (1997). Position-independent and high-level expression of human alpha-lactalbumin in the milk of

transgenic rats carrying a 210-kb YAC DNA. *Molecular Reproduction and Development, 47,* 157–163.
Fujiwara, Y., Takahashi, R. I., Miwa, M., Kameda, M., Kodaira, K., Hirabayashi, M., Suzuki, T., & Ueda, M. (1999). Analysis of control elements for position-independent expression of human alpha-lactalbumin YAC. *Molecular Reproduction and Development, 54,* 17–23.
Garner, I., & Colman, A. (1998). Therapeutic proteins from livestock. In A. J. Clark (Ed.), *Animal Breeding: Technology for the 21st Century* (pp. 64–71), London: CRC Press.
Genzyme Transgenics (1996). Production of recombinant antibodies in the milk of transgenic animals (abstract).
Gordon, J. W., Scangos, G. A., Plotkin, D. J., Barbosa, J. A., & Ruddle, F. H. (1980). Genetic transformation of mouse embryos by microinjection of purified DNA. *Proceedings of the National Academy of Sciences USA, 77,* 7380–7384.
Hammer, R. E., Pursel, V. G., Rexroad, C. E., Jr., Wall, R. J., Bolt, D. J., Ebert, K. M., Palmiter, R. D., & Brinster, R. L. (1985). Production of transgenic rabbits, sheep and pigs by microinjection. *Nature, 315,* 680–683.
Han, Z. S., Li, Q. W., Zhang, Z. Y., Xiao, B., Gao, D. W., Wu, S. Y., Li, J., Zhao, H. W., Jiang, Z. L., & Hu, J. H. (2007). High-level expression of human lactoferrin in the milk of goats by using replication-defective adenoviral vectors. *Protein Expression and Purification,* doi:10.1016/j.pep.2006.11.019.
Hiripi, L., Makovics, F., Halter, R., Baranyi, M., Paul, D., Carnwath, J. W., Bösze, Z., & Niemann, H. (2003). Expression of active human blood clotting factor VIII in mammary gland of transgenic rabbits. *DNA Cell Biology, 22,* 41–45.
Huszar, D., Balling, R., Kothary, R., Magli, M. C., Hozumi, N., Rossant, J., & Bernstein, A. (1985). Insertion of a bacterial gene into the mouse germ line using an infectious retrovirus vector. *Proceedings of the National Academy of Sciences USA, 82,* 8587–8591.
Jacobs, K., Shoemaker, C., Rudersdorf, R., Neill, S. D., Kaufman, R. J., Mufson, A., Seehra, J., Jones, S. S., Hewick, R., Fritsch, E. F., et al. (1985). Isolation and characterization of genomic and cDNA clones of human erythropoietin. *Nature, 313,* 806–810.
Jahner, D., Haase, K., Mulligan, R., & Jaenisch, R. (1985). Insertion of the bacterial *gpt* gene into the germ line of mice by retroviral infection. *Proceedings of the National Academy of Sciences USA, 82,* 6927–6931.
Keefer, C. L., Baldassarre, H., Keyston, R., Wang, B., Bhatia, B., Bilodeau, A. S., Zhou, J. F., Leduc, M., Downey, B. R., Lazaris, A., & Karatzas, C. N. (2001). Generation of dwarf goat (*Capra hircus*) clones following nuclear transfer with transfected and nontransfected fetal fibroblasts and *in vitro*-matured oocytes. *Biological Reproduction, 64,* 849–856.
Kim, N. Y., Kim, J. H., & Kim, H. J. (2005). Effect of low adapted temperature and medium composition on growth and erythropoietin (EPO) production by Chinese hamster ovary cells. *Archives of Pharmaceutical Research, 28,* 220–226.
Klinge, L., Straub, V., Neudorf, U., Schaper, J., Bosbach, T., Gorlinger, K., Wallot, M., Richards, S., & Voit, T. (2005). Safety and efficacy of recombinant acid alpha-glucosidase (rhGAA) in patients with classical infantile Pompe disease: Results of a phase II clinical trial. *Neuromuscular Disorders, 15,* 24–31.
Koles, K., van Berkel, P. H., Mannesse, M. L., Zoetemelk, R., Vliegenthart, J. F., & Kamerling, J. P. (2004a). Influence of lactation parameters on the N-glycosylation of recombinant human C1 inhibitor isolated from the milk of transgenic rabbits. *Glycobiology, 14,* 979–986.
Koles, K., van Berkel, P. H., Pieper, F. R., Nuijens, J. H., Mannesse, M. L., Vliegenthart, J. F., & Kamerling, J. P. (2004b). N- and O-glycans of recombinant human C1 inhibitor expressed in the milk of transgenic rabbits. *Glycobiology, 14,* 51–64.
Korhonen, V. P., Tolvanen, M., Hyttinen, J. M., Uusi-Oukari, M., Sinervirta, R., Alhonen, L., Jauhiainen, M., Janne, O. A., & Janne, J. (1997). Expression of bovine beta-lactoglobulin/human erythropoietin fusion protein in the milk of transgenic mice and rabbits. *European Journal of Biochemistry, 245,* 482–489.

Krantz, S. B. (1991). Erythropoietin. *Blood, 77,* 419–434.

Krimpenfort, P., Rademakers, A., Eyestone, W., van der Schans, A., van den Broek, S., Kooiman, P., Kootwijk, E., Platenburg, G., Pieper, F., Strijker, R., et al. (1991). Generation of transgenic dairy cattle using "*in vitro*" embryo production. *Biotechnology (NY), 9,* 844–847.

Limonta, J. M., Castro, F. O., Martinez, R., Puentes, P., Ramos, B., Aguilar, A., Lleonart, R. L., & de la Fuente, J. (1995). Transgenic rabbits as bioreactors for the production of human growth hormone. *Journal of Biotechnology, 40,* 49–58.

Lipinski, D., Jura, J., Kalak, R., Plawski, A., Kala, M., Szalata, M., Jarmuz, M., Korcz, A., Slomska, K., Jura, J., Gronek, P., Smorag, Z., Pienkowski, M., & Slomski, R. (2003). Transgenic rabbit producing human growth hormone in milk. *Journal of Applied Genetics, 44,* 165–174.

Lubon, H., & Paleyanda, R. K. (1997). Vitamin K-dependent protein production in transgenic animals. *Thrombosis and Haemostasis, 78,* 532–536.

Martin, S. L., Downey, D., Bilton, D., Keogan, M. T., Edgar, J., & Elborn, J. S. (2006). Safety and efficacy of recombinant alpha(1)-antitrypsin therapy in cystic fibrosis. *Pediatric Pulmonology, 41,* 177–183.

Massoud, M., Attal, J., Thepot, D., Pointu, H., Stinnakre, M. G., Theron, M. C., Lopez, C., & Houdebine, L. M. (1996). The deleterious effects of human erythropoietin gene driven by the rabbit whey acidic protein gene promoter in transgenic rabbits. *Reproduction Nutrition Development, 36,* 555–563.

McCreath, K. J., Howcroft, J., Campbell, K. H., Colman, A., Schnieke, A. E., & Kind, A. J. (2000). Production of gene-targeted sheep by nuclear transfer from cultured somatic cells. *Nature, 405,* 1066–1069.

McKee, C., Gibson, A., Dalrymple, M., Emslie, L., Garner, I., & Cottingham, I. (1998). Production of biologically active salmon calcitonin in the milk of transgenic rabbits. *Nature Biotechnology, 16,* 647–651.

Mercier, J. C., & Vilotte, J. L. (1993). Structure and function of milk protein genes. *Journal of Dairy Science, 76,* 3079–3098.

Nagy, A., Gocza, E., Diaz, E. M., Prideaux, V. R., Ivanyi, E., Markkula, M., & Rossant, J. (1990). Embryonic stem cells alone are able to support fetal development in the mouse. *Development, 110,* 815–821.

Niemann, H., Halter, R., Carnwath, J. W., Herrmann, D., Lemme, E., & Paul, D. (1999). Expression of human blood clotting factor VIII in the mammary gland of transgenic sheep. *Transgenic Research, 8,* 237–247.

Paleyanda, R. K., Velander, W. H., Lee, T. K., Scandella, D. H., Gwazdauskas, F. C., Knight, J. W., Hoyer, L. W., Drohan, W. N., & Lubon, H. (1997). Transgenic pigs produce functional human factor VIII in milk. *Nature Biotechnology, 15,* 971–975.

Park, J. K., Lee, Y. K., Lee, P., Chung, H. J., Kim, S., Lee, H. G., Seo, M. K., Han, J. H., Park, C. G., Kim, H. T., Kim, Y. K., Min, K. S., Kim, J. H., Lee, H. T., & Chang, W. K. (2006). Recombinant human erythropoietin produced in milk of transgenic pigs. *Journal of Biotechnology, 122,* 362–371.

Parker, M. H., Birck-Wilson, E., Allard, G., Masiello, N., Day, M., Murphy, K. P., Paragas, V., Silver, S., & Moody, M. D. (2004). Purification and characterization of a recombinant version of human α-fetoprotein expressed in the milk of transgenic goats. *Protein Expression and Purification, 38,* 177–183.

Riego, E., Limonta, J., Aguilar, A., Perez, A., de Armas, R., Solano, R., Ramos, B., Castro, F. O., & de la Fuente, J. (1993). Production of transgenic mice and rabbits that carry and express the human tissue plasminogen activator cDNA under the control of a bovine alpha S1 casein promoter. *Theriogenology, 39,* 1173–1185.

Rival-Gervier, S., Viglietta, C., Maeder, C., Attal, J., & Houdebine, L. M. (2002). Position-independent and tissue-specific expression of porcine whey acidic protein gene from a bacterial artificial chromosome in transgenic mice. *Molecular Reproduction and Development, 63,* 161–167.

Rudolph, N. S. (1999). Biopharmaceutical production in transgenic livestock. *Trends in Biotechnology, 17*, 367–374.

Salamone, D., Baranao, L., Santos, C., Bussmann, L., Artuso, J., Werning, C., Prync, A., Carbonetto, C., Dabsys, S., Munar, C., Salaberry, R., Berra, G., Berra, I., Fernandez, N., Papouchado, M., Foti, M., Judewicz, N., Mujica, I., Munoz, L., Alvarez, S. F., Gonzalez, E., Zimmermann, J., Criscuolo, M., & Melo, C. (2006). High level expression of bioactive recombinant human growth hormone in the milk of a cloned transgenic cow. *Journal of Biotechnology, 124*, 469–472.

Sanchez, O., Toledo, J. R., Rodriguez, M. P., & Castro, F. O. (2004). Adenoviral vector mediates high expression levels of human growth hormone in the milk of mice and goats. *Journal of Biotechnology, 114*, 89–97.

Schnieke, A. E., Kind, A. J., Ritchie, W. A., Mycock, K., Scott, A. R., Ritchie, M., Wilmut, I., Colman, A., & Campbell, K. H. (1997). Human factor IX transgenic sheep produced by transfer of nuclei from transfected fetal fibroblasts. *Science, 278*, 2130–2133.

Semeniuk, D. J., Boismenu, R., Tam, J., Weissenhofer, W., & Murgita, R. A. (1995). Evidence that immunosuppression is an intrinsic property of the α-fetoprotein molecule. *Advances in Experimental Medicine and Biology, 383*, 255–269.

Shen, W., Lan, G., Yang, X., Li, L., Min, L., Yang, Z., Tian, L., Wu, X., Sun, Y., Chen, H., Tan, J., Deng, J., & Pan, Q. (2006). Targeting the exogenous *htPAm* gene on goat somatic cell β-casein locus for transgenic goat production. *Molecular Reproduction and Development, 74*, 428–434.

Simons, J. P., Wilmut, I., Clark, A. J., Archibald, A. L., Bishop, J. O., & Lathe, R. (1988). Gene transfer to sheep. *Bio/Technology, 6*, 179–183 (abstract).

Soriano, P., Cone, R. D., Mulligan, R. C., & Jaenisch, R. (1986). Tissue-specific and ectopic expression of genes introduced into transgenic mice by retroviruses. *Science, 234*, 1409–1413.

Soulier, S., Hudrisier, M., Da Silva, J. C., Maeder, C., Viglietta, C., Besnard, N., & Vilotte, J. L. (2003). Substitution of the α-lactalbumin transcription unit by a CAT cDNA within a BAC clone silenced the locus in transgenic mice without affecting the physically linked Cyclin *T1* gene. *Genetics Selection Evolution, 35*, 239–247.

Stinnakre, M. G., Soulier, S., Schibler, L., Lepourry, L., Mercier, J. C., & Vilotte, J. L. (1999). Position-independent and copy-number-related expression of a goat bacterial artificial chromosome α-lactalbumin gene in transgenic mice. *Biochemistry Journal, 339 (Pt 1)*, 33–36.

Stromqvist, M., Houdebine, M., Andersson, J. O., Edlund, A., Johansson, T., Viglietta, C., Puissant, C., & Hansson, L. (1997). Recombinant human extracellular superoxide dismutase produced in milk of transgenic rabbits. *Transgenic Research, 6*, 271–278.

Subramanian, A., Paleyanda, R. K., Lubon, H., Williams, B. L., Gwazdauskas, F. C., Knight, J. W., Drohan, W. N., & Velander, W. H. (1996). Rate limitations in posttranslational processing by the mammary gland of transgenic animals. *Annals of the New York Academy of Sciences, 782*, 87–96.

Swartz, J. R. (2001). Advances in *Escherichia coli* production of therapeutic proteins. *Current Opinion in Biotechnology, 12*, 195–201.

Thomassen, E. A., van Veen, H. A., van Berkel, P. H., Nuijens, J. H., & Abrahams, J. P. (2005). The protein structure of recombinant human lactoferrin produced in the milk of transgenic cows closely matches the structure of human milk-derived lactoferrin. *Transgenic Research, 14*, 397–405.

Toledo, J. R., Sanchez, O., Segui, R. M., Garcia, G., Montanez, M., Zamora, P. A., Rodriguez, M. P., & Cremata, J. A. (2006). High expression level of recombinant human erythropoietin in the milk of non-transgenic goats. *Journal of Biotechnology, 123*, 225–235.

van Berkel, P. H., Welling, M. M., Geerts, M., van Veen, H. A., Ravensbergen, B., Salaheddine, M., Pauwels, E. K., Pieper, F., Nuijens, J. H., & Nibbering, P. H. (2002). Large scale production of recombinant human lactoferrin in the milk of transgenic cows. *Nature Biotechnology, 20,* 484–487.

Van Cott, K. E., Butler, S. P., Russell, C. G., Subramanian, A., Lubon, H., Gwazdauskas, F. C., Knight, J., Drohan, W. N., & Velander, W. H. (1999). Transgenic pigs as bioreactors: a comparison of gamma-carboxylation of glutamic acid in recombinant human protein C and factor IX by the mammary gland. *Genetic Analysis, 15,* 155–160.

Van Cott, K. E., Lubon, H., Gwazdauskas, F. C., Knight, J., Drohan, W. N., & Velander, W. H. (2001). Recombinant human protein C expression in the milk of transgenic pigs and the effect on endogenous milk immunoglobulin and transferrin levels. *Transgenic Research, 10,* 43–51.

van der Putten, H., Botteri, F. M., Miller, A. D., Rosenfeld, M. G., Fan, H., Evans, R. M., & Verma, I. M. (1985). Efficient insertion of genes into the mouse germ line via retroviral vectors. *Proceedings of the National Academy of Sciences USA, 82,* 6148–6152.

Van Hove, J. L., Yang, H. W., Wu, J. Y., Brady, R. O., & Chen, Y. T. (1996). High-level production of recombinant human lysosomal acid α-glucosidase in Chinese hamster ovary cells which targets to heart muscle and corrects glycogen accumulation in fibroblasts from patients with Pompe disease. *Proceedings of the National Academy of Sciences USA, 93,* 65–70.

Velander, W. H., Johnson, J. L., Page, R. L., Russell, C. G., Subramanian, A., Wilkins, T. D., Gwazdauskas, F. C., Pittius, C., & Drohan, W. N. (1992). High-level expression of a heterologous protein in the milk of transgenic swine using the cDNA encoding human protein C. *Proceedings of the National Academy of Sciences USA, 89,* 12003–12007.

Whitelaw, C. B. (2003). Transgenic livestock made easy. *Trends in Biotechnology, 22,* 157–159.

Whitelaw, C. B., Archibald, A. L., Harris, S., McClenaghan, M., Simons, J. P., & Clark, A. J. (1991). Targeting expression to the mammary gland: Intronic sequences can enhance the efficiency of gene expression in transgenic mice. *Transgenic Research, 1,* 3–13.

Wilmut, I., Schnieke, A. E., McWhir, J., Kind, A. J., & Campbell, K. H. (1997). Viable offspring derived from fetal and adult mammalian cells. *Nature, 385,* 810–813.

Wolf, E., Jehle, P. M., Weber, M. M., Sauerwein, H., Daxenberger, A., Breier, B. H., Besenfelder, U., Frenyo, L., & Brem, G. (1997). Human insulin-like growth factor I (IGF-I) produced in the mammary glands of transgenic rabbits: Yield, receptor binding, mitogenic activity, and effects on IGF-binding proteins. *Endocrinology, 138,* 307–313.

Wright, G., Carver, A., Cottom, D., Reeves, D., Scott, A., Simons, P., Wilmut, I., Garner, I., & Colman, A. (1991). High level expression of active human α1-antitrypsin in the milk of transgenic sheep. *Biotechnology (NY), 9,* 830–834.

Zinovieva, N., Lassnig, C., Schams, D., Besenfelder, U., Wolf, E., Muller, S., Frenyo, L., Seregi, J., Muller, M., & Brem, G. (1998). Stable production of human insulin-like growth factor 1 (IGF-1) in the milk of hemi- and homozygous transgenic rabbits over several generations. *Transgenic Research, 7,* 437–447.

Ziomek, C. A. (1998). Commercialization of proteins produced in the mammary gland. *Theriogenology, 49,* 139–144.

V
The Influence of Nutrition on the Production of Bioactive Milk Components

Insulin-Like Growth Factors (IGFs), IGF Binding Proteins, and Other Endocrine Factors in Milk: Role in the Newborn

Jürg. W. Blum and Craig R. Baumrucker

Abstract The role of colostrum and milk in the neonate has been chiefly recognized as a comprehensive nutrient foodstuff. In addition, the provision of colostrum—the first milk—for early immune capacity has been well documented for several species. Colostrum is additionally a rich and concentrated source of various factors that demonstrate biological activity *in vitro*. Three hypotheses have been proposed for the phenotypic function of these secreted bioactive components: (1) only mammary disposal, (2) mammary cell regulation, and (3) neonatal function [gastrointestinal tract (GIT) or systemic]. Traditionally, it was assumed that the development of the GIT is preprogrammed and not influenced by events occurring in the intestinal lumen. However, a large volume of research has demonstrated that colostrum (or milk-borne) bioactive components can basically contribute to the regulation of GIT growth and differentiation, while their role in postnatal development at physiological concentrations has remained elusive. Much of our current understanding is derived from cell culture and laboratory animals, but experimentation with agriculturally important species is taking place.

This chapter provides an overview of work conducted primarily in neonatal calves and secondarily in other species on the effects on neonates of selected peptide endocrine factors (hormones, growth factors, in part cytokines) in colostrum. The primary focus will be on insulin-like growth factors (IGFs) and IGF binding proteins (IGFBPs) and other bioactive peptides, but new interest and concern about steroids (especially estrogens) in milk are considered as well.

Keywords: endocrine factors · colostrum · milk · neonate · gastrointestinal tract · metabolism.

J. W. Blum
Veterinary Physiology, Vetsuisse Faculty, University of Bern, CH-3012 Bern, Switzerland. , Ph: +41-78-7211220, FAX: +41-26-4077297
e-mail: juerg.blum@physio.unibe.ch

Introduction

Colostrum and mature milk contain nutrients, minerals, trace elements, and (pre-) vitamins as well as nonnutrient (mostly bioactive) components, mammary epithelial cells, and their components as well as leukocytes. Nonnutrient substances include immunoglobulins (Ig), hormones, growth factors, releasing factors, cytokines, prostaglandins, enzymes, lactoferrin, transferrin, breakdown products of milk proteins, nucleotides, polyamines, and oligosaccharides. Several review papers (Blum, 2006; Blum & Baumrucker, 2002; Blum & Hammon, 2000; Campana & Baumrucker, 1995; Donovan & Odle, 1994; Gopal & Gill, 2000; Grosvenor et al., 1993; Molkentin, 2000; Oda et al., 1989; Schlimme et al., 2000) covered some of these factors in the past.

Important bovine colostral peptide endocrine factors are shown in Table 1. In bovine colostrum, many nonnutrients are derived from blood, as is the case for IgG_1, growth hormone (GH), prolactin (PRL), IGF-1, insulin, and glucagon. Other nonnutrient substances are produced in the mammary gland by lactocytes, such as some of the IGFBPs (Gibson et al., 1998). The greatest mass of bioactive proteins, peptides, and hormones is available to newborns in the first colostrum. There are species differences with respect to their contents: For example, bovine colostrum concentrations of IGFs are high (Blum & Hammon, 2000; Malven et al., 1987; Ronge & Blum, 1988; Vega et al., 1991), but components of the epidermal growth factor (EGF) family are in low concentration when compared to human and rat colostrum (Shing & Klagsbrun, 1984). Concentrations of nonnutrient

Table 1 Reported Endocrine Factors Found in Bovine Colostrum and Milk

EndocrineLigand	Colostrum	Milk	Source
IGF-1	1–3 µg/mL	10–50 ng/mL	Malven et al. (1987)
IGF-2	1.8 µg/mL	1–20 ng/mL	Vega et al. (1991)
IGFBPs	~3 µg/mL	~2 µg/mL	Puvogel et al. (2005)
EGF (likely betacellulin)	3 ng/mL	1.5 ng/mL	Iacopetta et al. (1992); Xiao et al. (2002)
Betacellulin	2.3 ng/mL	~2 ng/mL	Bastian et al. (2001)
TGF α	2.2–7.2 µg/mL	0–8.4 µg/mL	
TGF-β2	74 ng/mL; act: 150–1150 ng/mL		Pakkanen (1998)
TGF β 1&2	?	8 ng/mL	Cox and Burk (1991)
FGF (acidic)	?	~6 ng/mL	Rogers et al. (1995)
FGF (basic)	?	~20 ng/mL	Rogers et al. (1995)
Insulin	6–37 ng/mL	4–7 ng/mL	Malven et al. (1987)
Prolactin	500–800 ng/mL	6–8 ng/mL	Kacsóh et al. (1991)
Leptin	13.9 ng/mL	6.1 ng/mL	Pinotti and Rosi (2006)

peptides in cow's milk decline rapidly after the onset of lactation. Concentrations of many of these peptides are higher in colostrum than in blood (IgG_1, IGFs, insulin), while some are present in lower concentrations in colostrum than in blood (GH, glucagon). Nonnutrient substances are distributed among the casein, whey, and fat fractions. While the mechanism of appearance in milk of bioactive components synthesized by lactocytes would be expected to follow milk secretory pathways (Akers & Kaplan, 1989), the transfer mechanisms of components originating from blood or other mammary cells are not well understood, except for IgG. When cows are milked, the greatest concentrations of IGF-1, IGF-2, insulin, PRL, and other peptides are found in the first removed (cisternal) milk fraction, with gradually lower amounts in ensuing fractions of total delivery, followed by a rise toward the end of milking, a pattern that does not coincide with changes in dry matter content (Ontsouka et al., 2003). As a consequence, the availability of these components to the suckling calf is highly variable.

Intake of colostrum is essential for the survival of calves and many other species during the neonatal period as a result of lacking placental transfer of IgG (Butler, 1974). Colostrum components exert effects on the GIT, produce transient systemic metabolism and endocrine changes, and have long-lasting effects on immunoprotection as well as nutritional status. It has become apparent that colostrum yields effects on neonates also through nonnutrient (bioactive) substances. Sufficient provision of colostral immunoglobulins is well established to reduce mortality rates due to infections in calves and other species (Bush & Staley, 1980; Quigley & Drewry, 1998). Klagsbrun (1978) first demonstrated that milk contains factors that stimulate growth (mitogens) have the highest concentrations in colostrum. Some of the bioactive factors in colostrum, such as IGFs, enter the GIT in an active form and survive the digestive process (Koldovsky, 1989), and GIT interactions during intestinal development and systemic effects following absorption of such factors in the "open" GIT are possible. In fact, nonnutrient factors in colostrum modulate the GIT microbial population, have profound effects on the GIT (epithelial cell proliferation, migration, differentiation, apoptosis; protein synthesis and degradation; digestion, absorption; motility, immune system development, and function), and in part exert systemic effects outside the GIT on metabolism and endocrine systems, vascular tone and homeostasis, activity and behavior, and systemic growth. However, the value of many milk bioactive factors for neonates is still in question. Differences in milk composition between species and responses to milk-borne bioactive components may be expected due to different ontogenetic development of tissues and organs. However, in defense of formulas and replacers that are void of many milk bioactive substances, data accumulated over the last 20 to 30 years indicate no major developmental impairment of formula-fed children (Bernt & Walker, 1999), and after intake of colostrum, calves develop without unusual problems if fed milk replacers (Quigley & Drewry, 1998).

Effects of Colostrum and of Colostral Bioactive Substances on the Gastrointestinal Tract

General Aspects

Immediately after birth, the GIT is characterized by accelerated growth rate, changes in structural organization, increased metabolism, and digestive and absorptive functions, as has been shown for the calf (Baumrucker et al., 1994a; Bittrich et al., 2004; Blättler et al., 2001; Blum, 2005, 2006; Blum & Hammon, 2000; Bühler et al., 1998; David et al., 2003; Guilloteau et al., 1997; Hammon & Blum, 1997a; Kühne et al., 2000; Norrman et al., 2006; Sangild et al., 2001; Sauter et al., 2003, 2004; Schottstedt et al., 2005) and other species (Kelly, 1994; Burrin et al., 1995; Xu et al., 2002; Louveau & Gondret, 2004). During milk feeding, GIT growth is particularly enhanced. In rodents, crypt hyperplasia is high after birth and reduces about 10-fold from milk feeding to weaning (Cummins & Thompson, 2002).

Widdowson et al. (1976) first showed that colostrum intake provokes drastic morphological and functional changes of the GIT in neonates. Maturation of the GIT is modulated by ingested nutrients, regulatory substances (hormones, growth factors, cytokines, and neurotransmitters) produced and acting within the GIT wall (juxtacrine, autocrine, paracrine effects), produced in the GIT and released into the gut lumen (luminokines) or produced in and outside the GIT and circulating in blood (Blum, 2006). Levi-Montalcini and Cohen (1960) first suggested that an endocrine factor in the saliva of mice, which was later found to be epidermal growth factor (EGF) (Barnard et al., 1995), induced epidermal hyperplasia in the eyelid (precocious eyelid opening) and tooth eruption.

For colostral or milk protein factors to impact the neonate, the ingested factors must (1) not only survive the digestive functions of the GIT, but also retain biological activity, (2) have GIT receptors that respond to the surviving bioactive factor concentration in the GIT, or (3) be taken up to exert biological activity. Thus, the effects of GIT regulatory peptides are dependent not only upon the presence but also on the number and affinity of specific receptors or transporters and on postreceptor events. In fact, many of the endocrine factors occurring in bovine mammary secretions have been reported to have corresponding receptors in the GIT of the bovine species (Table 2). Because GIT site difference characteristics have been demonstrated in neonatal calves for IGF, GH, and insulin receptors (Cordano et al., 1998; Georgiev et al., 2003,; Georgieva et al., 2003; Ontsouka et al., 2004a, b, c), for and nuclear receptors (Krüger et al., 2005), the effects of colostrum and milk bioactive components are expected to be different in different regions of the GIT.

Many receptors in cells are currently defined as "orphan receptors" (Mohan & Heyman, 2003; Wise et al., 2004), which are family variations of known receptors that have not been associated with a ligand. This suggests either that these receptors are a molecular and synthesis event of no consequence or that the

Table 2 Receptors (mRNA or Protein) Reported to Be Present in the GIT of Bovine Species

Endocrine Receptor	Source
EGFR/TGF αR	No reports
IGF-IR	Georgiev et al. (2003); Baumrucker et al. (1994b)
IGF-IIR	Georgiev et al. (2003); Baumrucker et al. (1994b)
EpoR	Cloned; bone and kidney (Suliman et al., 1999)
TGF βR	Roelen et al. (1998)
FGFR	Bovine follicles and mammary (Berisha et al., 2004; Plath et al., 1998)
InsulinR	Georgiev et al. (2003)
ProlactinR	Scott et al. (1992)
Leptin	Chelikani et al. (2003)

ligand of consequence has not been identified, as shown by the identification of ligands of the family of peroxisome proliferator-activated receptors (Mahle, 2004). In colostrum or milk, the potential for discovery of new ligands for known or unknown receptors is high. The complex mix of bioactive components in colostrum and milk may also explain the difficulty in demonstrating a single-factor function, when functions may rely upon multiple receptors triggering multiple events.

Most of the studies on GIT receptors did not approach the cellular orientation. As an example, in rodents the expression of the EGF receptor was located on the basolateral surface of the intestinal epithelial cells (Montaner et al., 1999), while in more recent studies the EGF receptor was on the apical (brush border) membrane of the epithelium (Wallace et al., 2001). Differences in the abundance of mRNA coding for IGF-1, IGF-2, IGFBP-2, and IGFBP-3, for the IGF type-1 and IGF type-2 receptors, for the GH receptor, and for the insulin receptor in compartmentalized layers (fractions) of the jejunum and ileum of five-day-old calves have been demonstrated (Ontsouka et al., 2004b).

Studies in calves have focused on the GIT epithelium because it is the primary site of interactions with nutrients, nonnutrient substances, and microbes that are important for GIT function and health. The epithelial system (with respect to crypt depth and crypt cell proliferation rate, villus size) of the small intestine in preterm calves is much less responsive to feeding than in full-term calves and responds differently with respect to digestive enzymes (Bittrich et al., 2004). In full-term calves, "normal" colostrum intake compared with feeding milk replacer (or "formulas") enhanced the crypt cell proliferation rate but not the villus size (Blättler et al., 2001). Maximal compared to "normal" colostrum intake or feeding an extract of first-milked colostrum whey (that contained nonnutritive substances, such as IGF-1, etc.) did not affect the proliferation rate of crypt cells, but the villus size was increased, suggesting that the increased small intestinal villus size was the consequence of enhanced crypt hyperplasia, enhanced migration rate of crypt epithelial cells to villus tips, reduced apoptosis rate, or increased survival rate (Blättler et al., 2001; Roffler et al., 2003). The data showed that high amounts of colostrum enhance villus growth and therefore the potential for intestinal absorption, but crypt depth did not correlate with villus circumference and height (Bühler et al., 1998; Blättler et al., 2001). Negative correlations between crypt cell size and crypt cell proliferation rates suggested negative feedback control of small intestinal epithelial cell growth. Increased intake of colostrum also increased the absorptive capacity if xylose was used as a test substance (Baumrucker et al., 1994a; Hammon & Blum, 1997a; Sauter et al., 2004; Schottstedt et al., 2005).

Few studies in neonatal calves have compared the effects of feeding colostrum (containing nutritive and nonnutritive substances) on the GIT with the effects of a milk-derived formula (containing nutritive substances in comparable amounts as in colostrum, but without nonnutritive substances). When colostrum extracts (containing bioactive substances, such as IGF-1) were added to the formula, epithelial cell proliferation rates in the small intestine were higher in calves fed the colostrum extract than in controls (Roffler et al., 2003).

Only a small number of studies have been conducted on the ontogenetic changes in GIT lymphocyte populations in calves (David et al., 2003). Feeding colostrum rather than formula to calves reduced the number of proliferating cells in lymphoid follicles and of B-lymphocytes in Peyer's patches, but not apoptotic rates and the number of T-lymphocytes in ileum, suggesting that feeding colostrum (and therefore probably also its bioactive components, such as IgG_1) spared active immune responses (Norrman et al., 2003).

Epidermal Growth Factor Family

Members of the EGF peptide family share a common structure and bind to the EGF receptor (EGFR) (Barnard et al., 1995). Five related ligands have been identified: EGF, transforming growth factor-β (TGF-β), heparin binding EGF-like growth factor, amphiregulin, and betacellulin. EGF and TGF-β have been the most studied. Salivary glands are the major source of EGF and TGF-β in adults, but other endogenous sources are the kidneys, Brunner's glands, and Paneth cells of the small intestine (Dvorak et al., 1994). For the postnatal rodent, maternal milk is the major source of the intestinal EGF (Schaudies et al., 1989).

The EGF increases cell proliferation *in vitro* (Berseth, 1987) and *in vivo* when provided to rodents in high doses (Pollack et al., 1987). Physiological doses seem to have little impact. The infusion of EGF in rodents showed profound effects on cell proliferation and some effects upon crypt fission (Berlanga-Acosta et al., 2001). EGFR knockout mice survive for only a maximum of eight days after birth and suffer from impaired epithelial development in several organs, including the skin, lung, and GIT (Miettien et al., 1995), but the genetic background of the mice introduced variation in the phenotypic expression (Sibilia & Wagner, 1995). No reports of small intestinal EGFR in cattle have appeared.

Reports of EGF in bovine milk have indicated a much lower concentration (<2 ng/mL: Iacopetta et al., 1992; 2–3 ng/mL: Xiao et al., 2002) when compared to human milk (30–40 ng/mL), and betacellulin is present in bovine colostrum and milk at similar concentrations as EGF (Bastian et al., 2001). A search for EGF in expressed sequence tags (EST) databases or expressed mRNA databases for EGF and EGF-like messages is negative. Taken together, these data suggest that EGF in bovine species is not as significant when compared to rodents and humans and possibly pigs. Although negative results are not proof that it does not exist in cow's milk, it is now thought that the detection of EGF in bovine milk is rather a crossover of the antibodies to other EGF family members and that betacellulin is a likely candidate. Thus, while bovine mammary tissue synthesizes EGF (or other EGF family members) and TGF-β (Plaut, 1993) and secretes EGF or betacellulin (and perhaps TGF-β) into milk, the low concentration in light of the higher doses required for rodent intestinal effects suggests little opportunity for intestinal effects.

Growth Hormone, IGFs, and IGF Binding Proteins

Of great importance among milk bioactive components is the IGF system. It consists of three ligands, three corresponding high-affinity receptors, six IGFBPs that associate with the IGFs with high affinity, and five IGFBP-related proteins that bind to IGFs with approximately 10-fold lower affinity when compared with IGFBPs (Baxter, 2000). Within the bovine species, IGFs and IGFBPs are highly concentrated in the prepartum secretion and in first-milked colostrum and then rapidly decrease to low levels in mature milk. The detectable level and rank of specific IGFBPs in bovine mammary secretions are IGFBP-3 > IGFBP-2 ≈ IGFBP-4 > IGFBP-5, and the pattern of change during the lactation cycle is different from that of the circulation. The IGFBPs were thought previously to act principally by modulating IGF action by inhibiting or enhancing IGF binding to IGF receptors (IGFR). While circulating IGFBP-3 primarily originates from hepatic cells, this protein is also produced locally in many tissues (Zapf, 1995). In bovine mammary tissue, it is synthesized in lactocytes (Gibson et al., 1999) and especially during the involution and prepartum periods (Vega et al., 1991).

The, effects of IGF-1 on the GIT in neonatal pigs and calves have been studied. Based on these studies, IGFs can survive to a considerable extent in the small intestine (Xu et al., 2002) and especially in neonates. IGF-1 and IGF-2 effects are mediated by specific receptors that in the GIT are present in epithelial cells, fibroblasts, endothelia, and smooth muscles (Howarth, 2003). In calves IGFR numbers are different among different GIT sites and change depending on age and nutrition (Georgiev et al., 2003; Georgieva et al., 2003; Hammon & Blum, 2000; Ontsouka et al., 2004a, b, c). Parenteral IGF-1 administration enhances mucosal (epithelial), submucosal, and muscularis thickness;

longitudinal and cross-sectional GIT growth; and sodium absorption and sodium-dependent nutrient (glucose) absorption (Alexander & Carey, 1999). Recombinant human IGF-1 added to a formula increased intestinal villus growth, lactase activity, and lactase mRNA expression in artificially reared piglets (Houle et al., 2000). Oral IGF-I suppressed the proteolytic degradation of lactase and its precursor (Burrin et al., 2001). Feeding physiological amounts of IGF-1 or milk-borne IGF-1 increased lactase synthesis and reduced aminopeptidase activity (Burrin et al., 2001). Feeding pharmacological amounts of IGF-1 variably increased the proliferation rate of crypt cells, reduced the apoptosis rate of epithelial cells, and increased the villus size and protein synthesis, but the growth rate was not enhanced in transgenic pigs overexpressing IGF-1 (Burrin et al., 1999). There is no significant absorption of ingested IGF-1 or Long-R^3-IGF-I in calves and piglets (Vacher et al., 1995; Hammon & Blum, 1997; Donovan & Odle, 1994). Regulation of the GIT effects of ingested IGFs, IGFs produced within and outside the GIT, and IGFs circulating in blood is complex. It is further complicated by interactions with other endocrine systems (GH, insulin, and cortisol), modification (mostly inhibition) of IGF effects by IGFBPs, proteolysis of IGFBPs followed by IGF cleavage (Elmlinger et al., 1999), and possible interactions with lactoferrin (Baumrucker & Erondu, 2000).

The presence of mRNA of the GH receptor (GHR), of the IGF type 1 and 2 receptors (IGF-1R, IGF-2R), of the insulin receptor (IR), of IGF-1 and IGF-2, and of IGFBP 1-3 has been reported in all parts of the GIT of neonatal calves (Cordano et al., 1998; Pfaffl et al., 2002; Ontsouka et al., 2004a). There were marked differences in mRNA abundance among different GIT sites, suggesting variable mRNA synthesis and/or turnover rates and variable importance of these receptors and binding proteins for GIT growth and maturation. In the ileum or jejunum of five-day-old calves, mRNA levels differed between compartmentalized layers (fractions) containing villus tips, crypts, and lamina propria (Ontsouka et al., 2004b). Members of the somatotropic axis and of the IR were therefore not evenly expressed in different jejunal and ileal layers of neonatal calves. Higher mRNA levels of the GH-IGF-insulin system (except IGF-2R) in the small intestine of calves on day 5 than on day 1 suggested that this part of the GIT may be a main target for these colostral factors. The B_{max} of the IGF-1R, IGF-2R, and IR was measured in the small intestine and colon of neonatal calves; interestingly, the mRNA levels and maximal binding (B_{max}; used as an index of receptor numbers) of the IGF-1R, IGF-2R, and IR were negatively associated (Georgiev et al., 2003). The different studies also showed that the B_{max} of the IGF-2R decreased while that of the IR increased, whereas the B_{max} of IGF-1R did not change during the first postnatal week; the B_{max} values of IGF-1R, IGF-2R, and IR were modified by differences in feeding, and the B_{max} values of IGF-2R and IR at birth were lower in preterm than in full-term calves (Georgiev et al., 2003; Hammon & Blum, 2002).

In earlier studies, feeding recombinant human IGF-1 for seven days enhanced the incorporation of (3H)-thymidine into the DNA of isolated enterocytes *ex vivo* (Baumrucker et al., 1994a), but possible morphological effects were not studied. Feeding a formula (containing only traces of nonnutrient components) plus oral IGF-1 (milk derived from transgenic rabbits; amounts corresponding to those present in colostrum, i.e., 0.4 mg of IGF-1/L formula) for seven days had no significant effects on the proliferation of small intestinal crypt cells nor on villus growth (Roffler et al., 2003). Thus, ingested IGF-1, in physiological amounts as present in colostrum, is obviously not—or not solely—responsible for intestinal growth and development in neonatal calves.

Parenteral administration of GH for seven days enhanced crypt cell areas in neonatal calves but not the small intestinal villus size (Bühler et al., 1998), suggesting that GH does not rapidly stimulate neonatal intestinal growth and development.

Lactoferrin

Lactoferrin (Lf) is a multifunctional protein having several distinct biological activities, such as antimicrobial activity, inhibition/modulation of cytokine activation by neutrophils, immunomodulation, modification of apoptosis, and (colon) cancer prevention (Van der Strate et al., 2001). The Lf content of mammary secretions varies with development in both humans and cows, is elevated during involution, and is high in bovine colostrum (Nuijens et al., 1996; Muri et al., 2005). Feeding of Lf to calves decreased the villus size in the jejunum, enhanced the size of Peyer's patches in the ileum of calves, and exerted immunomodulatory effects by increasing plasma IgG levels (Prgomet et al., 2006). While Lf receptors have been characterized in the intestinal epithelium (Suzuki et al., 2001), additional studies have suggested these to be nucleolin (Legrand et al., 2004). Lf has a bipartite nuclear localization sequence (NLS) and can enter the nucleus of some cells (Garre et al., 1992). Nucleolin may be involved in Lf binding as well as translocation to the nuclear compartment (Legrand et al., 2004).

Bovine Lf (bLf) and IGFBP-3 binding has been reported for mammary cells (Baumrucker et al., 2003). Besides direct effects by Lf, there are important interactions among Lf, (all-*trans*)-retinoic acid (atRA), and IGFBP-3 (Baumrucker & Erondu, 2000). Thus, cells cultured in the presence of retinoids showed increased cell death, and all-*trans* retinoic acid (atRA) inhibited insulin- and EGF-stimulated cell proliferation (Purup et al., 2001; Cheli et al., 2003) and PARP-p85 apoptotic staining (Baumrucker et al., 2002). The application of bLf blocked apoptosis in these cells, altered atRA effects upon cell cycle progression, and allowed for continued cell proliferation (Baumrucker et al., 2002). A dose-dependent stimulation of bovine Lf (bLf) on reporter construct (RXR-α-luciferase) may be one of the bLf links to cell growth and apoptotic mechanisms (Baumrucker et al., 2003).

The IGFBP-3 binds to molecules other than IGFs (Fowlkes & Serra, 1996). With the recognition of an NLS, IGFBP-3 was shown to enter the nucleus of some cells and to interact with nuclear transcription factors (Jaques et al., 1997). Exogenous IGFBP-3 induced apoptosis in cell lines, but in some cases this action was not mediated by the type 1 IGF receptor (Valentinis & Baserga, 1996). In addition, IGFBP-3 binds to transferrin that is related to Lf (Weinzimer et al., 2001). Interestingly, Lf in bovine lactocyte membranes does bind IGFBP-3 (Gibson et al., 1998), and because bLf has an NLS sequence (Baumrucker et al., 1999), there is a link between IGFBP-3 and Lf. Thus, both genes also have a retinoic acid response element in their regulatory sequences.

Based on these premises, studies on interactions among Lf, vitamin A/retinoic acid, and the IGF system were performed in neonatal calves. Vitamin A and Lf can stimulate protein synthesis in neonates. However, newborn calves are vitamin A-deficient and have a low Lf status, but plasma vitamin A and Lf levels rapidly increase after the ingestion of colostrum, which normally contains relatively high amounts of Lf and vitamin A (Blum et al., 1997; Muri et al., 2005; Zanker et al., 2000a). On that basis, neonatal calves were fed a milk-based formula with or without vitamin A, Lf or vitamin A plus Lf, or colostrum to study protein synthesis in the jejunum and liver of neonatal calves (Rufibach et al., 2006). L-$[^{13}C]$-valine was intravenously administered to determine the fractional protein synthesis rate (FSR) in the jejunum and liver. There were no effects of vitamin A and Lf on intestinal and hepatic protein synthesis and no interactions between vitamin A and Lf, but the FSR of protein in the jejunum was significantly correlated with histomorphometrical traits of the jejunum, and the FSR of protein in the liver was significantly correlated with plasma albumin concentrations.

Because there is evidence that Lf and vitamin A interact with IGF binding proteins and thus influence the status and effects of IGF, the hypothesis was also tested that vitamin A and Lf influence the epithelial growth, development, and absorptive capacity of the small and large intestines and modulate intestinal immune tissues (Peyer's patches) (Schottstedt et al., 2005). The study showed that feed supplementation with vitamin A and Lf influenced growth of the ileum and colon. Interactions were observed between vitamin A and Lf on epithelial cell maturation, villus growth, and the size of follicles in Peyer's patches of neonatal calves.

Transforming Growth Factor-β

There are three known general mammalian isoforms of transforming growth factor-β (TGF-β: $β_1$, $β_2$, and $β_3$) in bovine mammary secretions (Roberts & Sporn, 1992), but there may be many more ligands of the TGF-β family (Massague, 1998). All three isoforms appear to be physiologically important (Roberts, 1998). These factors regulate a plethora of biological processes including wound healing, proliferation, apoptosis, immune response, and cell differentiation (Shi & Massague, 2003). TGF-β inhibits intestinal cells *in vitro*

(Barnard et al., 1995) and is thought to be involved in terminal differentiation of GIT epithelial cells, for which speak TGF effect mRNA expression in a villus gradient (Koyama & Podalsky, 1989). Interest is high in human milk TGF-β and stimulation of the maturation of the infant's immune system (Ogawa et al., 2004). Some suggestions of TGF-β's effect on IgG absorption in the neonatal calf have appeared (Hammer et al., 2004). The bovine TGF-β type 1 receptor has been cloned, and type 1 and 2 TGF-β receptor mRNA are present in the bovine intestine (Roelen et al., 1998). Because TGF-β ligands appear to be low in concentration in bovine mammary secretions (Table 1), their physiological role in the bovine neonate is unclear.

Erythropoietin

Erythropoietin (Epo) has also been detected in milk. Primarily produced in the kidneys, it plays a central role in the regulation of erythropoiesis. Epo is found in amniotic fluid at high levels; because the fetus swallows amniotic fluid, Epo may be one of several of the GIT development factors *in utero* (Teramo et al., 1987). The demonstration that Epo is synthesized (Juul et al., 2000) and secreted into human milk (King et al., 1998) inspired the identification of the Epo receptor in the intestinal cells and tissues of humans and rodents (Piatak et al., 1993). Evidence for Epo-trophic effects and protective function in the colon have appeared (Juul et al., 2001). Enterally provided Epo to the suckling rat appears to localize in the liver and other peripheral tissues (Miller-Gilbert et al., 2001). Stimulation of erythropoiesis in rodents has been reported (Bielecki et al., 1973), but no stimulation of erythropoiesis was found in human neonates (Juul, 2003). Thus, the erythropoiesis function for milk-borne Epo is controversial, and the main function may remain in the intestine as a trophic factor (Semba & Juul, 2002). Information concerning the presence of ruminant milk Epo has not yet been published.

Leptin

Leptin in mammals is produced mainly by white and brown adipocytes, but also by skeletal muscle, the stomach, placenta, and mammary gland (Chilliard et al., 2001). The mammary gland secretes leptin, and concentrations in colostrum are higher than in mature milk (Chilliard et al., 2001). Plasma leptin levels in human and rat neonates are high postnatally and then decrease; in rats and mice, concentrations increase transiently during the suckling period and are elevated in breastfed compared with formula-fed infants in some studies, possibly due in part to absorption of ingested colostral or milk leptin. Plasma leptin concentrations in growing preruminant calves and lambs are influenced by feeding intensity and are associated more with fat accretion rates than with fat mass (for details, see

Blum et al., 2005). Plasma leptin concentrations did not rise postnatally in lambs that were born small (Ehrhardt et al., 2003); after colostrum intake, plasma leptin in neonatal calves remained stable (whereas it decreased when calves were fed a formula) in one study (Blum et al., 2005) but transiently increased after colostrum intake in another study (Rérat et al., 2005). With respect to the effects of nutrition and interactions with hormones and metabolites, calves behaved differently from what is known in mature cattle (Blum et al., 2005).

That leptin is present in milk (Bonnet et al., 2002) and the discovery of the leptin receptor in the intestine (Ahima & Flier, 2000) stimulated research to elucidate its role in the neonatal intestine (Alavi et al., 2002). Intestinal studies with rodents show increased intestinal mass and changes in carbohydrate and amino acid absorption (Alavi et al., 2002). Studies in the piglet indicate that leptin feeding contributes to the development of the small intestine structure and function (Wolinski et al., 2003).

Steroids

In addition to cortisol, testosterone, and progesterone, estrogens are secreted with milk and have attracted much interest in recent years. Based on extractions and mass spectrum analysis of milk in Holstein–Friesian dairy cows, Malekinejad et al. (2006) detected estrone, estriol, 17 α-estradiol, and 17 β-estradiol (Table 3). All were shown to consist of 58–92% conjugated forms (sulfation or glucoronidation), as has been previously reported (McGarrigle & Lachelin, 1983). Conjugation is a means of making steroids more water-soluble for excretion (bile and other). Although conjugated steroids are no longer biologically active, if bacterial sulfatases or glucouronidases are present, they may become reactivated after deconjugation. Nonconjugated or free steroids have biological potencies of 17 β-estradiol >estrone > > estriol > 17 α -estradiol (Fritsche & Steinhart, 1999). Estrone is 5 to 10 times less potent than 17 β-estradiol, and the others may be considered to be inactive unless they occur in very high concentrations. However, estrone and estriol can be converted into catechol metabolites, which are carcinogenic (Li et al., 1985), while estriol has been reported to be protective (Follingstad, 1978) against tumor development. Thus, there are differences among steroid potencies in mammalian systems, and milk-borne steroids are conjugated for deactivation or metabolically converted to catechol derivatives that may or may not possess biological activity and carcinogenic potency within the mammalian system where production occurs. Because nonconjugated steroids pass through biological membranes, it is not surprising that they also appear in milk (Fritsche & Steinhart, 1999), but the recent report of most being conjugated (Malekinejad et al., 2006) suggests that the mechanisms of transfer of conjugated steroids into milk are unknown. Only a few reports have indicated that 17 β-estradiol is synthesized from provided androgens by mammary tissue

Table 3 Steroids in Cow's Milk (in g/mL)

Endocrine Ligand	Colostrum	Raw Milk[a]	Commercial Milk[b]	Source
Estrone	2000–4000[d]	9.2–118[c]	8–20[c]	Malekinejad et al. (2006); Janowski et al. (2002); Pope and Roy (1953)
17 αEstradiol	?	7–47[d]	nd[c]	Malekinejad et al. (2006)
17 βEstradiol	1500–2000[d] 1000[f]	6–221[c]	10–20[c]	Malekinejad et al. (2006); Janowski et al. (2002); Pope and Roy (1953)
Estriol	?	nd[c]	nd[c]	Malekinejad et al. (2006)
Progesterone	?	11,300[e]	2,100–11,000[e]	Fritsche and Steinhart (1999)
Testosterone	?	50–150[e]	10[ef]	Fritsche and Steinhart (1999)
Cortisol	1,590–4,400[d]	350[d]	710[d]	Butler and Des Bordes (1980); Shutt and Fell (1985)

[a] First to third trimester of pregnancy.
[b] 0–3.5% fat.
[c] Mass spectrometry.
[d] Radioimmunoassay (RIA).
[e] Enzyme linked immuno-sorbant assay (ELISA).
[f] Uterine growth bioassay.
nd Not detected.

in vitro (Janowski et al., 2002; Maule Walker et al., 1983), but the specific cell source and contribution to mammary secretion are unknown.

The GIT and liver have an important role in steroid metabolism. The 17 β-hydrosteroid dehydrogenase in the GIT and liver converts steroids (Mindnich et al., 2004) and reduces the amount of orally absorbed bioactive steroids in the first pass. There is a highly effective decrease (~500-fold) in the presence of 17 β-estradiol when the steroid is delivered via the GIT (Plowchalk & Teeguarden, 2002) when compared to the intravenous route. This well-known mechanism protects from the effects of orally administered or ingested steroids.

Systemic Effects of Colostral Bioactive Components on Metabolism and Endocrine Systems

It has been known for some time that the GIT of newborn mammalian species allows marked absorption of large proteins. This has been clearly demonstrated by the transfer of Ig into the systemic circulation (Quigley & Drewry, 1998). At some point, this "open" gut changes and large proteins, such as Ig, are no longer absorbed. At this point, the GIT is termed "closed." Importantly, the mammalian species exhibit different closure times for Ig after birth (calves and piglets: 24 to 48 hours; rodents: about day 16; humans: around 8 weeks). This differential "closure" concept is important to keep in mind when considering the "absorption" of bioactive components appearing in milk. Much of our understanding of the appearance and resulting phenotypic alterations that appear in the literature has been established with suckling rodents (Koldovsky, 1995) or other species that retain an "open" GIT for a considerable time after birth.

It is well known that colostrum intake changes the IgG status (Quigley & Drewry, 1998) of neonatal calves. Our studies in full-term calves (Ronge & Blum, 1988; Blum et al., 1997; Hadorn et al., 1997; Egli & Blum, 1998; Hammon & Blum, 1998, 1999, 2002; Hammon et al., 2000; Kühne et al., 2000; Rauprich et al., 2000a, b; Zanker et al., 2000b, 2001a, b; Nussbaum et al., 2002; Muri et al., 2005) have shown that after the intake of colostrum, there is a rise in blood plasma concentrations not only of IgG1, but also of Lf, total protein, albumin, and essential amino acids and a dramatic decrease in the glutamine/glutamate ratio. Plasma urea concentrations increase if high amounts of colostrum are fed, and plasma glucose concentrations increase with a delay of several days. Furthermore, there is a rise in plasma concentrations of glucose, triglycerides, phospholipids, total cholesterol, and essential fatty acids as well as of β-carotene, retinol, and α-tocopherol. On the other hand, colostrum intake does not markedly or immediately change plasma mineral, creatinine, lactate, and nitrate concentrations, but creatinine, lactate, nitrate, urate, and ascorbate concentrations rapidly decrease after birth. The very high plasma nitrate status of neonatal calves suggests markedly enhanced nitric oxide production (Blum et al., 2001), and we have furthermore shown the nitrosylation of tissue proteins, plasma albumin, and plasma IgG after the ingestion of colostrum in association with high activities of endothelial and inducible nitric oxide synthetases in some organs, such as in the small intestine and liver (Christen et al., 2007). Preterm calves that survive the first days of life exhibit relatively normal metabolic and endocrine changes (Bittrich et al., 2002).

Many proteins and peptides (lactalbumin, ovalbumin, and Lf within the first 24–48 hours after birth besides IgG_1) are absorbed intestinally and appear in the circulation of calves (Michanek et al., 1989; Muri et al., 2005). Lf even appears in the cerebrospinal fluid of neonatal calves (Talukder et al., 2003). On the other hand, there is nonexistent or only negligible absorption and (or) appearance in the systemic circulation of insulin, IGF-1, Long-R^3-IGF-1, and PRL within the first 24 hours after birth in calves, even when insulin and IGF-1 are administered in pharmacological amounts (Grütter & Blum, 1991; Baumrucker et al., 1994a; Hammon & Blum, 1998), demonstrating marked differences with respect to the absorption of peptides. There are also obvious species differences.

Nevertheless, during the first postnatal week(s), major changes in the blood levels of hormones can be observed in calves, as have been shown for PRL, adrenocorticotrophic hormone, GH, IGFs (and IGFBPs 1-3), insulin, glucagon, leptin, thyroxine (T4), 3.5.3'-triiodothyronine (T3), and cortisol (Baumrucker & Blum, 1994; Baumrucker et al., 1994b; Blum et al., 2005; Egli & Blum, 1998; Grütter & Blum 1991a, b; Hadorn et al., 1997; Hammon & Blum, 1998, 1999; Hammon et al., 2000; Kinsbergen et al., 1994; Kühne et al., 2000; Oda et al., 1989; Rauprich et al., 2000a, b; Ronge & Blum, 1988; Skaar et al., 1994; Nussbaum et al., 2002; Sauter et al., 2003; Sparks et al., 2003; Zanker et al., 2001). Although insulin and IGF-1 and probably many other peptide hormones are barely absorbed in the intestine in neonatal calves, it cannot be fully excluded that situations can potentially be found in which insulin and IGF-1 are absorbed, as in part reported for pigs (Xu et al., 2002). The situation is complex because neonatal calves are able to produce IGF-1 since IGF-1 mRNA is expressed in the liver, GIT, spleen, thymus, lymph nodes, and kidney (Cordano et al., 2000; Pfaffl et al., 2002; Ontsouka et al., 2004a). Based on the literature from humans, the sum of all sources of endogenous IGF-1 (saliva, bile, pancreatic secretions, GIT secretions) is around 100-fold what might appear in the GIT from any food source. Administration of bovine GH can basically increase plasma IGF-1 concentrations in neonatal calves (Coxam et al., 1989; Hammon & Blum, 1997), and there are significant changes to IGF binding proteins 1–3 in response to differences in feeding and endocrine treatments (Hammon & Blum, 1997b; Sauter et al., 2003). Although the GH-IGF-1 axis in neonatal calves is basically functional, the system is not fully mature because hepatic IGF-1 expression in neonatal calves is small (Cordano et al., 2000; Pfaffl et al., 2002), and GH effects on the IGF systems are markedly smaller in neonatal calves and calves that are a few weeks old than in older cattle (Ceppi & Blum, 1994; Hammon & Blum, 1997b). Lack of the expected variations of T4 and T3 levels despite marked differences in energy intake in several of our studies is in contrast to studies conducted by Grongnet et al. (1985). There is also no nutritional influence on GH and PRL levels. However, plasma concentrations of gastrin, glucose-dependent insulinotropic polypeptide, and cholecystokinin are increased, whereas concentrations of somatostatin and motilin decrease after colostrum intake (Guilloteau et al., 1997; Hadorn et al., 1997). Plasma cortisol concentrations were lower in colostrum-fed than formula-fed calves (Hammon & Blum, 1998).

There is no evidence for endocrine "imprinting" in calves by differences in feeding immediately postnatally (Zanker et al., 2001a), possibly because neonatal calves are born at a relatively mature stage. This may be different in species (rats, mice, dogs, cats) in which neonates are born relatively immature.

Conclusions and Outlook

With its nutrient and nonnutrient components, the intake of colostrum exerts marked effects on GIT development and function. Colostrum intake provides immunoprotection (passive immunity by Ig), thereby likely reducing the need for early active immune reactions, and is essential for the survival of neonates of many species, such as calves. Furthermore, there are important systemic effects on metabolism and on various endocrine systems due to the intake of nutrient and nonnutrient colostral components that contribute to survival in the stressful postnatal period.

The knowledge of higher concentrations of endocrine factors in colostrum than in mature milk suggests the possibility of a signal for the neonate.

Alternatively, long-term exposure to mature milk with lower concentrations may also have a lasting impact or an advantage on survival and growth. The evidence of GIT survival of IGFs and EGF from digestion and their effect on gut epithelial cells are becoming clearer. As for the absorption and systemic impact of bioactive factors, the evidence is rather meager.

For many years it has been known that the many different GIT cells are derived from common stem cell precursor cells located in intestinal crypts. Until recently, little was known about the events that commit the stem cells to other cell types of the GIT epithelium. While the enteroendocrine cell lines comprise only a small fraction of the total epithelium, they are the cells that secrete the gut hormones. It has become evident that the function of the GIT hormones was not only to modulate the function of other digestive organs like the intestine, pancreas, and stomach, but that gut hormones have been implicated in the regulation of other physiological processes such as appetite regulation by influence over the central nervous system and insulin secretion. It is possible that the exposure of the GIT to milk endocrine factors may alter the rate and quantitative distribution of GIT endocrine cells that could impact the appearance of these factors, which finally could impact the phenotype of the neonate.

The rapid expansion of molecular biology studies that involve the mechanism of cell signaling with *in vitro* experimentation has provided new insights into the possible regulation of cells and tissues. However, *in vivo* experimentation often shows that such mechanisms are not similarly observed. Mouse knockout or knock-in experiments often show little or no effect. The explanation for negative results may be compensatory responses by as yet unknown mechanisms—and it may be similar with various endocrine factors occurring in bovinesecretions and their impact on the bovine neonate.

Acknowledgment The studies of JWB have been supported by the Swiss National Science Foundation, CH-Bern (Grants 32-30188.90, 32-36140.92, 32-051012.97, 32-56823.99, 32-59311.99, 32-67205.01, 32-59311.01); by the Schaumann Foundation, D-Hamburg; by F. Hoffmann-La Roche, CH-Basel; by Novartis (formerly Ciba Geigy AG), CH-Basel, Switzerland; by Gräub AG, CH-Bern; and by the Swiss Federal Veterinary Office, CH-Liebefeld-Bern. The studies of CRB have been supported by the Penn State University Experiment Station and multiple USDA-NRI grants.

References

Ahima, R. S., & Flier, J. S. (2000). Leptin. *Annual Review of Physiology, 62*, 413–437.
Akers, R. M., & Kaplan, R. M. (1989). Role of milk secretion in transport of prolactin from blood into milk. *Hormones and Metabolic Research, 21*, 362–365.
Alavi, K., Schwartz, M. Z., Prasad, R., O'Connor, D., & Funanage, V. (1989). Regulation of intestinal epithelial cell growth by transforming growth factor type β. *Proceedings from the National Academy of Science, 86*, 1578–1582.

Alavi, K., Schwartz, M. Z., Prasad, R., O'Connor, D., & Funanage, V. (2002). Leptin: A new growth factor for the small intestine. *Journal of Pediatric Surgery, 37*, 327–330.

Alexander, A. N., & Carey, H. V. (1999). Oral IGF-I enhances nutrient and electrolyte absorption in neonatal piglet intestine. *American Journal of Physiology, 277*, G619–G625.

Barnard, J. A., Beauchamp, R. D., Russell, W. E., Dubois, R. N., & Coffey, R. J. (1995). Epidermal growth factor-related peptides and their relevance to gastrointestinal pathophysiology. *Gastroenterology, 108*, 564–580.

Bastian, S. E. P., Dunbar, A. J., Priebe, I. K., Owens, P. C., & Goddard, C. (2001). Measurement of betacellulin levels in bovine serum, colostrum and milk. *Journal of Endocrinology, 168*, 203–212.

Baumrucker, C. R., & Blum, J. W. (1994). Effects of dietary recombinant insulin-like growth factor on concentrations of hormones and growth factors in the blood of newborn calves. *Journal of Endocrinology, 140*, 15–21.

Baumrucker, C. R., Hadsell, D. L., & Blum, J. W. (1994a). Effects of dietary insulin-like growth factor I on growth and insulin-like growth factor receptors in neonatal calf intestine. *Journal of Animal Science, 72*, 428–433.

Baumrucker, C. R., Green, M. H., & Blum, J. W. (1994b). Effects of dietary rhIGF-I in neonatal calves on the appearance of glucose, insulin, D-xylose, globulins and γ-glutamyltransferase in blood. *Domestic Animal Endocrinology, 11*, 393–403.

Baumrucker, C. R., Gibson, C. A., Shang, Y., Schanbacher, F. L., & Green, M. H. (1999). Lactoferrin specifically binds to IGFBP-3 to cause competitive displacement of IGF from IGF:IGFBP-3 complexes and induces internalization and nuclear localization of lactoferrin:IGFBP-3 complexes in mammary epithelial cells. *Proceedings of the 81st Endocrine Annual Meeting*, San Diego, p. 404.

Baumrucker, C. R. & Erondu, N. E. (2000). Insulin–like growth factor (IGF) system in the bovine mammary gland and milk. *Journal of Mammary Gland Biology and Neoplasia, 5*, 53–64.

Baumrucker, C. R., Gibson, C. A., Zavodovskaya, M., Schanbacher, F. L., & Green, M. H. (2002). Interaction of lactoferrin and IGFBP-3 with the retinoid signaling system: Cell growth and apoptosis of mammary cells. *Proceedings of the 84th Endocrine Annual Meeting*, San Francisco, p. 362.

Baumrucker, C. R., Gibson, C. A., & Schanbacher, F. L. (2003). Bovine lactoferrin binds to insulin-like growth factor-binding protein-3. *Domestic Animal Endocrinology, 24*, 287–303.

Baxter, R. C. (2000). Insulin-like growth factor (IGF)-binding proteins: Interactions with IGFs and intrinsic bioactivities. *American Journal of Physiology, 278*, E967–E976.

Berisha, B., Sinowwatz, F., & Schams, D. (2004). Expression and localization of fibroblast growth factor (FGF) family members during the final growth of bovine ovarian follicles. *Molecular Reproduction and Development, 67*, 162–171.

Berlanga-Acosta, J., Playford, R. J., Mandir, N., & Goodlad, R. A. (2001). Gastrointestinal cell proliferation and crypt fission are separate but complementary means of increasing tissue mass following infusion of epidermal growth factor in rats. *Gut, 48*, 803–807.

Bernt, K. M., & Walker, W. A. (1999). Human milk as a carrier of biochemical messages. *Acta Paediatrica, 88 (Suppl)*, 27–41.

Berseth, C. L. (1987). Enhancement of intestinal growth in neonatal rats by epidermal growth factor in milk. *American Journal of Physiology, 253*, G662–G665.

Bielecki, M., Kazewska, M., Woijtiowicz, Z., & Gruszecki, W. (1973). The effect of orally administered erythropoietin on erythropoiesis in experimental animals. *Acta Physiologica Polonica, 24*, 351–356.

Bittrich, S., Morel, C., Philipona, C., Zbinden, Y., Hammon, H., & Blum, J. W. (2002). Physiological traits in preterm calves during their first week of life. *Journal of Animal Physiology and Animal Nutrition, 86*, 185–198.

Bittrich, S., Philipona, C., Hammon, H. M., Rome, V., Guilloteau, P., & Blum, J. W. (2004). Preterm as compared with full-term neonatal calves are characterized by morphological and functional immaturity of the small intestine. *Journal of Dairy Science, 87*, 1786–1795.

Blättler, U., Hammon, H. M., Morel, C., Philipona, C., Rauprich, A., Romé, V., le Huerou-Luron, I., Guilloteau, P., & Blum, J. W. (2001). Feeding colostrum, its composition and feeding duration variably modify proliferation and morphology of the intestine and digestive enzyme activities of neonatal calves. *Journal of Nutrition, 131,* 1256–1263.

Blum, J. W. (2005). Bovine gut development. In P. C. Garnsworthy (Ed.), *Calf and Heifer Rearing* (pp. 31–52). Nottingham, UK: Nottingham University Press.

Blum, J. W. (2006). Nutritional physiology of neonatal calves. *Journal of Animal Physiology and Animal Nutrition, 90,* 1–11.

Blum, J. W., & Baumrucker, C. R. (2002). Colostral and milk insulin-like growth factors and related substances: Mammary gland and neonatal (intestinal and systemic) targets. *Domestic Animal Endocrinology, 23,* 101–110.

Blum, J. W., & Hammon, H. (2000). Colostrum effects on the gastrointestinal tract, and on nutritional, endocrine and metabolic parameters in neonatal calves. *Livestock Production Science, 66,* 151–159.

Blum, J. W., Hadorn, U., Sallmann, H., & Schuep, W. (1997). Delaying the colostrum intake by one day impairs plasma lipid, essential fatty acid, carotene, retinol and alpha-tocopherol status in neonatal calves. *Journal of Nutrition, 127,* 2024–2029.

Blum, J. W., Zbinden, Y., Hammon, H. M., & Chilliard, Y. (2005). Plasma leptin status in young calves: Effects of preterm birth, age, glucocorticoid status, suckling, and feeding with an automatic feeder or by bucket. *Domestic Animal Endocrinology, 28,* 119–133.

Bonnet, M., Delavaud, C., Laud, K., Gourdou, I., Lerous, C., Djiane, J., & Chilliard, Y. (2002). Mammary leptin synthesis, milk leptin and their putative physiolgocial roles. *Reproduction Nutrition and Development, 42,* 399–413.

Bühler, C., Hammon, H., Rossi, G. L., & Blum, J. W. (1998). Small intestinal morphology in eight-day-old calves fed colostrum for different durations or only milk replacer and treated with long-R3-insulin-like growth factor I and growth hormone. *Journal of Animal Science, 76,* 758–765.

Burrin, D. G., Davis, T. A., Ebner, S., Schoknecht, P. A., Fiorotto, M. L., Reeds, P. J., & McAvoy, S. (1995). Nutrient-independent and nutrient-dependent factors stimulate protein synthesis in colostrum-fed newborn pigs. *Pediatric Research, 37,* 593–599.

Burrin, D. G., Fiorotto, M. L., & Hadsell, D. L. (1999). Transgenic hypersecretion of des(1-3) human insulin-like growth factor I in mouse milk has limited effects on the gastrointestinal tract in suckling pups. *Journal of Nutrition, 129,* 51–56.

Burrin, D. G., Stoll, B., Fan, M. Z., Dudley, M. A., Donovan, S. M., & Reeds, P. J. (2001). Oral IGF-I alters the posttranslational processing but not the activity of lactase-phlorizin hydrolase in formula-fed neonatal pigs. *Journal of Nutrition, 131,* 2235–2241.

Bush, L. J., & Staley, T. E. (1980). Absorption of colostral immunoglobulins in newborn calves. *Journal of Dairy Science, 63,* 672–680.

Butler, J. E. (1974). Immunoglobulins of the mammary secretions. In B. L. Larson (Ed.), *Lactation: A Comprehensive Treatise* (pp. 217–256). New York: Academic Press.

Butler, W. R., & Des Bordes, C. K. (1980). Radioimmunoassay technique for measuring cortisol in milk. *Journal of Dairy Science, 63,* 474–477.

Campana, W. M., & Baumrucker, C. R. (1995). Hormones and growth factors in bovine milk. In R. G. Jensen (Ed.), *Handbook of Milk Composition* (pp. 476–494). San Diego: Academic Press.

Ceppi, A., & Blum, J. W. (1994). Effects of growth hormone on growth performance, haematology, metabolites and hormones in iron-deficient veal calves. *Journal of Veterinary Medicine A, 41,* 443–458.

Cheli, F., Politis, I., Rossi, L., Fusi, E., & Baldi, A. (2003). Effects of retinoids on proliferation and plasminogen activator expression in a bovine mammary epithelial cell line. *Journal of Dairy Research, 70,* 367–372.

Chelikani, P. K., Glimm, D. R., & Kennelly, J. J. (2003). Short communication: Tissue distribution of leptin and leptin receptor mRNA in the bovine. *Journal of Dairy Science, 86*, 2369–2372.

Chilliard, Y., Bonnet, M., Delavaud, C., Faulconnier, Y., Leroux, C., Djiane, J., & Bocquier, F. (2001). Leptin in ruminants. Gene expression in adipose tissue and mammary gland, and regulation of plasma concentration. *Domestic Animal Endocrinology, 21*, 271–295.

Christen, S., Cattin, I., Knight, I., Wineyard, P. G., Blum, J. W., & Elsasser, T. (2007). High plasma nitrite/nitrate levels in neonatal calves are associated with high plasma levels of S-nitrosoalbumin and other S-nitrosothiols. *Experimental Biology and Medicine, 232*, 309–322.

Cordano, P., Hammon, H., & Blum, J. W. (1998). Tissue distribution of insulin-like growth factor-I mRNA in 8-day-old calves. In J. W. Blum, T. Elsasser, and P. Guilloteau (Eds.), *Symposium on Growth in Ruminants: Basic Aspects, Theory and Practice for the Future*, Bern, p. 288.

Cordano, P., Hammon, H. M., Morel, C., Zurbriggen, A., & Blum, J. W. (2000). mRNA of insulin-like growth factor (IGF) quantification and presence of IGF binding proteins, and receptors for growth hormone, IGF-I and insulin, determined by reverse transcribed polymerase chain reaction, in the liver of growing and mature male cattle. *Domestic Animal Endocrinology, 19*, 191–208.

Cox, D. A., & Burk, R. R. (1991). Isolation and characterisation of milk growth factor, a transforming-growth-factor-b2-related polypeptide from bovine milk. *European Journal of Immunology, 197*, 353–358.

Coxam, V., Bauchart, D., Durand, D., Davicco, M.-J., Opmeer, F., & Barlet, J. P. (1989). Nutrient effects on the hepatic production of somatomedin C (IGF1) in the milk-fed calf. *British Journal of Nutrition, 62*, 425–437.

Cummins, A. G., & Thompson, F. M. (2002). Effect of breast milk and weaning on epithelial growth of the small intestine of humans. *Gut, 51*, 748–754.

David, C. W., Norrman, J., Hammon, H. M., Davis, W. C., & Blum, J. W. (2003). Cell proliferation, apoptosis, and B- and T-lymphocytes in Peyer's patches of the ileum, in thymus and in lymph nodes of preterm calves, and in full-term calves at birth and on day 5 of life. *Journal of Dairy Science, 86*, 3321–3329.

Donovan, S. M., & Odle, J. (1994). Growth factors in milk as mediators of infant development. *Annual Review of Nutrition, 14*, 147–167.

Dvorak, B., Holubec, H., Lebouton, A. V., Wilson, J. M., & Koldovsky, O. (1994). Epidermal growth factor and transforming growth factor-α mRNA in rat small intestine: *In situ* hybridization study. *FEBS Letters, 352*, 291–295.

Egli, C. P., & Blum, J. W. (1998). Clinical, haematological, metabolic and endocrine traits during the first three months of life of suckling Simmentaler calves held in a cow-calf operation. *Journal of Veterinary Medicine A, 45*, 99–118.

Ehrhardt, R. A., Greenwood, P. L., Bell, A. W., & Boisclair, Y. R. (2003). Plasma leptin is regulated predominantly by nutrition in preruminant lambs. *Journal of Nutrition, 133*, 4196–4201.

Elmlinger, M. W., Grund, R., Buck, M., Wollmann, H. A., Feist, N., Weber, M. M., Speer, C. P., & Ranke, M. B. (1999). Limited proteolysis of the IGF binding protein-2 (IGFBP-2) by a specific serine protease activity in early breast milk. *Pediatric Research, 46*, 76–81.

Follingstad, A. H. (1978). Estriol, the forgotten estrogen? *Journal of the American Medical Association, 239*, 29–30.

Fowlkes, J. L., & Serra, D. M. (1996). Characterization of glycosaminoglycan-binding domains present in insulin-like growth factor-binding protein-3. *Journal of Biological Chemistry, 271*, 14676–14679.

Fritsche, S., & Steinhart, H. (1999). Occurrence of hormonally active compounds in food: A review. *European Food Research and Technology, 209*, 153–179.

Garre, C., Bianchi-Scarra, G., Sirito, M., Musso, M., & Ravazzolo, R. (1992). Lactoferrin binding sites and nuclear localization in K562(S) cells. *Journal of Cell Physiology, 153*, 477–482.

Georgiev, I. P., Georgieva, T. M., Pfaffl, M., Hammon, H. M., & Blum, J. W. (2003). Insulin-like growth factor and insulin receptors in intestinal mucosa of neonatal calves. *Journal of Endocrinology, 176*, 121–132.

Georgieva, T. M., Georgiev, I. P., Ontsouka, E., Hammon, H. M., Pfaffl, M., & Blum, J. W. (2003). Expression of insulin-like growth factors (IGF)-I and -II and of receptors for growth hormone, IGF-I, IGF-II, and insulin in the intestine and liver of pre- and full-term calves. *Journal of Animal Science, 81*, 2294–2300.

Gibson, C. A., Fligger, J. M., & Baumrucker, C. R. (1998). Specific insulin-like growth factor binding proteins-3 binding to membrane proteins of bovine mammary epithelial cells. *Endocrine Society, 80th Annual Meeting Abstract*, pp. P2–294.

Gibson, C. A., Staley, M. D., & Baumrucker, C. R. (1999). Identification of IGF binding proteins in bovine milk and the demonstration of IGFBP-3 synthesis and release by bovine mammary epithelial cells. *Journal of Animal Science, 77*, 1547–1557.

Gopal, P. K., & Gill, H. S. (2000). Oligosaccharides and glycoconjugates in bovine milk and colostrum. *British Journal of Nutrition, 84 (Suppl 1)*, S69–S74.

Grongnet, J. F., Grongnet-Pinchon, E., & Witowski, A. (1985). Neonatal levels of plasma thyroxine in male and female calves fed a colostrum or immunoglobulin diet or fasted for the first 28 hours of life. *Reproduction, Nutrition and Development, 25*, 537–543.

Grosvenor, C. E., Picciano, M. F., & Baumrucker, C. R. (1993). Hormones and growth factors in milk. *Endocrine Reviews, 14*, 710–728.

Grütter, R., & Blum, J. W. (1991a). Insulin and glucose in neonatal calves after peroral insulin and intravenous glucose administration. *Reproduction, Nutrition and Development, 31*, 389–397.

Grütter, R., & Blum, J. W. (1991b). Insulin-like growth factor I in neonatal calves fed colostrum or whole milk and injected with growth hormone. *Journal of Animal Physiology and Animal Nutrition, 66*, 231–239.

Guilloteau, P., le Huerou-Luron, I., Toullec, R., Chayvialle, J. A., Zabielski, R., & Blum, J. W. (1997). Gastrointestinal regulatory peptides and growth factors in young cattle and sheep. *Journal of Veterinary Medicine A, 44*, 1–23.

Hadorn, U., Hammon, H., Bruckmaier, R. M., & Blum, J. W. (1997). Delaying colostrum intake by one day has important effects on metabolic traits and on gastrointestinal and metabolic hormones in neonatal calves. *Journal of Nutrition, 127*, 2011–2023.

Hammer, C. J., Quigley, J. D., Ribeiro, L., & Tyler, H. D. (2005). Characterization of a colostrum replacer and a colostrum supplement containing IgG concentrate and growth factors. *Journal of Dairy Science, 87*, 106–111.

Hammon, H., & Blum, J. W. (1997a). Prolonged colostrum feeding enhances xylose absorption in neonatal calves. *Journal of Animal Science, 75*, 2915–2919.

Hammon, H., & Blum, J. W. (1997b). The somatotropic axis in neonatal calves can be modulated by nutrition, growth hormone, and Long-R^3-IGF-I. *American Journal of Physiology, 273*, E130–E138.

Hammon, H. M., & Blum, J. W. (1998). Metabolic and endocrine traits of neonatal calves are influenced by feeding colostrum for different durations or only milk replacer. *Journal of Nutrition, 128*, 624–632.

Hammon, H. M., & Blum, J. W. (1999). Free amino acids in plasma of neonatal calves are influenced by feeding colostrum for different durations or by feeding only milk replacer. *Journal of Animal Physiology and Animal Nutrition, 82*, 193–204.

Hammon, H. M., & Blum, J. W. (2002). Feeding different amounts of colostrum or only milk replacer modify receptors of intestinal insulin-like growth factors and insulin in neonatal calves. *Domestic Animal Endocrinology, 22*, 155–168.

Hammon, H. M., Zanker, I. A., & Blum, J. W. (2000). Delayed colostrum feeding affects IGF-I and insulin plasma concentrations in neonatal calves. *Journal of Dairy Science, 83,* 85–92.

Houle, V. M., Park, Y. K., Laswell, S. C., Freund, G. G., Dudley, M. A., & Donovan, S. M. (2000). Investigation of three doses of oral insulin-like growth factor-I on jejunal lactase phlorizin hydrolase activity and gene expression and enterocyte proliferation and migration in piglets. *Pediatric Research, 48,* 497–503.

Howarth, G. S. (2003). Insulin-like growth factor-I and the gastrointestinal system: Therapeutic indications and safety implications. *Journal of Nutrition, 133,* 2109–2112.

Iacopetta, B. J., Grieu, F., Horisberger, M., & Sunahara, G. I. (1992). Epidermal growth factor in human and bovine milk. *Acta Paediatrica Scandinavica, 81,* 287–291.

Janowski, T., Zdunczyk, S., Malecki-Tepicht, J., Baranski, W., & Ras, A. (2002). Mammary secretion of oestrogens in the cow. *Domestic Animal Endocrinology, 23,* 125–137.

Jaques, G., Noll, K., Wegmann, B., Witten, S., Kogan, E., Radulescu, R. T., & Havemann, K. (1997). Nuclear localization of insulin-like growth factor binding protein 3 in a lung cancer cell line. *Endocrinology, 138,* 1767–1770.

Juul, S. E. (2003). Enterally dosed recombinant human erythropoietin does not stimulate erythroopoiesis in neonates. *Journal of Pediatrics, 143,* 321–326.

Juul, S. E., Zhao, Y., Dame, J. B., Du, Y., Hutson, A. D., & Christensen, R. D. (2000). Origin and fate of erythropoietin in human milk. *Pediatric Research, 48,* 660–667.

Juul, S. E., Ledbetter, D. J., Joyce, A. E., Dame, C., Christensen, R. D., Zhao, Y., & DeMarco, V. (2001). Erythropoietin acts as a trophic factor in neonatal rat intestine. *Gut, 49,* 182–189.

Kacsóh, B., Toth, B. E., Avery, L. M., Deaver, D. R., Baumrucker, C. R., & Grosvenor, C. E. (1991). Biological and immunological activities of glycosylated and molecular weight variants of bovine prolactin in colostrum and milk. *Journal of Animal Science, 69 (Suppl),* 456.

Kelly, D. (1994). Colostrum, growth factors and intestinal development in pigs. In W.-B. Souffrat & H. Hagemeister (Eds.), *International Symposium on Digestive Physiology in Pigs* (pp. 151–166). Dummerstorf, Germany: EAAP Publication #80.

King, P. J., Sullivan, T. M., Roberts, R. A., Philipps, A. F., & Koldovsky, O. (1998). Human milk as a potential enteral source of erythropoietin. *Pediatric Research, 43,* 216–221.

Kinsbergen, M., Sallmann, H. P., & Blum, J. W. (1994). Metabolic, endocrine and hematological responses to normal feeding, total parenteral nutrition and fasting in one-week old calves. *Journal of Veterinary Medicine A, 41,* 268–282.

Klagsbrun, M. (1978). Human milk stimulates DNA synthesis and cellular proliferation in cultured fibroblasts. *Proceedings of the National Academy of Sciences USA, 75,* 5057–5061.

Koldovsky, O. (1989). Hormones in milk: Their possible physiological significance for the neonate. In E. Ledbenthal (Ed.), *Textbook of Gastroenterology and Nutrition in Infancy* (pp. 97–119). New York: Raven Press.

Koldovsky, O. (1995). Hormones in milk. *Vitamins and Hormones, 50,* 77–149.

Koyama, S., & Podalsky, D. K. (1989). Differential expression of transforming growth factors a and b in rat intestinal epithelial cells. *Journal of Clinical Investigation, 83,* 1768–1773.

Krüger, K., Blum, J. W., & Greger, D. L. (2005). Abundances of nuclear receptor and nuclear receptor target gene mRNA in liver and intestine of neonatal calves are differentially influenced by feeding colostrum and vitamin A. *Journal of Dairy Science, 88,* 3971–3981.

Kühne, S., Hammon, H. M., Bruckmaier, R. M., Morel, C., Zbinden, Y., & Blum, J. W. (2000). Growth performance, metabolic and endocrine traits, and absorptive capacity in neonatal calves fed either colostrum or milk replacer at two levels. *Journal of Animal Science, 78,* 609–620.

Legrand, D., Vigie, K., Said, E. A., Elass, E., Masson, M., Slomianny, M.-C., Carpentier, M., Briand, J.-P., Mazurier, J., & Hovanessian, A. G. (2004). Surface nucleolin participates in both the binding and endocytosis of lactoferrin in target cells. *European Journal of Biochemistry, 271,* 203–217.

Levi-Montalcini, R., & Cohen, S. (1960). Effects of the extract of the mouse submaxillary salivary glands on the sympathetic system of mammals. *Annals of the New York Academy of Sciences, 85,* 324–341.

Louveau, I., & Gondret, F. (2004). Regulation of development and metabolism of adipose tissue by growth hormone and the insulin-like growth factor system. *Domestic Animal Endocrinology, 27,* 241–255.

Mahle, Z. (2004). PPAR trilogy from metabolism to cancer. *Current Opinion in Clinical Nutrition and Metabolic Care, 7,* 397–402.

Malekinejad, H., Scherpenisse, P., & Bergwerff, A. A. (2006). Naturally occurring estrogens in processed milk and in raw milk (from gestated cows). *Journal of Agricultural Food Chemistry, 54,* 9785–9791.

Malven, P. V., Head, H. H., Collier, R. J., & Buonomo, F. C. (1987). Periparturient changes in secretion and mammary uptake of insulin and in concentrations of insulin and insulin-like growth factors in milk of dairy cows. *Journal of Dairy Science, 70,* 2254–2265.

Massague, J. (1998). TGF β signal transduction. *Annual Review of Biochemistry, 67,* 753–791.

Maule Walker, F. M., Davis, A. J., & Fleet, I. R. (1983). Endocrine activity of the mammary gland: Oestrogen and prostaglandin secretion by the cow and sheep mammary glands during lactogenesis. *British Veterinary Journal, 139,* 171–177.

McGarrigle, H. H. G., & Lachelin, G. C. L. (1983). Oestrone, oestradiol and oestriol glucosiduronates and sulphates in human puerperal plasma milk. *Journal of Steroid Biochemistry, 18,* 607–611.

Michanek, P., Ventorp, M., & Westrom, B. (1989). Intestinal transmission of macromolecules in newborn dairy calves of different ages at first feeding. *Research in Veterinary Science, 46,* 375–379.

Miettien, P. J., Berger, J. E., Meneses, J., Phung, Y., Pedersen, R. A., Werb, Z., & Derynck, R. (1995). Epithelial immaturity and multiorgan failure in mice lacking epidermal growth factor receptor. *Nature, 376,* 337–341.

Miller-Gilbert, A. L., Dubuque, S. H., Dvorak, B., Williams, C. S., Grille, J. G., Woodward, S. S., Koldovsky, O., & Kling, P. J. (2001). Enteral absorption of erythropoietin in the suckling rat. *Pediatric Research, 50,* 261–267.

Mindnich, R., Moller, G., & Adamski, J. (2004). The role of 17 β-hydroxysteroid dehydrogenases. *Molecular and Cellular Endocrinology, 218,* 7–20.

Mohan, R., & Heyman, R. A. (2003). Orphan nuclear receptor modulators. *Current Topics in Medical Chemistry, 3,* 1637–1647.

Molkentin, J. (2000). Occurrence and biochemical characteristics of natural bioactive substances in bovine milk lipids. *British Journal of Nutrition, 84 (Suppl 1),* S47–S53.

Montaner, B., Asbert, M., & Perez-Tomas, R. (1999). Immunolocalization of transforming growth factor- α and epidermal growth factor receptor in the rat gastroduodenal area. *Digestive Diseases and Sciences, 44,* 1408–1416.

Muri, C., Schottstedt, T., Hammon, H. M., Meyer, E., & Blum, J. W. (2005). Hematological, metabolic, and endocrine effects of feeding vitamin A and lactoferrin in neonatal calves. *Journal of Dairy Science, 88,* 1062–1077.

Norrman, J., David, C. W., Sauter, S. N., Hammon, H. M., & Blum, J. W. (2003). Effects of dexamethasone on lymphoid tissue in the gut and thymus of neonatal calves fed with colostrum or milk replacer. *Journal of Animal Science, 81,* 2322–2332.

Nussbaum, A., Schiessler, G., Hammon, H. M., & Blum, J. W. (2002). Growth performance and metabolic and endocrine traits in calves pair-fed by bucket or by automate starting in the neonatal period. *Journal of Animal Science, 80,* 1545–1555.

Oda, S., Satoh, H., Sugawara, T., Matsunaga, N., Kuhara, T., Katoh, K., Shoji, Y., Nihei, A., Ohta, M., & Sasaki, Y. (1989). Insulin-like growth factor-I, GH, insulin and glucagon concentrations in bovine colostrum and in plasma of dairy cows and neonatal calves around parturition. *Comparative Biochemistry and Physiology A*, *94*, 805–808.

Ogawa, J., Sasahara, A., Taketoshi, Y., Sira, M. M., Futatani, T., Kanegane, H., & Miyawaki, T. (2004). Role of transforming growth factor-b in breast milk for initiation of IgA production in newborn infants. *Early Human Development*, *77*, 67–75.

Ontsouka, E. C., Bruckmaier, R. M., & Blum, J. W. (2003). Fractionized milk composition during removal of colostrum and mature milk. *Journal of Dairy Science*, *86*, 2005–2011.

Ontsouka, E. C., Hammon, H. M., & Blum, J. W. (2004a). Expression of insulin-like growth factors (IGF)-1 and -2, IGF-binding proteins-2 and -3, and receptors for growth hormone, IGF type-1 and -2 and insulin in the gastrointestinal tract of neonatal calves. *Growth Factors*, *22*, 63–69.

Ontsouka, E. C., Philipona, C., Hammon, H. M., & Blum, J. W. (2004b). Abundance of mRNA encoding for components of the somatotropic axis and insulin receptor in different layers of the jejunum and ileum of neonatal calves. *Journal of Animal Science*, *82*, 3181–3188.

Ontsouka, C. E., Sauter, S. N., Blum, J. W., & Hammon, H. M. (2004c). Effects of colostrum feeding and dexamethasone treatment on mRNA levels of insulin-like growth factors (IGF)-I and -II, IGF binding proteins-2 and -3, and on receptors for growth hormone, IGF-I, IGF-II, and insulin in the gastrointestinal tract of neonatal calves. *Domestic Animal Endocrinology*, *26*, 155–175.

Pakkanen, R. (1998). Determination of transforming growth factor-beta 2 (TFG- β2) in bovine colostrum samples. *Journal of Immunoassay and Immuno-chemistry*, *19*, 23–37.

Pfaffl, M. W., Georgieva, T. M., Georgiev, I. P., Ontsouka, E., Hageleit, M., & Blum, J. W. (2002). Real-time RT-PCR quantification of insulin-like growth factor (IGF)-1, IGF-1 receptor, IGF-2, IGF-2 receptor, insulin receptor, growth hormone receptor, IGF-binding proteins 1, 2, and 3 in the bovine species. *Domestic Animal Endocrinology*, *22*, 91–102.

Piatak, M., Jr., Luk, K.-C., Williams, B., & Lifson, J. D. (1993). Quantitative competitive polymerase chain reaction for accurate quantitation of HIV DNA and RNA species. *Bio/Technology*, *14*, 70–78.

Pinotti, L., & Rosi, F. (2006). Leptin in bovine colostrum and milk. *Hormone and Metabolic Research*, *38*, 89–93.

Plaut, K. (1993). Role of epidermal growth factor and transforming growth factors in mammary development and lactation. *Journal of Dairy Science*, *76*, 1526–1538.

Plowchalk, D. R., & Teeguarden, J. (2002). Development of a physiologically based pharmacokinetic model for estradiol in rats and humans: A biologically motivated quantitative framework for evaluating responses to estradiol and other endocrine-active compounds. *Toxicological Science*, *69*, 60–78.

Pollack, P. F., Goda, T., & Colony, P. C. (1987). Effects of enterally fed epidermal growth factor on the small and large intestine of the suckling rat. *Regulatory Peptides*, *17*, 121–132.

Pope, G. S., & Roy, J. H. (1953). The oestrogenic activity of bovine colostrum. *Biochemical Journal*, *53*, 427–430.

Prgomet, C., Prenner, M. L., Schwarz, F. J., & Pfaffl, M. W. (2006). Effect of lactoferrin on selected immune system parameters and the gastrointestinal morphology in growing calves. *Journal of Animal Physiology and Animal Nutrition*, DOI:10.1111/j.1439-0396.2006.00649.x.

Purup, S., Jensen, S. K., & Sejrsen, K. (2001). Differential effects of retinoids on proliferation of bovine mammary epithelial cells in collagen gel culture. *Journal of Dairy Research*, *68*, 157–164.

Puvogel, G., Baumrucker, C. R., Sauerwein, H., Rühl, R., Ontsouka, E., Hammon, H. M., & Blum, J. W. (2005). Effects of an enhanced vitamin A intake during the dry period on retinoids, lactoferrin, IGF-system, mammary gland epithelial cell apoptosis and subsequent lactation in dairy cows. *Journal of Dairy Science*, *88*, 1785–1800.

Quigley, J. D., & Drewry, J. J. (1998). Nutrient and immunity transfer from cow to calf pre- and postcalving. *Journal of Dairy Science, 81*, 2779–2790.

Rauprich, A. B., Hammon, H. M., & Blum, J. W. (2000a). Effects of feeding colostrum and a formula with nutrient contents as colostrum on metabolic and endocrine traits in neonatal calves. *Biology of the Neonate, 78*, 53–64.

Rauprich, A. B., Hammon, H. M., & Blum, J. W. (2000b). Influence of feeding different amounts of first colostrum on metabolic, endocrine, and health status and on growth performance in neonatal calves. *Journal of Animal Science, 78*, 896–908.

Rérat, M., Zbinden, Y., Saner, R., Hammon, H., & Blum, J. W. (2005). *In vitro* embryo production: Growth performance, feed efficiency, and hematological, metabolic, and endocrine status in calves. *Journal of Dairy Science, 88*, 2579–2593.

Roberts, A. B. (1998). Molecular and cell biology of TGF-β. *Mineral and Electrolyte Metabolism, 24*, 111–119.

Roberts, A. B., & Sporn, M. B. (1992). Differential expression of the TGF-β isoforms in embryogenesis suggests specific roles in developing and adult tissues. *Molecular Reproduction and Development, 32*, 91–98.

Roelen, B. A., Van Eijk, M. J., Van Rooijen, M. A., Bevers, M. M., Larson, J. H., Lewin, H. A., & Mummery, C. L. (1998). Molecular cloning, genetic mapping, and developmental expression of a bovine transforming growth factor beta (TGF-β) type I receptor. *Molecular Reproduction and Development, 49*, 1–9.

Roffler, B., Fäh, A., Sauter, S. N., Hammon, H. M., Gallmann, P., Brem, G., & Blum, J. W. (2003). Intestinal morphology, epithelial cell proliferation, and absorptive capacity in neonatal calves fed milk-born insulin-like growth factor-I or a colostrum extract. *Journal of Dairy Science, 86*, 1797–1806.

Rogers, M. L., Belford, D. A., Francis, G. L., & Ballard, F. J. (1995). Identification of fibrobla growth factors in bovine cheese whey. *Journal of Dairy Research, 62*, 501–507.

Ronge, H., & Blum, J. W. (1988). Insulin-like growth factor I binding proteins in dairy cows, calves and bulls. *Acta Endocrinologica, 121*, 153–160.

Rufibach, K., Stefanoni, N., Rey-Roethlisberger, V., Schneiter, P., Doherr, M. G., Tappy, L., & Blum, J. W. (2006). Protein synthesis in jejunum and liver of neonatal calves fed vitamin A and lactoferrin. *Journal of Dairy Science, 89*, 3075–3086.

Sangild, P. T. (2001). Transitions in the life of the gut and brain. In J. E. Lindberg & B. Ogle (Eds.), *Digestive Physiology of Pigs* (pp. 3–17). Wallingford, UK: CABI Publishing.

Sauter, S. N., Ontsouka, E., Roffler, B., Zbinden, Y., Philipona, C., Pfaffl, M., Breier, B. H., Blum, J. W., & Hammon, H. M. (2003). Effects of dexamethasone and colostrum intake on the somatotropic axis in neonatal calves. *American Journal of Physiology, 285*, E252–E261.

Sauter, S. N., Roffler, B., Philipona, C., Morel, C., Rome, V., Guilloteau, P., Blum, J. W., & Hammon, H. M. (2004). Intestinal development in neonatal calves: Effects of glucocorticoids and dependence of colostrum feeding. *Biology of the Neonate, 85*, 94–104.

Schaudies, R. P., Grimes, J., Davis, D., Rao, R. K., & Koldovsky, O. (1989). EGF content in the gastrointestinal tract of rats: Effect of age and fasting/feeding. *American Journal of Physiology, 256*, G856–G861.

Schlimme, E., Martin, D., & Meisel, H. (2000). Nucleosides and nucleotides: Natural bioactive substances in milk colostrum. *British Journal of Nutrition, 84 (Suppl 1)*, S59–S68.

Schottstedt, T., Muri, C., Morel, C., Philipona, C., Hammon, H. M., & Blum, J. W. (2005). Effects of feeding vitamin A and lactoferrin on epithelium of lymphoid tissues of intestine of neonatal calves. *Journal of Dairy Science, 88*, 1050–1061.

Scott, P., Kessler, M. A., & Schuler, L. A. (1992). Molecular cloning of the bovine prolactin receptor and distribution of prolactin and growth hormone receptor transcripts in fetal and utero-placental tissues. *Molecular and Cellular Endocrinology, 89*, 47–58.

Semba, R. D., & Juul, S. E. (2002). Erythropoietin in human milk: Physiology and role in infant health. *Journal of Human Lactation, 18*, 252–261.

Shi, Y., & Massague, J. (2003). Mechanisms of TGF-β signaling from cell membrane to the nucleus. *Cell, 113*, 685–700.

Shing, Y. W., & Klagsbrun, M. (1984). Human and bovine milk contain different sets of growth factors. *Endocrinology, 115*, 273–282.

Shutt, D. A., & Fell, L. R. (1985). Comparison of total and free cortisol in bovine serum and milk or colostrum. *Journal of Dairy Science, 68*, 1832–1834.

Sibilia, M., & Wagner, E. F. (1995). Strain-dependent epithelial defects in mice lacking the EGF receptor. *Science, 269*, 234–238.

Skaar, T. C., Baumrucker, C. R., Deaver, D. R., & Blum, J. W. (1994). Diet effects and ontogeny of alterations of circulating insulin-like growth factor binding proteins in newborn dairy calves. *Journal of Animal Science, 72*, 421–427.

Sparks, A. L., Kirkpatrick, J. G., Chamberlain, C. S., Waldner, D., & Spicer, L. J. (2003). Insulin-like growth factor-I and its binding proteins in colostrum compared to measures in serum of Holstein neonates. *Journal of Dairy Science, 86*, 2022–2029.

Suliman, H. B., Logan-Henfrey, L., Majiwa, P. A. O., ole-Moiyoi, O., & Feldman, B. F. (1999). Analysis of erythropoietin and erythropoietin receptor genes expression in cattle during acute infection with *Trypanosoma congolense*. *Experimental Hematology, 27*, 37–45.

Suzuki, Y. A., Shin, K., & Lonnerdal, B. (2001). Molecular cloning and functional expression of a human intestinal lacctoferrin receptor. *Biochemistry, 40*, 15771–15779.

Talukder, M. J., Takeuchi, T., & Harada, E. (2003). Receptor-mediated transport of lactoferrin into the cerebrospinal fluid via plasma in young calves. *Journal of Veterinary Medical Science, 65*, 957–964.

Teramo, K. A., Widness, J. A., Clemons, G. K., Voutilainen, P., McKinlay, S., & Schwartz, R. (1987). Amniotic fluid erythropoietin correlates with umbilical plasma erythropoietin in normal and abnormal pregnancy. *Obstetrics and Gynecology, 69*, 710–716.

Vacher, P. Y., Bestetti, G., & Blum, J. W. (1995). Insulin-like growth factor I absorption in the jejunum of neonatal calves. *Biology of the Neonate, 68*, 354–367.

Valentinis, B., & Baserga, R. (1996). The IGF-I receptor protects tumor cells from apoptosis induced by high concentrations of serum. *Biochemical and Biophysical Research Communications, 224*, 362–368.

Van der Strate, B. W. A., Beljaars, L., Molema, G., Harmsen, M. C., & Meijer, D. K. F. (2001). Antiviral activities of lactoferrin. *Antiviral Research, 52*, 225–239.

Vega, J. R., Gibson, C. A., Skaar, T. C., Hadsell, D. L., & Baumrucker, C. R. (1991). Insulin-like growth factor (IGF)-I and -II and IGF binding proteins in serum and mammary secretions during the dry period and early lactation in dairy cows. *Journal of Animal Science, 69*, 2538–2547.

Wallace, L. E., Hardin, J. A., & Gall, D. G. (2001). Expression of EGF and ERBB receptor proteins in small intestinal epithelium. *Gastroenterology, 120*, A511.

Weinzimer, S. A., Gibson, T. B., Collet-Solberg, P. F., Khare, A., Liu, B., & Cohen, P. (2001). Transferrin is an insulin-like growth factor-binding protein-3 binding protein. *Journal of Clinical Endocrinology and Metabolism, 86*, 1806–1813.

Widdowson, E. M., Colombo, V. E., & Artavanis, C. A. (1976). Changes in the organs of pigs in response to feeding for the first 24 hours afer birth. *Biology of the Neonate, 28*, 272–281.

Wise, A., Jupe, S. C., & Rees, S. (2004). The identification of ligands at orphan G-protein coupled receptors. *Annual Review of Pharmacology and Toxicology, 44*, 43–66.

Wolinski, J., Biernat, M., Guilloteau, P., Westrom, B. R., & Zabielski, R. (2003). Exogenous leptin controls the development of the small intestine in neonatal piglets. *Journal of Endocrinology, 177*, 215–222.

Xiao, X., Xiong, A., Chen, X., Mao, X., & Zhou, X. (2002). Epidermal growth factor concentrations in human milk, cow's milk and cow's milk-based infant formulas. *Chinese Medical Journal, 115*, 451–454.

Xu, R. J., Sangild, P. T., Zhang, Y. Q., & Zhang, S. H. (2002). Bioactive compounds in porcine colostrum and milk and their effects on intestinal development in neonatal pigs. In *Biology of the Intestine in Growing Animals* (pp. 169–192). Amsterdam: Elsevier.

Zanker, I. A., Hammon, H. M., & Blum, J. W. (2000a). Beta-carotene, retinol and alpha-tocopherol status in calves fed the first colostrum at 0–2, 6–7, 12–13 or 24–25 hours after birth. *International Journal of Vitamin and Nutrition Research, 70,* 305–310.

Zanker, I. A., Hammon, H. M., & Blum, J. W. (2000b). Plasma amino acid pattern during the first month of life in calves fed the first colostrum at 0–2 h or at 24–25 h after birth. *Journal of Veterinary .Medicine A, 47,* 107–121.

Zanker, I. A., Hammon, H. M., & Blum, J. W. (2001a). Delayed feeding of first colostrum: Are there prolonged effects on haematological, metabolic and endocrine parameters and on growth performance in calves? *Journal of Animal Physiology and Animal Nutrition, 85,* 53–66.

Zanker, I. A., Hammon, H. M., & Blum, J. W. (2001b). Activities of γ-glutamyltransferase, alkaline phosphatase and aspartate-aminotransferase in colostrum, milk and blood plasma of calves fed first colostrum at 0–2, 6–7, 12–13 and 24–25 h after birth. *Journal of Veterinary Medicine A, 48,* 179–185.

Zapf, J. (1995). Physiological role of the insulin-like growth factor binding proteins. *European Journal of Endocrinology, 132,* 645–654.

Probiotics, Immunomodulation, and Health Benefits

Harsharn Gill[1] and Jaya Prasad[2]

Abstract Probiotics are defined as live microorganisms that, when administered in adequate amount, confer a health benefit on the host. Amongst the many benefits associated with the consumption of probiotics, modulation of the immune system has received the most attention. Several animal and human studies have provided unequivocal evidence that specific strains of probiotics are able to stimulate as well as regulate several aspects of natural and acquired immune responses. There is also evidence that intake of probiotics is effective in the prevention and/or management of acute gastroenteritis and rotavirus diarrhoea, antibiotic-associated diarrhoea and intestinal inflammatory disorders such as Crohn's disease and pouchitis, and paediatric atopic disorders. The efficacy of probiotics against bacterial infections and immunological disorders such as adult asthma, cancers, diabetes, and arthritis in humans remains to be proven. Also, major gaps exist in our knowledge about the mechanisms by which probiotics modulate immune function. Optimum dose, frequency and duration of treatment required for different conditions in different population groups also remains to be determined. Different probiotic strains vary in their ability to modulate the immune system and therefore efficacy of each strain needs to be carefully demonstrated through rigorously designed (randomised, double-blind, placebo-controlled) studies. This chapter provides an over view of the immunomodulatory effects of probiotics in health and disease, and discusses possible mechanisms through which probiotics mediate their disparate effects.

---

[1] Department of Primary Industries, Werribee, Victoria 3030, Australia
e-mail: Harsharn.Gill@dpi.vic.gov.au
[2] School of Molecular Sciences, Victoria University, PO Box 14428, Melbourne, Victoria 8001, Australia

Introduction

The human gastrointestinal tract harbors a diverse microflora representing several hundred different species. The colonization of the gastrointestinal tract begins immediately after birth. The colonization pattern is affected by factors such as mode of delivery, initial diet, and geographical location (Fanaro et al., 2003). In breastfed infants, between days 4 and 7, bifidobacteria become predominant, accumulating to 10^{10}–10^{11} CFU/g. Thus, nearly 100% of all bacteria cultured from stools of breastfed infants are bifidobacteria (Mitsuoka, 1996). During weaning, when an adult diet is consumed, the stools of infants shift to the Gram-negative bacillary flora of adults; bifidobacteria decrease by 1 log, the number of bacteroidaceae, eubacteria, peptostreptococcaceae, and usually Gram-positive clostridia outnumber bifidobacteria, which constitute 5–10% of the total flora. Lactobacilli, megaspherae, and veillonellae are often found in adult feces, but the counts are usually less than 10^7 CFU/g. In elderly persons, bifidobacteria decrease, clostridia significantly increase, as do lactobacilli, streptococci, and enterobacteriaceae (Woodmancey et al., 2004).

It is estimated that the gastrointestinal tract of an adult human contains 10^{13} bacteria, 10 times the number of eukaryotic cells in the body. The density of bacterial colonization increases progressively from the stomach (10^{3-4} CFU/g) to the colon (10^{10-11} CFU/g). Based on their effect on the intestinal environment, these bacteria can be grouped into three categories: beneficial bacteria, harmful bacteria, and bacteria exhibiting an intermediate property. Harmful bacteria are those that possess pathogenicity or transform food components into harmful substances (ammonia, amines, hydrogen sulfide, and indole from proteins) and include *Clostridium, Veillonella, Proteus,* and the Enterobacteriaceae family. Beneficial bacteria represented by *Bifidobacterium* and *Lactobacillus* suppress the harmful bacteria and exert many beneficial physiological effects. They have no harmful effect on the host. *Bacteroides, Eubacterium,* and anaerobic streptococci belong to the intermediate group. These bacteria do not show any virulence under normal conditions, but they may cause opportunistic infections when the host immunity or resistance is lowered (Ishibashi et al., 1997). Normally, a delicate balance exists among various communities of the intestinal flora and the harmful bacteria remain under check, leading to a healthy state. However, this balance can be altered as a result of many endogenous (nutrient availability, diet, diarrhea, etc.) and exogenous (antibiotic therapy, excessive hygiene, stress, aging, etc.) factors (Suskovic et al., 2001). Disturbances in the intestinal ecosystem are generally characterized by a remarkable increase in bacterial counts in the small intestine, by an increase in the numbers of aerobes, mostly enterobacteriaceae and streptococci, by the reduction or disappearance of bifidobacteria, and/or often by the presence of *Clostridium perfringens* (Mitsuoka, 1992). Recent studies have provided overwhelming evidence that the administration of probiotics could be effective in restoring intestinal microbial balance and gut homeostasis.

Probiotics are defined as live microorganisms that, when administered in adequate amounts, confer a health benefit on the host (FAO/WHO, 2001).

Intestinal Microflora and the Development of the Immune System

The immune system of a newborn is immunologically naïve and functionally immature (Kelly & Coutts, 2000). Exposure to antigens during early life is essential to drive the development of the gut mucosal immune system and to maintain immune homeostasis. Microbial antigens derived from the resident flora and the environment play a pivotal role in the maturation of gut-associated lymphoid tissue (Glaister, 1973; Moreau et al., 1978) and normal resistance to disease (Yamazaki et al., 1982). This has been clearly demonstrated in studies on germ-free mice. Germ-free animals have a poorly developed immune system; they have fewer IgA plasma cells and intraepithelial lymphocytes in the intestinal mucosa, and lower levels of immunoglobulins, compared to their conventionally reared counterparts (Gordon & Bruckner-Kardoss, 1961; Crabbe, 1968; Crabbe et al., 1970; Gordon & Pesti, 1971; Glaister 1973), and exhibit increased susceptibility to disease (Roach & Tannock, 1980; Yamazaki et al., 1982). However, the normal development of the immune system is restored when the germ-free animals are reared in a conventional environment or given normal intestinal microflora (Bauer et al., 1965; Crabbe et al., 1999). The role of the microbial flora in the development and regulation of host immunity is also highlighted by differences in the intestinal microflora and humoral immune responses of vaginally born versus Caesarean-delivered infants whose mothers received prophylactic antibiotics (Gronlund et al., 2000); microbial colonization in vaginally born infants is associated with the maturation of mucosal immune responses, especially circulating IgA- and IgM-secreting cells.

Another important role of intestinal microflora in the induction and maintenance of oral tolerance has also been demonstrated. GF mice fail to develop oral tolerance, whereas reconstitution of the gut microflora in GF mice at the neonatal stage, but not later, leads to the development of normal tolerance (Sudo et al., 1997). A reduced microbial exposure in Western societies has also been associated with an increased incidence of atopic and autoimmune disorders (Rook & Stanford, 1998).

Probiotics and Human Health

Consumption of probiotics is associated with a range of health benefits including stimulation of the immune system, protection against diarrheal disease and nosocomial and respiratory tract infections, lowering of cholesterol, attenuation of overt immunoinflammatory disorders (such as inflammatory bowel disease, allergies), and anticancer effects. The scope of this chapter is limited to the immunomodulatory effects of probiotics only. Readers are advised to see reviews by Guarner and Malagelada (2002), Gill and Guarner (2004), Sullivan and Nord (2006), and Quigley and Flourie (2007) for information on other health benefits of probiotics.

Probiotics and Modulation of Intestinal Microflora

Metchnikoff (1907) first proposed the hypothesis that colon bacteria adversely affect human health by "autointoxication." Further, he proposed that the longevity of Bulgarian peasants was due to the consumption of large quantities of fermented milk containing live beneficial microorganisms. As delineated in earlier sections, the intestinal microbiota impacts markedly on the immunology, biochemistry, physiology, and nonspecific disease resistance of the host (Gordon & Pesti, 1971). These observations have prompted the view that modification of the composition of the intestinal microbiota by means of dietary supplements might promote health (Goldin & Gorbach, 1992; Roberfroid, 1998). Most commonly used probiotic bacteria belong to the genera *Lactobacillus* and *Bifidobacterium* (Prasad et al., 1999). Lactobacilli are Gram-positive, nonspore-forming rods, catalase-negative, and usually nonmotile and do not reduce nitrate. The most frequently used lactobacilli species include *Lb. acidophilus, Lb. salivarius, Lb. casei, Lb. plantarum, Lb. fermentum,* and *Lb. brevis* (Mikelsaar et al., 1998). On the other hand, bifidobacteria are Gram-positive, nonspore-forming rods with distinct cellular bifurcating or club-shaped morphologies. The most commonly used species include *B. animalis, B. longum, B. bifidum,* and *B. infantis.*

In terms of the probiotic dose, it is generally thought that at least 10^9 CFU/day need to be ingested (Ouwehand et al., 2002). In a study aimed at determining the impact of consumption of *Bifidobacterium longum* (10^9 CFU/day) on the fecal flora of healthy adult human subjects, it was found that the fecal levels of lecithinase-negative clostridia were significantly reduced (Benno & Mitsuoka, 1992). In another study, intake of yogurt enriched with *B. longum* was found to significantly increase bifidobacteria counts in the feces of treated subjects compared with subjects given control yogurt (Bartram et al., 1994). Langhendries and co-workers (1995) reported the impact of consuming fermented infant formula containing viable bifidobacteria (10^6 CFU/g of *B. bifidum*) in full-term infants, wherein significant increases in resident bifidobacteria were observed. Fermented milk containing *Lb acidophilus* LA2 was consumed by human adult volunteers for seven days; the resident lactobacilli as well as bifidobacteria increased significantly in the feces (Hosoda et al., 1996). After investigating the impact of consumption of follow-up formula (NAN BF) containing *B. bifidum* strain Bb12 on fecal flora, the authors reported that the resident bifidobacteria significantly increased and the clostridia counts decreased (Fukushima et al., 1997). In a study involving adult human subjects consuming yogurt containing *Lb. acidophilus* and *B. bifidum*, Chen et al. (1999) reported that, after 10-day consumption, the subject fecal analysis displayed significant increase in the resident bifidobacteria and, at the same time, a significant drop in the coliforms counts. In a relatively long-term study (six-month preintervention period, six-month intervention, and three-month postintervention) on the effects of consuming *Lactobacillus rhamnosus* DR20 on the microecology of healthy human subjects, Tannock et al. (2000) reported that the strain DR20 was detected at different levels in different test subjects during the intervention period. The presence of DR20 among numerically predominant strains was related to the presence or absence of a stable indigenous population of lactobacilli during the control period. In addition, it was concluded that consumption of the DR20-containing milk product (1.6×10^9 CFU/day) transiently altered the lactobacilli and enterococci population of the feces of the majority of consumers. In another study, human subjects consumed *B. lactis* HN019 containing (3×10^{10} CFU/day) reconstituted milk for four weeks. At the end of the four weeks, the resident bifidobacteria and lactobacilli content increased significantly, and hence the probiotic was able to transiently impact the gut flora toward a beneficial effect. The probiotic counts in feces reached as high as 12.5×10^8 CFU/g (Gopal et al., 2003).

The effective probiotic dose for desired efficacy has received much attention recently, which probably is a result of both commercial (cost) and scientific interest. In order to determine the effective dose of *B. animalis* subsp. *lactis* Bb12, four doses (10^8, 10^9, 10^{10}, 10^{11} CFU/day) were given to adult volunteers in different groups. The fecal recovery of Bb12 increased significantly with increasing dose; however, the fecal bacterial composition was unaffected (Larsen et al., 2006). In another study, the effective dose of *B. lactis* HN019 that could influence the fecal flora of elderly (mean age: 69.5 years) human subjects was investigated. Three doses (5×10^9; 1×10^9, and 6.5×10^7 CFU/day) were administered in reconstituted milk. The probiotic intervention increased the number of resident bifidobacteria and reduced the enterobacteria. In addition, the enterococci and lactobacilli counts were increased. Even the lowest dose administered influenced the fecal microflora composition in elderly subjects (Ahmed et al., 2007). Elderly gut microecology appears to be more amenable for manipulation with bifidobacteria as the natural levels drop in the elderly. All these studies conclusively provide evidence that probiotic administration, though at a smaller level (10^9 CFU/day into 10^{14} CFU/g gut content) when compared to the total number of microbiota, can influence the intestinal microecology and deliver the desired health benefits.

Probiotics and Stimulation of the Immune System

Several animal and human studies have provided evidence that specific strains of lactic acid bacteria are able to stimulate as well as regulate several aspects of natural and acquired immune responses. It has also been shown that significant differences exist in the ability of bifidobacteria and lactobacilli strains to modulate the immune system and that the responses are dose-dependent. Several excellent reviews on the immunomodulatory effects of probiotics have been published in recent years; readers are advised to consult these for additional information (Gill, 1998, 2003; Erickson & Hubbard, 2000).

Immunological detection of probiotics and probiotic-derived products in the gut is performed by specialized membranous cells (M cells), overlying the Payer's patches and the epithelial cells. Dendritic cells, distributed throughout the subepithelium, have also been shown to have the ability to directly sample lumenal antigens. Antigens taken up by M cells are delivered to antigen-presenting cells (APCs) that process and present antigens to naïve T cells. APCs are able to discriminate between closely related microbes and their products through the expression of pattern-recognition receptors (e.g., TLRs and CD14) that recognize pathogen-associated molecular patterns (PRRs). The nature of cytokine secretion, phenotype, and state of activation of APCs determine whether T cells differentiate into T helper 1 (Th1), T helper 2 (Th2), or T regulatory (Treg) cells. Subsequent activation of Th1 cells leads to the production of IFN-γ, TNF-α, and IL-2 and is associated with the development of cell-mediated and cytoxic immunity; activated Th2 cells mainly secrete IL-4, IL-5, and IL-13, which promote antibody production and are associated with atopy; Treg cells secrete IL-10 and TGF-β, which downregulate activities of both Th1 and Th2 cells.

Innate (Nonspecific) Immunity

The innate responses constitute the first line of host defense and operate nonselectively against pathogens/abnormal antigens. The major cellular effectors of nonspecific immunity include epithelial cells, phagocytic cells (monocytes, macrophages, neutrophils), and natural killer cells (NK cells). Probiotics have been found to modulate the functions of all these cells.

Effect on Phagocytic and NK Cell Activity

Phagocytic cells are effective in eliminating microbial pathogens, whereas NK cells are crucial for defense against viral infections and cancers. The ability of probiotics to enhance the phagocytic activity of peripheral blood leucocytes (monocytes/macrophages and PMN) has been demonstrated in a number of human studies (Gill, 2003). Intake of *Lb. johnsonii* La1, *B. lactis* Bb12, *L. rhamnosus* HN001, or *B. lactis* HN019 resulted in the enhanced phagocytic capacity of peripheral blood leukocytes (PMN and monocytes) in healthy subjects (Schiffrin et al., 1995; Donnet-Hughes et al., 1999). The PMNs exhibited significantly greater improvement in phagocytic capacity compared with monocytes. The increases in phagocytic activity were dose-dependent (Donnet-Hughes, 1999) and were maintained for several weeks after cessation of probiotic intake (Schiffrin et al., 1995; Gill et al., 2001a, b). In another study, *Lactobacillus* GG was found to induce activation of neutrophils (increased the expression of phagocytosis receptors CR1, CR3, FcγRI, and FcαR) in healthy subjects but to inhibit milk-induced activation of neutrophils in milk-hypersensitive subjects (Pelto et al., 1998). An enhanced oxidative burst or microbicidal capacity of PMN cells in subjects fed probiotics or yogurt has also been demonstrated (Arunachalam et al., 2000; Mikes et al., 1995; Parra et al., 2004).

It has also been reported that probiotic intake is able to restore the age-related decline in phagocytic cell function (Gill, 2002). Aged subjects fed milk containing *Lb. rhamnosus* (HN001) or *Bifidobacterium lactis* (HN019) for three to six weeks exhibited significantly higher phagocytic activity than subjects fed milk without probiotics (Arunachalam et al., 2000; Gill et al., 2001a, b; Gill & Rutherfurd, 2001; Sheih et al., 2001). Importantly, subjects with relatively poor preintervention immunity status consistently showed greater improvement in phagocytic cell function than subjects with adequate preintervention immune function (Gill et al., 2001c). Furthermore, enhancement in phagocytic capacity was also age-related, with subjects older than 70 years exhibiting significantly greater improvements in immune function than those under 70 years (Gill et al., 2001a, b; Gill & Rutherfurd, 2001).

The augmentation of NK cell activity (*ex vivo*) and increases in the percentage of NK cells in the peripheral blood in healthy subjects following regular consumption of yogurt or milk containing probiotics have also been demonstrated (Gill et al., 2001b; Chiang et al., 2000; Sheih et al., 2001; Olivares et al., 2006). As with phagocytic activity, improvements in NK cell function in the elderly subjects, following intake of probiotics, were significantly correlated with age (Gill et al., 2001c). Similar observations regarding enhancement of phagocytic and NK cell function have been made in animals fed probiotics (Gill, 1998; Cross, 2002). Differences in the ability of live versus dead bacteria have also been reported.

It is important to note, however, that several studies have found no effect of probiotic intake on natural immune function (Spanhaak et al., 1998). Whether this has been due to the poor immunostimulatory ability of the probiotic strains used, suboptimal dose, probiotic viability, or some other reason is not known. Strain- and dose-dependent differences in the ability of LAB to modulate immune function are well documented (Donnet-Hughes et al., 1999; Gill, 1998).

Acquired Immunity

The acquired immunity comprises antibody- and cell-mediated responses and is characterized by its specificity and memory.

Consumption of specific probiotics has been shown to enhance antibody responses to natural infections and to systemic and oral immunizations (Isolauri et al., 1995; Majamaa et al., 1995; Kaila et al., 1992, 1995; Fukushima et al., 1998; Link-Amster et al., 1994; de Vrese et al., 2001). In a randomized, placebo-controlled study, Kaila et al. (1992) found significantly higher levels of specific mucosal and serum antibody responses in children with rotavirus following administration with *Lactobacillus* GG fermented milk compared with children receiving a placebo. It has also been demonstrated that viable probiotics are more efficient at stimulating rotavirus-specific immune response than the nonviable bacteria; the proportion of subjects exhibiting rotavirus-specific response at the convalescent stage was higher in the live group (10 out of 12 children) compared with the group given dead bacteria (2 out of 13) (Majamaa et al., 1995; Kaila et al., 1995).

Significantly superior antibody responses and seroconversion rates following parenteral or oral immunization in subjects given probiotics have also been demonstrated. Following immunization with a *Salmonella* vaccine in subjects given probiotics (*B. bifidum*, *L. acidophilus* La1, *Lactobacillus*), significantly higher specific serum IgA antibody and IgA-secreting cell responses were reported (Link-Amster et al., 1994; He et al., 2000). Consistent with these observations, a trend toward increased anti-*Salmonella* IgA levels in subjects receiving LGG and oral *Salmonella* vaccine was reported by Fang et al. (2000). The enhanced immunogenicity of a live rotavirus vaccine in infants given probiotics has also been observed; infants given oral rotavirus vaccine and *Lactobacillus* GG had significantly more IgA- and IgM-secreting cells compared with infants given vaccine only (Isolauri et al., 1995). In another investigation, supplementation with specific strains of probiotics was shown to enhance the efficacy of poliovirus vaccine (de Vrese et al., 2001). In a randomized, double-blind, placebo-controlled study, subjects given yogurt containing *L. rhamnosus* and *L. paracasei* had significantly higher virus-neutralizing antibody responses (mainly IgA) following vaccination with live attenuated polioviruses compared with subjects given placebo (chemically acidified milk). The levels of polio-specific serum IgG and IgA in volunteers consuming yogurt were also significantly increased (de Vrese et al., 2001). In another study, administration of a formula containing bifidobacteria to infants who were immunized against poliovirus several months prior to enrollment in the study was found to enhance total fecal IgA and anti-poliovirus fecal IgA (Fukushima et al., 1998). Similar effects of probiotic administration on antibody responses to a range of antigens and bacterial pathogens have been reported in several animal studies (Gill, 1998).

Together these observations suggest that specific strains of LAB exhibit potent adjuvant properties. The adjuvant effects of probiotics appear to be mediated through improved antigen presentation function: increased transport of antigenic materials across the gut mucosa and upregulation of antigen-presenting molecules and co-stimulatory molecules on immune cells (Heyman, 2001) and/or an increased number of B cells (De Simone et al., 1991). Thus, probiotics could be effective in improving the efficacy of oral and parenteral vaccines.

Cytokine Production

Cytokines comprise the largest and most pleiotropic group of immune response mediators. Initiation, maintenance, and resolution of both innate and acquired immune responses are regulated by cell-to-cell communication via cytokines.

The ability of probiotics to induce cytokine production by a range of immunocompetent cells may explain how they are able to influence both innate and acquired immune responses. Several studies have reported enhanced levels of IFN-γ, IFN-α, and IL-2 in healthy subjects given probiotics (de Simone et al., 1986; Solis-Pereyra & Lemonnier, 1991; Wheeler et al., 1997; Halpern et al., 1991; Aattouri & Lemonnier, 1997; Kishi et al., 1996; Arunachalam et al., 2000). Long-term consumption of yogurt has also been shown to increase the production of IL1β, IL-6, IL-10, IFN-γ, and TNF-α (Halpern et al., 1991; Aattouri & Lemonnier, 1997; Solis-Pereyra & Lemonnier, 1993; Miettinen et al., 1996). *In vitro*, LAB-induced production of IFN-γ, IL-1, TNF-α, IL-10, IL-12, IL-18, and TGF-β by mononuclear cells and DCs has also been demonstrated (Cross et al., 2002; Miettinen et al., 1998; Gill & Guarner, 2004; Lammers et al., 2003; Niers et al., 2005).

IL-12 and IL-18 induce IFN-γ production by T, B, and NK cells, while IFN-γ enhances the phagocyte capacity of phagocytic cells, induces MHC1 and MHCII expression on a variety of cells, potentiates antitumor cytotoxicity, stimulates helper T cell function, and improves the immunogenicity of vaccines (Nussler & Thomson, 1992). TNF-α, together with IFN-γ, increases the microbicidal capacity of macrophages and exerts cytotoxic effect against tumors. IFN-α plays an important role in early stages of host protection against viruses, bacteria, and cancers. IL-1 stimulates proliferation of T and B cells; IL-6 induces differentiation to antibody-secreting plasma cells; IL-2 stimulates proliferation and differentiation of B cells and NK cells and plays a role in the induction and regulation of T cell-mediated immune responses. IL-10 and TGF-β play an immunoregulatory role (Gill, 2003).

Probiotic-Induced Immunostimulation and Disease Resistance

Infectious Diseases

Infections with gastrointestinal and respiratory tract pathogens (bacteria and viruses) continue to be a major health problem worldwide. Several well-controlled studies have provided evidence that the administration of specific strains of probiotics could be effective in the prevention and/or treatment of infectious diarrhea (Table 1). A meta-analysis of studies published between 1966 and 2000 revealed that the administration of probiotics, compared to a placebo, was effective in reducing the duration of acute rotavirus diarrhea by 0.7 days (95% confidence interval: 0.3–1.2 days) and the frequency of diarrhea by 1.6 stools on day 2 of treatment (95% confidence interval: 0.7–2.6 fewer stools).

The results of several recent studies Table 1 have further shown that oral intake of probiotics is also effective against respiratory tract infections (Hatakka et al., 2001; Habbermann et al., 2001; Turchet et al., 2003; de Vrese et al., 2006). Several mechanisms by which probiotics mediate their protective effects have been suggested. However, their relative contribution remains unknown. The ability of probiotics to mediate protection at extraintestinal sites and against viral infections strongly suggests that probiotic-induced immune stimulation may be a major contributor.

An association between enhanced specific and nonspecific antibody responses (IgA-secreting cells and serum IgA) and a reduction in the duration of diarrhea in children hospitalized for acute viral diarrhea following the administration of probiotics have been reported in a number of studies (Kaila et al., 1992, 1995; Majamaa et al., 1995; Guandalini et al., 2000). An augmentation of immune responses (number of T-helper cells, NK cell activity, secretion of IFN-α and IFN-γ) and a reduction in the symptom score, duration of common cold episodes, and days with fever in subjects given probiotics during the winter/spring period have also been observed (De Vrese et al., 2006). Similarly, several animal studies have reported a positive relationship between enhanced immune responses (serum and mucosal antibodies, phagocytic cell function, and NK cell activity) and resistance to infection (*Salmonella, E. coli*, etc.) following oral administration with probiotics (Gill et al., 2001d; Shu & Gill, 2002).

Cancer

Studies in experimental animals have shown that supplementation with specific probiotic strains is effective in preventing the establishment, growth, and metastasis of chemically induced and transferrable tumors (Rafter, 2002; Capurso et al., 2006). In humans, probiotic supplementation has been shown to reduce the risk of colon cancer by inhibiting the transformation of pro-carcinogens to carcinogens, inactivating mutagenic compounds, and suppressing the growth of pro-carcinogenic bacteria. A negative association between the reduced incidence of cancer and the consumption of fermented dairy products, containing lactobacilli and bifidobacteria, has also been reported from a number of epidemiological and population-based case-control studies. However, there is little direct evidence of the antitumor efficacy of probiotics in

Table 1 Efficacy of Probiotics in the Prevention and Treatment of Diarrhea and Respiratory Diseases in Children: Some Examples

Probiotic Used	Study Population	Design	Outcome	Immune Effect	Reference
L. casei S train GG vs. placebo	Infants with diarrhea (82% due to rotavirus)	Double-blind, placebo-controlled	Reduction in number of motions/day (1.4 vs. 2.4; $P < 0.001$).	Not recorded	Isolauri et al. (1991)
L. case strain GG vs. placebo	Children with rotavirus diarrhea	Randomized, controlled	Reduction in duration of diarrhea (1.5 vs. 2.3 days; $P = 0.002$).	Not recorded	Isolauri et al. (1994)
LGG, L. casei subsp rhamnosus (Lactophilus) or S. thermophilus + L. delbruckii (Yalacta)	Children with acute rotavirus diarrhea	Randomized	Reduction in duration of diarrhea in LGG group (1.8 vs. 2.8 days in Lactophilus and 2.6 days in Yalacta groups).	Enhancement of rotavirus-specific IgA and specific antibody secreting cells	Majamaa et al. (1995)
L. reuteri	Infants with acute diarrhea	Randomized, placebo-controlled	Reduction in duration of diarrhea in L. reuteri group (1.7 vs. 2.9 days; $P = 0.07$).	Not recorded	Shornikova et al. (1997)
L. case strain GG vs. placebo	Children with diarrhea (unknown etiology)	Randomized, placebo-controlled	Reduction in duration of diarrhea in LGG group (7199 vs. 58.3 hours), reduction in the number of watery stools.	Not recorded	Guandalini et al. (2000)
B. bifidum, S. thermophilus vs. placebo	Children—prevention of diarrhea	Double-blind, placebo-controlled	Reduction in the incidence of diarrhea (8/26 vs. 2/29). $P < 0.035$ for rotavirus diarrhea.	Not recorded	Saavadra et al. (1994)
L. GG vs. placebo	Undernourished children (6 to 24 months old)—prevention of diarrhea	Randomized, placebo-controlled	Significantly fewer episodes of diarrhea in LGG group.	Not recorded	Oberhelman et al. (1999)

Table 1 (continued)

Probiotic Used	Study Population	Design	Outcome	Immune Effect	Reference
L. casei, S. thermophilus, L bulgaricus or *S. thermophilus, L bulgaricus* vs. placebo	Children (19 months old)—prevention of diarrhea	Randomized, blind, placebo-controlled	Reduction in the duration of diarrhea in *L. casei* group over the 6 months ($P = 0.009$).	Not recorded	Pedone et al. (1999)
L. rhamnosus strains 573L/1; 573L/2; 573L/3 or placebo	Children (2 months to 6 years old)—with infectious diarrhea	Randomized, blind, placebo-controlled	Reduction in the duration of rotavirus diarrhea (76 ± 35 hours vs. 115 ± 67 hours; $P = 0.03$).	Not recorded	Szymanski et al. (2006)
B. lactis Bb12 or *L. reuteri* ATCC 55730	Children (4–10 months old)—prevention of infections	Randomized, double-blind, placebo-controlled	Reduction in the number (0.31 in control group; 0.13 in *B. lactis* group and 0.02 in *L. reuteri* group) and duration of episodes of diarrhea.	Not recorded	Weizman et al. (2005)
B. lactis Bb12	Children (8 months or younger)	Multicenter, double-blind, controlled study	Reduction in the duration of episodes of diarrhea in probiotic group (5.1 ± 3.3 days vs. 7.0 ± 5.5 days).	Not recorded	Chouraqui et al. (2004)
Bifidobacterium Bb12 alone or with *S. thermophilus*	Children (6–36 months)—prevention of rotavirus diarrhea	Placebo-controlled	Prevention of symptomatic rotavirus infection.	No significant increase in antibody levels in treatment group, indicating no infection (30% of control group showed subclinical infection)	Phuapradit et al. (1999)

Table 1 (continued)

Probiotic Used	Study Population	Design	Outcome	Immune Effect	Reference
Lactobacillus GG	Children (1–6 years old)—prevention of diarrhea and respiratory infections	Randomized, double-blind, placebo-controlled	Reduction in the number of days of absence from day care center due to illness (4.9 vs. 5 days; $P < 0.03$). Reduction (17%) in the incidence of respiratory tract infections ($P = 0.05$).	Not recorded	Hatakka et al. (2001)
Verum (*Lactobacillus* and *Bifidobacterium* spp)	Prevention of common colds in healthy adults	Randomized, double-blind, controlled	Significant reduction in duration of episodes (7.0 vs. 8.9 in control, $P < 0.045$)	Significant increase in cytotoxic plus T suppressor cells ($CD8^+$) and T helper cells ($CD4^+$)	De Vrese et al. (2006)
Verum (*Lactobacillus* and *Bifidobacterium* spp)	Prevention of common cold in healthy adults	Randomized, double-blind, placebo-controlled	Reduction (13.6%) in incidence of virally induced infections ($P = 0.07$). Significant reduction (54%) in	Significant increase in T-lymphocytes including $CD4^+$ and $CD8^+$ cells as well as monocytes	Winkler et al. (2005)

human subjects. Rafter and colleagues (2007) reported a protective effect of synbiotic therapy in a randomized, double-blind, placebo-controlled study involving polypectomized patients and colon cancer patients. Synbiotic administration for 12 weeks resulted in a significant reduction in colorectal proliferation and the capacity of fecal water to induce toxicity of colonic cells, along with an improvement in epithelial barrier function in polypectomized patients. Furthermore, symbiotic therapy prevented an increase in IL-2 secretion by peripheral blood mononuclear cells in the polypectomized patients and enhanced the production of interferon-γ in cancer patients. The protective effects of probiotic supplementation against bladder cancer have also been demonstrated (Aso et al., 1995; Sawamura et al., 1994). It was also suggested that probiotic-induced stimulation of the immune system, as indicated by increases in the percentages of T-helper cell and NK cells, and augmentation of NK cell activity may play an important role in the suppression of tumor development. Several other mechanisms by which probiotics mediate anticancer effects have also been suggested (Rafter, 2002).

Probiotics and Attenuation of Immunoinflammatory Disorders

Inflammatory Bowel Disease (IBD)

Inflammatory bowel disease (IBD) consists of mainly two forms: Crohn's disease (CD) and ulcerative colitis (UC). Both diseases are chronic in nature and are characterized clinically by relapses and remissions. UC is characterized by inflammation with superficial ulcerations limited to the mucosa of the colon. Inflammation in CD patients is transmural with large ulcerations, and occasionally granulomas are observed. UC is generally confined to the large intestine, while CD shows a discontinuous pattern, potentially affecting the entire GI tract (Sheil et al., 2007). The etiology of the IBD is unknown. Results of several recent studies suggest that genetic factors and an abnormal host immune response to resident luminal bacteria are involved in the development and/or maintenance of IBD (Bonen & Cho, 2003; Mahida & Rolfe, 2004); CD is a Th1-mediated disease, whereas UC is a Th2-mediated disorder. Studies with animal models of IBD (genetically engineered and germ-free) have clearly demonstrated that the induction of intestinal inflammation is associated with the presence of enteric bacteria. The presence of enteric bacteria or their products in the inflamed tissue and alterations in patients with IBD have also been reported (Fedorak & Madsen, 2004). These observations have led to the evaluation of probiotic therapy as a means for modifying the luminal microbial environment and restoring immune homeostasis, for the management and treatment of IBD. The results of these interventions have been encouraging (Table 2). As per the criteria of evidence-based medicine, there is level 1 evidence to support the therapeutic use of probiotics for the treatment of

Table 2 Efficacy of Probiotics in the Prevention and Treatment of IBD: Some Examples

Probiotic Used	Study Population	Design	Outcome	Immune Effect	Reference
B. longum and synergy 1	Patients with active UC	Double-blind, randomized controlled trial	Reduction in sigmoidoscopy scores (scale 0–6) in the probiotic group (3.1) compared with placebo (3.2).	Significant reduction in the mRNA levels for human β-defensins 2, 3, and 4 after treatment ($p = 0.016, 0.038,$ and 0.008, respectively). Also, reduction in TNF-α and IL1-α after treatment ($p = 0.018$ and 0.023, respectively)	Furrie et al. (2005)
VSL#3*	Ambulatory patients with active UC (treatment and preventing relapse of IBD)	Open-label experiment	Induction of remission/response rate of 77% with no adverse events.	Not recorded	Bibiloni et al. (2005)
Fermented milk product containing *Lactobacillus* La-5 and *Bifidobacterium* Bb12	Patients with UC operated on with ileal-pouch-anal anastomosis and patients with ileorectal anastomosis	Open-label experiment	Reduction in the median endoscopic score of inflammation during intervention in the UC/IPAA patients.	Not recorded	Laake et al. (2005)
VSL#3	Patients with pouchitis (PADI score 7 or more)	Randomized and placebo-controlled	Remission was maintained for one year in 17 patients (85%) on VSL#3 and in one patient (6%) on placebo ($p < 0.0001$).	Not recorded	Mimura et al. (2004)

Table 2 (continued)

Probiotic Used	Study Population	Design	Outcome	Immune Effect	Reference
VSL#3	Patients with UC operated on with ileal-pouch-anal anastomosis.	Placebo-controlled	Effective in the prevention of the onset of acute pouchitis and improvement in the quality of life of patients with IPAA.	Significant reduction in mucosal mRNA expression levels of IL-1β, IL-8, and IFN-γ compared with placebo-treated patients. Increase in the number of polymorphonuclear cells	Gionchetti et al. (2003)
VSL#3	Patients with chronic pouchitis	A double-blind, placebo-controlled trial	Effective in the prevention of flare-ups of chronic pouchitis.	Not recorded	Gionchetti et al. (2000)
E. coli (Nissle, 1917)	Pateints with CD symptoms	Placebo-controlled trial with 24 patients	Reduction in relapse rate.	Not recorded	Malchow (1997)

CD = Crohn's disease; UC = ulcerative colitis.
*(Contains *Lactobacillus acidophilus*, *Lactobacillus plantarum*, *Lactobacillus casei*, *Lactobacillus delbrueckii*, *Bifidobacterium longum*, *Bifidobacterium infantis*, *Streptococcus salivaris*).

postoperative pouchitis, and levels 2 and 3 evidence to support the use of probiotics for the treatment of UC and CD (Fedorak & Madsen, 2004). The modulation or regulation of dysregulated immune responses has been suggested to be the primary mechanism by which probiotics mediate their beneficial effects (described in the next section); the ability of different probiotic strains to induce distinct mucosal cytokine profiles and modulate polarized Th1 and/or Th2 responses is well documented (Ghosh et al., 2004).

Allergies

Allergies represent an exaggerated and imbalanced immune response to an environmental or food antigen. Classical allergy is a type I hypersensitivity reaction. It is driven by the preferential activation of Th2 cells producing IL-4 and IL-5 cytokines and is characterized by an increased synthesis of IgE and activation and recruitment of eosinophils. Depending upon the mode of allergen entry, allergic reactions are commonly manifested by urticaria, rhinitis, vomiting, and/or diarrhea.

Recent studies have shown that insufficient or aberrant exposure to microbes, early in life, may be responsible for the rise in the prevalence of allergies in the westernized countries over the past 40 years. It has also been reported that differences (qualitative and quantitative) in the composition of neonatal gut microflora precede the development of allergy; children who developed allergy had fewer bifidobacteria and enterococci and higher levels of clostridia and *Staphylocuccus aureus* in their intestinal flora than nonallergic children (Bjorksten et al., 2001); differences in the composition of GI microbiota in individuals from industrialized versus nonindustrialized countries have also been observed (Drasar, 1974). The increased risk of developing allergic rhinoconjunctivitis in infants born by Caesarean section (C-section), compared with those delivered vaginally, further demonstrates the crucial role of indigenous microflora in shaping the development of the immune system (Renz-Polster et al., 2005).

To date, many studies have examined the efficacy of probiotic supplementation in the prevention and treatment of allergic disorders. These studies have shown that probiotic supplementation could have a beneficial effect in infants at high risk of atopy and those presenting with cow's milk allergy and atopic eczema and dermatitis (Table 3). It has also been shown that the benefits of probiotic supplementation during infancy extend beyond infancy (Kalliomaki et al., 2003; Lodinova-Zodnikova et al., 2003). Interestingly, however, the administration of probiotics in young adults or teenagers with birch pollen or apple allergy, before, during, and after pollen birch season, was found to be ineffective in alleviating the symptoms of allergy or reducing the use of medicine. This suggests that probiotic intervention, before or immediately after

Table 3 Efficacy of Probiotics in the Prevention and Treatment of Allergic Diseases: Some Examples

Probiotic Used	Study Population	Design	Outcome	Immune Effect	Reference
Probiotics + galacto-oligosaccharide	Mother and infant pairs (up to 6 months). Prevention of food allergy, eczema, asthma, and allergic rhinitis	Randomized, placebo-controlled trial	Reduction in the incidence of atopic diseases ($p < 0.052$), eczema ($p < 0.035$), and atopic eczema ($p < 0.025$)	Not recorded	Kukkonen et al. (2007)
L. casei shirota	Patients with allergic rhinitis triggered by Japanese cedar pollen	Randomized, double-blind, placebo-controlled study	Reduction in nasal symptom-medication score in the probiotic group	Not recorded	Tamura et al. (2007)
Lb gasseri TMC0356	Subjects with perennial allergic rhinitis	Controlled trial with 15 subjects showing high serum IgE levels and allergic symptoms	Not reported	Reduction in serum total IgE levels ($p < 0.05$). Significant increase in the proportion of Th1 cells [on days 14 ($p < 0.01$) and after 28 ($p < 0.05$)]	Morita et al. (2006)
Bifidobacterium longum BB536	Subjects with history of Japanese cedar pollinosis (JCPsis)	Randomized, double-blind, placebo-controlled trial	Significant improvements in eye symptoms in the probiotic group ($p = 0.0057$). Also, reduction in rhinorrhea and nasal blockage	Decrease in JCP-specific IgE levels	Xiao et al. (2006)

Table 3 (continued)

Probiotic Used	Study Population	Design	Outcome	Immune Effect	Reference
Lactobacillus rhamnosus and *Bifidobacteria lactis*	Children with established atopic dermatitis	Placebo-controlled	Improvement in AD only in food-sensitized children	Not recorded	Sistek et al. (2006)
Lb. fermentum VRI-033	Children (6–18 months) with moderate to severe atopic dermatitis (AD)	Randomized, double-blind, placebo-controlled trial	Significant reduction in the SCORAD index over time ($p < 0.03$)	Reduction in IgE level [35.7 (±6.0) in placebo group versus 31.8 (±4.3) in probiotic group]	Weston et al (2005)
Lactobacillus fermentum PCC trademark	Young children with moderate-to-severe atopic dermatitis (AD)	Randomized, placebo-controlled trial	Improvement in AD severity	Significant increase in IFN-γ production following stimulation with PHA and SEB at the end of the supplementation period (week 8: $P = 0.004$ and 0.046) as well as 8 weeks after cessation of supplementation (week 16: $P = 0.005$ and 0.021)	Prescott et al (2005)
Enterogermania (containing *Bacillus clausii*)	Adult subjects (mean age 22.3 years) with allergic rhinitis	Controlled trial involving 10 subjects	Symptoms not reported	Significant decrease in IL4 levels ($p = 0.004$); significant increase in IFN-γ ($p = 0.038$), TGF-β ($p = 0.039$), and IL10 ($p = 0.009$) levels	Ciprandi et al. (2005)

Table 3 (continued)

Probiotic Used	Study Population	Design	Outcome	Immune Effect	Reference
Lactobacillus GG (LGG), a mixture of four probiotic strains (MIX)	Infants with atopic eczema/dermatitis syndrome (AEDS) and food allergy	Randomized, double-blind, placebo-controlled	Reduction in SCORAD in IgE-sensitized infants. Reduction in AT in the LGG group, but not in other treatment groups	Increase in IgA levels in the probiotic group compared with the placebo group (LGG vs. placebo, $p = 0.064$; MIX vs. placebo, $p = 0.064$), after challenge, in subjects with IgE-associated CMA infants, increase in fecal IgA ($p = 0.014$), and decrease in TNF-α compared to placebo	Viljanen et al. (2005)
Enterogermania (containing *Bacillus clausii*)	Allergic children (mean age: 4.4 years) with recurrent respiratory infections	Controlled trial involving 10 children attending nursery school	Symptoms not reported	Significant reduction in IL-4 levels ($p < 0.01$) and a significant increase in IFN-γ ($p < 0.05$), IL-12 ($p < 0.001$), TGF-β ($p < 0.05$), and IL-10 ($p < 0.05$) levels	Ciprandi et al. (2004)
Lactobacillus rhamnosus 19070-2 and *Lactobacillus reuteri* DSM 122460	Children (1–13 years old) with atopic dermatitis	Double-blind, placebo-controlled, crossover study	Improvement in SCORAD	Reduction in serum eosinophil cationic protein levels ($P = 0.03$) in the probiotic group	Rosenfeldt et al. (2003)

Table 3 (continued)

Probiotic Used	Study Population	Design	Outcome	Immune Effect	Reference
Probiotics	Mother-infant pairs with history of atopic diseases	Double-blind, placebo-controlled study	Significant reduction in the risk of developing atopic eczema in probiotic group compared to placebo (15% and 47%, respectively; $P = 0.0098$)	Significant increase in TGF-β2 level in human milk in probiotic group (2885 pg/mL) vs. placebo (1340 pg/mL) $P = 0.018$)	Rautava et al. (2002)
Lactobacillus GG	Mother-infant pairs with history of atopic eczema	Randomized, double-blind, placebo-controlled study	Significant reduction in the incidence of atopic eczema ($p < 0.008$)	No effect on IgE levels	Kalliomaki et al. (2001)
Bifidobacterium lactis Bb-12 or Lactobacillus strain GG (ATCC 53103)	Infants (mean age: 4.6 months) with history of atopic eczema	Randomized, double-blind, placebo-controlled study	Reduction in SCORAD in the Bifidobacterium lactis Bb-12 group to 0 (0–3.8), and in the Lactobacillus GG group to 1 (0.1–8.7) vs. unsupplemented 13.4 (4.5–18.2)	Reduction in the concentration of soluble CD4 in serum and eosinophilic protein X in urine	Isolauri et al. (2000)
Lactobacillus GG (ATCC 53103)	Milk hypersensitive and healthy adult subjects	Double-blind, crossover study	Downregulation of immunoinflammatory response in milk-hypersensitive subjects	Significant reduction in the expression of CR1, Fc-γRI, and Fc-αR in neutrophils and CR1, CR3 and Fc-αR in monocytes	Pelto et al. (1998)

CMA = cow's milk allergy; SEB = Staphylococcus aureus enterotoxin B

birth, is more effective in inducing immunological tolerance, as the immune system is immature, compared with older children with a fully mature immune system (Ouwehand, 2007). Several mechanisms by which probiotics exert preventive/therapeutic antiallergy effects have been suggested. These include reduced immunogenicity of potential allergens through modification of their structure (Rokka et al., 1997), stabilization of the gut mucosal barrier, and restoration of immune system homeostasis through induction of regulatory innate and adoptive immune responses (Guarner et al., 2006).

Mechanisms by Which Probiotics Correct Immunological Disorders

A balance between Th1-Th2 is considered important for immune system homeostasis. Allergic disorders that are mediated by Th2 cells and IBD together with autoimmune disorders (e.g., type 1 diabetes) driven by Th1 cells were therefore considered the result of an imbalance between Th1-Th2 responses (Rook & Brunet, 2005). However, a parallel rise in the incidence of allergies and autoimmunity and IBD (in industrialized countries) in the past few decades and the simultaneous occurrence of Th1- and Th2-mediated disorders suggest that this simple assumption is unable to explain the underlying mechanisms (Guarner et al., 2006). Recent studies have shown that a defective Treg cell activity may be the central cause; patients with type 1 diabetes and multiple sclerosis, and individuals with predisposition to allergy, exhibit deficient Treg cell activity (Guarner et al., 2006).

Evidence from *in vitro* and *in vivo* studies suggests that probiotics may mediate their beneficial effects through induction of regulatory T cells, rather than skewing of Th1 or Th2 responses (Fig. 2). Treg cells suppress both Th1-and Th2-type immune responses through production of IL-10 and TGF-β. Increased levels of TGF-β in breast milk (Rautava et al., 2002) and elevated levels of IL-10 and TGF-β in atopic children following administration of probiotics have also been observed (Pessi et al., 2000; Isolauri et al., 2000). The ability of probiotics to induce regulatory DCs (Hart et al., 2004; Drakes et al., 2004) that drive the polarization of T cells toward Treg cells has been demonstrated (Di Giancinto et al., 2005). An association between the increased expression of IL-10 and the prevention of flare-ups of chronic UC (Cui et al., 2004) and a reduction in pro-inflammatory cytokines in tissue obtained from subjects with pouchitis following treatment with probiotics have also been observed (Lammers et al., 2005). Probiotics have been demonstrated *in vitro* to increase IL-10 synthesis and secretion (in macrophages and T cells) without significantly modifying pro-inflammatory cytokines in inflamed mucosa of patients with active ulcerative colitis (Pathmakanthan et al., 2004).

A strong support for the role of Treg cells is also provided by the results of recent animal studies. Di Giacinto et al. (2005) reported an increased number of Treg cells bearing surface TGF-β, following administration of probiotics, in an animal model of colitis. These cells were effective in conferring protection against colitis in a cell-transfer system. Importantly, the protective effect was dependent on TGF-β and IL-10 and was abolished by appropriate neutralizing antibodies. Furthermore, probiotics, whether delivered orally or subcutaneously, and bacterial DNA have been found to be effective in attenuating colitis and arthritis in mice (McCarthy et al., 2003; Sheil et al., 2004; Rachmilewitz et al., 2004). Chapat et al. (2004) showed that IFN-γ producing $CD8^+$ T cell- mediated ability of orally administered probiotic *L casei* to reduce skin inflammation due to contact sensitivity was Treg cell-dependent. In a recent study, probiotic administration was found to induce IL-10 production and prevent spontaneous autoimmune diabetes in the nonobese diabetic mouse (Calcinaro et al., 2005). This clearly suggests that the mechanisms by which probiotics mediate their effects are not restricted to the gut and are likely to be mediated by Treg cells. Once generated, Treg cells are able to move to other tissues (Rook & Brunet, 2005).

It has also been suggested that modulation of dendritic cell function by probiotics is a critical step that directs the polarization of naïve T cells to Treg cells (Braat et al., 2004). Different probiotics induce different DC activation patterns (expression of cytokines and maturation surface markers), with some strains exhibiting the ability to inhibit DC activation by other lactobacilli (Christensen et al., 2002); therefore, these are likely to exert different effects.

Thus, probiotics have been demonstrated to augment health benefits by influencing the gut flora composition and restoring the intestinal homeostasis. In addition, stimulation of the host immune system to enhance innate (macrophage and NK cell activity), humoral (pathogen-/vaccine-specific antibody and antibody-producing cells), and cell-based (Treg function) immunity has the potential to influence the general health of the world's population. Development of immunization procedures that avoid the use of needles and adjuvants is highly desirable, as it will reduce vaccine costs and make the large-scale implementation of immunization programs possible. Further research effort to understand the mechanisms by which the probiotics modulate the activity of macrophages and NK cells and enhance the immunogenecity of vaccines would help to identify potential candidate probiotics with superior properties.

References

Aattouri, N., & Lemonnier, D. (1997). Production of interferon induced by *Streptococcus thermophilus*: Role of $CD4^+$ and $CD8^+$ lymphocytes. *Journal of Nutritional Biochemistry*, 8, 25–31.

Ahmed, M., Prasad, J., Gill, H., Stevenson, L., & Gopal, P. (2007). Impact of consumption of different levels of *Bifidobacterium lactis* HN019 on the intestinal microflora of elderly human subjects. *Journal of Nutrition, Health, and Aging*, 11, 26–31.

Arunachalam, K., Gill, H. S., & Chandra, R. K. (2000). Enhancement of natural immune function by dietary consumption of *Bifidobacterium lactis* (HN019). *European Journal of Clinical Nutrition, 54,* 263–267.

Aso, Y., Akaza, H., Kotake, T., Tsukamoto, T., Imai, K., & Naito, S. (1995). Preventive effect of a *Lactobacillus casei* preparation on the recurrence of superficial bladder cancer in a double-blind trial. The BLP Study Group. *European Urology, 27*(2), 104–109.

Bartram, H. P., Scheppach, W., Gerlach, S., Ruckdeschel, G., Kelber, E., & Kasper, H. (1994). Does yogurt enriched with *Bifidobacterium longum* affect colonic microbiology and fecal metabolites in health subjects? *American Journal of Clinical Nutrition, 59,* 428–432.

Bauer, H., Paronetto, F., Burns, W., & Einheber, A. (1965). The non-specific enhancement of the immune response by the bacterial flora. Studies in germfree mice. Federation Proceedings.

Benno, Y., & Mitsuoka, T. (1992). Impact of *Bifidobacterium longum* on human fecal microflora. *Microbiology and Immunology, 36,* 683–694.

Bibiloni, R., Fedorak, R. N., Tannock, G. W., Madsen, K. L., Gionchetti, P., Campieri, M., De Simone, C., & Sartor, R. B. (2005). VSL#3 probiotic-mixture induces remission in patients with active ulcerative colitis. *American Journal of Gastroenterology, 100,* 1539–1546.

Bjorksten, B., Sepp, E., Julge, K., Voor, T., & Mikelsaar, M. (2001). Allergy development and the intestinal microflora during the first year of life. *Journal of Allergy and Clinical Immunology, 108,* 516–520.

Bonen, D. K., & Cho, J. H. (2003). The genetics of inflammatory bowel disease. *Gastroenterology, 124,* 521–536.

Braat, H., van den Brande, J., van Tol, E., Hommes, D., Peppelenbosch, M., & van Deventer, S. (2004). *Lactobacillus rhamnosus* induces peripheral hyporesponsiveness in stimulated $CD4^+$ T cells via modulation of dendritic cell function. *American Journal of Clinical Nutrition, 80,* 1618–1625.

Calcinaro, F., Dionisi, S., Marinaro, M., Candeloro, P., Bonato, V., Marzotti, S., Corneli, R. B., Ferretti, E., Gulino, A., Grasso, F., De Simone, C., Di Mario, U., Falorni, A., Boirivant, M., & Dotta, F. (2005). Oral probiotic administration induces interleukin-10 production and prevents spontaneous autoimmune diabetes in the non-obese diabetic mouse. *Diabetologia, 48,* 1565–1575.

Capurso, G., Marignani, M., & Fave, G. D. (2006). Probiotics and the incidence of colorectal cancer: When evidence is not evident. *Digestive and Liver Disease, 38 (Suppl 2),* S277–S282.

Chapat, L., Chemin, K., Dubois, B., Bourdet-Sicard, R., & Kaiserlian, D. (2004). *Lactobacillus casei* reduces $CD8^+$ T cell-mediated skin inflammation. *European Journal of Immunology, 34,* 2520–2528.

Chen, R. M., Wu, J. J., Lee, S. C., Huang, A. H., & Wu, H. M. (1999). Increase of intestinal *Bifidobacterium* and suppression of coliform bacteria with short-term yogurt ingestion. *Journal of Dairy Science, 82*(11), 2308–2314.

Chiang, B. L., Sheih, Y. H., Wang, L. H., Liao, C. K., & Gill, H. S. (2000). Enhancing immunity by dietary consumption of a probiotic lactic acid bacterium (*Bifidobacterium lactis* HN019): Optimization and definition of cellular immune responses. *European Journal of Clinical Nutrition, 54,* 849–855.

Chouraqui, J. P., Van Egroo, L. D., & Fichot, M. C. (2004). Acidified milk formula supplemented with *Bifidobacterium lactis*: Impact on infant diarrhea in residential care settings. *Journal of Pediatric Gastroenterology and Nutrition, 38,* 288–292.

Christensen, H. R., Frokiaer, H., & Pestka, J. J. (2002). Lactobacilli differentially modulate expression of cytokines and maturation surface markers in murine dendritic cells. *Journal of Immunology, 168,* 171–178.

Ciprandi, G., Tosca, M. A., Milanese, M., Caligo, G., & Ricca, V. (2004). Cytokines evaluation in nasal lavage of allergic children after *Bacillus clausii* administration: A pilot study. *Pediatric Allergy and Immunology, 15,* 148–151.

Ciprandi, G., Vizzaccaro, A., Cirillo, I., & Tosca, M. A. (2005). *Bacillus clausii* exerts immuno-modulatory activity in allergic subjects: A pilot study. *Allergie und Immunologie (Paris), 37,* 129–134.

Crabbe, J. C., Wahlsten, D., & Dudek, B. C. (1999). Genetics of mouse behavior: Interactions with laboratory environment. *Science, 284*(5420), 1670–1672.

Crabbe, P. A. (1968). The lymphoid tissue of human gastrointestinal mucous membrane. II. Its role. *Presse Medicine, 76,* 1875–1878.

Crabbe, P. A., Nash, D. R., Bazin, H., Eyssen, H., & Heremans, J. F. (1970). Immunohistochemical observations on lymphoid tissues from conventional and germ-free mice. *Laboratory Investigation, 22,* 448–457.

Cross, M. L. (2002). Microbes versus microbes: Immune signals generated by probiotic lactobacilli and their role in protection against microbial pathogens. *FEMS Immunology and Medical Microbiology, 34,* 245–253.

Cross, M. L., Mortensen, R. R., Kudsk, J., & Gill, H. S. (2002). Dietary intake of *Lactobacillus rhamnosus* HN001 enhances production of both Th1 and Th2 cytokines in antigen-primed mice. *Medical Microbiology and Immunology, 191,* 49–53.

Cui, H. H., Chen, C. L., Wang, J. D., Yang, Y. J., Cun, Y., Wu, J. B., Liu, Y. H., Dan, H. L., Jian, Y. T., & Chen, X. Q. (2004). Effects of probiotic on intestinal mucosa of patients with ulcerative colitis. *World Journal of Gastroenterology, 10,* 1521–1525.

De Simone, C., Salvadori, B. B., Negri, R., Ferrazzi, M., Baldinelli, L., & Vesely, R. (1986). The adjuvant effect of yogurt on production of gamma interferon by ConA-stimulated human peripheral blood lymphocytes. *Nutrition Reports International, 33,* 419–433.

De Simone, C., Rosati, E., Moretti, S., et al. (1991). Probiotics and stimulation of the immune response. *European Journal of Clinical Nutrition, 45 (Suppl),* 32–34.

de Vrese, M., Fenselau, S., Feindt, F., et al. (2001). Einfluss von Probiotika auf die immunantwort auf eine polioschluckimpfung [Effects of probiotics on immune response to polio vaccination]. *Proceedings of the German Nutrition Society, 3,* 7.

de Vrese, M., Winkler, P., Rautenberg, P., Harder, T., Noah, C., Laue, C., Ott, S., Hampe, J., Schreiber, S., Heller, K., & Schrezenmeir, J. (2006). Probiotic bacteria reduced duration and severity but not the incidence of common cold episodes in a double blind, randomized, controlled trial. *Vaccine, 24,* 6670–6674.

Di Giacinto, C., Marinaro, M., Sanchez, M., Strober, W., & Boirivant, M. (2005). Probiotics ameliorate recurrent Th1-mediated murine colitis by inducing IL-10 and IL-10-dependent TGF-β-bearing regulatory cells. *Journal of Immunology, 174,* 3237–3246.

Donnet-Hughes, A., Rochat, F., Serrant, P., et al. (1999). Modulation of nonspecific mechanisms of defense by lactic acid bacteria: Effective dose. *Journal of Dairy Science, 82,* 863–869.

Drakes, M., Blanchard, T., & Czinn, S. (2004). Bacterial probiotic modulation of dendritic cells. *Infectious Immunology, 72,* 3299–3309.

Drasar, B. S. (1974). Some factors associated with geographical variations in the intestinal microflora. *Society for Applied Bacteriology Symposium Series, 3,* 187–196.

Erickson, K. L., & Hubbard, N. E. (2000). Probiotic immunomodulation in health and disease. *Journal of Nutrition, 130 (2S Suppl),* 403S–409S.

Fanaro, S., Chierici, R., Guerrini, P., & Vigi, V. (2003). Intestinal microflora in early infancy: Composition and development. *Acta Paediatrics Supplement, 91,* 48–55.

Fang, H., Elina, T., Heikki, A., & Seppo, S. (2000). Modulation of humoral immune response through probiotic intake. *FEMS Immunology and Medical Microbiology, 29,* 47–52.

FAO/WHO (2001). Report of a joint FAO/WHO expert consultation on evaluation of health and nutritional properties of probiotics in food including powder milk with live lactic acid bacteria, Cordoba, Argentina, October 1–4, 2001. Available at http://www.fao.org/es/esn/food/foodandfood_probio_en.stm#contacts.

Fedorak, R. N., & Madsen, K. L. (2004). Probiotics and prebiotics in gastrointestinal disorders. *Current Opinion in Gastroenterology, 20,* 146–155.

Fukushima, Y., Li, S.-T., Hara, H., Terada, A., & Mitsuoka, T. (1997). Effect of follow-up formula containing bifidobacteria (NAN-BF) on fecal flora and fecal metabolites in healthy children. *Bioscience Microflora, 16,* 65–72.

Fukushima, Y., Kawata, Y., Hara, H., Mitsuoka, T., et al. (1998). Effect of a probiotic formula on intestinal immunoglobulin A production in healthy children. *International Journal of Food Microbiology, 42,* 39–44.

Furrie, E., Macfarlane, S., Kennedy, A., Cummings, J. H., Walsh, S. V., O'Neil, D. A., & Macfarlane, G. T. (2005). Synbiotic therapy (*Bifidobacterium longum*/Synergy 1) initiates resolution of inflammation in patients with active ulcerative colitis: A randomised controlled pilot trial. *Gut, 54,* 242–249.

Ghosh, S., van Heel, D., & Playford, R. J. (2004). Probiotics in inflammatory bowel disease: Is it all gut flora modulation? *Gut, 53,* 620–622.

Gill, H. S. (1998). Stimulation of the immune system by lactic cultures. *International Dairy Journal, 8,* 535–544.

Gill, H. (2003). Probiotics to enhance anti-infective defences in the gastrointestinal tract. *Best Practice & Research in Clinical Gastroenterology, 17,* 755–773.

Gill, H. S., & Guarner, F. (2004). Probiotics and human health: A clinical perspective. *Postgraduate Medical Journal, 80,* 516–526.

Gill, H. S., & Rutherfurd, K. J. (2001). Immune enhancement conferred by oral delivery of *Lactobacillus rhamnosus* HN001 in different milk-based substrates. *Journal of Dairy Research, 68,* 611–616.

Gill, H. S., Cross, M. L., Rutherfurd, K. J., & Gopal, P. K. (2001a). Dietary probiotic supplementation to enhance cellular immunity in the elderly. *British Journal of Biomedical Science, 58,* 94–96.

Gill, H. S., Rutherfurd, K. J., & Cross, M. L. (2001b). Dietary probiotic supplementation enhances natural killer cell activity in the elderly: An investigation of age-related immunological changes. *Journal of Clinical Immunology, 21,* 264–271.

Gill, H. S,, Rutherfurd, K. J., Cross, M. L., & Gopal, P. K. (2001c). Enhancement of immunity in the elderly by dietary supplementation with the probiotic *Bifidobacterium lactis* HN019. *American Journal of Clinical Nutrition, 74,* 833–839.

Gill, H. S., Shu, Q., Lin, H., Rutherfurd, K. J., & Cross, M. L. (2001d). Protection against translocating *Salmonella typhimurium* infection in mice by feeding the immuno-enhancing probiotic *Lactobacillus rhamnosus* strain HN001. *Medical Microbiology and Immunology, 190,* 97–104.

Gionchetti, P., Rizzello, F., Venturi, A., Brigidi, P., Matteuzzi, D., Bazzocchi, G., Poggioli, G., Miglioli, M., & Campieri, M. (2000). Oral bacteriotherapy as maintenance treatment in patients with chronic pouchitis: A double-blind, placebo-controlled trial. *Gastroenterology, 119,* 305–309.

Gionchetti, P., Rizzello, F., Helwig, U., Venturi, A., Lammers, K. M., Brigidi, P., Vitali, B., Poggioli, G., Miglioli, M., & Campieri, M. (2003). Prophylaxis of pouchitis onset with probiotic therapy: A double-blind, placebo-controlled trial. *Gastroenterology,.124,* 1202–1209.

Glaister, J. R. (1973). Factors affecting the lymphoid cells in the small intestinal epithelium of the mouse. *International Archives of Allergy and Applied Immunology, 45,* 719–730.

Goldin, B. R., & Gorbach, S. L. (1992). *Probiotics. The Scientific Basis.* New York: Chapman and Hall.

Gopal, P. K., Prasad, J., & Gill, H. S. (2003). Effect of consumption of *Bifidobacterium lactis* DR10 and galactooligosaccharides on the microecology of the gastrointestinal tract in human subjects. *Nutrition Research, 23,* 1313–1328.

Gordon, H. A., & Bruckner-Kardoss, E. (1961). Effect of the normal microbial flora on various tissue elements of the small intestine. *Acta Anatomica (Basel), 44,* 210–225.

Gordon, H. A., & Pesti, L. (1971). The gnotobiotic animal as a tool in the study of host microbial relationships. *Bacteriology Reviews, 35,* 390–429.

Gronlund, M. M., Arvilommi, H., Kero, P., Lehtonen, O. P., & Isolauri, E. (2000). Importance of intestinal colonisation in the maturation of humoral immunity in early infancy: A prospective follow up study of healthy infants aged 0–6 months. *Archives of Disease in Childhood. Fetal and Neonatal Edition, 83,* F186–F192.

Guandalini, S., Pensabene, L., Zikri, M. A., Dias, J. A., Casali, L. G., Hoekstra, H., Kolacek, S., Massar, K., Micetic-Turk, D., Papadopoulou, A., de Sousa, J. S., Sandhu, B., Szajewska, H., & Weizman, Z. (2000). Lactobacillus GG administered in oral rehydration solution to children with acute diarrhea: A multicenter European trial. *Journal of Pediatric Gastroenterology and Nutrition, 30,* 54–60.

Guarner, F., & Malagelada, J.-R. (2002). Gut flora in health and disease. *Lancet, 360,* 512–519.

Guarner, F., Bourdet-Sicard, R., Brandtzaeg, P., Gill, H. S., McGuirk, P., van Eden, W., Versalovic, J., Weinstock, J. V., & Rook, G. A. (2006). Mechanisms of disease: The hygiene hypothesis revisited. *National Clinical Practice in Gastroenterology and Hepatology, 3,* 275–284.

Habermann, W., Zimmermann, K., Skarabis, H., Kunze, R., & Rusch, V. (2001). The effect of a bacterial immunostimulant (human *Enterococcus faecalis* bacteria) on the occurrence of relapse in patients with chronic recurrent bronchitis. *Arzneimittelforschung, 51,* 931–937.

Halpern, G. M., Vruwink, K. G., Van De Water, J., Keen, C. L., & Gershwin, M. E. (1991). Influence of long-term yoghurt consumption in young adults. *International Journal of Immunotherapy, 7,* 205–210.

Hart, A. L., Lammers, K., Brigidi, P., Vitali, B., Rizzello, F., Gionchetti, P., Campieri, M., Kamm, M. A., Knight, S. C., & Stagg, A. J. (2004). Modulation of human dendritic cell phenotype and function by probiotic bacteria. *Gut, 53,* 1602–1609.

Hatakka, K., Savilahti, E., Ponka, A., Meurman, J. H., Poussa, T., Nase, L., Saxelin, M., & Korpela, R. (2001). Effect of long term consumption of probiotic milk on infections in children attending day care centres: Double blind, randomised trial. *British Medical Journal, 322*(7298), 1327.

He, F., Tuomola, E., Arvilommi, H., et al. (2000). Modulation of humoral immune response through probiotic intake. *FEMS Immunology and Medical Microbiology, 29,* 47–52.

Heyman, M. (2001). Symposium on "dietary influences on mucosal immunity." How dietary antigens access the mucosal immune system. *Proceedings of the Nutrition Society, 60,* 419–426.

Hosoda, M., Hashimoto, H., He, F., Morita, H., & Hosono, A. (1996). Effect of administration of milk fermented with *Lactobacillus acidophilus* LA-2 on fecal mutagenicity and microflora in the human intestine. *Journal of Dairy Science, 79,* 745–749.

Ishibashi, N., Yaeshima, T., & Hayasawa, H. (1997). Bifidobacteria: Their significance in human intestinal health. *Malaysian Journal of Nutrition, 3,* 149–159.

Isolauri, E., Juntunen, M., Rautanen, T., Sillanaukee, P., & Koivula, T. (1991). A human *Lactobacillus* strain (*Lactobacillus casei* sp GG) promotes recovery from acute diarrhoea in children. *Pediatrics, 88,* 90–97.

Isolauri, E., Aila, M., Mykkanen, H., Ling, W. H., & Salminen, S. (1994). Oral bacteriotherapy for viral gastroenteritis. *Digestive Diseases and Science, 39,* 2595–2600.

Isolauri, E., Joensus, J., Suomalainen, H., Luomala, M., & Vesikari, T. (1995). Improved immunogenicity of oral DxRRX reabsorbant rotavirus vaccine by *Lactobacillus casei* GG. *Vaccine, 13,* 310–312.

Isolauri, E., Arvola, T., Sutas, Y., Moilanen, E., & Salminen, S. (2000). Probiotics in the management of atopic eczema. *Clinical and Experimental Allergy, 30,* 1604–1610.

Kaila, M., Isolauri, E., Soppi, E., Virtanen, F., Laine, S., & Arvilommi, H. (1992). Enhancement of the circulating antibody secreting cell response in human diarrhoea by a human *Lactobacillus* strain. *Pediatric Research, 32,* 141–144.

Kaila, M., Isolauri, E., Saxelin, M., et al. (1995). Viable versus inactivated *Lactobacillus* strain GG in acute rotavirus diarrhoea. *Archives of Diseases in Childhood, 72,* 51–53.

Kalliomaki, M., Salminen, S., Arvilommi, H., Kero, P., Koskinen, P., & Isolauri, E. (2001). Probiotics in primary prevention of atopic disease: a randomized placebo-controlled trial. *Lancet, 357*(9262), 1076–1079.

Kalliomaki, M., Salminen, S., Poussa, T., Arvilommi, H., & Isolauri, E. (2003). Probiotics and prevention of atopic disease: 4-year follow-up of a randomized placebo-controlled trial. *Lancet, 361*(9372), 1869–1871.

Kelly, D., & Coutts, A. G. (2000). Early nutrition and the development of immune function in the neonate. *Proceedings of the Nutrition Society, 59*, 177–185.

Kishi, A., Uno, K., Matsubara, Y., Okuda, C., & Kishida, T. (1996). Effect of the oral administration of *Lactobacillus brevis* subsp. *coagulans* on interferon-a producing capacity in humans. *Journal of American College of Nutrition, 15*, 408–412.

Kukkonen, K., Savilahti, E., Haahtela, T., Juntunen-Backman, K., Korpela, R., Poussa, T., Tuure, T., & Kuitunen, M. (2007). Probiotics and prebiotic galacto-oligosaccharides in the prevention of allergic diseases: A randomized, double-blind, placebo-controlled trial. *Journal of Allergy and Clinical Immunology, 119*, 192–198.

Laake, K. O., Bjorneklett, A., Aamodt, G., Aabakken, L., Jacobsen, M., Bakka, A., & Vatn, M. H. (2005). Outcome of four weeks' intervention with probiotics on symptoms and endoscopic appearance after surgical reconstruction with a J-configurated ileal-pouch-anal-anastomosis in ulcerative colitis. *Scandinavian Journal of Gastroenterology, 40*, 43–51.

Lammers, K. M., Brigidi, P., Vitali, B., Gionchetti, P., Rizzello, F., Caramelli, E., Matteuzzi, D., & Campieri, M. (2003). Immunomodulatory effects of probiotic bacteria DNA: IL-1 and IL-10 response in human peripheral blood mononuclear cells. *FEMS Immunology and Medical Microbiology, 38*, 165–172.

Lammers, K. M., Vergopoulos, A., Babel, N., Gionchetti, P., Rizzello, F., Morselli, C., Caramelli, E., Fiorentino, M., d'Errico, A., Volk, H. D., & Campieri, M. (2005). Probiotic therapy in the prevention of pouchitis onset: Decreased interleukin-1β, interleukin-8, and interferon-γ gene expression. *Inflammatory Bowel Diseases, 11*, 447–454.

Langhendries, J. P., Detry, J., Van Hees, J., Lamboray, J. M., Darimont, J., Mozin, M. J., Secretin, M. C., & Senterre, J. (1995). Effect of a fermented infant formula containing viable bifidobacteria on the fecal flora composition and pH of healthy full-term infants. *Journal of Pediatric Gastroenterology and Nutrition, 21*, 177–181.

Larsen, C. N., Nielsen, S., Kaestel, P., Brockmann, E., Bennedsen, M., Christensen, H. R., Eskesen, D. C., Jacobsen, B. L., & Michaelsen, K. F. (2006). Dose-response study of probiotic bacteria *Bifidobacterium animalis subsp lactis* BB-12 and *Lactobacillus paracasei* subsp *paracasei* CRL-341 in healthy young adults. *European Journal of Clinical Nutrition, 60*, 1284–1293.

Link-Amster, H., Rochat, F., Saudan, K. Y., et al. (1994). Modulation of a specific humoral immune response and changes in intestinal flora mediated through fermented milk intake. *FEMS Immunology and Medical Microbiology, 10*, 55–64.

Lodinova-Zadnikova, R., Cukrowska, B., & Tlaskalova-Hogenova, H. (2003). Oral administration of probiotic *Escherichia coli* after birth reduces frequency of allergies and repeated infections later in life (after 10 and 20 years). *International Archives of Allergy and Immunology, 121*, 209–211.

Mahida, Y. R., & Rolfe, V. E. (2004). Host-bacterial interactions in inflammatory bowel disease. *Clinical Science, 107*, 331–341.

Majamaa, H., Isolauri, E., Saxelin, M., & Vesikari, T. (1995). Lactic acid bacteria in the treatment of acute rotavirus gastroenteritis. *Journal of Pediatric Gastroenterology and Nutrition. 20*, 333–338.

Malchow, H. A. (1997). Crohn's disease and *Escherichia coli*. A new approach in therapy to maintain remission of colonic Crohn's disease? *Journal of Clinical Gastroenterology, 25*, 653–658.

McCarthy, J., O'Mahony, L., O'Callaghan, L., Sheil, B., Vaughan, E. E., Fitzsimons, N., Fitzgibbon, J., O'Sullivan, G. C., Kiely, B., Collins, J. K., & Shanahan, F. (2003). Double blind, placebo controlled trial of two probiotic strains in interleukin 10 knockout mice and mechanistic link with cytokine balance. *Gut, 52,* 975–980.

Metchnikoff, E. (1907). The prolongation of life. Revised edition from 1907, translated by Mitchell. London: C. Heinemann; also in (1974) *Dairy Science Abstracts, 36,* 656.

Miettinen, M., Vuopio-Varkila, J., & Varkila, K. (1996). Production of human tumour necrosis factor alpha, interleukin-6, and interleukin-10 is induced by lactic acid bacteria. *Infectious Immunology, 64,* 5403–5405.

Miettinen, M., Matikainen, S., Vuopio-Varkila, J., Pirhonen, J., Varkila, K., Kurimoto, M., & Julkunen, I. (1998). Lactobacilli and streptococci induce interleukin-12 (IL-12), IL-18, and γ-interferon production in human peripheral blood mononuclear cells. *Infectious Immunology, 66,* 6058–6062.

Mikelsaar, M., Mander, R., & Sepp, E. (1998). Lactic acid microflora in the human microbial ecosystem and its development. In S. Salminen and A. von Wright (Eds.), *Lactic Acid Bacteria: Microbiology and Functional Aspects* (pp. 279–342), 2nd ed. New York: Marcel Dekker.

Mikes, Z., Ferenicik, M., Jahnova, E., et al. (1995). Hypocholesterolemic and immunostimulatory effects of orally applied *Enterococcus faecium* M-74 in man. *Folia Microbiologica, 40,* 639–646.

Mimura, T., Rizzello, F., Helwig, U., Poggioli, G., Schreiber, S., Talbot, I. C., Nicholls, R. J., Gionchetti, P., Campieri, M., & Kamm, M. A. (2004). Once daily high dose probiotic therapy (VSL#3) for maintaining remission in recurrent or refractory pouchitis. *Gut, 53,* 108–114.

Mitsuoka, T. (1992). Intestinal flora and aging. *Nutrition Reviews, 50,* 438–446.

Mitsuoka, T. (1996). Intestinal flora and human health. *Asia Pacific Journal of Clinical Nutrition, 5,* 2–9.

Moreau, M. C., Ducluzeau, R., Guy-Grand, D., & Muller, M. C. (1978). Increase in the population of duodenal immunoglobulin A plasmocytes in axenic mice associated with different living or dead bacterial strains of intestinal origin. *Infectious Immunology, 21,* 532–539.

Morita, H., He, F., Kawase, M., Kubota, A., Hiramatsu, M., Kurisaki, J., & Salminen, S. (2006). Preliminary human study for possible alteration of serum immunoglobulin E production in perennial allergic rhinitis with fermented milk prepared with *Lactobacillus gasseri* TMC0356. *Microbiology and Immunology, 50,* 701–706.

Niers, L. E., Timmerman, H. M., Rijkers, G. T., van Bleek, G. M., van Uden, N. O., Knol, E. F.,

Nussler, A. K., & Thomson, A. W. (1992). Immunomodulatory agents in the laboratory and clinic. *Parasitology, 105 (Suppl),* S5–S23.

Oberhelman, R., Gilman, R. H., Sheen, P., Taylor, D., Black, R., Cabrera, L., Lescano, A., Mesa, R., & Madico, G. (1999). A placebo controlled trial of *Lactobacillus* GG to prevent diarrhoea in undernourished Peruvian children. *Journal of Paediatrics, 134,* 15–20.

Olivares, M., Paz Diaz-Ropero, M., Gomez, N., Sierra, S., Lara-Villoslada, F., Martin, R., Miguel Rodriguez, J., & Xaus, J. (2006). Dietary deprivation of fermented foods causes a fall in innate immune response. Lactic acid bacteria can counteract the immunological effect of this deprivation. *Journal of Dairy Research, 73,* 492–498.

Ouwehand, A. C. (2007). Antiallergic effects of probiotics. *Journal of Nutrition, 137 (Suppl 2),* 794S–797S.

Ouwehand, A., Isolauri, E., & Salminen, S. (2002). The role of the intestinal microflora for the development of the immune system in early childhood. *European Journal of Nutrition, 41 (Suppl 1),* I32–I37.

Parra, M. D., Martinez de Morentin, B. E., Cobo, J. M., Mateos, A., & Martinez, J. A. (2004). Daily ingestion of fermented milk containing *Lactobacillus casei* DN114001 improves innate-defense capacity in healthy middle-aged people. *Journal of Physiology and Biochemistry, 60,* 85–91.

Pathmakanthan, S., Li, C. K., Cowie, J., & Hawkey, C. J. (2004). *Lactobacillus plantarum* 299: Beneficial *in vitro* immunomodulation in cells extracted from inflamed human colon. *Journal of Gastroenterology and Hepatology, 19*, 166–173.

Pedone, C., Bernabeu, A., Postaire, E., Bouley, A., & Reinert, P. (1999). The effect of supplementation with milk fermented by *Lactobacilus casei* (strain DN-114 001) on acute diarrhoea in children attending day care centres. *International Journal of Clinical Practice, 53*, 179–184.

Pelto, L., Isolauri, E., Lilius, E. M., Nuutila, J., & Salminen, S. (1998). Probiotic bacteria down-regulate the milk-induced inflammatory response in milk-hypersensitive subjects but have an immunostimulatory effect in healthy subjects. *Clinical and Experimental Allergy, 28*, 1474–1479.

Pessi, T., Sutas, Y., Hurme, M., & Isolauri, E. (2000). Interleukin-10 generation in atopic children following oral *Lactobacillus rhamnosus* GG. *Clinical and Experimental Allergy, 30*, 1804–1808.

Phuapradit, P., Varavithya, W., Vathanophas, K., Sangchai, R., Podhipak, A., Suthutvoravut, U., Nopchinda, S., Chantraruksa, V., & Haschke, F. (1999). Reduction of rotavirus infection in children receiving bifidobacteria-supplemented formula. *Journal of the Medical Association of Thailand, 82 (Suppl 1)*, S43–S48.

Prasad, J., Gill, H. S., Smart, J. B., & Gopal, P. K. (1999). Selection and characterisation of *Lactobacillus* and *Bifidobacterium* strains for use as probiotics. *International Dairy Journal, 8*, 993–1002.

Prescott, S. L., Dunstan, J. A., Hale, J., Breckler, L., Lehmann, H., Weston, S., & Richmond, P. (2005). Clinical effects of probiotics are associated with increased interferon-γ responses in very young children with atopic dermatitis. *Clinical and Experimental Allergy, 35*, 1557–1564.

Quigley, E. M., & Flourie, B. (2007). Probiotics and irritable bowel syndrome: A rationale for their use and an assessment of the evidence to date. *Neurogastroenterology and Motility, 19*, 166–172.

Rachmilewitz, D., Katakura, K., Karmeli, F., Hayashi, T., Reinus, C., Rudensky, B., Akira, S., Takeda, K., Lee, J., Takabayashi, K., & Raz, E. (2004). Toll-like receptor 9 signaling mediates the anti-inflammatory effects of probiotics in murine experimental colitis. *Gastroenterology, 126*, 520–528.

Rafter, J. (2002). Scientific basis of biomarkers and benefits of functional foods for reduction of disease risk: Cancer. *British Journal of Nutrition, 88 (Suppl 2)*, S219–S224.

Rafter, J., Bennett, M., Caderni, G., Clune, Y., Hughes, R., Karlsson, P. C., Klinder, A., O'Riordan, M., O'Sullivan, G. C., Pool-Zobel, B., Rechkemmer, G., Roller, M., Rowland, I., Salvadori, M., Thijs, H., Van Loo, J., Watzl, B., & Collins, J. K. (2007). Dietary synbiotics reduce cancer risk factors in polypectomized and colon cancer patients. *American Journal of Clinical Nutrition, 85*, 488–496.

Rautava, S., Kalliomaki, M., & Isolauri, E. (2002). Probiotics during pregnancy and breast-feeding might confer immunomodulatory protection against atopic disease in the infant. *Journal of Allergy and Clinical Immunology, 109*, 119–121.

Renz-Polster, H., David, M. R., Buist, A. S., Vollmer, W. M., O'Connor, E. A., Frazier, E. A., & Wall, M. A. (2005). Caesarean section delivery and the risk of allergic disorders in childhood. *Clinical and Experimental Allergy, 35*, 1466–1472.

Roach, S., & Tannock, G. W. (1980). Indigenous bacteria that influence the number of *Salmonella typhimurium* in the spleen of intravenously challenged mice. *Canadian Journal of Microbiology, 26*, 408–411.

Roberfroid, M. B. (1998). Prebiotics and synbiotics: Concepts and nutritional properties. *British Journal of Nutrition, 80*, S197–S202.

Rokka, T., Syvaoja, E. L., Tuomine, J., et al. (1997). Release of bioactive peptides by enzymatic proteolysis of *Lactobacillus* GG fermented UHT milk. *Milchwissenschaft, 52*, 675–678.

Rook, G. A., & Brunet, L. R. (2005). Old friends for breakfast. *Clinical and Experimental Allergy, 35*, 841–842.

Rook, G. A., & Stanford, J. L. (1998). Give us this day our daily germs. *Immunology Today, 19*, 113–116.

Rosenfeldt, V., Benfeldt, E., Nielsen, S. D., Michaelsen, K. F., Jeppesen, D. L., Valerius, N. H., & Paerregaard, A. (2003). Effect of probiotic *Lactobacillus* strains in children with atopic dermatitis. *Journal of Allergy and Clinical Immunology, 111*, 389–395.

Saavedra, J. M., Bauman, N. A., Oung, I., Perman, J. A., & Yolken, R. H. (1994). Feeding of *Bifidobacterium bifidum* and *Streptococcus thermophilus* to infants in hospital for prevention of diarrhoea and shedding of rotavirus. *Lancet, 344*(8929), 1046–1049.

Sawamura, A., Yamaguchi, Y., Toge, T., et al. (1994). Enhancement of immuno-activities by oral administration of *Lactobacillus casei* in colorectal cancer patients. *Biotherapy, 8*, 1567–1572.

Schiffrin, E. J., Rochar, F., Link-Amster, H., et al. (1995). Immunomodulation of human blood cells following the ingestion of lactic acid bacteria. *Journal of Dairy Science, 78*, 491–497.

Sheih, Y. H., Chiang, B. L., Wang, L. H., Liao, C. K., & Gill, H. S. (2001). Systemic immunity-enhancing effects in healthy subjects following dietary consumption of the lactic acid bacterium *Lactobacillus rhamnosus* HN001. *Journal of the American College of Nutrition, 20(2 Suppl)*, 149–156.

Sheil, B., McCarthy, J., O'Mahony, L., Bennett, M. W., Ryan, P., Fitzgibbon, J. J., Kiely, B., Collins, J. K., & Shanahan, F. (2004). Is the mucosal route of administration essential for probiotic function? Subcutaneous administration is associated with attenuation of murine colitis and arthritis. *Gut, 53*, 694–700.

Sheil, B., Shanahan, F., & O'Mahony, L. (2007). Probiotic effects on inflammatory bowel disease. *Journal of Nutrition, 137(3 Suppl 2)*, 819S–824S.

Shornikova, A. V., Casas, I. A., Isolauri, E., et al. (1997). *Lactobacillus reuteri* as a therapeutic agent in acute diarrhoea in young children. *Journal of Pediatric Gastroenterology and Nutrition, 24*, 399–404.

Shu, Q., & Gill, H. S. (2002). Immune protection mediated by the probiotic *Lactobacillus rhamnosus* HN001 (DR20) against *Escherichia coli* O157:H7 infection in mice. *FEMS Immunology and Medical Microbiology, 34*, 59–64.

Sistek, D., Kelly, R., Wickens, K., Stanley, T., Fitzharris, P., & Crane, J. (2006). Is the effect of probiotics on atopic dermatitis confined to food sensitized children? *Clinical and Experimental Allergy, 36*, 629–633.

Solis Pereyra, B., & Lemonnier, D. (1991). Induction of $2'-5'$ A synthetase activity and interferon in humans by bacteria used in dairy products. *European Cytokine Network, 2*, 137–140.

Solis Pereyra, B., & Lemonnier, D. (1993). Induction of human cytokines by bacteria in dairy foods. *Nutrition Research, 13*, 1127–1140.

Spanhaak, S., Havenaar, R., & Schaafsma, G. (1998). The effect of consumption of milk fermented by *Lactobacillus casei* strain Shirota on the intestinal microflora and immune parameters in humans. *European Journal of Clinical Nutrition, 52*, 899–907.

Sudo, N., Sawamura, S., Tanaka, K., Aiba, Y., Kubo, C., & Koga, Y. (1997). The requirement of intestinal bacterial flora for the development of an IgE production system fully susceptible to oral tolerance induction. *Journal of Immunology, 159*, 1739–1745.

Sullivan, A., & Nord, C. E. (2006). Probiotic lactobacilli and bacteraemia in Stockholm. *Scandinavian Journal of Infectious Disease, 38*, 327–331.

Suskovic, J., Kos, B., Goreta, J., & Matosic, S. (2001). Lactic acid bacteria and Bifidobacteria in synbiotic effect. *Food Technology and Biotechnology, 39*, 227–235.

Szymanski, H., Chmielarczyk, A., Strus, M., Pejcz, J., Jawien. M., Kochan, P., & Heczko, P. B. (2006). Colonisation of the gastrointestinal tract by probiotic *L. rhamnosus* strains in acute diarrhoea in children. *Digest of Liver Diseases, 38 (Suppl 2)*, S274–S276.

Tamura, M., Shikina, T., Morihana, T., Hayama, M., Kajimoto, O., Sakamoto, A., Kajimoto, Y., Watanabe, O., Nonaka, C., Shida, K., & Nanno, M. (2007). Effects of probiotics on allergic rhinitis induced by Japanese cedar pollen: Randomized double-blind, placebo-controlled clinical trial. *International Archives of Allergy and Immunology, 143,* 75–82.

Tannock, G. W., Munro, K., Harmsen, H. J. M., Welling, G. W., Smart, J., & Gopal, P. K. (2000). Analysis of the fecal microflora of human subjects consuming a probiotic product containing *Lactobacillus rhamnsous* DR20. *Applied and Environmental Microbiology, 66,* 2578–2588.

Turchet, P., Laurenzano, M., Auboiron, S., & Antoine, J. M. (2003). Effect of fermented milk containing the probiotic *Lactobacillus casei* DN-114001 on winter infections in free-living elderly subjects: A randomised, controlled pilot study. *Journal of Nutrition, Health and Aging, 7,* 75–77.

Viljanen, M., Savilahti, E., Haahtela, T., Juntunen-Backman, K., Korpela, R., Poussa, T., Tuure, T., & Kuitunen, M. (2005). Probiotics in the treatment of atopic eczema/dermatitis syndrome in infants: A double-blind placebo-controlled trial. *Allergy, 60,* 494–500.

Weizman, Z., Asli, G., & Alsheikh, A. (2005). Effect of a probiotic infant formula on infections in child care centers: Comparison of two probiotic agents. *Pediatrics, 115,* 5–9.

Weston, S., Halbert, A., Richmond, P., & Prescott, S. L. (2005). Effects of probiotics on atopic dermatitis: A randomised controlled trial. *Archives of Diseases in Childhood, 90,* 892–897.

Wheeler, J. G., Shema, S. J., Bogle, M. L., et al. (1997). Immune and clinical impact of *Lactobacillus acidophilus* on asthma. *Annals in Allergy, Asthma and Immunology, 79,* 229–233.

Winkler, P., De Vrese, M., Laue, C.,, & Schrezenmeir, J. (2005). Effect of dietary supplement containing probiotic bacteria plus vitamins and minerals on common cold infections and cellular immune parameters. *International Journal of Clinical Pharmacology and Therapeutics, 43,* 318–326.

Woodmansey, E. J., McMurdo, M. E., Macfarlane, G. T., & Macfarlane, S. (2004). Comparison of compositions and metabolic activities of fecal microbiotas in young adults and in non-antibiotic-treated elderly subjects. *Applied and Environmental Microbiology, 70,* 6113–6122.

Xiao, J. Z., Kondo, S., Yanagisawa, N., Takahashi, N., Odamaki, T., Iwabuchi, N., Miyaji, K., Iwatsuki, K., Togashi, H., Enomoto, K., & Enomoto, T. (2006). Probiotics in the treatment of Japanese cedar pollinosis: A double-blind placebo-controlled trial. *Clinical and Experimental Allergy, 36,* 1425–1435.

Yamazaki, S., Kamimura, H., Momose, H., Kawashima, T., & Ueda, K. (1982). Protective effect of *Bifidobacterium*-monoassociation against lethal activity of *Escherichia coli*. *Bifidobacteria Microflora, 1,* 55.

Potential Anti-Inflammatory and Anti-Infectious Effects of Human Milk Oligosaccharides

C. Kunz and S. Rudloff

Abstract There is increasing evidence of the local effects within the gastro intestinal tract and the systemic functions of human milk oligosaccharides (HMO). In addition to the vast majority of *in vitro* data, animal studies underline the high potential of HMO to influence very different processes. HMO probably influence the composition of the gut microflora through effects on the growth of bifidus bacteria. Whether the concomitant low number of pathogenic microorganisms in breastfed infants is also caused by HMO is an intriguing question that still has yet to be proven. Due to the similarity of HMO to epithelial cell surface carbohydrates, an inhibitory effect on the adhesion of pathogens to the cell surface is most likely. If this could be shown in humans, HMO would provide a new way to prevent or treat certain infections. It would also indicate supplementing infant formula based on cow's milk with HMO, as those oligosaccharides are either not detectable or present only in low numbers in bovine milk. As some HMO can be absorbed and circulate in blood, systemic effects may also be influenced. Due to their similarities to selectin ligands, HMO have been tested in *in vitro* studies demonstrating their anti-inflammatory abilities. For example, it has been shown that sialic acid-containing oligosaccharides reduce the adhesion of leukocytes to endothelial cells, an indication for an immune regulatory effect of certain HMO. We cover these topics after a short introduction on the structures of HMO, with a particular emphasis on their blood group and secretor specificity.

Concentrations of Oligosaccharides in Human Milk

Human milk is a rich source of complex oligosaccharides synthesized within the mammary gland. The concentration of this fraction, however, varies widely due to the lactational stage decreasing from high amounts of up to 50 g/L or more in

---

C. Kunz
Institute of Nutritional Sciences, Justus Liebig University Giessen, Wilhlemstr 20,
35392 Giessen, Germany
e-mail: clemens.kunz@uni-giessen.de

colostrum to an average of 10–15 g/L in mature milk (Kunz et al., 1999). The interindividual variability can partly be explained by the genetic variance of the donors leading to different blood group–specific components as well as a distinguished set of milk oligosaccharides.

The major components among complex oligosaccharides are lacto-N-tetraose (0.5–1.5 g/L) and their monofucosylated derivatives, lacto-N-fucopentaose I or II (up to 1.7 g/L) (Table 1). Lacto-N-tetraose and its monofucosylated derivatives add up to 50–70% of the total complex carbohydrates. Of the sialylated components, the content of sialyllactose (NeuAcα2-6Lac and NeuAcα2-3Lac) is about 1.0 g/L followed by isomers of monosialylated lacto-N-tetraose and disialylated lacto-N-tetraose (Table 2). The total amount of complex oligosaccharides in mature milk is between 10–15 g/L. In bovine milk, only small amounts of oligosaccharides are detectable, with sialyllactose as the major component.

Table 1 Structures of Selected Neutral Human Milk Oligosaccharides for Cell Culture Studies

Abbreviation	Compound	Structure
LNT	Lacto-N-tetraose	Galβ1-3GlcNAcβ1-3Galβ1-4Glc
LNH	Lacto-N-hexaose	Galβ1-4GlcNAcβ1-6 Galβ1-4Glc 　　　　　　　　　　　3 　　　　　　　　　　　| 　　　　　　　　Galβ1-4GlcNAcβ1
LNFP I	Lacto-N-fucopentaose I	Galβ1-3GlcNAcβ1-3Galβ1-4Glc 　　　　　2 　　　　　| 　　　　Fucα1
LNFP II	Lacto-N-fucopentaose II	Galβ1-3GlcNAcβ1-3Galβ1-4Glc 　　　　　4 　　　　　| 　　　　Fucα1
LNFP III	Lacto-N-fucopentaose III	Galβ1-4GlcNAcβ1-3Galβ1-4Glc 　　　　　3 　　　　　| 　　　　Fucα1
LNDFH I	Lacto-N-difucohexaose I	Galβ1-3GlcNAcβ1-3Galβ1-4Glc 　　　　2　　4 　　　　|　　| 　　Fucα1 Fucα1
LNDFH II	Lacto-N-difucohexaose II	Galβ1-3GlcNAcβ1-3Galβ1-4Glc 　　　　4　　　　　　　3 　　　　|　　　　　　　| 　　Fucα1　　　　　Fucα1
2'FL	2'-Fucosyllactose	Fucα1-2Galβ1-4Glc
3-FL	3-Fucosyllactose	Galβ1-4Glc 　3 　| 　Fucα1

Potential Anti-Inflammatory and Anti-Infectious Effects 457

Table 2 Structures of Selected Acidic Human Milk Oligosaccharides

Abbreviation	Compound	Structure
SL	N-acetyl-neuraminyl lactose	Mixture of 3' SL and 6'SL (80%); remainder primarily 6'-sialyllactosamin (less than 5% lactose)
3'-SL	3'-Sialyllactose	Neu5Acα2-3Galβ1-4Glc
6'-SL	6'-Sialyllacotse	Neu5Acα2-6Galβ1-4Glc
LST a	LS-tetrasaccharide a	Galβ1-3GlcNAcβ1-3Galβ1-4Glc 3 | Neu5Acα2
LST b	LS-tetrasaccharide b	Galβ1-3GlcNAcβ1-3Galβ1-4Glc 6 | Neu5Acα2
LST c	LS-tetrasaccharide c	Galβ1-4GlcNAcβ1-3Galβ1-4Glc 6 | Neu5Acα2
DSLNT	Disialyl-lacto-N-tetraose	Galβ1-3GlcNAcβ1-3Galβ1-4Glc 3 6 | | Neu5Acα2 Neu5Acα2

Structural Considerations: Neutral Oligosaccharides and Lewis Blood Group Secretor Status

The presence of different neutral oligosaccharides in human milk depends on the activity of specific enzymes in the lactating gland. An α1-2-fucosyltransferase is expressed in about 77% of all Caucasians who are classified as secretors. Therefore, oligosaccharides in milk from these women are characterized by the presence of α1-2-fucosylated components, e.g., 2'-fucosyllactose (Fucα1-2Galβ1-4Glc), lacto-N-fucopentaose I (Fucα1-2Galβ1-3GlcNAcβ1-3Galβ1-4Glc), or more complex oligosaccharides, all possessing Fucα1-2Galβ1-3GlcNAc residues (Table 3).

In Lewis (a+b−) individuals, another fucosyltransferase attaches Fuc residues in α1-4 linkages to a subterminal GlcNAc residue of type 1 chains. Therefore, in "Lewis (a+b−), nonsecretor milk," the major fucosylated oligosaccharide is lacto-N-fucopentaose II (Galβ1-3[Fucα1-4]GlcNAcβ1-3Galβ1-4Glc) (Table 3). This characteristic component is found in about 20% of the population.

Table 3 Blood Group Specificity and Enzyme Activity in the General Population

Blood Groups	Enzyme Activity	% Population
Secretors	α1-2 Fucosyltransferase	70
Lewis a	α1-4 Fucosyltransferase	20
Lewis b	α1-2 Fucosyltransferase	70
	α1-4 Fucosyltransferase	
Lewis a⁻b⁻	α1-3 Fucosyltransferase	10

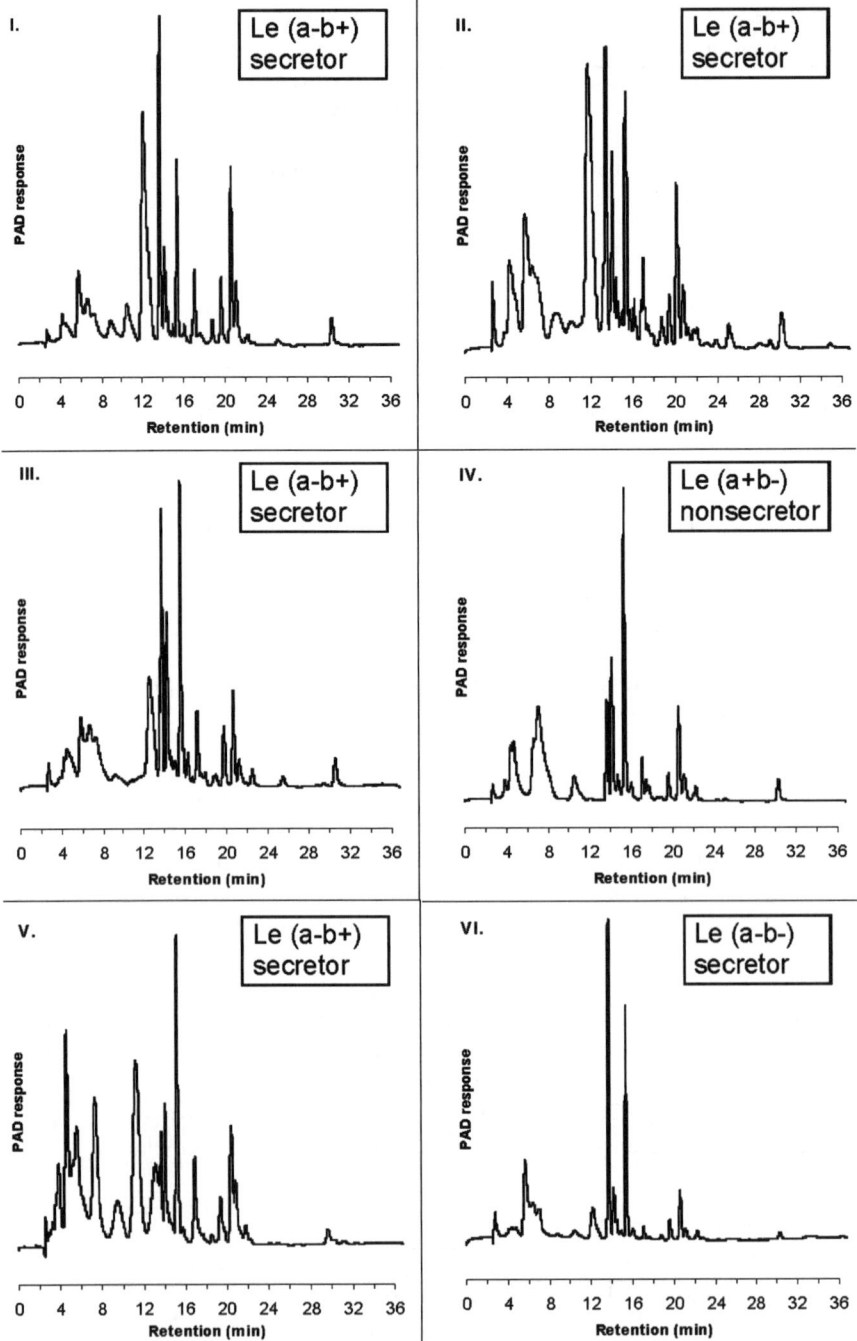

Fig. 1 HMO profile after high-pH anion exchange chromatograpy with pulsed amperometric detection (from Kunz et al., 1996). The figure demonstrates the large difference in the total amount and in the variety of HMO in women with a different Lewis blood group specificity

In Lewis (a−b+) donors, who represent about 70% of the population, both fucosyltransferases–the secretor gene and the Lewis gene–dependent form–are expressed. Here one of the major milk oligosaccharides is lacto-N-difucohexaose I (Fucα1-2Galß1-3[Fucα1-4]GlcNAcß1-3Galß1-4Glc) (Table 3).

In about 5% of the population who belong to blood group Lewis (a−b−), a third type of milk oligosaccharide was found carrying Fuc α1-3 linked to GlcNAc in type 2 chains. The major oligosaccharide in the milk of these donors is lacto-N-fucopentaose III (Galß1-4[Fucα1-3]GlcNAcß1-3Galß1-4Glc) (Table 3).

This different enzyme activity has a marked effect on the pattern and amount of oligosaccharides in milk (Figure 1). Whether this obvious difference has an impact on infants' health either immediately or in later life has not been investigated so far. Only one study has tried to identify 2-fucosyl-lactose as the decisive components in milk that should be responsible for a lower number of certain infections in breastfed infants (Morrow et al., 2004). However, the study was not designed to address this question as a primary outcome. It was a retrospective study in which the incidence of specific diseases had been linked to the presence of an individual component, i.e., 2-fucosyl-lactose. Major drawbacks of the study are its retrospective design, the presence of more than 100 other oligosaccharides in milk, and the presence of other anti-infectious components in milk that have not been investigated. Therefore, the interpretation of the study remains largely speculative.

Milk Oligosaccharides and the Gut

Oligosaccharides as Growth Factors for Bifidobacterium bifidum

A bifidus predominance of the intestinal flora of breastfed infants was reported by Moro more than 100 years ago (1900). He concluded that human milk contains a growth factor for these microorganisms. By using a "Bifidum mutant" (*Bifidobacterium bifidum subspecies Pennsylvanicum, B. bifidum subsp. Penn.*), György et al. (1953) referred to a mixture of oligosaccharides containing N-acetyl glucosamine (GlcNAc) they called gynolactose as the Bifidus factor. Many *in vitro* studies have demonstrated that GlcNAc-containing oligosaccharides are able to enhance the growth of *B. bifidum subsp. Penn.*, whereas other N-containing sugars showed less growth-promoting activity. Some of these *in vitro* data have not been verified in following experiments. In this context, it is necessary to recognize that the Bifidus strain used in György's group is a laboratory strain that does not regularly occur in the feces of breastfed infants. In addition, there is no evidence that the monosaccharide GlcNAc itself is a good growth factor for microbes in infants. However, *in vitro* studies clearly demonstrated a positive effect of GlcNAc, if this monosaccharide was present within an oligosaccharide structure.

In addition to oligosaccharides, there are several glycoconjugate fractions, i.e., glycoproteins or glycolipids, which may also have a bifidogenic effect. For the fraction of free oligosaccharides in human milk, it is likely that they have more specific effects both locally within the gastrointestinal tract and systemically.

Compared to the relevance of bifidobacteria as a marker for the breastfed infant's microflora, there are several other well-documented differences in the intestinal flora between human milk-fed and formula-fed infants. As Adlerberth (1997) thoroughly discussed, it has consistently been observed that breastfed infants have lower counts of clostridia and enterococci and higher numbers of staphylococci compared to formula-fed infants, even if the population density of bifidobacteria was similar in both groups. In addition, it is noteworthy that the influence of breastfeeding on bacteroides and lactobacilli seems to be very weak, if at all detectable.

As human milk is still the gold standard for the production of infant formula, new products like those containing prebiotics on the market were shown to be suitable to increase the number of bifidobacteria in the infants' gut; however, no data have been published so far with regard to the growth inhibition of pathogenic microorganisms. In human milk, oligosaccharide structures are present and prevent the adhesion of certain microorganisms by acting as soluble receptor analogs due to a specific monosaccharide composition and the linkages between those.

This raises the important issue that it is not the mere presence of growth factors in milk that determines bacterial colonization and leads to a higher number of bifidobacteria and lactobacilli, but that more specific mechanisms are responsible for the concomitant low numbers of pathogenic bacteria in the gut flora of breastfed infants. As will be discussed below, certain complex oligosaccharides in human milk might be involved in these processes.

Oligosaccharides as Inhibitors of Pathogen Adhesion to Epithelial Cells

The decisive factor in the pathophysiology of infectious diseases such as diarrhea seems to be the ability of microorganisms to adhere to the mucosal surface and their subsequent spreading, colonization, and invasion (e.g., for *Escherichia coli, Helicobacter jejuni, Shigella* strains, *Vibrio cholerae,* and *Salmonella* species) (Kunz et al., 1997; Karlsson, 1995). Bacterial adhesion is a receptor-mediated interaction between structures on the bacterial surface and complementary (carbohydrate) ligands on the mucosal surface of the host. Many examples from *in vitro* studies demonstrate the high potential of milk oligosaccharides to interfere with these specific host pathogen interactions. As it is still not possible to produce large amounts of complex milk-type oligosaccharides for clinical studies, *in vivo* data are still lacking. However, some oligosaccharides have recently been tested as antiadhesive drugs in animals. For example, the intranasal or intratracheal administration of either oligosaccharides or neoglycoproteins in rabbits and rat pups markedly reduced experimental pneumonia caused by *S. pneumonia* (Idänpään-Heikkilä et al., 1997).

In another recent study, *H. pylori*-positive rhesus monkeys were treated with 3'sialyllactose alone or in combination with one of the commonly used antiulcer drugs, bismuth subsalicylate (a gastric surface coating agent) or omeprazole (a proton pump inhibitor) (Mysore et al., 1999). Of the six monkeys given milk oligosaccharides, only two were permanently cured, and a third animal was transiently cleared, while three of the animals remained persistently colonized. According to the authors, the antiadhesive therapy is safe and can cure or reduce *H. pylori* colonization in rhesus monkeys.

Clinical experience with oligosaccharides as antiadhesive drugs is still very limited. Ukkonen et al. (2000) investigated the efficacy of 3'-sialyllacto-N-neotetraose given intranasally for the prophylaxis of acute otitis media. In this randomized, double-blind, placebo-controlled study, 507 healthy children had been assigned either to the acidic milk oligosaccharide or to placebo as intranasal sprays twice daily for three months. Although the study failed to reduce the incidence of nasopharyngeal colonization with *Streptococcus pneumoniae* and *Hemophilus influenzae* as well as that of acute otitis media, this may be, for example, due to the fact that during natural infection, bacteria can express multiple lectins with diverse specificities, the inhibition of which may require a cocktail of oligosaccharides (Sharon & Ofek, 2000).

Besides their potential in pathophysiological situations such as inflammation, milk-type oligosaccharides could also exert specific functions in physiological events, e.g., in the intestinal cell maturation. Present data suggest that the cell adhesion molecule MAdCAM-1 is involved in the homing of lymphocyte subsets to the intestinal tract (Berg et al., 1992; McEcer, 1994). It is strongly expressed in high endothelial cells in Peyers' patches and in mesenteric but not peripheral lymph nodes. In addition, it also contains a mucin-like domain that functions as a ligand for L-selectin, which allows the rolling of leukocytes along an MAdCAM-1–coated surface. Therefore, MAdCAM-1 has the structural requirements to contribute to both the rolling (by low-affinity binding) and the firm adhesion by very high-affinity binding through its interactions with integrins. In terms of infant nutrition, it is intriguing to speculate that oligosaccharides may interfere with this process, resulting in an increased leukocyte/lymphocyte accumulation in the mucosa, thus supporting phagocytosis or antibody production.

Recent observations indicate that HMO not only influence systemic processes such as leukocyte endothelial cell or leukocyte platelet interactions (Bode et al., 2004a, b), but also induce intestinal cellular processes (Kuntz et al., 2004, 2007). These oligosaccharides seem to affect the gut in at least two different ways: as growth factors influencing normal gut development and maturation and as anti-inflammatory and immune components modulating the intestinal immune system.

Systemic Effects: Leukocyte-Endothelial Cell Interactions

Human milk is a rich source of oligosaccharides with structural similarities to the binding determinants of selectin ligands (Rudloff et al., 2002). Since oligosaccharides characteristic of those in human milk are also present in the urine of human milk-fed infants, it can be assumed that an intestinal absorption occurs (Kunz et al., 2000; Rudloff et al., 1996). These data were furthermore supported by *in vitro* studies on Caco-2 cells demonstrating that nHMO are transported across the intestinal epithelium by receptor-mediated transcytosis and paracellular pathways, whereas acidic HMO cross the intestinal lining via paracellular routes (Gnoth et al., 2001). The estimated amount of oligosaccharides present in the circulation of human milk-fed infants is comparable to serum concentrations after a dosage of carbohydrate drugs currently under investigation and may thus be high enough to have an impact on immune reactions.

Excessive leukocyte infiltration causes severe tissue damage in a variety of inflammatory diseases (Lasky, 1995; Carden & Granger, 2000). The initial step leading to leukocyte extravasation is mediated by selectins on activated endothelium and their oligosaccharide ligands on leukocytes (Springer, 1994). As HMO contain binding determinants for selectins, they exert the potential to affect leukocyte rolling and adhesion to endothelial cells.

It has recently been shown that the adhesion of monocytes, lymphocytes, or neutrophils isolated from human peripheral blood passing over TNF-α-activated endothelial cells (HUVEC, human umbilical vein endothelial cells) was reduced by up to 50% using sialylated HMO (Bode et al., 2004a). These effects were even more pronounced than those achieved by soluble sialyl-Lewis x, a physiological binding determinant for selectins. Several active components within the oligosaccharide fraction of human milk were identified, e.g., 3'-sialyl-lactose and 3'-sialyl-3-fucosyl-lactose. These results indicate that specific oligosaccharides in human milk may serve as anti-inflammatory components and might therefore contribute to the lower incidence of inflammatory diseases in human milk-fed infants.

The most intriguing question generated by the results of the present study is whether a reduced leukocyte adhesion in human milk-fed infants can be regarded as a benefit or a disadvantage. The lower incidence of infectious and inflammatory diseases in human milk-fed infants compared to formula-fed infants supports the beneficial effects of human milk. Taking the pathogenic understanding for a prevalent chronic disorder such as neonatal necrotizing enterocolitis (NEC) as a model, oligosaccharides from human milk might at least act on two different levels: (1) bacterial colonization and the local adhesion of pathogens to the intestinal surface are key events in the pathogenesis of NEC. HMO were reported to modify bacterial colonization and to prevent the attachment of several pathogens to the intestinal epithelium (Sharon & Ofek, 2000). HMO might therefore influence the onset of the disease; and (2) an excessive leukocyte infiltration, followed by the production of reactive oxygen species, was shown to play a major role in the progression of NEC (Musemeche et al., 1991). The resulting tissue necrosis leads to a breakdown of the epithelial barrier and is followed by bacterial translocation, sepsis, and multiorgan failure in severe cases. As observed in the present study, acidic oligosaccharides isolated from human milk were able to reduce both leukocyte rolling and adhesion and might therefore be regarded as beneficial. The combination of both mechanisms could be a possible explanation for the lower incidence of NEC in human milk- versus formula-fed infants (Wright et al., 1989; Lucas & Cole, 1990; Koloske, 2001). However, it should be stressed that the inhibition of leukocyte adhesion at sites of inflammation in newborn infants *in vivo* through HMO to an extent that has been observed in *in vitro* studies needs to be further investigated.

Conclusions and Perspective

Human milk contains a large variety of oligosaccharides with the potential to modulate the gut flora, to affect different gastrointestinal activities, and to influence inflammatory processes. Whether these functions occur *in vivo* in the human milk-fed infant has not been proven yet. One reason is that larger amounts of milk oligosaccharides needed for clinical studies are not available.

Currently, many studies are comparing the effects of plant-derived prebiotic digosachaides with those of milk oligosaccharides, although no structural similarity exists between the two kinds of carbohydrates. In addition, the effect on the growth of bifidobacteria may be influenced by HMO as well as on other major pathogenic microbes. As human milk is still the gold standard for the production of infant formula, new products like those containing prebiotics have been shown to be suitable to increase the number of bifidobacteria in the infants' gut; however, no data have been published so far with regard to the growth inhibition of pathogenic microorganisms that can be demonstrated for breastfed infants.

Studies in humans with infant formula supplemented milk oligosaccharides are still lacking, although there is currently a high level of activity among food companies to produce enough oligosaccharides by different techniques. It can be assumed that soon the first individual milk oligosaccharides may be available for clinical studies. This raises many questions such as

- which component(s) should be added,
- how much should be added to an infant formula,
- what is their metabolic fate, or
- which infants should receive it?

Another intriguing question relates to the susceptibility of infants to infectious diseases depending on the amount and type of oligosaccharides they receive from their mother's milk. There are large differences in the oligosaccharide pattern; the total amount an infant receives per day depends on the mother's blood group and secretor status. Therefore, the question to be addressed in the future is whether this difference leads to more infections due to the intake of lower amounts of specific oligosaccharides, e.g., in infants from mothers who are Lewis-negative compared to milk from mothers with Lewis a or Lewis b status.

References

Adlerberth, I. (1997). The establishment of a normal intestinal microflora in the newborn infant. In L. A. Hanson and R. B. Yolken (Eds.), *Probiotics, Other Nutritional Factors, and Intestinal Flora* (pp. 63–78). 42nd Nestlé Nutrition Workshop in Beijing (China). New York: Raven Press.

Beachey, E. H. (1981). Bacterial adherence: Adhesin-receptor interactions mediating the attachment of bacteria to mucosal surfaces. *Journal of Infectious Diseases, 143*, 325–345.

Berg, E. L., Magnani, J., Warnock, R. A., Robinson, M. K., Butcher, E. C. (1992). Comparison of L-Selectin and E-Selectin Ligand Specificities: The L-Selection Can Bind the E-Selection Ligands Sialyl Le^x and Sialyl Le^a. *Biochemical and Biophysical Research Communicaitons, 184*, 2, 1048–1055.

Inhibition of monocyte, lymphocyte, and neutrophil adhesion to endothelial cells by human milk oligosaccharides. *Thrombosis and Haemostasis, 92*, 1402–1410.

Bode, L., Rudloff, S., Kunz, C., Strobel, S., & Klein, N. (2004b). Human milk oligosaccharides reduce platelet-neutrophil complex formation leading to a decrease in neutrophil ß2 integrin expression. *Journal of Leukocyte Biology, 76*, 1–7.

Carden, D. L., & Granger, D. N. (2000). Pathophysiology of ischemia-reperfusion injury. *Journal of Pathology, 190*, 255–266.

Engfer, M. B., Stahl, B., Finke, B., Sawatzki, G., & Daniel, H. (2000). Human milk oligosaccharides are resistant to enzymatic hydrolysis in the upper gastrointestinal tract. *American Journal of Clinical Nutrition, 71*, 1589–1596.

Gibson, G., & Roberfroid, M. B. (1995). Dietary modulaton of the human colonic microbiota: Introducing the concept of prebiotics. *Journal of Nutrition, 125*, 1401–1412.

Gnoth, M. J., Kunz, C., Kinne-Saffran, E., & Rudloff, S. (2000). Human milk oligosaccharides are minimally digested *in vitro*. *Journal of Nutrition, 130*, 3014–3020.

György, P. (1953). A hitherto unrecognized biochemical difference between human milk and cows milk. *Pediatrics, 11*, 98–108.

Idänpään-Heikkilä, I., Simon, P. M., Zopf, D., Vullo, T., Cahill, P., Sokol, K., & Tuomanen, E. (1997). Oligosaccharides interfere with the establishment andprogression of experimental pneumococcal pneumonia. *Journal of Infectious Diseases, 176*, 704–712.

Kannagi, R. (2002). Regulatory roles of carbohydrate ligands for selectins in the homing of lymphocytes. *Current Opinion in Structural Biology, 12*, 599–608.

Karlsson, K. A. (1995). Microbial recognition of target-cell glycoconjugates. *Current Opinion in Structural Biology, 5*, 622–635.

Kosloske, A. M. (2001). Breast milk decreases the risk of neonatal necrotizing enterocolitis. Adv Nutr Res, 10,123–137.

Kuntz, S., Henkel, C., Rudloff, S., & Kunz, C. (2003). Effects of neutral oligosaccharides from human milk on proliferation, differentiation and apoptosis in intestinal epithelial cells. *Proceedings of the Annual Meeting of Experimental Biology*, San Diego.

Kunz, C. et al. (1996).

Kunz, C., Rudloff, S., Hintermann, A., Pohlentz, G., Egge, H. (1996). High-pH anion exchange chromatography with pulsed amperometric detection and molar response facotr of human milk oligosacchrides. *J Chromatogr B Biomed Appl, 685*, 211–221.

Kunz, C., Rudloff, S., Schad, W., & Braun, D. (1999). Lactose-derived oligosaccharides in the milk of elephants: Comparison with human milk. *British Journal of Nutrition, 82,* 391–399.

Kunz, C., Rudloff, S., Baier, W., Klein, N., & Strobel, S. (2000). Oligosaccharides in human milk. Structural, functional and metabolic aspects. *Annual Review of Nutrition, 20,* 699–722.

Lasky, L. A. (1995). Selectin-carbohydrate interactions and the initiation of the inflammatory response. *Annual Review of Biochemistry, 64,* 113–139.

Lucas, A., Cole, T. J. (1990). Breast milk and neonatal necrotizing enterocolits. *Lancet, 336,* 1519–1523.

McEver, R. P. (1994). Selectins. *Curr Opin Immunol, 6,* 75–84.

Moro, E. (1900). Morphologische und bakteriologische Untersuchungen über die Darmbakterien des Säuglings: Die Bakteriumflora des normalen Frauenmilchstuhls [in German]. *Jahrbuch Kinderh, 61,* 686–734.

Morrow, A. L., Ruiz-Palacios, G. M., Altaye, M., Jiang, X., Guerrero, M. L., Meinzen-Derr, J. K., Farkas, T., Chaturvedi, P., Pickering, L. K., & Newburg, D. S. (2004). Human milk oligosaccharides are associated with protection against diarrhea in breast-fed infants. *Journal of Pediatrics, 145,* 297–303.

Mysore, J. V., Wigginton, T., Simon, P. M., Zopf, D., Heman-Ackah, L. M., & Dubois, A. (1999). Treatment of *Helicobacter pylori* infection in rhesus monkeys using a novel anti-adhesion compound. *Gastroenterology, 117,* 1316–1325.

Ofek, I., & Sharon, N. (1990). Adhesins as lectins: Specificity and role in infection. *Current Topics in Microbiology and Immunology, 151,* 91–114.

Rudloff, S., Pohlentz, G., Diekmann, L., Egge, H., & Kunz, C. (1996). Urinary excretion of lactose and oligosaccharides in preterm infants fed human milk or infant formula. *Acta Paediatrics, 85,* 598–603.

Sharon, N., & Ofek, I. (2000). Safe as mother's milk: Carbohydrates as future anti-adhesion drugs for bacterial diseases. *Glycoconjugate Journal, 17,* 659–664.

Springer, T. A. (1990). Adhesion receptors of the immune system. *Nature, 346,* 425–434.

Tungland, B. C., & Meyer, P. D. (2002). Dietary fiber and human health. *Comprehensive Reviews in Food Science and Food Safety, Vol. 2.*

Ukkonen, P., Varis, K., Jernfors, M., Herva, E., Jokinen, J., Ruokokoski, E., Zopf, D., & Kilpi, T. (2000). Treatment of acute otitis media with an antiadhesive oligosaccharide: A randomised, double-blind, placebo-controlled trial. *Lancet, 356,* 1398–1402.

Wright, A. L., Holberg, C. J., Martinez, F. D., et al. (1989) Breast feeding and lower respiratory tract illness in the first year of life. Group Health Medical Associates. *BMJ, 299,* 946–949.

On the Role of Breastfeeding in Health Promotion and the Prevention of Allergic Diseases

L. Rosetta and A. Baldi

Abstract Based on animal models, we specify the major role of different bioactive milk components known to participate significantly in neonatal health promotion and in protection against a large number of infectious diseases and the development of allergies and asthma.

Introduction

The increasing prevalence of allergic disease during the past 30 years, particularly in industrialized countries, was shown in several recent epidemiological studies among both adults and children (Butland et al., 1997; Hill et al., 1999; Chan-Yeung et al., 2000). Such an alarming issue leads us to question the mechanisms that confer a predisposition to allergies and to examine the possibility to decrease or at least limit the occurrence of allergic episodes or try to control the risk factors likely to predispose individuals to allergies later in life (Kramer, 1988; Kjellman & Nilsson, 1999). Prevention is always preferred, even if significant progress can be initiated to control the frequency of episodes or the severity of symptoms (Exner & Greinecker, 1996; Prescott, 2003).

Allergy is defined as an immune-mediated inflammatory response to common environmental allergens that are otherwise harmless (Douglass & O'Hehir, 2006). Although allergic diseases are mainly determined by a combination of genetic and environmental factors, the processes that lead to allergies are initiated early in life (Forsyth et al., 1985). Consequently, the main public health measures to decrease the prevalence of allergic diseases will

---

L. Rosetta
CNRS UPR 2147, 44 rue de , Amidal Mouchez, 75044 PARIS, France, Ph: +33 4 43435644, Fax: +33443145630
e-mail: lyliane.rosetta@ivry.cnrs.fr

act on environmental factors, not only atmospheric pollution, but also the food, quality of housing, and lifestyle of infants and children (Fiocchi et al., 1984).

In the prevention of allergy, in atopic or nonatopic families, one aspect is the very early initiation of an immunomodulatory environment in infants, and the key period following birth seems to be a determinant (Halken, 2004). Many authors have highlighted the possible role of the mother's health and diet during pregnancy related to the occurrence of allergies in the infant, for example, for cow's milk allergy (Baylis et al., 1983), and the risk of intrauterine exposure to allergens. Consequently, asthma and gene-environment interaction exposure during the intrauterine period were investigated (Becker, 2005). Various degrees of risk linked to gender and birth order have also been mentioned (Businco et al., 1983). Parental smoking was diagnosed as an aggravating factor (Cogswell et al., 1987). During infancy, the main atopic symptoms are atopic dermatitis, recurrent wheezing, and adverse reaction to food, particularly cow's milk protein and chicken egg. We will examine the possible protective effect of breast milk against allergic diseases and try to explain contradictory results (Duchen et al., 2000; Harris et al., 2001).

Preventive programs set up to decrease the risk of allergic diseases in babies at risk of atopy have focused on the first six months of life (Bruno et al., 1993). Various studies have questioned the difference in the risk of allergic disease between cow's milk protein and soy milk protein in early life; the results have not been conclusive. The only quasi-certitude was that breast milk can confer a small advantage of long-term protection against respiratory infection (Burr et al., 1993). Other authors have compared the effects of different infant feeding to exclusively breastfed infants (Chandra et al., 1989; Chandra, 1997, 2000). Breastfeeding has been shown to confer immunological protection during a critical period in life (Hoppu et al., 2001). Exclusive breastfeeding for at least four to six months and late solid food introduction (after four to six months) are associated with fewer atopic diseases, food allergies, eczema, and wheezing. The major health advantage of breastfeeding that has been clearly demonstrated remains in the protection of the infant from certain infections in early life (Golding et al., 1997).

The immature immunocompatibility system is influenced by the composition of diet; either milk composition or the timing of introduction of weaning food seems capital. The contact with potential allergens during the first year of life (Halken, 2003) seems particularly determinant for the predisposition of children in the onset of allergic diseases (Gruskay, 1982).

This chapter examines the actual knowledge on the role of breastfeeding in the prevention of allergic diseases. We focus on the modification to the composition of human milk at different stages of lactation, i.e., colostrum and early lactation, mature milk at midlactation, milk composition and characteristics during late lactation, and the possible preventive effect of its components on the development of allergic disease (Kemp & Kakakios, 2004). We briefly describe the different attitudes toward breastfeeding today in industrialized and developing countries (Kannan et al., 1999) and the influence of public health policy and support to women after delivery in order to encourage the choice to breastfeed newborns (Haque et al., 2002).

Bioactive Components of Breast Milk

Breast milk provides a wide variety of proteins that have a unique amino acid composition and highly desirable physicochemical properties. Proteins in human milk are an important source of amino acids for the growth of breastfed infants. In addition, they display several extranutritional properties to promote the development and healthfulness of the neonate. The physiological significance of milk proteins includes the modulation of digestive and gastrointestinal functions, hemodynamic regulation (modulation), enhancement of immune function, and modulation of intestinal microflora (Schanbacher et al., 1998; Lönnerdal, 2003). While several bioactivities are directly related to native proteins, others are latent until the proteolytic breakdown of milk proteins generates a number of peptides endowed with various biological properties. Once they are liberated in the body, bioactive peptides may act as regulatory components with hormone-like activity (Meisel, 1997; Schanbacher et al., 1997). Proteolysis in the mammary gland can be attributed to native proteases, such as the plasmin/plasminogen system (Baldi et al., 1996; Politis, 1996; Fantuz et al., 2001). Proteins are subjected to hydrolytic breakdown during gastric processing and later upon exposure of the proteins with indigenous or intestinal bacteria-derived enzymes in the gut (Baldi et al., 2005).

The physiological role of bioactive milk proteins and peptides is supported by other nonnutrient compounds such as lipids, oligosaccharides, and complex carbohydrates: These multifunctional components exert synergistic effects in order to promote the healthfulness of the newborn (Lönnerdal, 2000).

Early Nutrition and Prevention of Infections

It is well recognized that breastfeeding can reduce the incidence and severity of respiratory and gastrointestinal infections in the neonate in both developing and developed countries (Chien & Howie, 2001). Furthermore, the value of breast milk may extend well beyond weaning and may have significant effects on the mature stages of life. Human neonates that are breastfed for up to six months have fewer health-related problems later in life than formula-fed infants: several epidemiological studies have associated breastfeeding with reduced incidence of type 1 diabetes, celiac disease, inflammatory bowel disease, rheumatoid arthritis, asthma, and eczema (Field, 2005). The physiological and protective effects of the bioactive components of human milk have not been obtained from studies in infants but rather from *in vitro* data or studies performed with laboratory animals. The precise role of milk protein peptides on mucosal immunity in humans must be determined.

Antimicrobial Components of Human Milk

Breast milk contains a wide range of antimicrobial components that exert their protective role on both the lactating mammary gland and the suckling neonate at a time when its immune system is still immature. Human milk provides a multi-layered defense system: pathogens can be inactivated directly by antimicrobial lipids and proteins, such as secretory antibodies, lactoferrin, and lysosyme, and can be prevented from binding to cellular receptors and co-receptors by oligosaccharides and carbohydrates (Isaacs, 2005). Secretory antibodies provide protection against antigens and pathogens that the mother has been exposed to; their protective effect is particularly evident in diarrheal infections (Telemo & Hanson, 1996; Lönnerdal, 2000). Glycoconjugates and oligosaccharides seem to protect the infant by blocking pathogens or toxin binding to the cell surface (Newburg, 1996). The lipids in human milk exert antiviral, antibacterial, and antiprotozoal activity only after digestion in the gastrointestinal tract (Isaacs et al., 1990). A multitude of proteins in human milk have inhibitory activity against pathogenic bacteria, virus, and fungi. Among these, lactoferrin shows broad-spectrum antimicrobial activity against a number of microorganisms, including *Escherichia coli* (Dionysius et al., 1993), *Bacillus* spp. (Oram & Reiter, 1968), *Salmonella typhimurium*, *Shigella dysenteriae,* and *Streptococcus mutans* (Batish et al., 1988; Payne et al., 1990). *In vivo* studies on young calves demonstrated that the administration of lactoferrin, combined with the lactoperoxidase system, reduced the number of CFU (colony forming units) of *E. coli* in digesta and feces (Van Leeuwen et al., 2000). Moreover, mice fed milk supplemented with bovine lactoferrin had a reduced proliferation of *Clostridium* species in the gut (Teraguchi et al., 1995). Li and Mine (2004) showed that the administration of lactoferrin before the intravenous injection of bacterial toxin lipopolysaccharides (LPS) in piglets reduced the mortality compared to BSA- (bovine serum albumin) treated animals. Pecorini et al. (2003) tested the antimicrobial activity of recombinant porcine lactoferrin, expressed in *Pichia pastoris* yeast, on *E. coli* strains, isolated from piglets with neonatal and weaning diarrhea, showing a 30% inhibitory effect of recombinant lactoferrin at a concentration of 0.1 mg/mL on bacterial growth compared with the control (0 mg/mL). The antibiotic effect of lactoferrin is also mediated by the generation of a potent peptide, lactoferricin, after proteolytic digestion (Bellamy et al., 1992). *In vitro* studies have shown that bovine lactoferrin peptides exhibited antibacterial activity against a number of Gram-positive and Gram-negative bacteria (Hoek et al., 1997). A novel antimicrobial peptide, derived from bovine lactoferrin and designated lactoferrampin, exhibited candidicidal activity and was active against *B. subtilis, E. coli,* and *Pseudomonas aeruginosa* (Van der Kraan et al., 2004).

Probiotic Support of Intestinal Microflora

One of the earliest documented differences between breastfed and artificially fed infants was the microbial populations in their gut (Balmer & Wharton, 1989). Breastfed infants have a reduced number of potentially pathogenic bacteria, such as *E. coli, Bacteroides, Campylobacter,* and *Streptococci,* and an increased number of *Bifidobacteria* and *Lactobacilli,* whereas the microflora composition of formula-fed infants closely resembles that of the adult gut. It is generally believed that the acidic microenvironment produced by lactic acid bacteria inhibits the proliferation of pathogenic bacteria, thereby enhancing the defense against infection (Lönnerdal, 2003; Newburg, 2005). Several factors in milk are known to have a prebiotic activity. Milk, and particularly colostrum, contains a family of oligosaccharides that are potent promoters of *Bifidobacteria* growth (Newburg, 1996). Bifidobacterial proliferation is also enhanced by κ-casein peptides and lactoferrin. Several studies showed that lactoferrin is able to stimulate the growth of bacteria belonging to *Lactobacillus* and *Bifidobacterium* genera (Liepke et al., 2002; Kim et al., 2004). This growth-promoting effect is dependent on the strain and may be related to the presence of lactoferrin binding proteins on the surface and in the cytosolic fraction of sensitive bacterial cells (Kim et al., 2004). *In vitro* studies on the ability of recombinant porcine lactoferrin expressed in *P. pastoris,* bovine lactoferrin, and human lactoferrin to stimulate the growth of four *Lactobacillus* strains showed that *Lb. casei ssp casei* was most stimulated by low concentrations of both proteins, whereas other strains were not affected (Pecorini et al., 2005). It is worthy to note that the effect of human milk bioactive compounds on the growth of a beneficial intestinal microflora is the result of a synergistic interaction among the different (peptide and nonpeptide) functional components of milk (Schanbacher et al., 1998).

Immunomodulatory Milk Peptides: The Dichotomy Between Suppression and Induction

Breast milk contains a multitude of unique components and nutrients that provide the exclusive protein supply for the newborn: All human neonates depend upon milk proteins until they enter the fifth month of their life. Besides the nutritional aspects, it has been well recognized that bioactive components in human milk influence the immune status of the neonate; the bioactivities of milk not only provide protection but also "educate" the infant immune system in the early postnatal period (Baldi et al., 2005). Human neonates are born with an immature acquired immune system, and many of the innate components of mucosal immunity are not fully developed. The first few months are very critical because neonates are exposed to a large number of microorganisms, foreign proteins, and chemicals, and their immune system should develop oral tolerance to nutrient molecules and avoid tolerance to pathogen-derived antigens (Field, 2005; Baldi et al., 2005). The term "immunomodulation" was purposely adopted to indicate that suppression of the immune system or induction toward pathogen-derived antigens may be required in certain instances (oral tolerance). The successful development of tolerance contributes to lower incidences of food-related allergies in breastfed infants (van Odijk et al., 2003).

Mounting evidence suggests that several peptides generated by the hydrolysis of milk proteins can regulate the overall immune function of the neonate. Pecquet et al. (2000) reported that mice fed with either bovine β-lactoglobulin tryptic hydrolysates or fractions of the hydrolysate were tolerized against β-lactoglobulin. Prioult et al. (2004) showed that *Lb. paracasei* peptidases were capable of further hydrolyzing tryptic-chymotryptic peptides of β-lactoglobulin. Furthermore, they reported that these peptides repressed lymphocyte proliferation and upregulated interleukin-10 (IL-10) production. Hydrolysis of casein with pepsin and trypsin or additionally with enzymes derived from *Lb. casei* strain GG generates molecules capable of inhibiting lymphocyte proliferation (Sutas et al., 1996). Thus, indigenous enzymes in the gut together with enzymes of bacterial origin have been proven beneficial in the downregulation of hypersensitivity reactions to ingested proteins in human neonates.

The main mechanism of protection against pathogen-derived antigens provided by mucosal immunity is mediated through IgA-producing cells and secretory IgA that neutralize and thus prevent the entry of potentially harmful antigens in the host. Thus, stimulation of the local immune response can be effective in preventing certain diseases caused by microorganisms entering the host through the oral route. A great number of studies demonstrate the ability of milk peptides to enhance the immune function (for references, see Clare & Swaisgood, 2000; Hill et al., 2000). Politis and Chronopoulou provide an extensive discussion of this topic in the present book. Politis et al. (2003) also showed that the effect of casein digests on immune parameters may depend upon the immune system's maturity. The effect of two peptides, corresponding to residues 191–193 and 63–68 of bovine β-casein, were studied on membrane-bound urokinase plasminogen activator (uPA) and expression of major histocompatability complex (MHC) class II antigens by porcine blood neutrophils. Results indicated that both peptides reduced the amount of u-PA present on the cell membrane as well as the expression of MHC class II antigens only at the time of weaning and not four weeks later. Thus, β-casein synthetic peptides appear to be effective only at a time that coincides with the immaturity of the immune system; they are not effective when the immune system has presumably gained its full functionality.

The Prevention of Allergic Disease in Childhood

Predictive Factors of and Risk Factors for Allergy

Allergy is an excessive immune-mediated response to various allergens; atopic subjects have a genetic predisposition to develop immunoglobulin E antibodies (IgE) in response to exposure to allergens (Arshad, 2001). In families where one or both parents are tested positive to atopy by skin-prick tests, their children seem more likely to develop allergic diseases (Prescott & Tang, 2005), although the nature of the relationship between childhood wheezing and atopy remains uncertain (Kurukulaaratchy et al., 2006). The most common allergic disorders are asthma, rhinitis, eczema, and urticaria. The form of allergic diseases can change from infancy to childhood and adulthood for the same individual, according to his or her exposure to allergens and to the previous development of the disease. In infancy, the most common symptoms are atopic dermatitis, gastrointestinal symptoms, and recurrent wheezing. Association with different symptoms is often seen; for example, asthma can be associated with wheezy bronchitis, allergic rhinitis, and/or conjunctivitis (Aberg et al., 1989). A number of epidemiological studies based on a large cohort of children have highlighted the apparent increase in the prevalence of allergic diseases among children. This trend was mainly observed in industrialized countries (Butland et al., 1997; Wilson et al., 1998; Habbick et al., 1999; Kull et al., 2002; Chulada, 2003; Kummeling et al., 2005, 2007).

Two hypotheses have recently been proposed to explain the increase in the prevalence of allergy observed in developed countries. One hypothesis suggests a more frequent exposure to environmental pollution either through a change in the level and frequency of food allergens, via possible chemical contamination of vegetables fruits or water, or through the increase in air pollution by sulfur dioxide, nitrogen oxides, ozone, or diesel exhaust particles and tobacco smoke. The second hypothesis suggests that the change in lifestyle with larger urban populations and less contact with animals, livestock, poultry, and pets is likely to contribute to the development of atopy, and atopic sensitization is associated with allergic disorders in children (Arshad, 2001; McGeady, 2004). In addition, prenatal exposure to food allergens seems likely to influence fetal immune responses associated with eczema and allergic sensitization during childhood. Some food ingredients, mainly the composition in fatty acid—associated with a higher maternal intake of margarine and vegetable oils—are positively associated with higher odds ratio of allergic diseases in young children and negatively associated with high maternal fish intake. The authors of this study concluded that mothers' intake of foods rich in n-6 polyunsaturated fatty acid during pregnancy may increase the risk of allergic disease in their offspring, while foods rich in n-3 polyunsaturated fatty acid may decrease the risk (Sausenthaler et al., 2007). Parental smoking and particularly smoking of the mother during pregnancy and/or during breastfeeding are always associated with a significantly increased risk of childhood wheezing (Zacharasiewicz et al., 1999) and a higher risk of asthma. The severity and frequency of episodes of asthma in these children are related to the magnitude of exposure (Halken, 2004). Other environmental allergens have the same impact. In a recent epidemiological study carried out in Menorca, Spain, on prenatal exposure to dichlorodiphenyldichloroethylene (DDE), a widely used insecticide, it was shown that prenatal exposure to organochlorine compounds was related to an increased risk of asthma for the infant. DDE measured in the umbilical cord was associated with an increase in wheezing at 4 years of age for children who were followed since birth (Sunyer et al., 2005).

Sensitization to food allergens, which can start during the prenatal period, continues during early life with the sensitization to inhalent allergens. High levels of gene-environment interactions have been reported, especially in atopic asthma and its related traits (Martinez, 2007). For many years, house dust mites were suspected to have the highest responsibility in the prevalence of asthma, but different interventional studies to protect high-risk babies against house dust mites were recently inconclusive (Marks et al., 2006; Woodcock et al., 2004). The development of allergic diseases seems to be the combination of a genetic predisposition with an early sensitization to various allergens, and the timing and magnitude of different environmental factors seem to have a complicated interaction with genetic factors (Epton et al., 2007). The phenotypic expression of the disease can be facilitated by the presence of nonspecific adjuvant factors like tobacco smoke, infections, or urban air pollution (Thestrup-Pedersen, 2003; Halken, 2004; Sunyer et al., 2006a). Overall, allergic rhinitis decreased with geographical latitude, but a large sample of adults recently recruited throughout the world using a standardized allergen protocol has shown the main association of the prevalence of allergic rhinitis with language groups, reflecting unknown genetic and cultural risk factors (Wjst et al., 2005).

The prevalence of allergic diseases and sensitization seems to be inversely related to socioeconomic level (Almqvist et al., 2005), but the proportion of children sensitized to cats or dogs was not higher in households with or without pets (Arshad, 2001) and no relationship with feather bedding compared to synthetic bedding was shown in recent comparative studies (Nafstad et al., 2002).

The association of atopic dermatitis in infancy with reduced neonatal macrophage inflammatory protein-1β levels suggests a link with immature immune responses at birth. There is a possible impairment of cutaneous adaptation to extrauterine life in eczema and atopic dermatitis (Sugiyama et al., 2007). Among other associated factors, body composition (Oddy, 2004) and obesity have been identified as possible additional risk factors for respiratory allergy when combined with environmental factors like urban lifestyle compared to farm life (Radon & Schulze, 2006). In the KOALA birth cohort study, the authors observed reduced microbial exposure in urban children, which can influence gut microbiota composition. This is known to precede the manifestation of atopy (Penders et al., 2006).

Protective Factors: Clinical and Epidemiological Aspects

Many attempts have been made to test possible protective or preventive measures to avoid or at least decrease the incidence of allergic diseases in infants and children. Wilson et al. (1998) showed that the duration of exclusive breastfeeding and the time of introduction of solid food can modulate the prevalence of respiratory allergic diseases. The third National Health and Nutrition Examination Survey (NHANES III), conducted from 1988 to 1994, concluded that breastfeeding might delay the onset of allergy or actively protect children younger than 24 months of age (Chulada, 2003). A number of issues regarding the physiological mechanism by which breastfeeding protects against asthma and recurrent wheezing remain unresolved. Nevertheless, it was shown that strictly breastfed children are protected against allergies during the period of breastfeeding and shortly after cessation, particularly in reducing the prevalence of asthma and recurrent wheezing in children exposed to environmental smoke tobacco.

The composition of milk varies from early to midlactation, e.g., from the secretion of colostrum during the period of initiation to mature milk, after some weeks of lactation. As shown earlier in this chapter, milk composition influences the composition of gut microbiota in breastfed infants (see also the chapter by Kunz and Rudloff and that by Blum and Baumrucker in this book). The immunomodulatory components in breast milk affect the immune development of the infant, and optimal transfer of passive immunity IgG and IgA is on the day of delivery, at the maximum during the prelacteal period. Recent research set up to compare the composition of colostrum from allergic and nonallergic women has shown a large variation in the concentration of neutral oligosaccharides in milk samples collected during days 2–4 postpartum (Sjogren et al., 2006). Although the authors found a trend between a higher consumption of neutral oligosaccharides and the later development of infant allergy, the concentration in neutral oligosaccharides was not directly correlated with mother's allergy or with allergy development in children up to 18 months. TGF-β is the dominant cytokine in colostrum and was shown to have anti-inflammatory properties and to be able to downregulate IgE synthesis together with IL-10 (Bottcher et al., 2003). Particular cytokine patterns in mother's milk may influence the development of atopic diseases in breastfed infants (Rigotti et al., 2006). TGF-β was significantly less secreted in mature milk than in the colostrum of allergic mothers and less in allergic mothers than in nonallergic mothers. No change in IL-10 was measured in the same samples. In the KOALA study, breast milk samples were collected at one month postpartum to analyze the level of different cytokines (TGF-β, IL-10, IL-12) and soluble CD14. The results were analyzed according to mother's allergic status, and the authors found no evidence of an association between milk cytokine levels and the manifestation of wheezing and allergic sensitization of their 2-year-old infants (Snijders et al., 2006). Otherwise, in the same cohort, a nonsignificant trend toward a reduced risk of eczema in the first year of life was associated with increased duration of breastfeeding (Snijders et al., 2007). In the Melbourne atopy cohort, a higher level of long-chain omega-3 fatty acid in the colostrum of mothers with high-risk infants did not confer protection against the development of atopy for them (Stoney et al., 2004). Earlier prospective longitudinal studies already pointed out that being breastfed was associated with lower rates of recurrent wheezing at 6 years (3.1 vs. 9.7%; $P < 0.01$) for nonatopic children (Wright, 1995). In industrialized countries, very few children are exclusively breastfed for more than a few weeks postpartum, generally not longer than the duration of maternity leave. Recent work has focused on the duration of exclusive breastfeeding necessary to confer a significant protection against different types of allergy to children. A clinical trial has shown that exclusive breastfeeding for at least four months had a preventive effect on atopic dermatitis in infants at risk of atopic disease during the first year of life (Schoetzau, 2002). The risk of atopic dermatitis during the first year of life was related to maternal atopic dermatitis and lower concentrations of macrophage inflammatory protein-1β in cord blood, but not to viral or bacterial infection during pregnancy or breastfeeding. Paternal hay fever was associated negatively with the development of atopic dermatitis in the infant (Sugiyama et al., 2007).

The main conclusion about the role of exclusive breastfeeding was that it confers protection against a large number of infectious diseases, and we should stress the major role of colostrum in this matter (Hanson, 1999). This is particularly crucial in developing countries where exclusive breastfeeding provides a high level of protection against acute respiratory infections and diarrhea and consequently has a direct effect on the survival of the children (Arifeen et al., 2001). Breast milk is recognized to be the best food for the newborn, but the six-month duration of exclusive breastfeeding recommended by WHO since 2002 has been debated on the basis of its inadequacy to meet energy requirements of the six-month-old infant (Reilly & Wells, 2005). In developing countries, some women breastfeed for about 2.5 years, but exclusive breastfeeding is often seen for the first three to four months postpartum only. Apart from the energy requirement of the infant, other positive and negative aspects of exclusive breastfeeding should be taken into account. Both protective and allergy-promoting effects of breastfeeding have been reported. It seems that some immunomodulatory components vary according to the mother's allergy status. Maternal exposure to food allergens or outdoor allergens may increase the risk of allergies in their breastfed babies. There is some evidence that breast milk is a nonnegligible way to ingest organochlorines during infancy in areas where insecticides are intensively sprayed. A follow-up of children in an area with high levels of organochlorine compounds has shown that from birth to age 4, the mean DDE level among children with artificial feeding decreased by 72%, while among breastfed children it increased by 53% (Sunyer et al., 2006b).

Although the benefit of breastfeeding for many aspects of infant health, either immediately or later in life, is not disputable, the evidence for its protective effect on allergic disease continues to be inconclusive (Schack-Nielsen & Michaelsen,

2006). Even results from a study carried out in poor suburbs of Cape Town, South Africa, suggest a protective effect of prolonged breastfeeding on the development of allergic diseases, particularly hay fever, in children born to nonallergic parents. This protective effect was not found in children with an allergic predisposition (Obihara, 2005).

Breast milk provides passive protective factors that lessen the expression of neonatal allergic and infectious diseases and have a powerful effect on intestinal development and host defense (Walker, 2004). On the other hand, the effect of introducing the baby to small amounts of allergen through breast milk can be more harmful than beneficial. Those infants breastfed by mothers exhibiting positive challenges to specific foods manifest allergic responses themselves following the first breastfeeding after the mother's challenge (Wilson, 1983). The manipulation of maternal diet during early lactation may influence the risk of allergy for their breastfed infant: avoidance of cow's milk and peanuts in a nursing mother's diet may help to escape allergic diseases in her six-month-old infant. Bruno et al. (1993) showed that the amount of bovine milk protein found 12 hours after nursing mother's ingestion in her breast milk is more likely to induce an IgE-type response in her high-risk breastfed infant than direct ingestion of bovine milk, which will trigger the production of an IgG-type response. There was a significant linear inverse association between breastfeeding duration and allergic diseases in children without atopic parents, but not in children with an allergic predisposition. Nonatopic parents are defined as belonging to families without a history of asthma, eczema, or hay fever. A birth cohort study in New Zealand showed that children who were breastfed for more than four weeks increased the likelihood of skin test responses to common allergens at the age of 13 years, and more than doubled the risk of diagnosed asthma in mid-childhood, with effects persisting into adulthood (Sears et al., 2002).

Conclusions

Preventive strategies should include both genetic and environmental factors, as an infant with one or both parents being atopic are more at risk than an infant without atopic parents. In high-risk children, breastfeeding should be recommended for the first four to six months and solid food should not be introduced before four months. Nursing mothers should avoid food allergens during pregnancy and lactation to protect their infant more efficiently against atopy (Stanaland, 2004). Avoidance of food allergens and smoking during pregnancy and early lactation should certainly be recommended (Zieger, 1989). In families with atopic parents, preventive action may be decided in agreement with the pediatrician. Initiation of breastfeeding should be encouraged for all mothers immediately after delivery, and public health recommendations should stress the importance of giving colostrum for the infant's well-being.

References

Aberg, N., Engstrom, I., & Lindberg, U. (1989). Allergic diseases in Swedish school children. *Acta Paediatrica Scandinavica, 78*(2), 246–252.

Almqvist, C., Pershagen, G., & Wickman, M. (2005). Low socioeconomic status as a risk factor for asthma, rhinitis and sensitization at 4 years in a birth cohort. *Clinical and Experimental Allergy, 35*(5), 612–618.

Arifeen, S., Black, R. E., Antelman, G., Baqui, A., Caulfield, L., & Becker, S. (2001). Exclusive breastfeeding reduces acute respiratory infection and diarrhea deaths among infants in Dhaka slums. *Pediatrics, 108*(4), E67.

Arshad, S. H., Tariq, S. M., Matthews, S., & Hakim, E. (2001). Sensitization to common allergens and its association with allergic disorders at age 4 years: A whole population birth cohort study. *Pediatrics, 108*(2), E33.

Baldi, A., Savoini, G., Cheli, F., Fantuz, F., Senatore, E., Bertocci, L., & Politis, I. (1996). Changes in plasmin-plasminogen-plasminogen activator system in milk from Italian Fresian herds. *International Dairy Journal, 6*, 1045–1053.

Baldi, A., Politis, I., Pecorini, C., Fusi, E., Chronopoulou, R., & Dell'Orto, V. (2005). Biological effects of milk proteins and their peptides with emphasis on those related to the gastrointestinal ecosystem. *Journal of Dairy Research, 72 (Spec No.)*, 66–72.

Balmer, S. E., & Wharton, B. A. (1989). Diet and faecal flora in the newborn: Breast milk and infant formula. *Archives of Disease in Childhood, 64*, 1672.

Batish, V. K., Chander, H., Zumdegeni, K. C., Bhatta, K. L., & Singh, R. S. (1988). Antibacterial activity of lactoferrin against some common food-borne pathogenic organisms. *Australian Journal of Dairy Technology, 5*, 16–18.

Baylis, J. M., Leeds, A. R., & Challacombe, D. N. (1983). Persistent nausea and food aversions in pregnancy. A possible association with cow's milk allergy in infants. *Clinical Allergy, 13*(3), 263–269.

Becker, A. B. (2005). Primary prevention of allergy and asthma is possible. *Clinical Reviews in Allergy and Immunology, 28*(1), 5–16.

Bellamy, W., Takase, M., Yamauchi, K., Wakabayashi, H., Kawase, K., & Tomita, M. (1992). Identification of the bactericidal domain of lactoferrin. *Biochimica et Biophysica Acta, 1121*, 130–136.

Bottcher, M. F., Jenmalm, M. C., & Bjorksten, B. (2003). Cytokine, chemokine and secretory IgA levels in human milk in relation to atopic disease and IgA production in infants. *Pediatric Allergy and Immunology, 14*(1), 35–41.

Bruno, G., Milita, O., Ferrara, M., Nisini, R., Cantani, A., & Businco, L. (1993). Prevention of atopic diseases in high risk babies (long-term follow-up). *Allergy Proceedings, 14*(3), 181–186; discussion, 186–187.

Burr, M. L., Limb, E. S., Maguire, M. J., Amarah, L., Eldridge, B. A., Layzell, J. C., & Merrett, T. G. (1993). Infant feeding, wheezing, and allergy: A prospective study. *Archives of Disease in Childhood, 68*(6), 724–728.

Businco, L., Marchetti, F., Pellegrini, G., Cantani, A., & Perlini, R. (1983). Prevention of atopic disease in "at-risk newborns" by prolonged breast-feeding. *Annals of Allergy, 51* (2 Pt 2), 296–299.

Butland, B. K., Strachan, D. P., Lewis, S., Bynner, J., Butler, N., & Britton, J. (1997). Investigation into the increase in hay fever and eczema at age 16 observed between the 1958 and 1970 British birth cohorts. *British Medical Journal, 315*(7110), 717–721.

Chan-Yeung, M., Manfreda, J., Dimich-Ward, H., Ferguson, A., Watson, W., & Becker, A. (2000). A randomized controlled study on the effectiveness of a multifaceted intervention program in the primary prevention of asthma in high-risk infants. *Archives of Pediatrics & Adolescent Medicine, 154*(7), 657–663.

Chandra, R. K. (1997). Five-year follow-up of high-risk infants with family history of allergy who were exclusively breast-fed or fed partial whey hydrolysate, soy, and conventional cow's milk formulas. *Journal of Pediatric Gastroenterology and Nutrition, 24*(4), 380–388.

Chandra, R. K. (2000). Food allergy and nutrition in early life: Implications for later health. *Proceedings of the Nutrition Society, 59*(2), 273–277.

Chandra, R. K., Singh, G., & Shridhara, B. (1989). Effect of feeding whey hydrolysate, soy and conventional cow milk formulas on incidence of atopic disease in high risk infants. *Annals of Allergy, 63*(2), 102–106.

Chien, P. F., & Howie, P. W. (2001). Breast milk and the risk of opportunistic infection in infancy in industrialised and non-industrialised settings. *Advances in Nutritional Research, 10*, 69–104.

Chulada, P. C., Arbes, S. J., Jr., Dunson, D., & Zeldin, D. C. (2003). Breast-feeding and the prevalence of asthma and wheeze in children: Analyses from the Third National Health and Nutrition Examination Survey, 1988–1994. *Journal of Allergy and Clinical Immunology, 111*(2), 328–336.

Clare, D. A., & Swaisgood, H. E. (2000). Bioactive milk peptides: A prospectus. *Journal of Dairy Science, 83*, 1187–1195.

Cogswell, J. J., Mitchell, E. B., & Alexander, J. (1987). Parental smoking, breast feeding, and respiratory infection in development of allergic diseases. *Archives of Disease in Childhood, 62*(4), 338–344.

Dionysius, D. A., Grieve, P. A., & Milne, J. M. (1993). Forms of lactoferrin: Their antibacterial effect on enterotoxigenic *Escherichia coli. Journal of Dairy Science, 76*, 2597–2606.

Douglass, J. A., & O'Hehir, R. E. (2006). Diagnosis, treatment and prevention of allergic disease: The basics. *The Medical Journal of Australia, 185*(4), 228–233.

Duchen, K., Casas, R., Fageras-Bottcher, M., Yu, G., & Bjorksten, B. (2000). Human milk polyunsaturated long-chain fatty acids and secretory immunoglobulin A antibodies and early childhood allergy. *Pediatric Allergy and Immunology, 11*(1), 29–39.

Epton, M. J., Town, G. I., Ingham, T., Wickens, K., Fishwick, D., Crane, J., & NZA2CS (2007). The New Zealand Asthma and Allergy Cohort Study (NZA2CS): Assembly, demographics and investigations. *BMC Public Health, 7*, 26.

Exner, H., & Greinecker, G. (1996). [Prevention of allergy.] *Wiener medizinishe Wochenschrift (1946), 146*(15), 406–411.

Fantuz, F., Polidori, F., Cheli, F., & Baldi, A. (2001). Plasminogen activation system in goat milk and its relation with composition and coagulation properties. *Journal of Dairy Science, 84*, 1786–1790.

Field, C. J. (2005). The immunological components of human milk and their effect on immune development in infants. *Journal of Nutrition, 135*, 1–4.

Fiocchi, A., Borella, E., Di Donna, C., Lista, G. L., Turrini, D., & Riva, E. (1984). [Respiratory atopy and season of birth: Epidemiologic considerations.] *La Pediatria Medica e Chirurgica, 6*(5), 673–676.

Forsyth, B. W., McCarthy, P. L., & Leventhal, J. M. (1985). Problems of early infancy, formula changes, and mothers' beliefs about their infants. *Journal of Pediatrics, 106*(6), 1012–1017.

Golding, J., Emmett, P. M., & Rogers, I. S. (1997). Does breast feeding have any impact on non-infectious, non-allergic disorders? *Early Human Development, 49 (Suppl)*, S131–142.

Gruskay, F. L. (1982). Comparison of breast, cow, and soy feedings in the prevention of onset of allergic disease: A 15-year prospective study. *Clinical Pediatrics, 21*(8), 486–491.

Habbick, B. F., Pizzichini, M. M., Taylor, B., Rennie, D., Senthilselvan, A., & Sears, M. R. (1999). Prevalence of asthma, rhinitis and eczema among children in 2 Canadian cities: The International Study of Asthma and Allergies in Childhood. *Canadian Medical Association Journal, 160*(13), 1824–1828.

Halken, S. (2003). Early sensitisation and development of allergic airway disease—Risk factors and predictors. *Paediatric Respiratory Reviews, 4*(2), 128–134.

Halken, S. (2004). Prevention of allergic disease in childhood: Clinical and epidemiological aspects of primary and secondary allergy prevention. *Pediatric Allergy and Immunology, 15 (Suppl 16)*, 4–5, 9–32.

Haque, M. F., Hussain, M., Sarkar, A., Hoque, M. M., Ara, F. A., & Sultana, S. (2002). Breast-feeding counselling and its effect on the prevalence of exclusive breast-feeding. *Journal of Health, Population, and Nutrition, 20*(4), 312–316.

Harris, J. M., Cullinan, P., Williams, H. C., Mills, P., Moffat, S., White, C., & Newman Taylor, A. J. (2001). Environmental associations with eczema in early life. *British Journal of Dermatology, 144*(4), 795–802.

Hill, H. S., Doull, F., Rutherford, K. J., & Cross, M. L. (2000). Immunoregulatory peptides in bovine milk. *British Journal of Nutrition, 84 (Suppl 1)*, S111–S117.

Hoek, K. S., Milne, J. M., Grieve, P. A., Dionysius, D. A., & Smith, R. (1997). Antibacterial activity of bovine lactoferrin-derived peptides. *Antimicrobial Agents and Chemotherapy, 41*, 54–59.

Hoppu, U., Kalliomaki, M., Laiho, K., & Isolauri, E. (2001). Breast milk—Immunomodulatory signals against allergic diseases. *Allergy, 56 (Suppl 67)*, 23–26.

Isaacs, C. E. (2005). Human milk inactivates pathogens individually, additively and synergistically. *Journal of Nutrition, 135*(135), 1286–1288.

Isaacs, C. E., Kashyap, S., Heird, W. C., & Thormar, H. (1990). Antiviral and antibacterial lipids in human milk and infant formula feeds. *Archives of Disease in Childhood, 65*, 861–864.

Kannan, S., Carruth, B. R., & Skinner, J. (1999). Infant feeding practices of Anglo American and Asian Indian American mothers. *Journal of the American College of Nutrition, 18*(3), 279–286.

Kemp, A., & Kakakios, A. (2004). Asthma prevention: Breast is best? *Journal of Paediatrics and Child Health, 40*(7), 337–339.

Kim, W. S., Ohashi, M., Tanaka, T., Kumura, H., Kim, G. Y., Kwon, I. K., Goh, J. S., & Shimazaki, K. (2004). Growth-promoting effects of lactoferrin on *L. acidophilus* and *Bifidobacterium* spp. *Biometals, 17*, 279–283.

Kjellman, N. I., & Nilsson, L. (1999). Is allergy prevention realistic and beneficial? *Pediatric Allergy and Immunology, 10*(12 Suppl), 11–17.

Kramer, M. S. (1988). Does breast feeding help protect against atopic disease? Biology, methodology, and a golden jubilee of controversy. *Journal of Pediatrics, 112*(2), 181–190.

Kull, I., Wickman, M., Lilja, G., Nordvall, S. L., & Pershagen, G. (2002). Breast feeding and allergic diseases in infants—A prospective birth cohort study. *Archives of Disease in Childhood, 87*(6), 478–481.

Kummeling, I., Thijs, C., Penders, J., Snijders, B. E., Stelma, F., Reimerink, J., Koopmans, M., Dagnelie, P. C., Huber, M., Jansen, M. C., de Bie, R., & van den Brandt, P. A. (2005). Etiology of atopy in infancy: The KOALA Birth Cohort Study. *Pediatric Allergy and Immunology, 16*(8), 679–684.

Kummeling, I., Stelma, F. F., Dagnelie, P. C., Snijders, B. E., Penders, J., Huber, M., van Ree, R., van den Brandt, P. A., & Thijs, C. (2007). Early life exposure to antibiotics and the subsequent development of eczema, wheeze, and allergic sensitization in the first 2 years of life: The KOALA Birth Cohort Study. *Pediatrics, 119*(1), e225–e231.

Kurukulaaratchy, R. J., Matthews, S., & Arshad, S. H. (2006). Relationship between childhood atopy and wheeze: What mediates wheezing in atopic phenotypes? *Annals of Allergy, Asthma and Immunology, 97*(1), 84–91.

Li, E. W., & Mine, Y. (2004). Immunoenhancing effects of bovine glycomacropeptide and its derivatives on the proliferative response and phagocytic activity of human macrophage-like cells, U937. *Journal of Agricultural Food Chemistry, 52*, 2704–2708.

Liepke, C., Adermann, K., Raida, M., Mägert, H. J., Forssmann, W. G., & Zucht, H. D. (2002). Human milk provides peptides highly stimulating the growth of bifidobacteria. *European Journal of Biochemistry, 269*, 712–718.

Lönnerdal, B. (2000). Breast milk: A truly functional food. *Nutrition, 16*, 509–511.
Lönnerdal, B. (2003). Nutritional and physiologic significance of human milk proteins. *American Journal of Clinical Nutrition, 77*, 1537S–1543S.
Marks, G. B., Mihrshahi, S., Kemp, A. S., Tovey, E. R., Webb, K., Almqvist, C., Ampon, R. D., Crisafulli, D., Belousova, E. G., Mellis, C. M., Peat, J. K., & Leeder, S. R. (2006). Prevention of asthma during the first 5 years of life: A randomized controlled trial. *Journal of Allergy and Clinical Immunology, 118*(1), 53–61.
Martinez, F. D. (2007). Gene-environment interactions in asthma: With apologies to William of Ockham. *Proceedings of the American Thoracic Society, 4*(1), 26–31.
McGeady, S. J. (2004). Immunocompetence and allergy. *Pediatrics, 113*(4 Suppl), 1107–1113.
Meisel, H. (1997). Biochemical properties of bioactive peptides derived from milk proteins: Potential nutraceuticals for food and pharmaceutical applications. *Livestock Production Science, 50*, 125–138.
Nafstad, P., Nystad, W., & Jaakkola, J. J. (2002). The use of a feather quilt, childhood asthma and allergic rhinitis: A prospective cohort study. *Clinical and Experimental Allergy, 32*(8), 1150–1154.
Newburg, D. S. (1996). Oligosaccharides and glycoconjugates in human milk: Their role in host defence. *Journal of Mammary Gland Biology and Neoplasia, 1*, 271–283.
Newburg, D. S. (2005). Innate immunity and human milk. *Journal of Nutrition, 135*, 1308–1312.
Oddy, W. H., Sherriff, J. L., de Klerk, N. H., Kendall, G. E., Sly, P. D., Beilin, L. J., Blake, K. B., Landau, L. I., & Stanley, F. J. (2004). The relation of breastfeeding and body mass index to asthma and atopy in children: A prospective cohort study to age 6 years. *American Journal of Public Health, 94*(9), 1531–1537.
Oram, J. D., & Reiter, B. (1968). Inhibition of bacteria by lactoferrin and other iron-chelating agents. *Biochimica et Biophysica Acta, 170*, 351–365.
Payne, K., Davidson, P. M., & Oliver, S. P. (1990). Influence of bovine lactoferrin on the growth of Listeria monocytogenes. *Journal of Food Protection, 53*, 468–472.
Pecorini, C., Martino, P., Reggi, S., Fogher, C., & Baldi, A. (2003). Evaluation of biological effects of recombinant porcine lactoferrin expressed in *Pichia pastoris* yeast. Proceedings of the 6th International Conference on Lactoferrin: Structure, Function and Applications, Capri, Italy p70.
Pecorini, C., Savazzini, F., Martino, P. A., Fusi, E., Fogher, C., & Baldi, A. (2005). Heterologous expression of biologically active porcine lactoferrin in *Pichia pastoris* yeast. *Veterinary Research Communications, 29 (Suppl 2)*, 379–382.
Pecquet, S., Bovetto, L., Maynard, F., & Fritsche, R. (2000). Peptides obtained by tryptic hydrolysis of bovine β-lactoglobulin induce specific oral tolerance in mice. *Journal of Allergy and Clinical Immunology, 105*, 514–521.
Penders, J., Thijs, C., van den Brandt, P. A., Kummeling, I., Snijders, B., Stelma, F., Adams, H., van Ree, R., & Stobberingh, E. E. (2007). Gut microbiota composition and development of atopic manifestations in infancy: The KOALA birth cohort study. *Gut 56* (5), 661–667.
Politis, I. (1996). Plasminogen activator system: Implications for mammary cell growth and involution. *Journal of Dairy Science, 79*, 1097–1107.
Politis, I., Voudouri, A., Bizelis, I., & Zervas, G. (2003). The effect of various vitamin E derivatives on the urokinase-plasminogen activator system of ovine macrophages and neutrophils. *British Journal of Nutrition, 89*, 259–265.
Prescott, S. L. (2003). Early origins of allergic disease: A review of processes and influences during early immune development. *Current Opinion in Allergy and Clinical Immunology, 3*(2), 125–132.
Prescott, S. L., & Tang, M. L. (2005). The Australasian Society of Clinical Immunology and Allergy position statement: Summary of allergy prevention in children. *Medical Journal of Australia, 182*(9), 464–467.

Prioult, G., Pecquet, S., & Fliss, I. (2004). Stimulation of interleukin-10 production by acidic β-lactoglobulin-derived peptides hydrolyzed with *Lactobacillus paracasei* NCC2461 peptidases. *Clinical and Diagnostic Laboratory Immunology, 11*, 266–271.

Radon, K., & Schulze, A. (2006). Adult obesity, farm childhood, and their effect on allergic sensitization. *Journal of Allergy and Clinical Immunology, 118*(6), 1279–1283.

Reilly, J. J., & Wells, J. C. (2005). Duration of exclusive breast-feeding: Introduction of complementary feeding may be necessary before 6 months of age. *British Journal of Nutrition, 94*(6), 869–872.

Rigotti, E., Piacentini, G. L., Ress, M., Pigozzi, R., Boner, A. L., & Peroni, D. G. (2006). Transforming growth factor-beta and interleukin-10 in breast milk and development of atopic diseases in infants. *Clinical and Experimental Allergy, 36*(5), 614–618.

Sausenthaler, S., Koletzko, S., Schaaf, B., Lehmann, I., Borte, M., Herbarth, O., von Berg, A., Wichmann, H. E., Heinrich, J., the LISA Study Group (2007). Maternal diet during pregnancy in relation to eczema and allergic sensitization in the offspring at 2 y of age. *American Journal of Clinical Nutrition, 85*(2), 530–537.

Schanbacher, F. L., Talhouk, R. S., & Murray, F. A. (1997). Biology and origin of bioactive peptides in milk. *Livestock Production Science, 50*, 105–123.

Schanbacher, F. L., Talhouk, R. S., Murray, F. A., Gherman, L. I., & Willett, L. B. (1998). Milk-borne bioactive peptides. *International Dairy Journal, 8*, 393–403.

Sears, M. R., Greene, J. M., Willan, A. R., Taylor, D. R., Flannery, E. M., Cowan, J. O., Herbison, G. P., & Poulton, R. (2002). Long-term relation between breastfeeding and development of atopy and asthma in children and young adults: A longitudinal study. *Lancet, 360*(9337), 901–907.

Snijders, B. E., Damoiseaux, J. G., Penders, J., Kummeling, I., Stelma, F. F., van Ree, R., van den Brandt, P. A., & Thijs, C. (2006). Cytokines and soluble CD14 in breast milk in relation with atopic manifestations in mother and infant (KOALA Study). *Clinical and Experimental Allergy, 36*(12), 1609–1615.

Snijders, B. E., Thijs, C., Kummeling, I., Penders, J., & van den Brandt, P. A. (2007). Breastfeeding and infant eczema in the first year of life in the KOALA birth cohort study: A risk period-specific analysis. *Pediatrics, 119*(1), e137–e141.

Stoney, R. M., Woods, R. K., Hosking, C. S., Hill, D. J., Abramson, M. J., & Thien, F. C. (2004). Maternal breast milk long-chain n-3 fatty acids are associated with increased risk of atopy in breastfed infants. *Clinical and Experimental Allergy, 34*(2), 194–200.

Sugiyama, M., Arakawa, H., Ozawa, K., Mizuno, T., Mochizuki, H., Tokuyama, K., & Morikawa, A. (2007). Early-life risk factors for occurrence of atopic dermatitis during the first year. *Pediatrics, 119*(3), e716–e723.

Sunyer, J., Torrent, M., Muñoz-Ortiz, L., Ribas-Fitó, N., Carrizo, D., Grimalt, J., Antó, J. M., & Cullinan, P. (2005). Prenatal dichlorodiphenyldichloroethylene (DDE) and asthma in children. *Environmental Health Perspectives, 113*(12), 1787–1790.

Sunyer, J., Jarvis, D., Gotschi, T., Garcia-Esteban, R., Jacquemin, B., Aguilera, I., Ackerman, U., de Marco, R., Forsberg, B., Gislason, T., Heinrich, J., Norbäck, D., Villani, S., & Künzli, N. (2006a). Chronic bronchitis and urban air pollution in an international study. *Occupational and Environmental Medicine, 63*(12), 836–843.

Sunyer, J., Torrent, M., Garcia-Esteban, R., Ribas-Fitó, N., Carrizo, D., Romieu, I., Antó, J. M., & Grimalt, J. O. (2006b). Early exposure to dichlorodiphenyldichloroethylene, breastfeeding and asthma at age six. *Clinical and Experimental Allergy, 36*(10), 1236–1241.

Sutas, Y., Soppi, E., Korhonen, H., Syvaoja, E. L., Saxelin, M., Rokka, T., & Isolauri, E. (1996). Suppression of lymphocyte proliferation in *in vitro* bovine casein hydrolyzed with *Lactobacillus casei* GG-derived enzymes. *Journal of Allergy and Clinical Immunology, 98*, 216–224.

Telemo, E., & Hanson, L. A. (1996). Antibodies in milk. *Journal of Mammary Gland Biology and Neoplasia, 1*, 243–249.

Teraguchi, S., Shin, K., Ozawa, K., Nakamura, S., Fukuwatari, Y., Tsuyuri, S., Namihira, H., & Shimanura, S. (1995). Bacteriostatic effect of orally administered bovine lactoferrin on proliferation of Clostridium species in the gut of mice fed bovine milk. *Applied and Environmental Microbiology, 61*, 501–506.

Thestrup-Pedersen, K. (2003). Atopic eczema. What has caused the epidemic in industrialised countries and can early intervention modify the natural history of atopic eczema? *Journal of Cosmetic Dermatology, 2*(3–4), 202–210.

Van der Kraan, M. I. A., Groenink, J., Nazmi, K., Veerman, E. C. I., Bolscher, J. G. M., & Amerongen, A. V. N. (2004). Lactoferrampin: A novel antimicrobial peptide in the N1-domain of bovine lactoferrin. *Peptides, 25*, 177–183.

Van Leeuwen, P., Oosting, S. J., Mouwen, J., & Verstegen, M. W. A. (2000). Effects of a lactoperoxidase system and lactoferrin, added to a milk replacer diet, on severity of diarrhoea, intestinal morphology and microbiology of digesta and faeces in young calves. *Journal of Animal Physiology and Animal Nutrition, 83*, 15–23.

van Odijk, J., Kull, I., Borres, M. P., Brandtzaeg, P., Edberg, U., Hanson, L. A., Host, A., Kuitunen, M., Olsen, S. F., Skerfving, S., Sundell, J., & Wille, S. (2003). Breastfeeding and allergic disease: A multidisciplinary review of the literature (1966–2001) on the mode of early feeding in infancy and its impact on later atopic manifestations. *Allergy, 58*(9), 833–843.

Wilson, A. C., Forsyth, J. S., Greene, S. A., Irvine, L., Hau, C., & Howie, P. W. (1998). Relation of infant diet to childhood health: Seven year follow up of cohort of children in Dundee infant feeding study. *British Medical Journal, 316*(7124), 21–25.

Wilson, W. H. (1983). Recurrent acute otitis media in infants—Role of immune complexes acquired in utero. *Laryngoscope, 93*(4), 418–421.

Wjst, M., Dharmage, S., André, E., Norback, D., Raherison, C., Villani, S., Manfreda, J., Sunyer, J., Jarvis, D., Burney, P., & Svanes, C. (2005). Latitude, birth date, and allergy. *PLoS Medicine, 2*(10), e294.

Woodcock, A., Lowe, L. A., Murray, C. S., Simpson, B. M., Pipis, S. D., Kissen, P., Simpson, A., Custovic, A., & NMAaAS Group (2004). Early life environmental control: Effect on symptoms, sensitization, and lung function at age 3 years. *American Journal of Respiratory and Critical Care Medicine, 170*(4), 433–439.

Zacharasiewicz, A., Zidek, T., Haidinger, G., Waldhor, T., Suess, G., & Vutuc, C. (1999). Indoor factors and their association to respiratory symptoms suggestive of asthma in Austrian children aged 6–9 years. *Wien Klin Wochenschr, 111*(21), 882–886.

Zieger, R. S., Heller, S., Mellon, M. H., Forsythe, A. B., O'Connor, R. D., Hamburger, R. N. & Schatz, M. (1989). Effect of combined maternal and infant food-allergen avoidance on development of atopy in early infancy a randomized study. *Journal of Allergy and Clinical Immunology, 84*(1), 72–89.

Subject Index

ABCG5 and *ABCG8* transporters
  identification and expression, 73
ACACA and *FASN* genes, 70
ACCA and *FASN* gene regulation, 80–83
ACCα isoenzyme expression, 74
ACE inhibitory activity, 304–305
ACE inhibitory and antihypertensive peptides
  cheeses from, 306–308
  commercial whey product from, 308, 310–312
ACE inhibitory peptides from fermented milk product, 304
Acetyl-CoA carboxylase (ACC), 70
Acquired immunity and probiotics, 427–428
Acyl Carrier Protein (ACP), 345
Acyl- CoA: ACP transacylase, 29
Acyl glycerol phosphate acyl transferase (AGPAT or LPAAT), 70, 349
Acyltransferases regulation, 84–85
Adhesin Hap, (Haemophilus influenzae) 174
Adipophilin and TIP47 proteins, 139
AGPAT, *see* Acyl glycerol phosphate acyl transferase
AIP deactivation
  lipopolysaccharide of, 154
  negative feedback for milk secretion, 153–154
A-Linolenic Acid (ALA), 344, 346
Alkaline Phosphatase (AlP) in milk, biological roles, 153
Allergic disease in childhood prevention
  clinical and epidemiological aspects, 474–477
  predictive factors of and risk factors for, 472–474
Allergy, 467–468
All-trans-β-carotene, 111
Alzheimer's disease, 244

Angiotensin-converting enzyme and inhibitors, 295, 296
Angiotensin-Converting Enzyme Inhibitors (ACEI) from milk proteins, 299–301
Angiotensin II vasoconstrictor, 296
Anteiso-methyl-fatty acids, 9
Anti-apoptotic Bcl-2 family members, 220–221
Antibody response, 334, 428, 474
Antibody therapy against *Streptococcus mutans*, 320, 324, 328
Antibody therapy in *Clostridium difficile* infection, 321–322, 328–329
α-1 Antiprypsin (hα1AT), 355, 368, 371, 376–377
Antithrombin III, 355, 366, 383
Antitumor peptides, 283–284
Apoptosis, 219–220
Apoptosis in cancerous cells, 181
Apoptosis inducing factor (AIF), 223
Apoptosis signal-regulating kinase-1 (ASK-1), 223
Arachidonic Acid, 346
Atopic dermatitis, 205
ATP-binding cassette (ABC) transporters, 73
Autophagy and cell death, 229–230

Bcl-2 and Bcl-xl overexpression, 228
Bcl-2 family and homology (BH) domains, 221
β-Cryptoxanthin, 111
bioreactors, posttranslational modifications, 356, 367
β-Lactoglobulin (β-lg) hydrolysis, 256
Bovine antibodies, 319
  clinical effect against rotavirus infection, 321, 327, 330–331

Bovine *(cont.)*
  in infections by *Escherichia coli*,
    321, 331–332
  therapy in *Clostridium difficile* infection,
    328–329
  treatment of *Cryptosporidium parvum*-
    induced diarrhea, 321, 325,
    329–330
Bovine colostral whey proteins, 333
Bovine colostrum contents, 113
Bovine Immunoglobulin Concentrate
    (BIC), 320
Bovine mammary gland, co-expression of
    FABP and CD36, 73
Bovine MFGM, 133–134
Bovine MFGM protein 2-DE
    separation, 133
Bovine milk
  and auto-oxidation, 119
  fat globules, 115
  fat, triglyceride structure, 348–351
  fatty acids and forage species, 30–33
  and nutrition effect on TFA and
    bioactive lipid, 29–30
  vitamin A (retinol) concentration, 110
Bovine neonatal Fc Receptor (bFcRn), 334
Branch-chain fatty acid concentrations in
    bovine milk fat, 10
Breast milk, bioactive components, 469
Butyrophilin protein in MFGM, 138

Campestanol and Campesterol
    in milk, 114
Carbonic anhydrase glycosylated
    enzyme, 139
Carboxylation, 365
Carotenoids
  biological effects on signaling
    pathways, 112
Carotenoids in cow's milk, 110–114
Casein-derived antimicrobial peptides,
    275, 278–279
Caspase-activated DNase (CAD), 222
Caspase-Independent pathways, 223–224
Caspases, cysteine proteases, 222
CD14
  discovery of, 196
  expression regulation of, 198–199
  host response to bacterial ligands,
    role of, 199
  infant health, relevance of, 202–204
  molecular characteristics of, 196
  pathological diseases, relevance of, 205

  as PRR, 200–202
  soluble forms of, 197–198
cdk inhibitor (CKI) p21 expression, 182
Cell death pathway autophagy, 230
Cell growth regulation, 182
T Cell-independent antigen
    polyvinylpyrrolidone (PVP), 243
CHD risk and TFA contents, 14
Chronic diseases
  and nutrition, 3, 4
  trans fatty acids, 10
Chylomicron-derived retinol ester
    hydrolysis, 118
CLA concentrations in milk and grass
    intake, 30
CLA-enriched butter contents, 15–16
Cleavage activating protein (SCAP), 92
Clostridium Difficile, 328
clotting factors blood, human proteins genes,
    355, 365, 373–374
Coenzyme A, 345
Colostral bioactive components effects on
    metabolism and endocrine
    systems, 407–409
Colostrinin (CLN), 241
Colostrum, 395–405, 407–409, 456, 468, 471,
    475–477
Colostrum and colostral bioactive
    substances effects on GI tract,
    397–400
Conjugated Linoleic Acid (CLA), 344
Copper P-type ATPase role in Cu
    translocation to milk, 150
Copper transporting P-type ATPases, 150
Cows feed, low-forage diet supplemented
    with soybean and fish oils study, 80
Cow's milk
  vitamin D content in, 113
  vitamin E content in, 112–113
Cream-derived membrane vesicles
    in milk, 146
Cryptosporidium Parvum, 329
Cryptosporidium parvum-induced
    diarrhea antibody treatment,
    321–324, 329–330
Cytokine production, 428–429
Cytoplasmic lipid droplets (CLDs),
    135–136
  approach, 130
Cytotoxic T lymphocytes (CTL), 224

Dairy products, fat-soluble vitamin
    content in, 111

Subject Index

Death domain protein RAIDD/
    CRADD, 223
Death-inducing signaling complex
    (DISC), 223
Defensins and cathelicidins, 280–283
De novo fatty acid synthesis, 29
    acetyl-CoA carboxylase gene *(ACAC)*,
        73–74
    FA esterification of, 76–77
    fatty acid synthase gene *(FASN)*, 74–75
    stearoyl-CoA desaturase, 75–76
DGAT expression, 77
Diacyl glycerol acyl transferase (DGAT),
    70, 349
Dietary PUFA ruminal biohydro-
    genation, 28
4,4-Dimethyloxazoline (DMOX) fatty acid
    derivatives, 18
Direct *in vivo* transfection, 361–362
Docosahexaenoic Acid (DHA), 346, 348
Dry matter intake (DMI), 116
Duodenal/Intravenous infusion of specific
    fatty acids, 88–90

Echidna (Tachyglossus aculeatus),
    monotreme mammal, 350–351
Eicosapentaenoic Acid (EPA), 346
Embryonic stem (ES) cells, 362
Enterohemorrhagic E. Coli, 332
Enteropathogenic E. Coli, 331
Enterotoxigenic E. Coli, 331
Epidermal growth factor family, 400–401
Epithelial cell and secretion mechanisms of
    lipid globules, 131
Erythropoietin, 355–356, 362, 378, 405
Escherichia Coli, 331, 350, 375, 382, 460, 470
Essential amino acid cysteine in milk and
    infants, 151–152
Even-chain *iso* acids, 9
Extracellular Superoxide Dismutase (hEC-
    SOD), 355, 377

FABP and CD36 co-expression, 73
FABP types expression in FA
    metabolism, 73
FA pair ratios in goat studies, 84
Fat-soluble vitamins in animal nutrition and
    milk, 116–119
Fatty acid binding proteins (FABP), 73
Fatty acid composition, 344–348
Fatty acid synthase (FAS) enzymes, 70
Fatty acid synthesis by bacteria in rumen, 29
Forage-based diet, 78

Forage conservation method, 36–37
Forage-to-concentrate ratio of diet, 38
Fortified milk, 113
FOSHU system in Japan and FOSHU
    products with antihypertensive
    effect, 312–313

Gain-of-function transgenic approach, 351
Gaucher's disease, 205
GH-IGF-insulin system, 402
Glutamic acid, 365, 374
γ-Glutamyl-transpeptidase (GGT), 151
Glutathione peroxidase (Gp) LM-associated
    enzyme, 151
Glycerol 3-Phosphate Acyl Transferase
    (GPAT), 70, 349
glycosylation, 363–365, 378–379, 381,
    383–384
Goats feed, hay/corn silage diets with
    vegetable lipids study, 81
Graft-*versus*-host reaction (GvH), 242
Gram+ and Gram- pathogens, 172
Granzyme A-activated DNase, 224
Growth Hormone, 355, 374, 401

Helicobacter Pylori, 332
Helicobacter pylori, oral therapy with bovine
    antibodies, 332
Heparan sulfate (HS) proteoglycans, 171
hEPO, *see* Erythropoietin
hGH, *see* Growth Hormone
hI, *see* Insulin
hIGF-1, *see* Insulin-like Growth Factor-1
Highly lipophilic food microconstituents
    (HLFMs), 110
hIL 2, *see* Interleukin-2
hLF, *see* Lactoferrin
human milk oligosaccharides (HMO)
    profile, 458
hPC, *see* Protein C
htPA, *see* Tissue Plasminogen Activator
Human α-defensin-5 (hD-5), 281
Human α-lactalbumin made lethal to tumor
    cells (HAMLET)
    and apoptosis, 226–227
    autophagy, 227–230
    Bcl-2 family and p53, 227
    and breastfeeding, 232–233
    nuclear interactions of, 231
    structure of, 218–219
    in vivo effects, 231–232
Human β-defensin-1 (hBD-1), 281
Human β-defensin-2 (hBD-2), 281

Human butyrophilin expression, 135
Human CAP18 protein (hCAP18), 281
Human CD14 sequence, 196
Human colostral MFGM proteins,
    annotated database, 135
Human health and functional milks, 119
Human Lf, 168
Human MFGM, 134–135
Human MFGM by 2-DE separation, 134
Human milk
  and carotenoids, 112
  erythropoietin (Epo), 405
  lactoferrin (Lf), 403–404
  leptin, 405–406
  oligosaccharides, 455–459
    as growth factors for bifidobacterium
      bifidum, 459–460
    as inhibitors of pathogen adhesion to
      epithelial cells, 460–461
    leukocyte-endothelial cell
      interactions, 462–463
  sCD14 levels, 198
  steroids, 406–407
  transforming growth factor-β, 404–405
Human neutrophil-derived-α-peptide
    (hNP1-3), 281
Human oral prophylactic studies, 323–324
Human recombinant proteins and transgenic
    livestock
  acid α-glucosidase (hαGLU), 380
  alpha-1-antitrypsin (hα$_1$AT), 376–377
  antithrombin (hAT-III), 383
  blood clotting factors, 373
  C1 inhibitor (hC1INH), 384
  collagen type I (hCOL), 385
  erythropoietin (hEPO), 378
  extracellular superoxide dismutase (hEC-
    SOD), 377–378
  α-Fetoprotein (hαFP), 381
  fibrinogen (hFIB), 380–381
  growth hormone (hGH), 374–375
  insulin-like growth factor-1 (hIGF-1),
    382–383
  interleukin-2 (hIL-2), 381–382
  lactoferrin (hLF), 375–376
  nerve growth factor beta (hNGF-β),
    384–385
  protein C (hPC), 374
  serum albumin (hSA), 375
  tissue-nonspecific alkaline phosphatase
    (hTNAP), 379
  tissue plasminogen activator (htPA), 379
Human therapeutic studies, 325–327

Human tumor cells *in vitro* studies, 20

IgA1 protease, 174
IIBC, *see* Immunized Bovine Colostrum
IL-18 production and inhibition of
    angiogenesis, 180
Immune cells regulation of recruitment, 178
Immune system development, 422–423
Immune whey protein concentrate, 328
Immunized Bovine Colostrum, 330
Immunomodulatory milk peptides, 471–472
Inhibitor of apoptosis (IAP) protein, 222
Inner Cell Mass (ICM), 362
Insulin, 355, 396
Insulin-like Growth Factor-1, 355, 382, 396
Interleukin-2, 355, 381
In vivo Δ9 desaturase activity, 84
Ions and organic substances
    transporatation, 149–150

Jun N-terminal kinase (JNK) pathways, 223

Kappacin microbial peptide, 279
Kawasaki disease, 205
K29R polymorphism, 165

Lactadherin human glycoprotein, 139
Lactating mammary cells, 130
Lactoferrin (Lf), 355, 362, 365, 375–376,
    403–04, 470–471
  antimicrobial activities of, 172–175
  binding molecules on microorganisms, 170
  commercial and clinical applications of,
    183–184
  3D structure of, 168
  gene structure and regulation of,
    164–165
  glycosylation of, 168
  in human milk, 163
  inflammation modulation, 176
  iron binding and release, 169
  mammalian binding proteins, 171
  mechanisms of action in cancer, 180
  as modulator of inflammation, 176–178
  peptide clusters location and putative
    functions, 166–167
  receptors in microorganisms and
    mammals, 169–170
  sequence characteristics of, 165
  stimulation of host immune responses by,
    178–179
  synthesis and localization of, 164
  and tumorigenesis and metastasis, 179

Subject Index

Lacto-N-tetraose, 456
Lactotripeptide, mode of action, 304–305
Lf gene polymorphisms, 165
Linoleic Acid, 344, 346
Lipid metabolism in rumen, 27–28
Lipogenic genes structure of
    lipoprotein lipase in
    ruminant species, 71
Lipopolysaccharide (LPS), 154
Lipoprotein lipase regulation, 79–80
Lipoprotein lipases (LPL) in
    lactatingmammary
    tissue, 118
Lipoprotein triacylglycerol hydrolysis, 71
Livestock species as bioreactors
  cattle, 369–370
  rabbits and pigs, 367
  sheep and goats, 369
Low-milk fat syndrome in cows, 86
LPAAT, *see* Lysophosphatidic Acid
    Acyltransferase
LPL transcripts in Bovines mammary
    tissue, 71
Lutein, 111
T Lymphocytes effects of Lf, 178–179
Lysophosphatidic Acid Acyltransferase, 349

major fatty acid positional distribution in
    triglycerides, 350
Mammalian inteleukin-1β-converting
    enzyme (ICE-3), 222
Mammary Epithelial Cells, in vitro studies on
    effect of fatty acids on lipogenesis,
    90–91
Mammary lipogenesis, 70
  gene characterization and
    mechanisms, 70–71
  gene expression regulation by
    dietary factors, 77–79
  by nutrition, 85–88
Mammary tissue
  and cell types, 72
  metabolic pathway for de novo FA
    synthesis in, 70
Medium-chain FA synthesis, 74–75
MFD diets and cows studies, 79
Microbial lipid synthesis, 28–29
Microbiota-host immune system
    interactions, 199
Milk
  Alkaline Phosphatase (AlP) biological
    roles, 153

  antibacterial peptides *in vivo* studies,
    284–285
  antyhypertensive peptides, 297–298,
    302–304
  cis-9, trans-11 conjugated
    linoleic acid (CLA)
    concentrations in, 50
  enzymes biological role, 144
  FA desaturation ratios and stearoyl-CoA
    desaturase (SCD) activity
    relationships, 85
  physical phases, 145
  and plant-derived sterols/
    phytosterols, 114
  the whole book is about milk proteins!
  protein analysis by electrophoresis,
    131–132
  as source of highly lipophilic
    microconstituents, 110–114
  as vehicle for highly lipophilic food
    microconstituents, 113–116
Milk composition and inclusion of oils/
    oilseeds in diet
  medium-chain SFA in bovine milk
    concentration, 39
  polyunsaturated fatty acids, 43
  trans fatty acids and conjugated linoleic
    acid, 43, 44
Milk fat
  branch-chain fatty acids, 9–10
  butyrate in, 8, 9
  CLA and TFA content, fresh pasture
    impact on, 30
  composition modification, 343–354
  conjugated linoleic acid, 19–22
  it is not clear to what it wants to point,
    too general
  monounsaturated fatty acids (MUFA), 6, 8
  PUFA concentrations and forage
    legumes, 34
  and seminatural grasslands, 35
  sphingomyelin, 22–23
  synthesis in ruminant mammary
    epithelial cell, 69
  time-dependent changes, 51
  trans fatty acids, 10–18
Milk fat-depressing (MFD) diets, 78–79
Milk fat globule membrane
    (MFGM), 129–130
  double-extraction method, 135
  protiens by proteomic tools, 136–138
  proteomic analysis, 133

Milk fatty acid composition and impact of concentrate supplements in diet, 37–38
Milk lipids metabolic origins mammary de novo fatty acid synthesis, 23–24
Milk oxidative stability, 120
Milk oxidative state and glutathione, 151
Milk peptides
    and immune function, 256–260
    and immune system, 267
    immunomodulating activity *in vivo*, 260–262
    induction/suppression of, 254–256
    origin of, 254
Milk ratio of myristoleic acid to myristic acid (cis-9 C14:1/C14:0), 84
Milk serum lipoprotein membrane (MSLM), 143
Milk serum lipoprotein membranes (MSLM), model for origin, 146–147
Milk short- and medium-chain fatty acid, relationships between, 82, 83
Mini-Mental State Examination (MMSE) scale, 247
Mitochondria and Bcl-2 family, 220–221
Mitochondrial outer membrane permeabilization (MOMP), 220
Monocyte/Macrophage and PMN activation by Lf, 179
Monounsaturated fatty acids (MUFA), 6, 8, 343
Mouse MFGM, 135–136
MSLM and MFGM phosphatase role, 154–155
MSLM/MFGM ratio for milk protein, 148
Mucin 1 glycosylated transmembrane protein, 138–139
MUFA, *see* Monounsaturated Fatty Acids
MUFA-rich diet, 6, 8
Multiple sclerosis, 205

Na^+/K^+ ATPase pump, 150
National nutritional guidelines, 3
NEFA utilization for milk lipid synthesis, 72
Nematode caenorhabditis elegans, 344
Neonate and intake of colostrum, 395–397
N-ethylmalmeimide (NEM), 76
Neurodegenerative disorders, 244
NK cell cytotoxicity, 180
Nonsteroidal Anti-inflammatory Drugs, 333–334

NSAIDs, *see* Nonsteroidal Anti-inflammatory Drugs
Nucleolin protein, 171
Nucleoside triphosphate pyrophosphohydrolase (NTPPPH), 155
Nucleotide metabolizing enzymes, 156–158

Odd-chain anteiso, 9
Odd-chain iso acids, 9
Oleic Acid, 344
oral candidiasis treatment, 320
Oral therapy with bovine antibodies against *Helicobacter pylori*, 332
β–Oxidation of FA in mitochondria, 74

Palmitic Acid, 348
p53 and resistance to cell death, 224–226
Pattern recognition receptors (PRR), 195
Peptide fractionation profile, 258–261
Peptide hBD-1, 282
Phagocytic and NK cell activity effect, 426–427
not enough specific therefore not informative
Phosphorus metabolizing enzymes, 152
Plasma protein ceruloplasmin and Cu transport, 50
Poly-N-acetyllactosamine antennae, 168
Polyunsaturated fatty acids (PUFA), 6, 343, 346
Polyunsaturated fatty acids (PUFA) synthesis pathways, 346
Position effect, transgene integration, 361
Posttranslational modifications recombinant proteins from milk, 363–365
Proline-rich polypeptide (PRP) biological properties, 246
Probiotics
    and attenuation of immunoinflammatory disorders allergies, 440–441
    inflammatory bowel disease (IBD), 440
    and human health, 423
    induced immunostimulation and disease resistance
        cancer and, 439
        infectious diseases and, 429, 439
    mechanisms for correction of immunological disorders, 441–443
    and modulation of intestinal microflora, 423–425
    and stimulation of immune system, 425–426

Programmed Cell Death (PCD), 219, 222–223
Proline-rich polypeptide (PRP)
   clinical application of, 244–246
   in clinical trials, 246–247
   immunological effects of, 241–244
Pronuclear microinjection, 361
Prostaglandin inhibitor Ro 20-5720, 242
Protein C, 355, 374
Protein identification by PMF and MS/MS analysis, 134
Proteomic analysis, 130–132
Proteomics and biological research, 130–131
PUFA, see Polyunsaturated Fatty Acids
PUFA biohydrogenation in ensiled grass, 36–37
Pyrophosphate-related minerals formulas, 155

Radical productions, 149–150
Raw bovine milk content, 112
Reactive oxygen species overproduction inhibition, 177–178
recombinant human milk proteins, 355–391
Recombinant proteins from milk
   posttranslational modifications of, 363–365
   purification of, 366
Redox regulating enzymes, 148–149
Reduced glutathione (GSH) in milk, 151
Renin-angiotensin system (RAS), 296
Retinol and xenobiotic oxidations, 111
Retinol-binding protein (RBP), 119
Retinol equivalents (RE) per gram of fat, 112
Retroviral transfection, 361
Rheumatoid arthritis, 205
Rotavirus
   infective agent, 321–324, 327, 330, 423, 427–431
   placebo-controlled study, 330
RRR-α-tocopheryl acetate, 117
Rumen biohydrogenation process, 343, 347
Rumen, 18:2 n-6 and 18:3 n-3 metabolism pathways, 27
Ruminal biohydrogenation pathways, 28
Ruminant milk, TFA and bioactive lipids, 3–4

Saturated fatty acids (SFA), 3–8
SCD, see Stearoyl-CoA Desaturase
SCD activity nutritional regulation, 83–84
SCD gene expression in mammary gland and adipose tissue, 75
sCD14 levels in human breast milk during lactation, 197–198
Serine proteases Granzyme A and B, 224

Sheep red blood cells (SRBC), 242
Signaling pathways mediating nutritional regulation of gene expression, 91
Sitostanol and Sitosterol in milk, 114
Somatic cell counts (SCC) in cows' milk, 116
Sperm-mediated gene transfer, 362
Spontaneously hypertensive rats (SHR), 298
Stanols in milk, 114
Staphylococcus aureus, 170
Stearoyl-CoA Desaturase, 75–76, 344
Sterol regulatory binding protein-1 (SREBP-1), 91–93
Stigmasterol in milk, 114
Streptococcus mutans, 320
Streptococcus uberis, 170
Systemic inflammatory response syndrome, 205
Systemic lupus erythematosus, 205

TFA as pro-inflammatory, 12
TFA content of foods and legislation, 10–11
Th-1 cytokine-dominant environment in peripheral blood, 179
Therapeutic products for human patients and clinical trials, 366
Tissue-nonspecific AP (TNAP), 154, 379
Tissue Plasminogen Activator, 355, 379
TLR4-dependent and -independent signaling pathways, 179
α-Tocopherol, 112
   form of vitamin E, 112
   stereoisomers in milk, 117
α-Tocopherol transfer protein (TTP), 117
Toxic milk mouse illness, 150
Toxoplasma gondii, 170
Trans-10 C18:1 and fat yield curvilinear response curve, 86
Trans-10 C18:1 isomer synthesis in rumen, 87
TRANSFAIR study, 12
Trans fatty acids (TFA), 3, 4
Transgene, agalactic phenotype, 348
Transgenic animals production
   gene construct, 357–358
   livestock species choice, 358–360
   targeted tissue, 356–357
Transgenic livestock animals production methods, 360–363
Trans octadecadienoic acids in bovine milk fat, 17
"Trans-11 pathway", 87

Triglyceride biosynthesis, 349
Triglyceride structure of bovine milk fat, 348–351
trans fatty acid, intake in European Countries, 12
Tumor-induced angiogenesis in mice, 180
Tumor suppressor p53, 224
Type II cell death, 227

Urokinase plasminogen activator (u-PA), 257

Vitamin A activity of full-fat cow's milk, 112
Vitamin A (retinol) concentration of bovine milk, 110
Vitamin A role in, biological processes, 111
Vitamin content of milk and dietary manipulation, 118
Vitamin D content in cow's milk, 113
Vitamin E content in cow's milk, 112
Vitamin E supplementation in dairy cows milk, 117
Vitamins A and E and β-Carotene content of colostrum, 113
Voltage-dependent anion channel (VDAC), 221

Weaning Piglet Model, 256
Whey protein–derived antimicrobial peptides, 272–274
Wilson's disease in humans, 150
    See also Toxic milk mouse illness
Xanthine oxidase cytosolic enzyme, 139

Printed in the United States of America

SF
251
.B56
2008

Bioactive components of milk.

$139.00 35010000542199

DATE			

South University
709 Mall Blvd
Savannah, GA 31406

BAKER & TAYLOR

4-07-09